Ship Operation Technology

Manfred Pfaff

Ship Operation Technology

Reference Book and Guidebook

Springer

Manfred Pfaff
Abfallstromkontrolle
Bezirksregierung Detmold
Detmold, Germany

ISBN 978-3-658-32728-6 ISBN 978-3-658-32729-3 (eBook)
https://doi.org/10.1007/978-3-658-32729-3

This book is a translation of the original German edition „Schiffsbetriebstechnik" by Pfaff, Manfred, published by Springer Fachmedien Wiesbaden GmbH in 2018. The translation was done with the help of artificial intelligence (machine translation by the service DeepL.com). A subsequent human revision was done primarily in terms of content, so that the book will read stylistically differently from a conventional translation. Springer Nature works continuously to further the development of tools for the production of books and on the related technologies to support the authors.

Responsible Editor: Thomas Zipsner
This Springer imprint is published by the registered company Springer Fachmedien Wiesbaden GmbH part of Springer Nature.
The registered company address is: Abraham-Lincoln-Str. 46, 65189 Wiesbaden, Germany

Foreword

On January 1, 2015, the world merchant fleet comprised 50,422 ships[1] with >300 GT[2] and a deadweight tonnage of 1661 million dwt.[3] Bulk carriers formed the main part of this, followed by crude oil tankers.

To ensure that these ships can navigate the world's oceans safely, a functioning ship operation technology is of particular importance.

In addition to a few brief explanations of shipbuilding and the basics of the various propulsion systems, this textbook "Ship Operation Technology" deals with auxiliary systems and auxiliary machinery, power units, life-saving equipment, ship safety, and much more - in other words, with what makes the hull into a ship in the first place. In this respect, this book complements other literature dealing with the actual design and construction of ships and with the propulsion technology of ships.

The characteristics and significance of the various systems, equipment, plants, and machines presented here for the overall "ship" system are examined and simple calculations are carried out using "rules of thumb."

Where necessary, the essential legal and technical standards on which the abovementioned devices are based are mentioned and interpretations or dimensioning are carried out on the basis of these regulations as examples.

This is intended to enable the reader to determine roughly the data required for the design, such as heat exchanger area, rough calculation of drive power, rope and chain diameter, and much more, using simple means. Questions, such as which type of light guidance is required for which watercraft, will be answered. Notes on environmental protection and ship safety round off the work.

The appendix contains tables that may be helpful for ship operation practice.

Our sincere thanks go to the Mechanical Engineering editorial staff of Springer Vieweg Verlag, namely Dipl.-Eng. Mr. Thomas Zipsner, Ms. Ellen-Susanne Klabunde,

---

[1]See Naval Command (2015) Annual Report, Rostock 2015, p. 31.

[2]Abbreviation for gross tonnage, a measure for ship measurement.

[3]"Deadweight tonnage" = total carrying capacity of a merchant ship.

and Ms. Imke Zander. The many tips and suggestions from Mr. Zipsner and Ms. Klabunde, as well as the constant support with words and deeds from the editorial team have made a decisive contribution to the success of this book.

Furthermore, I would like to thank the numerous companies and private individuals who supported me extensively in my research for this book. Especially, I would like to mention my daughter Ramona as well as the shipping company AIDA Cruises and Mr. Michael Grund from the company Ocean Clean.

And now, good luck with reading and applying the book, for further education, studies, and of course in your job.

Lage, June 2017 Dr.-Eng. Manfred Pfaff

Contents

1 Job Description of the Ship Operation Technician

Admission to the duties of technical ship officer requires a completed course at a technical college or a university of applied sciences in ship operation technology or ship operation as well as a corresponding valid certificate of competency.[1]

Depending on the position held within the officer's crew, the following certificates of competency are required for the technical watch officer:

- The certificate of competency "technical watch officer on ships of unlimited machine capacity"
- The certificate of competency "second technical officer"

On the internet site of the German Navy (http://www.marine.de), the use of the ship operation technician is described as follows:

> Ship operation technicians are the specialists responsible for the operation and maintenance of the modern operating equipment, for example, the refrigeration, air-conditioning and environmental protection systems, the fire extinguishing, fuel and spraying systems as well as the cranes, lifts, and hoists that ensure the operational capability and utilizability of a ship or boat.

The main tasks are:

- Monitoring and operation of the ship's technical systems and equipment such as bilge system (pump outboard water that has penetrated), fire extinguishing equipment, fuel transfer and storage facilities, fresh water and distillate treatment plant, refrigeration, air-conditioning, and environmental protection systems;

[1] For more details, see [1].

M. Pfaff, *Ship Operation Technology*, https://doi.org/10.1007/978-3-658-32729-3_1

- Carrying out care, maintenance, and repair work (e.g., replacement of assemblies of the controlling and regulating equipment);
- Commanding a ship's protection squad;
- Implementation of immediate measures in the areas of leakage, fire, and NBC defense as well as in rescue and recovery services;
- Diving service in secondary use.

The "chief" of the operations technicians on board of naval units is the ship's technical officer (STO). In the German Navy, the propulsion officer (PO) and his staff (known as *stoker* in naval jargon) are responsible for the propulsion system.

On board of civilian ships, the executive engineer is called "chief." He has the obligationfor the responsibility and maintenance of all technical equipments and their operation. This also includes, for example, the supply of the necessary operating materials, water, working materials, spare parts, tools, and other goods required for the technical operation of the ship, which he determines in consultation with the ship's command in good time before the start of the voyage and he takes care of their organization.

Technical ship officers are also responsible for the training of employees in the field of technical ship operation. They are also responsible for the implementation of occupational safety and fire protection as well as the corresponding safety instruction in the operating rooms and, last but not least, for compliance with environmental protection regulations.

They advise the captain and the other nautical officers on all matters of technical ship operation and the use of the ship machinery.

Due to the extensive technical knowledge from "A for propulsion systems" to "Z for cylinder covers," and due to the ship's captain's expectation towards the chief or the STO and PO that the ship with its aggregates and technical equipment is always ready for operation and safe, the role of the ship's technical personnel can also be seen as the "caretaker" of the unit. Figure 1.1 shows a partial view of a ship's technical control station, the main workplace of the "chief."

Increasing requirements for energy efficiency, maritime environmental protection, and ship safety as well as increasing complexity of the systems characterize the challenge for ship operation technicians. The ship's engineer and the ship operation technician are required to have interdisciplinary knowledge regarding the monitoring and maintenance of ship machinery systems. They are responsible for the functioning of the technology on board a ship. General knowledge in the fields of physics, mathematics, electrical engineering, and mechanical engineering as well as in-depth knowledge in the fields of system engineering, machine dynamics, refrigeration and air-conditioning technology, and ship electronics is therefore required.

Fig. 1.1 Partial view of a ship's technical control station. *Photo* AIDA

Reference

1. http://www.deutsche-flagge.de/de/befaehigung/ausbildung/ausbildungsstaetten. Accessed: 28. Apr. 2017

2 Rules and Regulations

The construction and operation of ships, boats and other watercraft are subject to extensive standards, laws and other regulations. The following explanations will try to shed some light on the jungle of regulations and to show their basic systematics. An overview of the most important regulations to be applied in the construction of German ships has been drawn up by the Employer's Liability Insurance Association for Transport and Traffic[1] ("Berufsgenossenschaft für Transport und Verkehrswesen"; see Annex 1).

2.1 Hierarchy of Laws and Standards

Living together in a community is regulated by the state through laws and regulations. In Germany, our Constitutional Law (CL) takes precedence over all other laws and regulations. In addition to the directly applicable laws and ordinances, the federal government can also issue so-called administrative regulations. Their content is initially only binding on the administration, i.e. the enforcement authorities. In order for them to have direct effect for the citizen, they must be implemented by individual order by the relevant competent authority.

But now Germany is also integrated into the European Union (EU). The EU also has the possibility to regulate coexistence in this community of states. To this end, it may adopt EU regulations. These are directly applicable in each Member State. However, the EU can also issue EU directives. In order to have direct legal effect in a Member State, they must be transposed into national law by means of appropriate legislative procedures.

[1] See http://www.deutsche-flagge.de/de/download/bau-und-ausruestung/neu-und-umbau/uebersicht/rechtsvorschriften.

M. Pfaff, *Ship Operation Technology*, https://doi.org/10.1007/978-3-658-32729-3_2

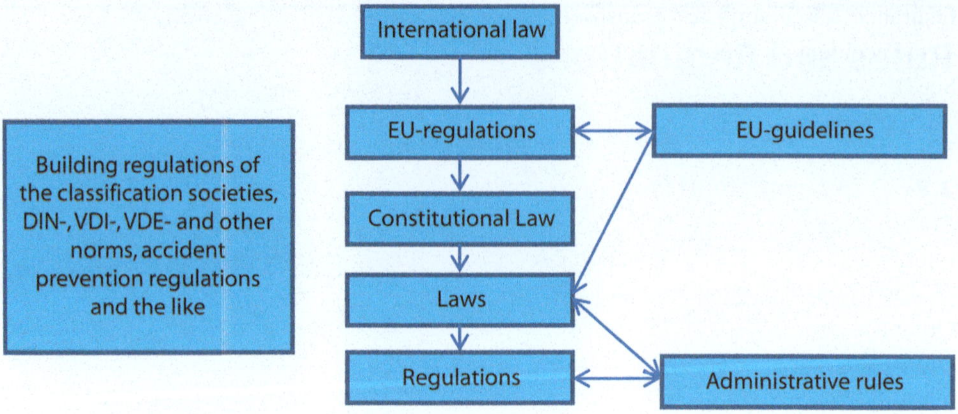

Fig. 2.1 Overview "hierarchy of standards"

In addition, there are regulations under international law. Sources of international law are bilateral or multilateral international treaties, customary international law and general legal principles (see also Fig. 2.1).

In addition to laws, regulations and other legal principles, requirements for shipbuilding and ship operation are found in the most diverse *technical regulations*. First and foremost are certainly the construction regulations of the classification societies such as DNV GL,[2] ABS (American Bureau of Shipping) or in the field of the German Navy there are the German Navy Construction Regulations for Ships of the German Navy ("Bauvorschriften für Schiffe der Deutschen Marine, BV-Hefte") issued by the Federal Office for Equipment, Information Technology and Utilization of the German Armed Forces ("Bundesamt für Ausrüstung, Informationstechnik und Nutzung der Bundeswehr," BAAINBw for short). In this book, reference is often still made to the former construction regulations of the former Germanische Lloyd, which have now been replaced by construction regulations of the classification society DNV GL.[3] Nevertheless, they still contain valuable suggestions and aids for the planning and dimensioning of a hull and its equipment.

There are also provisions in DIN, VDI and VDE standards[4] and in regulations and leaflets of the professional associations. The regulations of the Employer's Liability

[2]Merger of the classification societies *Det Norske Veritas* (*DNV*) and *Germanischer Lloyd* (*GL*).

[3]See below: https://rules.dnvgl.com/ServiceDocuments/dnvgl/#!/industry/1/Maritime/1/DNV%20GL%20rules%20for%20classification:%20Ships%20(RU-SHIP).

[4]Technical standards are issued by the German Institute for Standardization ("Deutsches Institut für Normung," DIN), the Association of German Engineers ("Verein Deutscher Ingenieure," VDI) and the Association of Electrical Engineering Electronics Information Technology ("Verband der Elektrotechnik Elektronik Informationstechnik e. V.," VDE).

Insurance Association for Transport and Traffic are of particular importance here, such as the accident prevention regulation for the sea ("Unfallverhütungsvorschrift See - UVV See").

2.2 International Law Regulations

International law and international conventions in the field of shipping are concluded by the maritime roof organizations, above all the IMO (International Maritime Organisation) and the ILO (International Labour Organisation), both organs of the United Nations. The main purpose of this is to develop standards for ship safety and maritime environmental protection. The following factors should be mentioned in this context **main marine engineering** Agreements are *SOLAS* (International Convention for the Safety of Life at Sea), *MARPOL* (Marine Pollution - International Convention for the Prevention of Pollution from Ships) with its six annexes and the *Collision Avoidance Rules, CAR* (*Kollisionsverhütungsregeln, KVR*).[5]

2.2.1 SOLAS

Already on January 20, 1914 representatives from 13 countries agreed to the first "International Convention for the Safety of Life at Sea." This international convention for the safety of human life at sea lays down fundamental requirements for ship safety [3, p. 27 f.].

Some important aspects of ship safety are listed below, for which requirements are defined in the SOLAS Convention.[6]

One aspect is leakage prevention measures. For example, the hull is divided by watertight bulkheads. A bulkhead is part of the structural protection of ships, consisting of stiffened vertical partitions which divide the hull into watertight compartments for safety and at the same time give it strength. Depending on the arrangement of the walls - along or across the midship axis - we speak of longitudinal or transverse bulkheads. This constructive protective measure increases the sink safety in case of leakage.

It also contains regulations on the nature of these watertight bulkheads, on openings in these bulkheads, in the outer skin and on the installation of double bottoms.

Requirements concerning machinery and electrical installations mainly refer to sufficient reserve and emergency power sources. Furthermore, according to SOLAS, electrical installations must be designed in such a way that sufficient protection against accidents caused by electric current is provided.

[5]Erles, N.-G., in: [2, p. 1037 ff.].

[6]Supplementary see also: http://www.imo.org.

There are also regulations on steering gear, as its safe operation is essential for the safe manoeuvrability of the ship.

A further regulatory component of this agreement are measures for fire protection, fire detection and firefighting.

Chapter III contains requirements concerning the number, type and nature of life-saving and distress signalling devices to be carried on board (inter alia lifeboats, portable radios, life jackets, line-throwing devices, signal ammunition).

Explanations inter alia about radiotelephony and navigation equipment and the Automatic Identification System (AIS) complete the content of this regulation.

In addition, special measures are prescribed for high-speed craft.

In response to existing terrorist threats, there are regulations to enhance the security of ships and port facilities ("ISPS Code" - International Ship and Port Facility Security Code).

2.2.2 MARPOL

MARPOL - **M**arine **Pol**lution. Environmental protection is also becoming increasingly important in the maritime sector.

The MARPOL Convention belongs to the most important regulations of the IMO.[7] This international convention for the prevention of pollution from ships formulates shipbuilding, operational and technical requirements for maritime environmental protection [3, p. 26 f.].

It was adopted as early as 1973, modified in 1978 and has been continuously expanded ever since. It now comprises a total of six special regulations (Annexes I–VI), e.g. on protection against oil pollution, on the transport of packaged pollutants or on the prevention of ship-generated waste [1, p. 18].

The most recent extension of MARPOL (Annex VI) regulates the emissions of air pollutants; in Rule 16, this Annex also describes requirements for ship waste incineration plants (a).

2.2.3 Collision Avoidance Rules

The **Collision Avoidance Rules (CAR)**[8] - officially "International Regulations for Preventing Collisions at Sea, 1972" (*Conventions on the International Regulations*

[7]"International Maritime Organization.".

[8]For CPR see: https://www.elwis.de/schifffahrtsrecht/seeschifffahrtsrecht/kvr/.

for Preventing Collisions at Sea COLREGs) - constitute international maritime law. They provide the basic legal framework for regulating the safety and ease of navigation on the high seas and associated waters. The CARs are intended to prevent ship collisions and apply to all ships, including pleasure craft.

The CARs are the basic rules of maritime (not inland waterway) transport and contain the following regulatory areas:

- Part A: General (Rules 1–3, inter alia scope, responsibility and general definitions),
- Part B: Avoidance and driving rules,
- Part C: Lights and Signal Bodies (Rules 20–31, inter alia Lights, Navigation Lights and Signal Bodies),
- Part D: Sound and light signals (Rules 32–37, inter alia representation of sound signals, sound signalling devices),
- Part E: Exemptions (Rule 38),

by which **Annexes I–IV** which provides further details on the arrangement and technical design of lights and signal bodies, on additional signals for vessels fishing close together, on technical details of acoustic signalling equipment and on emergency signs.

2.3 European Union regulations

For the ship's engineer, with regard to European requirements, the **Marine Equipment Directive** is of importance. In 1996, the European Union developed the Maritime Equipment Directive (MED) as Directive 96/98/EC to ensure the free movement of marine equipment within the EU. This directive came into force on January 1, 1999.

Annex A of the Directive lists the marine equipment covered by the Directive. For marine equipment of Annex A.1 there are internationally harmonised testing standards; for marine equipment of Annex A.2 these internationally harmonised testing standards do not yet exist.

Since the entry into force of this Directive, marine equipment listed in Annex A.1 is subject to an EC conformity assessment procedure and must be approved by a notified body. It can then be installed and used throughout the EU area without further national approval. The manufacturer must affix a symbolised steering wheel as a mark of conformity, the identification number of the notified body and the last two digits of the year of marking on the device.

The marine equipment referred to in Annexes A.1 and A.2 and the corresponding testing standards are subject to continuous changes, which are implemented annually in new versions of the Marine Equipment Directive. On April 9, 2015, the EU Commission adopted a further amendment to the Marine Equipment Directive (Directive 96/98/EC) and published it in the Official Journal of the EU (L95 of April 10, 2015) as Directive 2015/559/EU. This directive came into force in Germany on April 30, 2016.[9]

2.4 German Laws and Regulations

The abovementioned international and European law regulations must, in order to have direct legal effect in Germany, be implemented in German territorial waters ("12-sm zone"[10]) and on ships flying the German flag must be transposed into German law by national legislative acts.

For example, the Collision Prevention Rules have become applicable to Germany through the "Regulation on the International Regulations for the Prevention of Collisions at Sea of 1972."

With the *Ship Safety Act* are - besides other international and EU law regulations[11] - amongst others SOLAS and MARPOL have been implemented into German law.

The *Ship Safety Regulation* is the national implementation of SOLAS in German law as a supplement to the Ship Safety Act. It deals with safety standards of ships, their equipment and crew. In this respect, this Regulation serves the effective application of the Ship Safety Act in addition to maritime safety, including the protection of seafarers' health and safety at work and environmental protection. It contains, among other things, regulations on the design of steam boiler systems, radio systems and marine equipment.

The Employer's Liability Insurance Association for Transport and Traffic ("Berufsgenossenschaft für Transport und Verkehrswesen," BG) is a so-called state official ("Beliehene des Staates"). It is responsible to a large extent for the implementation of the abovementioned regulations in the field of ship safety and marine environmental protection. In addition, it is responsible for statutory accident prevention in accordance with the German Social Code. It covers the prevention of occupational accidents, occupational diseases and work-related health hazards. For this purpose, the BG issues separate regulations, such as accident prevention regulations, guidelines, information sheets, manuals and guides [2, p. 1041].

The BG's accident prevention regulations are of a statutory nature; violations of these regulations may be sanctioned by the BG with fines.

[9]http://www.bsh.de/de/schifffahrt/berufsschifffahrt/schiffsausruestungsrichtlinie/.

[10]The unit sm = Seemeile; 1 sm = 1852 m.

[11]See the annexes to the Ship Safety Act.

2.5 Construction Regulations of the Classification Societies

The classification societies (in Germany e.g. DNV GL or Bureau Veritas) also issue construction regulations which contain regulations on the required strength of the hull and the safe functioning of the technical systems. The aforementioned Federal Office for Equipment, Information Technology and Utilization of the German Armed Forces ("Bundesamt für Ausrüstung, Informationstechnik und Nutzung der Bundeswehr," BAAINBw) is working on further-reaching construction regulations for German naval ships that take into account the special requirements of a warship.

The construction regulations are ordinarily also a contractual component between the client of the ship and the shipyard.

Compliance with the construction regulations of the classification societies is important in that, after construction and successful trial run, this society issues a class certificate which indicates the ship's permissible cruising range. If the ship is to retain its so-called class, it must be presented at regular intervals for surveys to check that the ship is in proper condition; it is said that "the ship renews its class." In principle, the granting of the class certificate is comparable to the registration of a car, the regular inspection of the ship to maintain the class, the presentation of the car at the technical inspection agency.

2.6 Technical Standards

International and German technical standards are issued by associations and clubs. In principle, such sets of rules are not binding standards, but can be declared directly binding by contract, e.g. between shipowner and shipyard, by binding declaration in laws or regulations, but also by official orders to the party concerned (usually the shipowner). In this respect they have more the character of recommendations. These rules and regulations reflect the current state of the art.

Internationally important organizations for standardization are ISO (International Organization for Standardization) and in the field of electrical engineering the IEC (International Electrotechnical Commission).

In Germany technical standards are issued by the *German Institute for Standardization* (*"Deutsches Institut für Normung,"* DIN), by the *Association of German Engineers* (*"Verein Deutscher Ingenieure,"* VDI) and by the *Association for Electrical Engineering, Electronics, Information Technology e. V.* (*"Verband der Elektrotechnik Elektronik Informationstechnik e. V.,"* VDE). About the *Ship and Marine Technology Standards Body* (*"Normenstelle Schiffs- und Meerestechnik,"* NSMT)[12] at DIN, the current status of standardization with regard to relevant standards for the shipping industry can be enquired about.

References

1. AIDA: AIDAcares – Nachhaltigkeitsbericht (2009)
2. Bernhardt, F., Meier-Peter, H. (eds.): Handbuch Schiffsbetriebstechnik. Seehafen Verlag, Hamburg (2008)
3. Verband für Schiffbau und Meerestechnik e. V.: Schiffstechnik und Schiffbautechnologie. Seehafen Verlag, Hamburg (2006)

3 Hulls, Cordage, Superstructures, Anchor Gear, Corrosion Protection, and Deck Coverings

The shape, dimensions, and other structural elements of the hull depend on the intended use of the ship. It is not easy to simply classify ships according to any one principle; a differentiation can be made according to the type of propulsion: sail or machine, within the machine-powered vehicles according to the type of power generation: for example, diesel engines or diesel-electric power generation, combined diesel and gas turbine propulsion. Even steam engines for power generation can still be found from time to time. However, a distinction can also be made according to area of operation: Inland navigation or maritime shipping (e.g., territory and coastal shipping, open sea). Differentiation can be made according to the type of load: Liquid substances in tankers, bulk carriers or general cargo vessels, or passenger ships or ferries.

Other differentiating features are directly aimed at the hull forms: Glider or displacer, monohull, double hull (catamaran, see Fig. 3.1, or SWATH—Small Waterplane Area Twin Hull), or vehicle with three hulls (trimaran).[1]

However, this book does not discuss these aspects and hull design issues in detail and refers to the relevant literature.

3.1 Hull

A *boats-* or rather *hull* is the part of a boat or ship that gives it buoyancy. The *hull* is the finished, buoyant hull without the included technology. In inland navigation, the hull is also known as *ship's hull* [32].

[1]Supplementary also Lehmann, E. in [1, p. 876 ff.].

M. Pfaff, *Ship Operation Technology*, https://doi.org/10.1007/978-3-658-32729-3_3

Fig. 3.1 Double hull vehicle

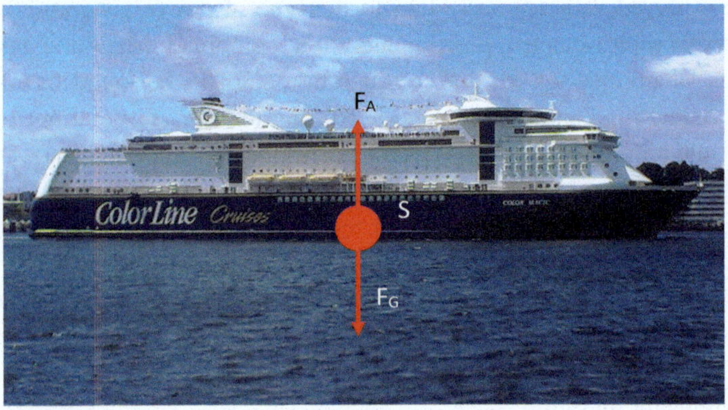

Fig. 3.2 Buoyancy ship. F_A buoyancy; F_G weight force ship; S focus

The underwater part of the hull is called *underwater hull* whose shape, when viewed from the side, is known as *lateral plan*. The *lateral plan* (from lateral: sideways) is therefore the lateral projection of the underwater surface.

3.1.1 Buoyancy and Stability

The *Archimedean principle* is important for the **buoyancy** of ships: The buoyancy is equal to the weight of the displaced liquid.

The vector of the weight force of the ship F_G acts perpendicular to the waterline through the center of gravity S of the vehicle (see [31]. The water displaced by the hull (displacement volume) has the weight force $F_{G\,\text{Water}}$. According to the Archimedean principle, this volume of water generates a buoyancy force F_A, which is equal to the weight of the displaced water. For a floating ship (also for diving submarines) applies

$$F_G = F_A. \tag{3.1}$$

Thus, the vehicle is in a floating equilibrium position as both vectors with the same amount act directly in opposite direction and thus become zero.

$$F_A = F_{G\,\text{Water,displaced}}, \tag{3.2}$$

$$F_{G\,\text{Water,displaced}} = V_{\text{Water,displaced}} \cdot \rho_{\text{Water}} \cdot g \tag{3.3}$$

where g the acceleration of gravity and ρ_{Water} the density of the water, which depends on its salinity (see Appendix 2).

The following symbols are used in shipbuilding:

- Underwater volume (displacement volume) of the hull ∇ $V_{\text{Water, displaced}}$,
- Total weight (displacement weight) of the vessel Δ_F F_G,
- Outer skin factor (shell plating coefficient)[2] a_H.

Thus, the formula for calculating the mass of the displaced water is

$$\Delta_F = \nabla\ (\text{m}^3) \cdot \rho_{\text{Water}}\ (\text{kg/m}^3) \cdot g\ (\text{m/s}^2) \text{ in Newton.} \tag{3.4}$$

For approximate calculations, the density values for water can be taken from Appendix 2.

In the design, the mass of the displaced water (= total mass of the ship) is also calculated; the so-called design equation is

$$\Delta_F = \text{Light Ship Weight} + \text{Deadwight} = L_{\text{PP}} \cdot B \cdot C_{\text{B}} \cdot \rho_{\text{Water}} \cdot (1 + a_H). \tag{3.5}$$

"C_{B}" is the so-called block coefficient. It indicates the ratio of the volume of the underwater hull to the circumscribing cuboid; it indicates the displacement and thus the carrying capacity of the ship.

$$C_{\text{B}} = \frac{\nabla}{L_{\text{PP}} \cdot B \cdot T} \tag{3.6}$$

"L_{pp}" is the length between the perpendiculars, that is, the distance between the center line of the rudder stock and the intersection of the waterline and the fore stern on the construction waterline.

[2]Takes into account the actual displacement compared to the "displacement on frames," often estimated at 0.003 for rough calculations.

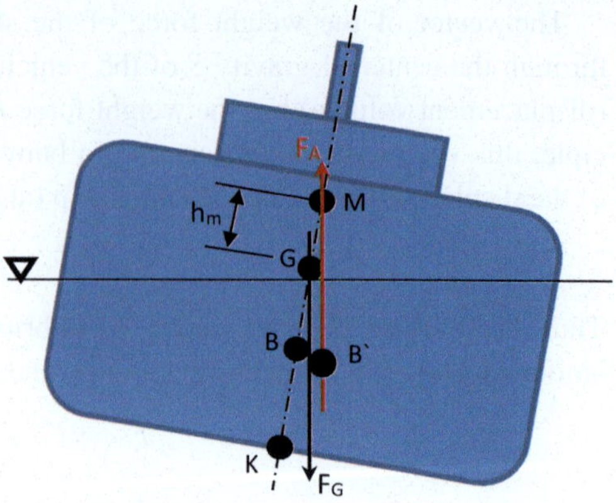

Fig. 3.3 Metacenter/stability. F_A buoyancy; F_G weight force; B, B' mold center on the underwater hull; G center of gravity; K keel; M metacentesis; h_m metacentric height

The smaller C_B the "slimmer" the ship. Fast ships usually have a small C_B value. The block coefficient is also known as *fullness*.[3]

With regard to the **stability** of watercraft, the following is important: In order to be reliable in motion, a hull must not only have sufficient freeboard, but also the ability to straighten up—keyword: metacentric height. The vertical distance between the vector of the ship's weight force and the buoyancy force determines the straightening moment. The buoyancy force runs through the metacentric height. For sufficient stability of the vessel, the metacenter of a dimensionally stable vessel must, therefore, be above its center of gravity (see Fig. 3.3).

Example for buoyancy

A ship is on the high seas, where the density of seawater is 1.03 g/cm³. The ship then enters the port. The density of the harbor water is only 1.00 g/cm³. After the ship has unloaded 600 tons of cargo, it is just as deep in the water as on the high seas. What is the ship's mass without the cargo?

Simplification: Instead of the weight force, the mass is used.

A ship floats when the mass of displaced water is equal to the mass of the ship. It always dives so deeply until this state of balance is established. As the density of seawater is greater than the density of fresh water (harbor water), the same volume of seawater also weighs more than fresh water and the ship does not dive as deep.

When the ship enters the harbor, it has to displace more water than in the sea because the mass of the ship does not change—it dives deeper. After unloading, it emerges again, the depth and thus the displaced volume decreases again.

The following equations therefore apply:

[3] For a more detailed study of the survey, see also [8] and [6, Chap. 1 General principles].

1. Seawater, load the ship:

Mass of ship + Mass of load = Mass of displaced salt water,

$$m_{\text{ship}} + m_L = m_{SW}, \tag{3.7}$$

$$m_{\text{ship}} + m_L = \rho_{SW} \cdot V. \tag{3.8}$$

2. Fresh water, ship unloaded:

Mass of ship = Mass of displaced harbor water,

$$m_{\text{ship}} = m_{HW}, \tag{3.9}$$

$$m_{\text{ship}} = \rho_{HW} \cdot V. \tag{3.10}$$

3. As the immersion depth remains the same after unloading, the following applies (buoyancy is equal to the mass of the displaced liquid $m = \rho/V$): Volume of displaced seawater before unloading = volume of displaced fresh water after unloading. In this respect, V = constant.

In Eqs. 3.8 and 3.10, the volume and mass of the ship appear as unknown quantities. As the displaced water volume is the same in both cases, these we convert them to V, equate them, and calculate the mass of the ship:

$$\frac{m_{\text{ship}} + m_L}{\rho_{SW}} = \frac{m_{\text{ship}}}{\rho_{HW}},$$

$$m_{\text{ship}} \cdot \rho_{HW} + m_L \cdot \rho_{HW} = m_{\text{ship}} \cdot \rho_{SW},$$

$$m_{\text{ship}} = \frac{m_L \cdot \rho_{HW}}{\rho_{SW} - \rho_{HW}},$$

$$m_{\text{ship}} = \frac{600\ \text{t} \cdot 1.00\ \text{g/cm}^3}{\left(1.03\ \text{g/cm}^3 - 1.00\ \text{g/cm}^3\right)} = 20{,}000\ \text{tons}.$$

Answer: The ship has a mass of 20,000 tons. ◄

3.1.2 Important Designations and Principal Dimensions

Figure 3.4 shows a few common **designations** on the ship.

Fig. 3.4 Common designations on the ship: *1 bug* is the front part of the hull; *2 bow bead*—also known as bulge bow—serves to improve the flow characteristics, lowers the required propulsion power, and thus reduces fuel consumption; *3* the *anchor* keeps the ship in the water, for example, when ship is lying in the roadstead, it also serves the ship's safety in case of maneuverability; *4 starboard* is the right-hand side of the ship (indicated by a green light at night or in poor visibility), and *port side* is the left-hand side of the ship (indicated by a red light), as seen from the stern to the bow; *5 stern* means the rear (aft) part of the vehicle; *6* the *chimney* is necessary for the discharge of exhaust gases from propulsion and other combustion systems; *7* the *superstructures* and *deckhouses* designate all superstructures above the upper deck [15, p. 47]; *8 upper deck,* also *main deck,* is the deck that closes the hull at the top; in shipbuilding terms, the main deck is also the deck in which the upper strength members of the hull are located, which run the full length of the ship; *9 bridge wing* (usually the widest part of the ship); *10 bridge*

Besides speed, the **main dimensions** determining the ship's design include [10, p. 855, 13, p. 2 f.].[4]

- Length, width, depth, and side height to freeboard deck,[5]
- Block coefficient (see Sect. 3.1.1),
- The ratio of the main frame area in relation to the molding edge to the rectangle of width and depth. The *molding edge* is the inner edge of the outer skin. *Main frame* is the frame at the greatest width of the ship [34],

[4]For further details, see [6]; also: [8, H. Definition], [6, Chap. 2 General arrangement design], http://www.risp-duisburg.de/files/technik.pdf.

[5]The freeboard deck (also known as the survey deck) is normally the uppermost continuous deck exposed to the weather and the sea, which has fixed closing devices for all openings in its exposed part and below which all openings in the sides of the ship are fitted with fixed watertight closing devices (see http://www.uni-protokolle.de/lexikon/freiborddeck.html; International Convention on Load Lines 1966, Annex I).

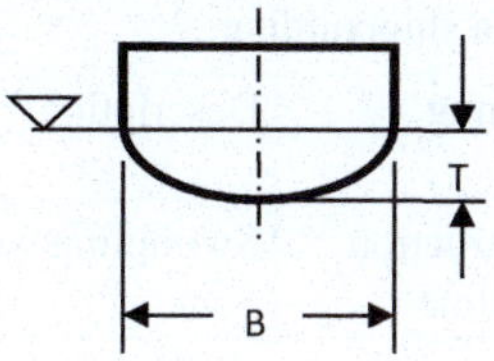

- Spantareal curve, especially the bow bead and the position of the shoulders.

With regard to the main dimensions, the information listed in Table 3.1 is usual.

In particular, the *length specifications* are further differentiated, such as LC, L*, and LS; these are special length specifications from the construction regulations of the classification societies.

3.2 Superstructure, Deckhouse, Chimney

Superstructures are structures on the freeboard deck, which extend from ship to ship or whose side plating is indented by $<0.04\ B$[6] from the outer skin. So-called effective superstructures are those that extend amidships in the range of $0.4\ L$[7] and $>0.15\ L$. The side plating as the outer skin and the deck as the belt deck are structurally connected to the hull to form a complete structure (see Fig. 3.5).

Deckhouses, on the other hand, are structures above the belt deck, whose side plating $>0.04\ B$ is indented from the outer skin; they are fitted on the hull (see Fig. 3.6).

On seagoing vessels, exhaust pipes are generally routed upward out of the ship[8] and terminate above the uppermost deck to allow an unobstructed discharge of the exhaust gases. The cladding of the exhaust pipes is the chimney jacket, and the entire component is the **chimney** (see Figs. 3.6 and 3.7). The chimney has the following tasks:

- Improvement of exhaust gas discharge (to prevent passengers and crew from being affected by soot and odor and to prevent soot particles from contaminating the deck),
- Avoid touching the hot surfaces of the exhaust gas pipes,
- Visual aspect; the chimney is often painted in the colors of the shipping company, and usually the logo of the shipping company can be found on it (see Fig. 3.7a and b).

---

[6] B = construction width (see Table 3.1).

[7] L = length between perpendiculars (see Table 3.1).

[8] In order to avoid infrared detection, exhaust gases from naval vessels are usually discharged from the side or aft of the hull, just above or even below the water surface. This type of exhaust gas routing is also common on recreational boats and yachts.

Table 3.1 Usual main dimensions in shipbuilding

Abbreviation	Engl. Designation	Meaning	Description
KWL	CWL	Construction waterline	Swimming waterline at summer freeboard
HL	AP	Rear perpendicular	Mostly rudder axle
VL	FP	Front perpendicular	Cut of front with the KWL
LAD		Length to deck	From foremost to aftmost fixed point (front edge stem to rear edge stern stem at deck level)
LÜA	LOA	Overall length	Important for the pier occupancy; measure from outermost forward part of ship (e.g., for tall ships, nock of bowsprit) to outermost aft end of ship (e.g., the nock of flagpoles protruding from the rear)
LWL		Length in the swimming waterline	(KWL; front edge of forestern—rear edge of sternpost in the KWL including rudder blade)
LZDL	LPP	Length between the perpendiculars	Length from aft perpendicular (AP), which is the axis of rotation of rudder, and the forward perpendicular (FP) as the intersection the design waterline with the front edge of stem (length between the pendulums)
B		Construction width	Usually width of the construction waterline (B_{DWL}); it is the largest width of the construction waterline, measured on molding edge
BÜA	BOA	Overall width	Width between the extreme points of ship to port side and starboard measured in the middle of the ship
D or also H		Page height	Height of the molding edge side desk of uppermost continuous decks above the base at half-length between the perpendiculars
F		Freeboard	Measured from KWL to top edge of deck covering to the side of ship at half ship length
T		Design depth	Measured at lower edge of the ground flange of steel ships halfway between the perpendiculars (LPP)
Tg		Largest depth	
V		Displacement of ship at frames	

Fig. 3.5 MSC ARMONIA—superstructures extend over almost the entire length of the ship

Fig. 3.6 Deckhouse of a container ship seen from astern

3.3 Nonintegrated Foundations

On ships, there are a large number of foundations for the installation of machines and devices. They are intended to transfer the masses and forces to the spatial structure and enable the components to be securely fastened. Their design aims at high degree of rigidity in order to keep deformations of the units to be supported below harmful limits.

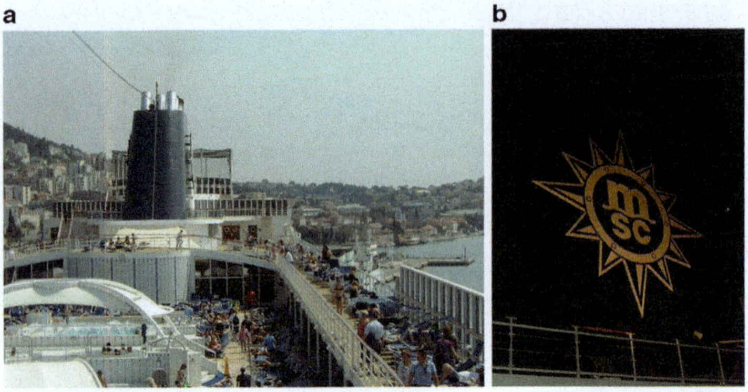

Fig. 3.7 Chimney of the MSC ARMONIA (**a** and **b**)

In addition, vibrations that may be caused by the machine or device are to be prevented by the foundation on the ship's structure, which can lead to annoyance for crews and passengers through vibration excitation in the ship's structures (frequency 1–80 Hz). Whole-body vibrations also include low-frequency vibrations (<1 Hz), to which the crew and passengers are exposed; they can cause them so-called seasickness (kinetosis).

In Germany, the "Noise and Vibration Occupational Health and Safety Ordinance"[9] ("Lärm- und Vibrations-Arbeitsschutzverordnung") for the protection of employees, including crew members of ships flying the German flag, stipulates **exposure limit and trigger values for vibration**[10] (vibration acceleration related to an 8-h working day—a_8) under Section 9 [4]:

1. For hand-arm vibrations, the following value shall apply
   a) The exposure limit value and
   b) The trigger value $a_8 = 2.5$ m/s^2.
2. For whole-body vibrations, the following value shall apply
   a) The exposure limit value in *x*, *y*, and *z* direction and
   b) The trigger value $a_8 = 0.5$ m/s^2.

The *x* and *y* coordinates give the direction of the oscillation in the horizontal direction, and the *z* coordinate in the vertical direction.

Employees on board of seagoing vessels are excluded from the scope of this regulation only where there is legislation containing equivalent provisions. Here, for example,

[9]Ordinance of March 6, 2007 (BGBl. I, p. 261), last amended by Article 3 of the Ordinance of July 19, 2010 (BGBl. I, p. 960).

[10]The exposure limit value is the absolute limit for exposure and, if the exposure action value is reached, the employer must carry out further investigations/measures to limit exposure.

the Germanischer Lloyd (GL) construction Regulations "I Ship Technology" can be used again. Under item I-1-1, "F. Vibration" is explained[11]:

Vibration limits on ships can be defined in several ways. Unless the application of other national or international regulations or standards is mandatory, the following guidelines and regulations are recommended:

For vibration loads to which the crew is exposed

- Measuring method, evaluation: according to ISO 6954, ed. 2000
- Limit values: according to ISO 6954, differentiated by ship type and location
- Ships under German flag: Guidelines and accident prevention regulations of the BG Verkehr [2]
- Vibration nuisance to passengers: GL class symbol "Harmony Class," GL Regulations on Rating Noise and Vibrations for Comfort, Cruise Ships (I-1-16)
- Vibrations of machines, devices, and other equipment:
- GL Regulations for Machinery and Equipment (I-1-2), Section 1.

These remarks show that there are indeed different standards for assessing whether or not a vibration is to be considered a nuisance. In individual cases, this depends on the contractual agreements between customer and shipyard or also on legal regulations or requirements of the classification societies.

In addition, the vibrations of the machines and devices can also lead to local crack formation due to alternating stress in the components of the ship's structure. Therefore, vibration-relevant units are usually installed in a vibration-damped manner (see Figs. 3.8 and 3.9). A selection of vibration dampers is shown as an example in Appendix 3.[12]

Nonintegrated foundations are defined as special components specially manufactured to accommodate machines or device and then integrated into the basic structure "ship" (see Fig. 3.10). In these cases, the units to be founded are therefore **not directly** connected to the actual ship's structure.

Integrated foundations, on the other hand, are those where parts of the normal ship's structure—possibly also in reinforced design—serve directly to accommodate assemblies (see Fig. 3.11).

The foundations can be classified as follows [15, p. 48]:

- Foundations for main engines,
- Foundations for deck support machinery, and
- Foundations for auxiliary machinery and other device.

---

[11]See also Appendix 4.

[12]On the above, among others [26].

Fig. 3.8 Air compressor for starting compressed air on foundation with vibration damper (nonintegrated foundation)

Fig. 3.9 Elastic mounting of a propulsion motor

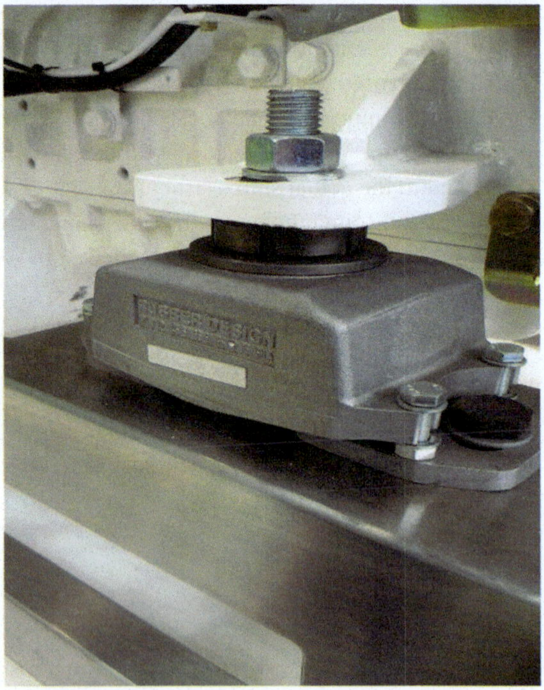

Fig. 3.10 Storage or installation of a fire pump (undamped installation)

Fig. 3.11 Foundation of a main propulsion engine

3.3.1 Excursus on Vibrations

Due to piston movements in the propulsion diesel engines, imbalances in rotary piston machines (e.g., pumps, blowers), etc., these units are set into oscillation (vibration). Unless they are decoupled by damping elements, these are transferred to the ship's structure.

An important property of an oscillating movement is its *frequency,* that is, the number of oscillations performed per second. For the frequency, use the symbol *f,* the SI unit is *hertz* (Hz), where 1 Hz is one oscillation per second (see also Fig. 3.12). Linked to the frequency is the *period of oscillation T*:

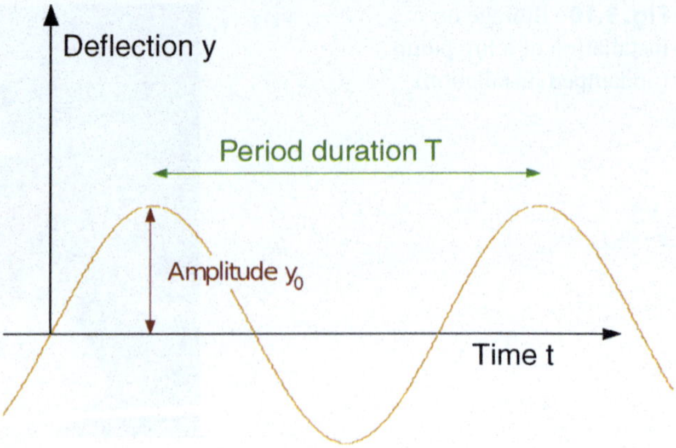

Fig. 3.12 Harmonic oscillation. *Graphic* Debenben, CC0 1.0

$$T = \frac{1}{f}. \tag{3.11}$$

A movement that is repeated at regular intervals is called *periodic oscillation or periodic motion*. If this oscillation is excited regularly, it is called a *forced oscillation*.

At this movement, the deflection of an oscillating mass particle from its original position can be expressed as a function of time:

$$y(t) = y_0 \cdot \cos(\omega t + \varphi). \tag{3.12}$$

Here y_0 is the maximum deflection of the oscillating mass particle from its rest position and is *amplitude* (often *m* is also used as index for "maximum"); ω is the so-called angular velocity or angular frequency (in the case of a forced vibration caused by the excitation force) and φ is the phase angle. The term $(\omega\, t+\varphi)$ is called *phase of the movement*.

The *angular frequency* or *angular velocity* is described by the following equation:

$$\omega = \frac{2\pi}{T} = 2\omega f. \tag{3.13}$$

The SI unit for the angular frequency is the radian per second (for consistency reasons φ must therefore also be given in radians).

However, the relevant tables not only ask for the amplitude as a limit value for the reasonableness of vibrations, but also for the vibration acceleration as a limit value. The acceleration *a* is generally the first derivation of speed *v* after the time *t*:

$$a(t) = \frac{\mathrm{d}v(t)}{\mathrm{d}t}. \tag{3.14}$$

The *vibration velocity v* is now again derived from the first derivative of the functional equation for the harmonic oscillation according to time:

$$v(t) = \frac{\mathrm{d}y(t)}{\mathrm{d}t} = \frac{\mathrm{d}}{\mathrm{d}t}\left[y_0 \cos(\omega t + \phi)\right], \tag{3.15}$$

$$v(t) = -\omega y_0 \sin(\omega t + \varphi). \tag{3.16}$$

This results in the *vibration acceleration,* also called *acceleration amplitude*:

$$a(t) = -\omega^2 y_0 \cos(\omega t + \varphi) = \omega^2 y(t). \tag{3.17}$$

If the acceleration of a particle is known as a function of time, Newton's law can also be used to determine the force that gives the particle this acceleration. That is interesting, for example, for the question of which force a vibrating mass will effect on its foundation:

$$F = m \cdot a = -(m \cdot \omega^2) \cdot y. \tag{3.18}$$

The result, a repulsive force proportional to the deflection but with opposite sign, is the so-called Hook's law for spring force:

$$F_F = -k \cdot y. \tag{3.19}$$

Here is the spring constant $k = m\,\omega^2$. Thus, it is immediately apparent that in a harmonic oscillation, the force is linearly proportional to the amplitude y. The angular frequency ω of the harmonic oscillation of the particle thus depends on the spring constant k and the mass m:

$$\omega = \sqrt{\frac{k}{m}}. \tag{3.20}$$

For the period duration, one thus receives:

$$T = 2\pi \cdot \sqrt{\frac{m}{k}}. \tag{3.21}$$

Now harmonic continuous oscillations may have to be damped in order not to exceed the specified limits. Our oscillating spring-mass system must be damped by an external force (e.g., by a rubber buffer). We then talk about damped harmonic oscillations. We assume that a force F_G is required for the compression of the rubber damper, which is proportional to the magnitude of the velocity of our mass; it then applies:

$$F_G = -D \cdot v, \tag{3.22}$$

where negative sign shows that force will work against force of oscillating mass. D is the *damping coefficient* (sometimes also b, as further used in the following). The force exerted by the spring on the weight is $F_F = -ky$. Furthermore, we assume that the gravitational force on the weight in relation to F_G and F_F is negligible. According to Newton's second law then applies:

$$-Dv - ky = ma. \tag{3.23}$$

By inserting and rearranging, one obtains:

$$m\frac{\mathrm{d}^2y}{\mathrm{d}t^2} + b\frac{\mathrm{d}y}{\mathrm{d}t} + ky = 0. \tag{3.24}$$

The solution of this differential equation is [9, p. 19]:

$$y(t) = y_0 \mathrm{e}^{-bt/2m} \cos(\omega' + \varphi). \tag{3.25}$$

In Eq. 3.25, "e" is the Eulerian number, ω' is the angular frequency of the damped system; it is given by

$$\omega' = \sqrt{\left(\frac{k}{m}\right) - \left(\frac{b^2}{4\,m^2}\right)}. \tag{3.26}$$

As mentioned above, imbalances of rotating systems lead to vibrations, which are transmitted via the components (problem of resonances!).[13] In order to explain some basic terms, here we consider the simplest model for bending vibrations, namely a single disk with mass m mounted centrally on a massless shaft, with rigid bearing and without damping. This system can be regarded to be a spring-mass system with one degree of freedom and performs forced vibrations caused by unbalance.

A *center of gravity eccentricity* e (= deviation from the shaft center) results in a rotating centrifugal force of magnitude $F_u = m\ e\ \omega^2$ with an angular velocity $\omega = 2\pi f$ where the product of mass times eccentricity is called *unbalance* $U = me$. The solution of the equations of motion results in a bent, *stationary shaft* where the deflection of the shaft center results in:

$$w = e\eta^2/(1 - \eta^2). \tag{3.27}$$

Here η is the so-called frequency ratio, that is, the ratio of the excitation frequency ω to the natural frequency ω_0 so: $\eta = \omega/\omega_0$. The *natural frequency* ω_0 is a very important parameter of any vibratory system and is calculated in this case from the simple relationship $\omega_0^2 = k/m$ where in the example considered k is the bending stiffness of the shaft. In general, k is the spring constant of spring-mass system. For rotating machines, the corresponding frequency is usually specified as *critical speed*:

$$n_{\text{crit}} = 30 \cdot \omega_0/\pi\ [\text{min}^{-1}]. \tag{3.28}$$

It is easy to see that for $\eta \Rightarrow 1$ or $\omega \Rightarrow \omega_0$, the deflection becomes infinitely large ($w \Rightarrow \infty$), one speaks of resonance.

Although in real systems, the deflection does not go to infinity because frictional influences (⇒ damping) limit the amplitudes to a finite dimension, this state must either be avoided completely or at least be passed through quickly during start-up and

[13] Follow [28].

shutdown processes of a machine in order to prevent overloading of the components or rubbing of the rotors in the housing.

The state of movement $\eta < 1$ is described as "subcritical," where the shaft center *W* is located within the center of gravity *S*, that is, the centrifugal force excitation and the shaft deflection are in the same phase.

With "supercritical motion state" $\eta > 1$, on the other hand, the center of gravity is *S* is located within the shaft center *W*, which means that deflection and excitation are in opposite phases.

For ($\eta \Rightarrow \infty$), the center of gravity moves to the origin, the shaft center rotates with the eccentricity *e*.

Sample task

An electric motor of the total mass *m* and an operating speed *n* is mounted on an elastic base. It runs with an imbalance corresponding to a mass m_u with an eccentricity *e*.

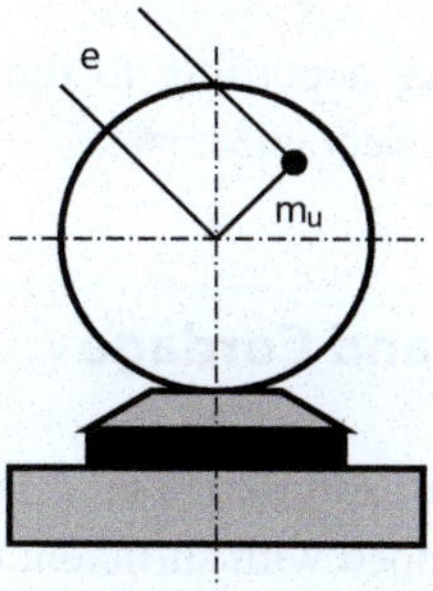

Given: $m_{tot} = 40$ kg, $m_u = 0.7$ kg, $e = 0.4$ mm, $n = 3000$ min^{-1}, $\eta = 4$

Wanted: a) Spring constant *k*, so that for undamped vibrations, a natural frequency ratio results from $\eta = \omega/\omega_0 = 4$,

b) Force applied to the foundation in the operating condition,

c) Check whether the vibration acceleration in the operating condition reaches critical values with respect to existing limit values

Solution[14]

a)

$$\omega/\omega_0 = 4, \text{ consequential: } \omega_0 = \omega/4,$$

$$n = 3000 \text{ min}^{-1} = 50 \text{ s}^{-1},$$

$$\omega = 2\pi f = 2\pi \cdot 50 \text{ s}^{-1} = 314 \text{ s}^{-1},$$

$$\omega_0 = 314 \text{ s}^{-1}/4 = 78.5 \text{ s}^{-1},$$

$$k = m_{tot} \cdot \omega_0^2 = 40 \text{ kg} \cdot (78.5 \text{ s}^{-1})^2 = 246{,}490 \text{ kg/s}^2.$$

[14]For further details on this task, see also [27].

b) The amount of force that the elastic support opposes the vibrations is proportional to the maximum amplitude that this support experiences; the proportionality factor is the spring constant:

$$|F| = k \cdot y,$$

where

$$y = e \cdot \frac{m_u \cdot \eta^2}{m_{\text{tot}}(1 - \eta^2)}, \quad \text{“}e\text{” is the eccentricity,}$$

therefore

$$y = 0.0004\ \text{m}[(0.7\ \text{kg} \cdot 4^2)/(40\ \text{kg}(1 - 16))] = |0.008\ \text{mm}|,$$

consequential

$$|F| = 246{,}490\ \text{kg/s}^2 \cdot 0.000008\ \text{m}, \quad F = 1.84\ \text{N}.$$

c)

$$a = \omega_0^2 \cdot y = 78.5^2\ \text{s}^{-2} \cdot 0.008\ \text{mm} = 49\ \text{mm/s}^2.$$

This means that all limit values according to the relevant regulations are complied with (see above explanations in Sect. 3.3). ◀

3.4 Anchor Gear, Lines, and Cordage

In order to keep the ship at sea, for example, in the roadstead, in a fixed position or safely moored at the pier, it must be equipped with sufficient *anchor gear* and *mooring* lines.

3.4.1 Anchor Gear

The *anchor gear* consists of an *anchor* and the *anchor chain*. The chain is connected to the anchor by means of the anchor shackle (often a swivel shackle). On smaller ships, the anchor chain can be replaced by a shorter chain lead (a few meters of chain directly at the anchor) and an anchor hawser (anchor line). As the weight of the anchor hawser or chain plays a significant role in the holding force of the anchor gear, occasionally, when using only hawser, lines with incorporated lead weights are used to compensate for the lower weight of the hawser to the chain. Riding weights can also be attached to the hawser for this purpose. However, these measures are more applicable to recreational craft. Only anchors and chains are used in maritime shipping.

With regard to the type of anchor, a distinction is made according to the anchoring reason. This is based on the knowledge that, depending on the anchor base, the optimum *anchor shape* should be selected.

Historical findings provide witness to ancient *anchor stones* (see Fig. 3.13).

Fig. 3.13 Anchor stone

The *anchor* commonly used on seagoing vessels today is the **patent anchor** (see Fig. 3.14), which is the best compromise solution in terms of durability for the different anchor bases, although there are nuances in the design with regard to its detail design (e.g., Hall anchor, Inglefield anchor, and others).

This anchor achieves its holding force by digging into the ground when the anchor chain is pulled almost horizontally. To ensure this horizontal pull, sufficient chain must be inserted when anchoring, depending on wind and sea conditions (see below).

Patent anchors belong to the stockless anchors. This means that at the upper end of the shaft, below the ring, they have no stick at right angles to the sparks. A further feature of the patent anchors is the rotating sparks.

The essential components of the patent anchor are shown in Fig. 3.15.

Besides the patent anchor, the *stock* or *admiralty anchor* (see Fig. 3.16) is still used for smaller fishing vessels, but especially (probably for reasons of nostalgia) for sailing ships.

Fig. 3.14 Patent anchor in anchor hull

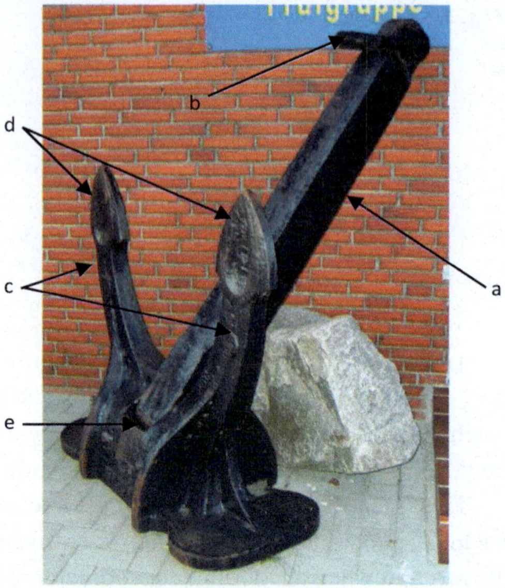

Fig. 3.15 Patent anchor. The *patent anchor* consists of the shaft (**a**) and the chain shackle (**b**). The two arms (**c**) with the hands or spade (**d**) are connected to the shaft by a bolt (**e**)

Fig. 3.16 Admiralty or stock anchor

The *chain* itself is composed of lengths from 27 m, which are connected by shackles. Each shackle is marked with a light color, to make it easier[15] to count the number of shackles that have run out when inserting the chain during anchoring; in this way, it is possible to determine how much anchor chain has already been inserted. The anchor

[15]The sailor understands *chain* as the lowering of the anchor chain.

Fig. 3.17 Anchor and mooring winch with anchor chain

chain runs through the anchor hawse to the ship's deck, from there further through a mechanical chain stopper over the traction wheel (the chain sprocket) of the anchor winch and further into the chain box.

In addition to securing the anchor by means of a chain stopper, the anchor winch also has a band brake for locking.

The wire rope stopper shown in Fig. 3.17 (see red arrow) is used for additional securing of anchor and chain outside the hunting ground.

The anchor chain or hawser must be inserted (extended) long enough so that the anchor remains flat on the ground even under tensile load via the anchor chain, otherwise it will be torn out of the ground and will no longer hold. If this is not the case, the ship shall be laid up for a *short day*. This should only be the case immediately prior to lifting (breaking out of the ground) and hoisting (pulling up), as the holding force of the anchor is then only low. In this respect, it is roughly assumed that three to five times the water depth should be attached to the chain or anchor hawser; in heavy sea and/or extreme wind conditions, this figure is correspondingly higher. The construction regulation of the classification societies may make different assumptions (see Sect. 3.4.2).

Seagoing vessels usually have between 10 and 12 shackles (1 shackle today corresponds to 27 m—see above) chain length per side (port and starboard); in this respect, the total chain length to be carried is about 600 m [25].

The length of the inserted chain is, therefore, indicated in shackles. For example, the instructions for anchoring can be: "5 shackles in the water" or "3 shackles on the capstan." It depends mainly on the water depth at the anchorage and the condition of the ground (silt, sand, shill, stones, vegetation, etc.). For the first few meters, the chain lies on the seabed and then leads in an arc (similar to an "e-function") upward to the

Fig. 3.18 Anchor chain

ship. The weight of the chain lying on the ground contributes significantly to the holding force. The bay in which the chain hangs from the ship to the ground is slightly stretched when force is applied to the ship for a short period of time, for example, by pitching movements in rough seas or swell, and thus acts as a shock absorber against force peaks, which could otherwise cause damage to the anchor capstan or break the chain. When the force is reduced, the weight of the chain or hawser sinks back to the bottom and the ship is pulled back to its original position. The greater the expected forces on the ship (due to wind pressure, current, sea state, ice drift), the more chain has to be inserted. If a lighter hawser is used instead of the chain, it should be correspondingly longer.

In general, stay chains are used as anchor chains (see Fig. 3.18).

3.4.2 Design or Dimensioning of Anchor and Chain

In maritime shipping today, the anchor chain is designed according to the regulations of the classification societies, such as the former GL.[16] The type of construction (normal chain or stay chain), strength, length, etc. are defined here using tables. The size, weight,

[16]See in more detail [8, Chap. 1 Hull, para. 18]. Note on the former GL Construction Regulations: These may differ from the construction regulations of other classification societies. Although new construction regulations have been issued by DNV · GL—the successor organization of GL—the old regulations still provide valuable information on planning, equipping, and dimensioning of hulls and components..

Fig. 3.19 Freeboard mark of the classification society **B**ureau **V**eritas. *TF* fresh water (tropics); *F* fresh water (summer); *T* tropics; *S* summer; *W* winter; *WNA* winter in the North Atlantic; *h* effective height from the summer low load line to the uppermost deckhouse $= a + \sum hi$; *a* distance, measured midships, from the summer low load line to the main deck side (m), $\Sigma h\ i$ sum of the deck heights of superstructures and deckhouses on the main deck with a width greater than *B*/4, measured at the center of the ship (m), any deck crack shall not be taken into account; if the main deck is locally interrupted or stepped in the area of the lowest superstructure or the lowest deckhouse, measurements shall be taken from the ideal deck to the center of the ship; *A* lateral area (m^2) of hulls, superstructures, and deckhouses, which are positioned above the low load line and within the length *L* and whose width is greater than *B/4*; *B* construction width, usually width of the construction waterline (B_{DWL}), it is the largest width of the construction waterline, measured on molding edge. *Photo* Wualex

and number of anchors can also be taken from these tables. The decisive factor for the design of the chain and anchor is the so-called "equipment number" (EN).

In inland navigation, the anchor chains and the anchor are designed to hold the ship against the current when anchoring.

The formula for calculating the EN for determining the equipment in accordance with the former GL Construction Regulation is based on the assumption that the flow velocity of the water is 2.5 m/s, the wind speed 25 m/s, and the **length of the anchor chain 6 to 10 times the water depth**. It is assumed that under normal circumstances, the ship will anchor with only one anchor and one anchor chain.

The equipment code Z_1 for *anchor* and *chains* is calculated according to Eq. 3.29[17]:

$$Z_1 = \text{EN} = D^{2/3} + 2hB + (A/10), \tag{3.29}$$

with *D* displacement on frames [t] in seawater with a density of 1.025 t/m^3 on summer freeboard[18] (freeboard mark, see Fig. 3.19).

[17]See [8, Chap. 1 Hull, para. 18, p. 18–1]; this was replaced by [6, Chap. 11, Hull equipment, supporting structure and appendages].

[18]For reasons of stability, a ship may not be loaded deeper than its freeboard marking. For further information on the freeboard markings, see also See Employer's Liability Insurance Association (Berufsgenossenschaft) Hamburg, "D2 Leaflet—On the meaning of the freeboard marks," in the version dated September 18, 1987.

An overview of the relevant standards relating to anchor chains can be found at the "Round Steel Chains" Standards Committee.[19]

Example for dimensioning anchor and chain

A cruise ship is being built at your shipyard. Determine the required number and type of anchors as well as the required anchor chain according to the relevant GL Construction Regulation. The following data material is provided by the design department:

$$D = 71{,}100 \text{ t}$$
$$h = 28 \text{ m}$$
$$B = 30 \text{ m}$$
$$A = 6{,}720 \text{ m}^2$$

Inserting the above data into the equation for the EN (Eq. 3.29) "EN $= D^{2/3} + 2hB + (A/10)$" delivers: EN$=$4068.

Construction regulation under **Table 1** of DNV·GL Regulation "Rules for Classification—Ships, Part 3 Hull, Chap. 11 Hull equipment, supporting structure, and appendages" follows (see Appendix 5): Two anchors with a weight of 12.3 t each are required; the total length of the anchor chain (stay chain) must be at least 687.5 m. Two of the three anchors must be connected to the chains, and a third anchor must be in reserve. ◀

3.4.3 Line and Cordage

On ships, line and cordage (see Fig. 3.20 for the assembly of cordage) are used for different purposes and made of different materials. The most important applications are as mooring lines and towing hawsers. However, according to their intended use, a distinction is made between cordage and lines for *standing* as well as *running rigging.* **Standing rigging** is the term used to describe ship's cordage (generally for sailing ships), which are firmly attached at both ends on deck or in the rigging (e.g., for bracing masts such as shrouds, stays, guy ropes, and also stretch ropes). The name is explained by the fact that these are usually not moved during maneuvers. The **running rigging** is the cordage that is usually shorn and moved by blocks (e.g., flag lines, halyards, and sheets, and also mooring lines, towing lines, and anchor lines).

[19]Round Steel Chain Standards Committee (Normenausschuss Rundstahlketten, NRK) at DIN German Institute for Standardization (Deutsches Institut für Normung e. V.) (www.din.de).

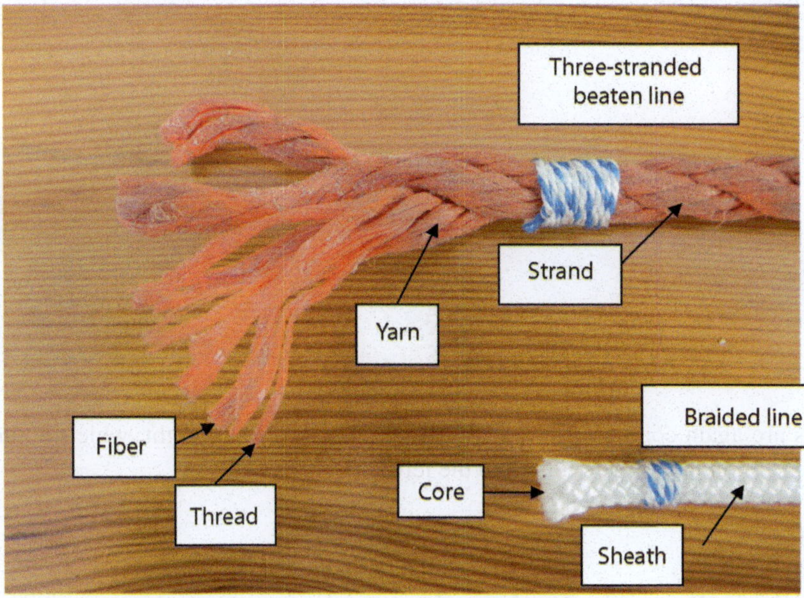

Fig. 3.20 Structure of cordage

Fig. 3.21 Twisted three-stranded cordage

Cordage[20] is the generic term for all *twisted* and *braided* cordage made of natural and synthetic fibers. A special group of the cordage forms the *Hercules cordage,* which has a soul (core) of wire rope.

Twisted cordage (see Fig. 3.21) is based on the twisting together of rope strands that are alternately turned to the left and right. Thus, in the most common ropes, three strands are twisted clockwise, which in turn are twisted anticlockwise from a multitude of yarns.

---

[20]The following (essentially) from [33].

Fig. 3.22 Braided cordage (mooring line)

The yarns are again composed of individual yarns turned to the right, which in turn are composed of individual fibers turned to the left.

Terminology

- **Fiber** = basic element of a thread
- **Thread** = composed of several twisted fibers
- **Yarn** = composed of several threads twisted in opposite directions
- **Cardel** or **ribbon** = composed of several yarns twisted in opposite directions
- **Rope** = composed of several oppositely twisted cardels
- **Hawser** = composed of several ropes twisted in opposite directions

Braided cordage (see Fig. 3.22) is available in three versions [22]:

- **Square mesh (Squareline)**: Eight strands are interlaced with each other. Half of these are turned to the left and half to the right, resulting in a roughly square cross section of the cordage. This type of braiding has the advantage that such ropes are very stretchy and easy to handle (often used as mooring lines in maritime shipping—see also Fig. 3.23).
- **Hollow braid:** This line consists only of load-bearing fibers and is therefore very light and easy to splice.
- **Core-sheath braid:** The load-bearing part of the line is surrounded by a sheath, which protects the core from abrasion and weathering.

Further distinguishing criteria can be found in Table 3.2.

Types of cordages

Natural fiber cordages (made of hemp, manila, sisal, coconut, cotton, or flax) are hardly used in seafaring because of the high wear and tear and the expensive, time-consuming production (exception: traditional shipping). Today, cordages made of synthetic fibers or wires have become widely used in commercial shipping. Synthetic fiber cordages have

Fig. 3.23 Lying alongside the pier

Table 3.2 Distinguishing criteria for cordage according to its diameter

Very strong cordage	φ>22 mm	Hawser
Weak cordage	φ up to 22 mm	Line
Thin cordage	φ up to 6 mm	Fastening material (also Hüsing, Schiemannsgarn)

Table 3.3 Types of cordages and their use

Type of cordage	Common trade names	Use
Natural fiber		Largely replaced by synthetic fiber cordages; use on traditional ships
Polyamide	Perlon, Nylon	Mooring lines, flag lines, towing hawsers
Polypropylene	Toplon, Polyprop	Floatable; lifeline, mooring lines
Polyester	Trevira, Diolen, Dacron	Mooring lines, anchor line, traps, pods
Polyethylene	Marlex	Mooring lines, towing hawser
Wire cordages		As standing rigging for shrouds and stay

the great advantage of being rot-resistant, but can be vulnerable to chemicals. The following materials are widely used (see also Table 3.3, for information on cordage materials and their application)[21]:

- Polypropylene (PP)—generally used for mooring lines,
- Polyamide (PA),
- Polyester (PES),
- Aramid and Dederon,
- Stainless steel material is now widely used for wire cordages (see Fig. 3.24).

[21]For color coding of fiber cordages, see Appendix 9.

Fig. 3.24 Wire rope construction. *Picture* Tachymètre, CC BY-SA 3.0

Table 3.3 clearly shows the types of cordages and their use.

External influences, exposure to chemicals, and also the handling of the cordage can lead to a *loss of strength,*[22] which can be adopted as follows:

- Through splices: approximately 10%,
- Sutured eye: approximately 20%,
- Through nodes: approximately 60–65%,
- By external heating due to friction,
- Due to internal heating because of the start of work (high bar), and
- Due to heat exposure because of solar radiation or heating.

Various *knots* is used for mooring the ship to bollards, rings, poles, etc. to connecting lines, etc. (colloquially also *sailor knots*; see also Appendix 8). Knots and stakes must be easy to insert, hold securely under load, and be easily and quickly released (without load).

As a rule, the pier can be fixed to the longitudinal side (see Fig. 3.23). To ensure that the ship lies safely, it is moored ashore with one or more *forelines*, *stern lines*, *fore* and *aft spring lines*, depending on its size. Large ships are sometimes stabilized between fore/aft line and spring line with an additional *chest* or *cross line* and with *head* and *tail line*. Figure 3.23 clearly shows two forelines (on a bollard) and four forespring lines (lines leading from front to aft).

The mooring lines are controlled by a *cluse* in the ship and occupied on bollards. On larger ships, the lines are often also pulled through by means of a cable winch.

For the question of the number, length, and breaking strength of mooring lines, the relevant construction regulations of the classification societies formulate the corresponding requirements, for example, under Table 1 of the DNV•GL Regulation "Rules for Classification—Ships, Part 3 Hull, Chap. 11 Hull equipment, supporting structure and

[22]Estimated breaking loads for cordages, see tables in Appendix 6.

appendages" contains instructions for the selection of anchors, towing hawsers, and mooring lines. Their number, length, and breaking strength are determined by *EN* (see also table in Appendix 5; [24]).

According to the abovementioned construction regulation, both wire and fiber ropes as well as ropes consisting of steel wire and fiber strands can be used for towing and mooring hawsers. The breaking forces given in Table 1 apply to steel and fiber ropes.

If mooring hawsers with breaking forces above 490 kN are specified in Table 1 of the abovementioned regulation, hawsers with lower breaking forces can also be provided if the number of hawsers is increased so that the product of breaking force × number of hawsers according to Table 1 is not fallen short of. However, the breaking strength of the individual hawser should not be less than 490 kN. The number of hawsers may also be reduced and the breaking strength of the hawsers increased at the same time, provided that the product of breaking strength × number of hawsers according to Table 1 is not undershot. However, there should be at least six mooring hawsers on board.

Regardless of the breaking load recommended in Table 1, the diameter of a fiber rope should not be smaller than 20 mm [24].

Example for the dimensioning of mooring lines

The fictitious ship in Sect. 3.4.2 must also be equipped with mooring lines. The customer would like to use cordages made of *PP*. Determine what is required according to the relevant construction regulation of DNV • GL. However, here for "*h*" 40 m and for "*A*" 7120 m^2 must be assumed.

The number of equipment for the recommended selection of hawsers and for determining the design load on fixed towing, mooring, and mooring equipment and their substructures is calculated in the same way as in Eq. 3.29; however, different values for "*h*" and "*A*" are used (they may be higher—as in this example. This takes into account the wind attack on the surface vessel: while a ship lying at anchor can swing around it, a vessel lying at the pier cannot give way to the wind pressure).

$$\mathrm{EN} = D^{2/3} + 2hB + (A/10). \tag{3.30}$$

With the given data, an EN = 4828 results.

From Table 1 of the Construction Regulation (see Appendix 5) "Rules for Classification—Ships, Part 3 Hull, Chap. 11 Hull equipment, supporting structure and appendages," seven mooring hawsers with a length of 200 m each result; the breaking load must be 686 kN.

Note: Between the **rope grade** and the **minimum breaking strength**, there is a relatively complex relationship, which is explained in more detail in DIN EN 12385-4.

The rope breaking load of wire ropes is determined by the diameter (*d*) in mm, the fill factor (*f*), the stranding factor (*k*), and the tensile strength (*Rm*) of the steel.

The diameter (d) is the largest outer diameter of the rope (edge measurement). The fill factor determines the proportion of the steel cross section in the total cross section. The stranding factor depends on the design; ropes always have strength, which is about 5–15% lower than the strength of the sum of the individual wires.

The minimum breaking force (MBF), that is, the minimum force the rope must achieve in the tensile test, is then calculated using the following formula:

$$F_{\min} = \frac{f \cdot Rm \cdot k \cdot d^2 \cdot \pi}{4} \text{ [N]}. \tag{3.31}$$

For the frequently used parallel lay ropes with steel core and the rope grade, for example, the MBF according to DIN-EN 12385 is: $F_{\min} = 630\, d^2$ [29]. ◀

Example for the design of a towing hawser

A harbor tug (see Fig. 3.25) should be equipped with a wire hawser as a precursor. The bollard pull of the tug is 30 t. The maximum rope angle to the towed tug shall be 45°. Select the required wire rope (see also Fig. 3.25, where the tension angle is clear).

The bollard pull is the horizontal force exerted by the tug. However, as the towing hawser runs from the tug to the bow of the annex at an angle α of maximum 45°, a so-called force triangle must be drawn for this case, as forces are vectors:

Fig. 3.25 Towing hawsers. *Photo* Buonasera, CC BY-SA 3.0

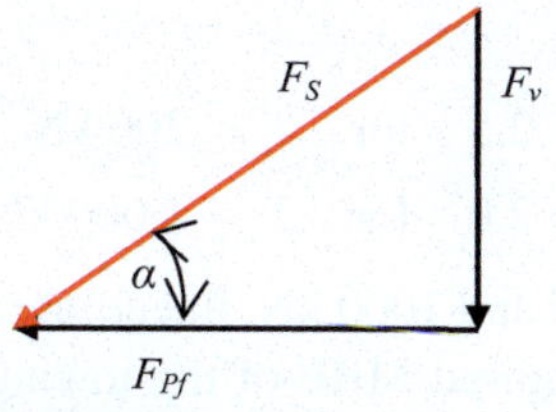

The pulling force in the towing line, therefore, has the horizontal component of the bollard pull (F_{bp}) and a vertically directed component (F_v). From the vector addition follows:

$$\vec{F_{tl}} = \vec{F_v} + \vec{F_{bp}}\,. \tag{3.32}$$

This vector addition can be controlled with the angle function

$$\cos\alpha = \frac{\vec{F_{bp}}}{\vec{F_{tl}}} \tag{3.33}$$

describe:

$$F_{tl} = \frac{F_{bp}}{\cos\alpha} = \frac{30\text{t}}{0.071} = 42\ \text{t}.$$

This corresponds to a force of $F_{tl} = 412$ kN. Taking into account a safety factor $f = 1.3$ for dynamic rope loads from wind and wave impact, the required breaking load strength for the towing hawser is 536 kN.

According to the catalog of, for example, the manufacturer Carl Stahl GmbH Munich,[23] a rope diameter of $d = 32$ mm must be selected (round strand rope, constr. 6×36 WS-IWRC, DIN EN 12385-4 (DIN 3064)).

According to No. C 4.3 of Section 25—Tugs—of the former GL Construction Regulations "I Ship Technology 1 Seagoing Vessels," the required MBF F_{min} of the towing hawser is to be determined on the basis of the design force T (the design force T corresponds to the towing force required by the operator or the bollard pull, if the pulling force is not defined—see Section 25, No. 2.1 of this construction regulation) and the factor K for the practical value as follows:

$$F_{min} = K \cdot T \tag{3.34}$$

[23]See Appendix 7 "DIN Ropes from Carl Stahl Munich."

with

$$K = 2.5 \quad \text{for} \quad T \leq 200\ \text{kN} \quad \text{and}$$
$$K = 2.0 \quad \text{for} \quad T \geq 1{,}000\ \text{kN}$$

(for values of T between 200 and 1000 kN, linear interpolation must be required).

According to GL, the required MBF of the towing hawser is to be determined as follows: As $T = 30{,}000\ \text{kg} \cdot 9.81\ \text{m/s}^2 = 294\ \text{kN}$, the K-value must first be interpolated linearly. ◄

Excursus on linear interpolation [7, p. 195 f.]

The linear interpolation, founded by Isaac Newton, is the simplest and probably the most widely used in practice. Here two given data points (x_0, f_0) and (x_1, f_1) are connected by a distance.

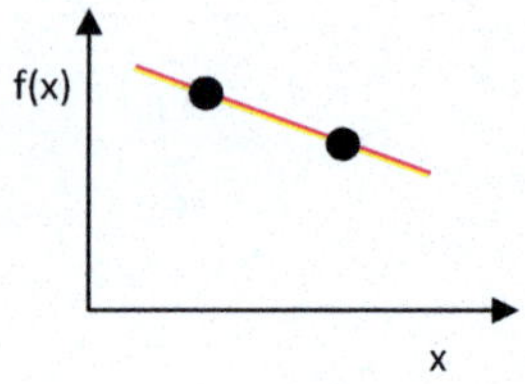

It is valid:

$$f(x) = f_0 + \frac{f_1 - f_0}{x_1 - x_0}(x - x_0) = f_0 \frac{x_1 - x}{x_1 - x_0} + f_1 \frac{x - x_0}{x_1 - x_0}. \tag{3.35}$$

Using the above equation with

$$f(x) = K$$
$$f_0 = 2.5$$
$$f_1 = 2.0$$
$$x_0 = 200\ \text{kN}$$
$$x_1 = 1{,}000\ \text{kN}$$
$$x = 294\ \text{kN}$$

follows for $K_{294} = 2.4$, thus

$$F_{\min} = 2.4 \cdot 294\ \text{kN}$$
$$F_{\min} = 706\ \text{kN}.$$

A comparison of the two results shows that, according to GL, a towing hawser with a higher MBF should be selected for our tug; according to the catalog of Carl Stahl Munich, a rope of rope grade $Rm = 1770$ with diameter 36 mm would therefore have to be selected. In this respect, it is important which load assumptions are made in individual cases or according to which construction regulation the design of a component is to be carried out.

3.4.4 Bollards

The mooring line or towing line is laid on bollards on board. The so-called rope friction is used (see Fig. 3.26) in order to keep the ship safe with only a few loops of the rope around the bollard.

On board, bollards are usually of welded construction, with a cast or welded thickening on top and are available in pairs. A double bollard on board is not only used to cover the mooring line (eight-shaped with a head strike—see Fig. 3.27), but can also be used as a brake by making use of the rope friction to slow down the ship with the mooring line passed first, usually the projection. For this purpose, it is covered with only a few turns, the loose end being sensitively lowered by hand. This is also known as shrinking.

A variant of the double bollard is the double cross bollard (Fig. 3.28). It is mainly used on smaller watercraft.

On a single bollard (see Fig. 3.29), a line is laid by means of bowline, weaving hitch or fixed (spliced) eye.

Example of rope friction at the bollard

The "Euler-Eytelwein formula," also called the rope friction formula, was developed by Leonhard Euler (1707–1783) and Johann Albert Eytelwein (1764–1848).

If a rope is wrapped around a bollard and pulled at one end of the rope, holding the other end with less force is sufficient to prevent the rope from slipping around

Fig. 3.26 Rope friction

Fig. 3.27 Head strike on double bollard

Fig. 3.28 Double cross bollards

Fig. 3.29 Single bollard

the bollard. This is because along the circumference of the bollard in contact with it, static frictional forces develop tangentially, which support holding.

For the ratio of pulling force F_z and the holding force F_h applies:

$$F_z \leq F_h \cdot e^{\mu_H \cdot \alpha}, \tag{3.36}$$

where α is the wrap angle (**in radians**), with which the rope is wrapped around the round object, and μ_H is the static friction coefficient.

If the rope slips on the bollard, the static friction coefficient μ_H is replaced by the coefficient of sliding friction μ_G.

As can be seen, the forces increase very rapidly with the angle of wrap. A steel cable, which is connected via a steel bollard with $\mu_H \approx 0.15$ to hold a ship, only needs about $\boldsymbol{F_h} \approx \mathbf{40\%}$ of the force to hold the force $\boldsymbol{F_z}$ that wants to cause a movement. With three wraps, 5.9% is already sufficient.

Calculation example: A ship exerts force $F_z = 50$ kN on the protrusion. The ship should be stopped manually (assumption $F_h = 0.3$ kN). The mooring line is made of PP ($\mu_H = 0.2$). How many full wraps are required around a bollard diameter 40 cm?

From the above equation, the following is calculated by converting and inserting for α in radians

$$\alpha = \frac{\ln \frac{F_z}{F_h}}{\mu_H}, \tag{3.37}$$

$$\alpha = 25.6.$$

By converting from radians to degrees with

$$\alpha^\circ = \frac{360^\circ \cdot 25.6}{2\pi} = 1{,}468^\circ. \tag{3.38}$$

By division with 360°, you get the number of wraps = 4. ◄

3.5 Ladders, Stairs, Railing

To get from one deck to the other, these heights are bridged by ladders, fixed ladders, and stairs. Both their constructive and structural design as well as the required strength are subject to technical regulations. The following sections provide an overview of the structural designs of these facilities.

3.5.1 Ladders

Basic provisions for ladders can be found in particular in the following regulations:

- BGV[24] D 19 "Watercraft with a type approval for inland waterways,"
- BGV D 36 "Accident prevention regulation for ladders and steps,"
- Safety rules for crampons and step irons (ZH 1/542),
- DIN EN 131-1 "Ladders; designations, types, functional dimensions,"
- DIN EN 131-2 "Ladders; requirements, testing, marking,"
- DIN 4567 "Ladders; assessment bases for ladders for special professional use,"

[24] BGV stands for "Berufsgenossenschaftliche Vorschrift" (Employer's Liability Insurance Association Regulation).

- DIN 24532 "Vertical fixed steel ladders,"
- DIN 83200 "Ladders on ships; overview, installation,"
- DIN 83202 "Fixed ladders on ships."

With regard to the cargo space, ladders on ***inland vessels*** is explained in Section 5 of the BGV D 19:

1. Vessels whose holds are accessed shall have at least one, or if the length of the hold exceeds 20 m, at least two fixed ladders per hold, which shall be diagonally offset.
2. Ladders and stairs shall allow safe entry and exit, also from the gangway. Single ladders must be provided with safety devices to prevent them from slipping and falling over.

Fixed metal ladders meet this requirement if they are guided in an escape route or have safe transitions at the point of interruption. They shall otherwise comply with this requirement if they

- DIN 83200 "Ladders on ships; overview, installation,"
- DIN 83202-1 "Fixed ladders on ships; lightweight design,"
- DIN 83202-2 "Fixed ladders on ships; medium-weight type,"
- DIN 83202-3 "Fixed ladders on ships; heavy construction"

correspond to this.

Rung gangways meet this requirement if they are guided in an escape route or have safe transitions at the point of interruption. In addition, they meet this requirement if they comply with ISO 9519 "Shipbuilding and marine engineering; wall and metal rungs" [11].

3.5.1.1 Strength of Ladders

The required strength of ladders results from the static calculation. The Employers' Liability Insurance Association Regulation (ELIAR; Berufsgenossenschaftliche Vorschrift, BGV) D 36 provides corresponding information:

> As a rule, the static calculation is based on a vertical force from 1500 N, which acts perpendicularly in the most unfavorable static position of the ladder in the use position. A safety factor of 1.75, related to the yield strength, must be taken into account in the static calculation of ladders and steps made of metal.
>
> For permissible bending stresses for ladders, see DIN EN 131-2 "Ladders; requirements, testing, marking."

Deflection

The deflection is determined according to DIN EN 131-2 "Ladders; requirements, testing, marking."

Table 3.4 Span and permissible deflection of ladders

Span L (mm)	Permitted deflection f (mm)
≤5000	$5L^2 \cdot 10^{-6}$
>5000 to ≤ 12,000	$0.043L - 90$

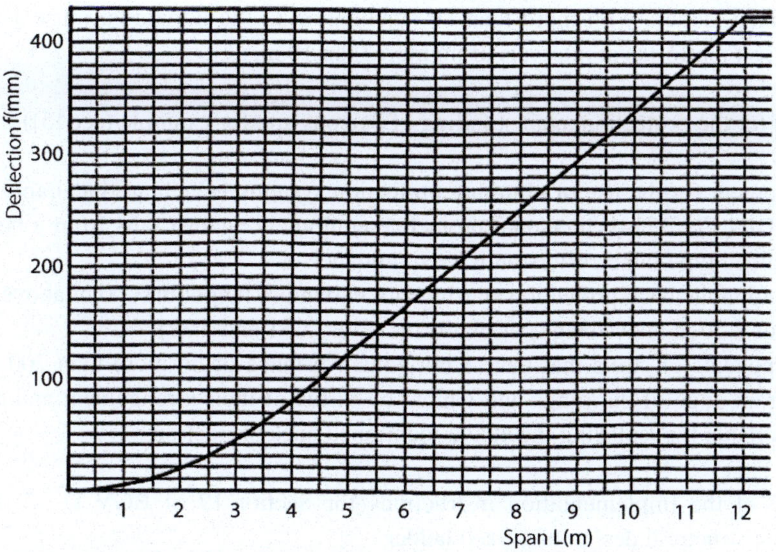

Fig. 3.30 Permissible deflection of conductors

The requirement for protection against excessive deflection is met if the deflection f as a function of the span L does not exceed the following values according to Table 3.4.

The span L is the ladder length minus a protrusion at each end of the ladder of 200 mm.

In Fig. 3.30, the permissible deflection f as a function of span L is shown.

Measures against excessive bending, especially on ladders with a length of more than 12 m, are, for example, spar supports or bracing.

3.5.2 Fixed Ladders

The difference between a ladder and a fixed ladder lies in the fact that a ladder is placed at an angle of about 65°–75° to the horizontal, while a fixed ladder is set up vertically.

Provisions for fixed ladders are formulated in Section 15 of BGV D 36 [5]:

1. Fixed ladders are only permitted if the installation of a staircase is not operationally possible or not necessary due to the low risk of accidents.
2. Fixed ladders must be firmly attached.

3. Fixed ladders must have a holding device at their point of exit.
4. Fixed ladders with possible fall heights of more than 5 m must, as far as operationally possible, be equipped with devices to protect against the fall of persons.
5. Fixed ladders with fall heights of more than 10 m must be equipped with devices that allow the use of fall arresters.
6. On fixed ladders with more than 80° inclination to the horizontal must be provided with rest platforms at intervals of no more than 10 m.

Devices for the protection against falling of persons are, for example (see [5]).

- Equipment for the use of climbing protection systems, which are inevitably activated (see DIN EN 353–1 "Personal protective equipment against falls from a height; Part 1: Guided type fall arresters with fixed guidance"),
- A continuous back protection starting at a maximum height of 3.00 m above the stand area or 2.20 m above stages or platforms (Figs. 3.31 and 3.32),
- Components or struts that have a horizontal distance of no more than 700 mm from the front edge of the rungs and which, by virtue of their arrangement and condition, are suitable for replacing the above-mentioned back protection.

Figure 5 of the implementation instructions for Section 15 of BGV D 36 (Fig. 3.31) shows the structural design of a fixed ladder.

3.5.3 Stairs

Stairs can be found both in the exterior and interior of ships. Especially on cruise ships, stairs of special design are often installed. Figure 3.33 shows stair sections on the AIDAmar.

Basic regulations for stairs are found in particular in the following regulations:

- DIN EN 13056:2000 "Stairs with angles of inclination from 30° to < 45°,"
- DIN EN 790:1994 "Stairs with angles of inclination from 45° to 60°,"
- DIN 83214 "Stairs and stair railings for exterior and interior use in seagoing vessels—Basic requirements,"
- DIN 83215 "Stairs and stair railings for the exterior and interior of seagoing vessels – Stairs,"
- (Stairs according to these standards are used in the exterior and interior of ships and of floating equipment of the German Navy; they are not intended for the passenger area on passenger ships. The individual parts, step depth, run width, and clear passage dimensions are specified. In addition, specifications are made, among other things, on the payload, the anti-slip properties of the steps, and the installation of the stairs.)

Fig. 3.31 Fixed ladder with back protection

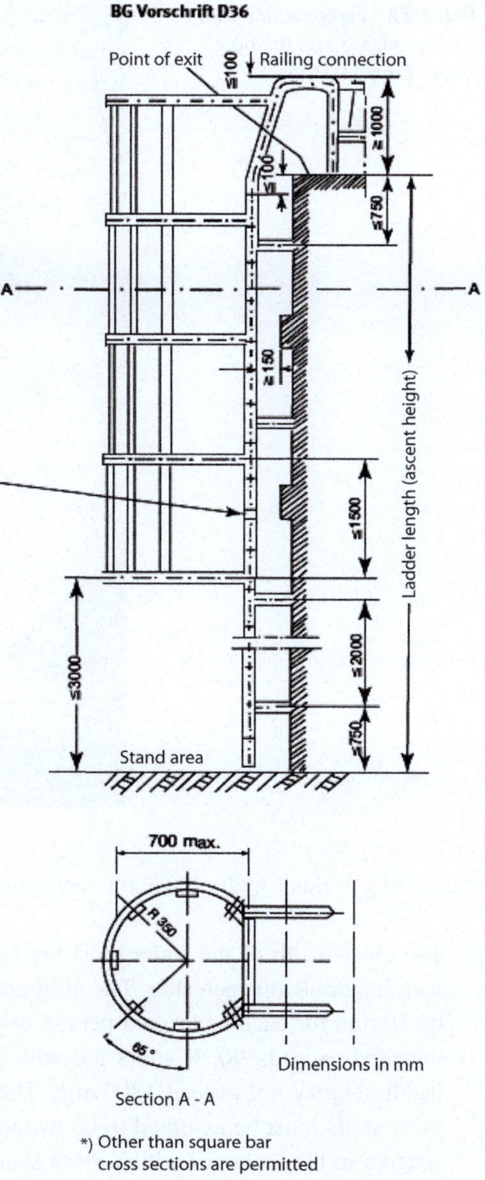

- DIN 83217 "Stairs and railings in cargo tanks of ships—Basic requirements."
- Essential requirements according to this standard: distance between supports of 1500 mm, railing height at stairs 1000 mm,
- Directive 2009/45/EC of the European Parliament and of the Council of May 6, 2009 on safety rules and standards for passenger ships.

Fig. 3.32 Fixed ladder with back protection on the boat deck of the AIDAmar

According to this EU directive, the following minimum requirements apply to stairs:

1. The clear width of the stairs must not be less than 900 mm. Stairs must be provided with handrails on each side. The minimum clear width of the stairs must be increased by 10 mm for each additional person if the number of the persons for whom they are intended exceeds 90. If stairs are wider than 900 mm, the clear width between the handrails may not exceed 1800 mm. The total number of persons to be evacuated by these stairs must be assumed to be two thirds of the crew and the total number of passengers in the areas for which these stairs are intended. The width of stairways shall at least comply with the standard adopted by IMO Resolution A.757(18).
2. All stairs intended for more than 90 persons must be arranged in the longitudinal direction of the ship.
3. Doorways and corridors and intermediate landings forming part of escape routes shall have the same dimensions as stairways.
4. The vertical extension of the stairs must not exceed 3.5 m without the presence of a landing and the angle of inclination of the stairs must not exceed 45°.

a b

Fig. 3.33 **a, b** Stairs in the ship AIDAmar

5. The stairway forecourt areas on each deck level must have a floor space of at least 2 m^2 and, if they are intended for more than 20 persons, must have and must be 1 m^2 larger for every additional 10 persons, but need not be larger than 16 m^2 in total with the exception of those stairway forecourt areas where there is direct access from public spaces to the stairwell.

Basically, the following applies to stairs (see Fig. 3.34):

a) Should run fore and aft, and
b) Be at least as wide as the openings or other traffic routes to which they lead.

If stairs are part of a workplace, which is always assumed in commercial shipping, they are also subject to the requirements of the Workplace Ordinance (Arbeitsstättenverordnung, ArbStättV).

The BG Transport and Traffic, Ship Safety Division, states under "D.1 Mechanical and electrical equipment" with status 02/2012 under 3.2 [17]:

Fig. 3.34 Stairs according to the above requirements

Stairs in service rooms should be made of steel according to DIN 83206. If possible, they should be arranged in the longitudinal direction of the ship. Stairs in machinery spaces with up to four steps and over shaft lines may be installed in the transverse direction of the ship.

The inclination of the staircase, in relation to the horizontal, must not exceed 60°. The clear height above stairs, measured at the leading edge of the step, must be 2.00 m. If the maximum height between two landings is greater than 3.70 m, they must be divided by landings. The tread area in front of stairs and landings may not be less than 600 mm × 600 mm. The step height should not exceed 230 mm, and the step width 140 mm. The step spacing must be completely even. The top step is to be designed according to DIN 83206 with 250-mm width. Dirt traps (mudguards) in accordance with DIN 83208 must be installed under the stairs, which are located above open spaces and traffic routes. There must be no skirting boards or edges at the top of the stairs to avoid the risk of tripping. Compare DIN 83204, 83205, 83206, 83207, and 83208.

On ships with a gross tonnage of more than 6000, stairs shall be provided in service rooms in which transverse thrust units are arranged. On ships with a gross tonnage of up to 6000, stairs are to be arranged where possible. A fixed ladder can be placed up to 3.00 m above the upper edge of the transverse beam channel.

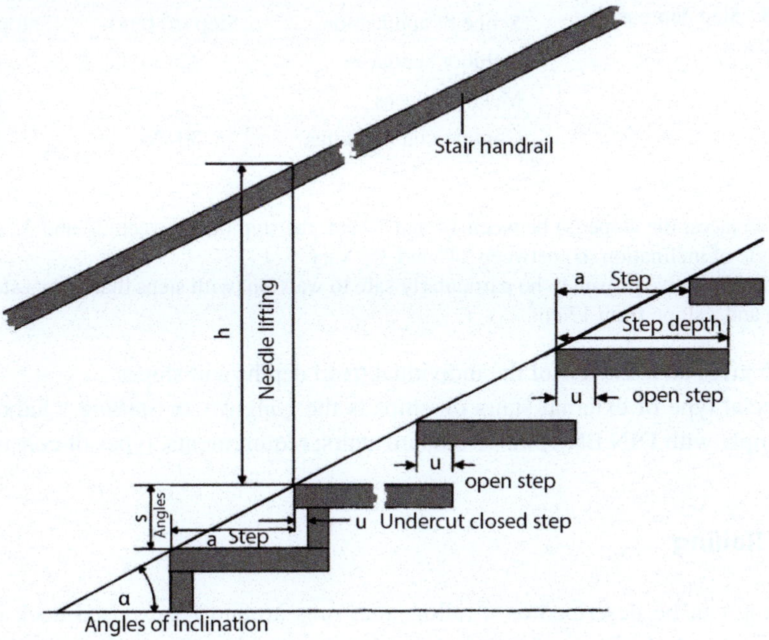

Fig. 3.35 Dimensions and designations of a staircase

Further requirements for stairs (see also Fig. 3.35) can be found in the Workplace Directive (WD; Arbeitsstättenrichtlinie, ASR) "A1.8—Traffic routes."[25] The following requirements are formulated under no. 4.5:

1. Staircases must be designed in such a way that they can be safely and easily climbed. This is achieved by means of sufficiently large, level, nonslip, recognizable, and load-bearing tread surfaces at regular intervals corresponding to the step size.
2. The slopes and treads of a staircase connecting two floors must not differ from each other. The steps should be illuminated with high contrast and, if possible, without disturbing glare for the user (see ASR A3.4 "Lighting").
3. Taking into account the risk of accidents, stairs with straight flights are to be preferred to those with spiral flights or spiral flight sections. Spiral and spindle staircases are not permitted on the first escape route (see ASR A2.3 "Escape routes and emergency exits, escape and rescue plan").
4. For stairs … the relationship between step length (SL), step (a), and slope (s) is the step size $2 \cdot s + a = \mathrm{SL}$. To ensure that a staircase is easy to walk on, the step length should be between 59 and 65 cm.

[25]Workplace Directives can be downloaded under [16].

Table 3.5 Step depth and slope of stairs

Scope of application	Step (*a*) (cm)	Slope (*s*) (cm)
Outdoor staircases	32–30	14–16
Meeting places	31–29	15–17
Commercial buildings	30–26	16–19

In workplaces, the slope (*s*) between 14 and 19 cm, the step (*a*) between 26 and 32 cm, and the angle of inclination (α) between 24° and 36° vary.

Staircases have proven to be particularly safe to walk on, with steps that have a step from 29 cm and a slope from 17 cm.

Table 3.5 gives an overview of the individual tread depths and slopes.

A special type of external stairs on ships is the *gangway* or *stelling*. Outboard stairs must comply with DIN EN 1502 “Outboard stairs; requirements, types of construction.”

3.5.4 Railing

The railing can be described as a railing that runs around an exposed deck or around deck openings. A distinction is made between closed, open, fixed, removable, and folding railings.

A *closed railing* is called a bulwark or entrenchment. This consists of a slab corridor running around the ship above the shear corridor. The upper end is formed by the railing profile. On superstructure decks, this profile is often covered with a teak strip, and on sailing ships and yachts, it is also used on the main deck.

An *open railing* (Fig. 3.36) consists of a series of vertical supports and horizontal intermediate bars; the upper end is formed by the railing profile.

Fig. 3.36 Railing

Height of railing and bulwark

In Section 21 "O. Railing" of the Classification and Construction Regulations I Ship Technology—1 seagoing vessel—of GL,[26] there are basic requirements for **railing**. According to these, their height must be at least 1.0 m above deck. The height below the lowest railing passage must be $\leq$230 mm, and the distance between the other passages must be $\leq$380 mm (see Fig. 3.36). With regard to the railing construction, reference is made to DIN 81702 or equivalent standards.

The **height of the bulwark** must be at least 1.0 m according to Section 6 "K. Bulwark" of the abovementioned GL Construction Regulation.

Regarding the **inland navigation**, the DIN EN 711 "Inland navigation vessels; Railings for decks; Requirements, types of construction" is relevant.

According to Section 11.02 *Protection against falls and crashes,* the regulation "Minimum technical requirements for vessels on the Rhine and on inland waterways of zones 1, 2, 3, and 4 for vessels applying for a certificate (Annex II to the Regulation on Inspection of Inland Navigation Vessels),"[27] the following rules must apply to the **inland navigation** transport:

> Outer edges of decks and those working areas where the height of fall may exceed 1 m must be fitted with bulwarks or hatch coamings, each at least 0.70 m high, or with railings in accordance with European standard EN 711, consisting of handrail, intermediate knee-level drawbar and baseboard. In the case of gangways, a baseboard and a continuous handrail must be provided at the hatch coaming. If gangway railings are available that cannot be folded down, the handrail at the hatch coaming can be dispensed with.
>
> In working areas where the height of fall exceeds 1 m, the inspection body may require appropriate facilities and equipment for safe working.

According to Annex 1 No. B. 24 "Safety requirements for new and existing passenger ships engaged on domestic voyages," Commission Directive 2010/36/EU of July 1, 2010, the railing must be at least 1.10-m high.

According to the Workplace Directive (ASR) A2.1 "Protection against falls from a height and falling objects, entering danger zones," the following railing heights apply in accordance with No. 5.1:

> The defenses must be at least 1.00 m high. In the case of parapets, the height of the reinforcement may be reduced to 0.80 m if the depth of the reinforcement is at least 0.20 m and the depth of the parapet provides equivalent protection against falls.
>
> If the fall height is more than 12 m, the height of the bracing must be at least 1.10 m.

[26]Stand 2012.

[27]In the version of the amendment dated June 16, 2014 (BGBl. I, p. 748).

Example for determining the height of the railing

On a cruise ship, the sun deck 15 m lies above the water level. Which railing height should be chosen?

Answer: 1.10 m. As GL's Construction Regulation requires at least 1.0 m, but the ship is also a place of work, and as the present case involves a fall height of>12.0 m, the regulation of ASR A2.1 is relevant because it is more specific. ◀

Notes on statics

For the design of the **bulwark**, binding statements are made in accordance with GL Construction Regulations in Section 6 "K. Bulwark."

The thickness of the bulwark must not therefore be less than

$$t = \left(0.75 - \frac{L}{1{,}000}\right) \cdot \sqrt{L} \text{ at ship length } L \leq 100 \text{ m} \tag{3.39}$$

and

$$t = 0.65 \cdot \sqrt{L} \quad \text{at ship length } L > 100 \text{ m,} \tag{3.40}$$

where for *L*, no greater value than 200 m must be used.

The bulwark, which is particularly exposed to the sea wash from the front in the fore ship area, shall have the thickness of the forecastle plating in accordance with a special calculation specification in accordance with Section 16, B.1 of this construction regulation.

The bulwark shall be supported at every second frame. As a rule, the section modulus of the support connected to the deck should not be less than

$$W \geq 4 \cdot p \cdot e \cdot l^2 \text{ [mm}^3\text{]} \tag{3.41}$$

with

e Column spacing in meters,
l Column length in meters

The load p is applied with at least 15 kN/m^2.[28]

With regard to the statics of the railing, the following must be observed: The railing support is to be considered statically like a clamped rod. This is subject to bending load in the horizontal direction by people leaning against it. Wind loads must also be taken into account additionally, as far as the railing is infill. In accordance with DIN 1055

[28]Further details can be found in the abovementioned construction regulation.

Fig. 3.37 Detail of the foot formation (clamping) of a railing support

"Load assumptions for buildings," vertical loads must also be assumed from the dead weight of the railing, depending on how the railing is mounted on the hull of the ship. With regard to the particularly relevant horizontal loads, there are usage categories with three different impact loads in this DIN 21. For private houses, for example, a horizontal load of 0.5 kN/m (corresponds to approximately 50 kg/m of railing) applies. In public buildings, the load is 1.0 kN/m, and for buildings with crowds of people, 2.0 kN/m. The loads must be applied to the uppermost point or the handrail. For a cruise ship, comparable to a building with crowds of people, 2.0 kN/m should therefore be taken into account.

Based on the horizontal force to be selected, the distances between the railing supports and the bending moment at the foot of the railing support (as already mentioned above, the foot of the railing support is to be regarded as a restraint; see Fig. 3.37), which must be introduced into the hull by design, must then be determined.

The wall thickness of the profiles used also results from the static calculation.

Example for dimensioning a railing support

An open railing, as shown in Fig. 3.37, should be designed. The distance between the railing supports should be 2 m; the railing support itself should be made of flat steel. The shear force Q of 2 kN/m will be considered. Height of the railing is 1.1 m. The steel used is normal shipbuilding steel. The cross section of the flat steel must be determined.

The static system is a clamped support with horizontal force acting on the upper end.

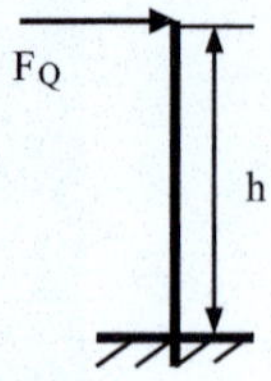

The lateral force on the support F_Q is calculated by multiplying the line load by the distance between supports s:

$$\begin{aligned} F_Q &= Q \cdot s \\ F_Q &= 2 \text{ kN/m} \cdot 2 \text{ m} \\ F_Q &= 4 \text{ kN.} \end{aligned} \tag{3.42}$$

The maximum bending moment of a cantilever arm at the clamping is calculated with the equation

$$M_b = F_Q \cdot h \tag{3.43}$$

with "h" the railing height to

$$\begin{aligned} M_b &= 4 \text{ kN} \cdot 1.1 \text{ m} \\ M_b &= 4.4 \text{ kNm.} \end{aligned}$$

The bending stress σ_B thus occurring at the clamping is calculated from the quotient of the bending moment and the axial section modulus of the column cross section. The existing bending stress must be less than the bending stress permissible for the material $\sigma_{B\text{permissible}}$. This can be found in various standard sheets. GL Construction Regulations I Ship Technology, 1 Seagoing Ships, Section 2 A gives a yield strength of R_{eH} of at least 235 N/mm^2 for normal strength shipbuilding steel. It can be used here as a permissible bending stress.

$$\sigma_B = \frac{M_B}{W} \leq \sigma_{B,\text{permissible}}. \tag{3.44}$$

With the given values for W, the result is obtained by switching over:

$$W = 18{,}723 \text{ mm}^3.$$

The axial section modulus for a rectangular cross section is determined as follows:

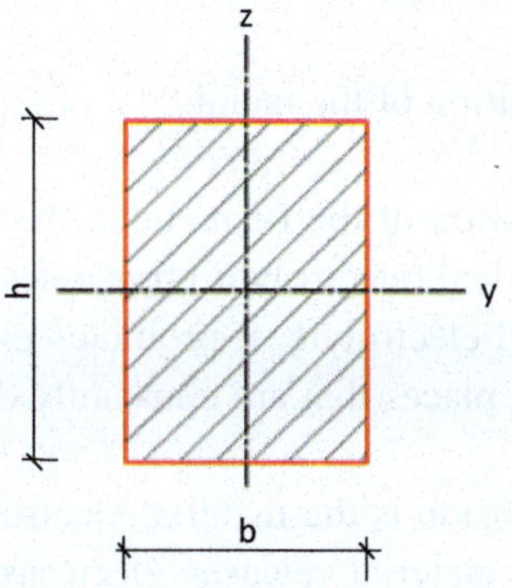

For a rectangle with the width b parallel to the y-axis and height h, the section modulus with respect to the horizontal axis is

$$W_y = \frac{b \cdot h^2}{6}. \tag{3.45}$$

For the same rectangle, the section modulus with respect to the vertical axis

$$W_z = \frac{h \cdot b^2}{6}. \tag{3.46}$$

In this respect, a flat steel 110×10 mm S235JR+AR DIN EN 10058 (with $W_y = 20{,}167$ mm^3) would be suitable. ◀

3.6 Corrosion Protection and Deck Coverings/floors

Wind, weather, and sea are taking their toll on the ship. Metallic parts tend to corrode, and the deck can become slippery due to water and possibly ice. In this respect, requirements must be made with regard to corrosion protection for reasons of occupational safety and for the protection of passengers.

3.6.1 Corrosion Protection

Corrosion[29] is a damage to a material caused by chemical attack; the attack occurs from the surface.

[29]For the definition, see [13, *p. 527*]; [10, *p. 55 f.*].

Corrosion is divided into two main areas:

1. Pure oxidation and
2. Electrochemical decomposition of the metal.

Pure oxidation is the combination of the Fe metal with oxygen in the air. The oxidation is "supported" by electrochemical (electrolytic) processes.

As seawater is a very good electrolyte, it promotes electrolytic decomposition wherever it is present. Particularly, places that are constantly exposed to air and seawater corrode well.

The actual "scene" of corrosion is the material/electrolyte interface. During the interfacial reactions, the metallic material releases electrons and is thereby dissolved (so-called anodic metal dissolution): $M \rightarrow M^+ + e^-$. At the same time, a cathodic metal deposition ($M^+ + e^- \rightarrow M$), a cathodic hydrogen evolution ($2\,H^+ + 2e^- \rightarrow H_2$), or cathodic hydroxyl ion formation ($1/2\,O_2 + 2\,H_2O + 2e^- \rightarrow 2\,OH^-$) takes place on the workpiece. Which of these reactions prevails depends on the environment and the metals involved.

The corrosion can also be caused by different *local elements*. Here, two different metals lie directly next to each other, an electrolyte is added (e.g., bronze plain bearing in steel housing in seawater). This is called *electrochemical corrosion* [13]. Together with the electrolyte solution, the two metals in contact form a short-circuited galvanic element, *the local element*. As with every galvanic element, electrons flow from the less noble metal to the nobler metal. The less noble metal (e.g., iron) goes into solution in the form of ions. At the more noble metal (e.g., the bronze bush), cations are discharged from the solution (in the case of seawater hydrogen ions). This means that the less noble metal is destroyed over time. The degree of electrochemical reactions depends on the electrochemical voltage series.[30] The greater the distance between the standard electrochemical potentials between the more noble and the less noble material, the greater and stronger the decomposition of the less noble metal.

During *rust formation,* the reaction of ferrous metals and oxygen to form iron oxides, metal dissolution, and cathodic hydroxyl formation take place in highly oxygenated environments (rainwater, seawater, flowing waters). *Rust* is the corrosion product of iron or steel by oxidation with oxygen.

The term **corrosion protection** refers to all measures to prevent corrosion. They must therefore be suitable for preventing the corrosion mechanisms described above [3].

An effective corrosion protection of the underwater hull consists of *coating*—passive corrosion protection—and a *cathodic protection* with galvanic sacrificial anodes or external current anodes—active corrosion protection.

[30]See, for example, [30]; for electrochemical voltage series of various elements, see Appendix 10.

3.6.1.1 Cathodic Corrosion Protection

Cathodic corrosion protection is always used in metal shipbuilding. Its structure and principle correspond to that of a galvanic element. The effect of cathodic corrosion protection is based on the compensation of oxidation-related currents on the metal surface by protective current. The task of cathodic corrosion protection in shipbuilding is to compensate for the corrosive influences on the underwater hull, that is, to protect the steel of the ship's hull in the area of the damage and defects of the coating system as well as the uncoated metal surfaces (e.g., propellers, shafts) from corrosion. The protective current can—as already mentioned—be generated either by galvanic sacrificial anodes or by external current protection anodes and direct current source.

Cathodic protection with galvanic sacrificial anodes

The principle is based on the fact that the underwater hull to be protected is connected to a less noble metal in a short circuit, whereby the steel becomes the cathode and the less noble metal becomes the sacrificial anode. Corrosion protection with sacrificial anodes (Fig. 3.38) cannot be adapted to the required protective current. Sacrificial zinc anodes represent only partial protection, that is, in the case of larger, exposed outer skin surfaces, the protection is not sufficient. A comparable full protection with sacrificial anodes, as achieved with potential-controlled external current anodes, is not justifiable for reasons of weight and cost. Openings in the outer skin, such as sea chests, bow thrusters, and also the ballast water tanks must be additionally protected by galvanic anodes. When the sacrificial anodes have worn away, they have to be replaced; this is usually after about 2 years.

Fig. 3.38 Sacrificial anodes in the stern tube and rudder blade area

Design of material requirements for galvanic sacrificial anodes for external protection (according to VG standard 81256-1)

At the beginning of the calculation of the material requirements is the determination of the size of the area to be protected. As far as the corrosion protection of the underwater surface of ships is concerned, the area is usually to be taken from the building documents. If this is not the case, the size of the area can be calculated with sufficient accuracy using the following equation:

$$A_U = L_{\mathrm{CWL}} \cdot (B_{\mathrm{CWL}} + 2T_{\mathrm{CWL}}) \cdot \delta \tag{3.47}$$

with:

A_U	Underwater area (wetted surface of ship; m^2),
L_{CWL}	Length at construction waterline in meter (m),
B_{CWL}	Width at construction waterline at molding edge of frame at 0.5 L_{CWL} (m),
T_{CWL}	Construction depth at 0.5 L_{CWL} obtained at the base (m),
δ	Full displacement ratio(see [34]),

where $\delta = B \cdot T$ is, with B being the width of the ship on main frame and T the depth.

The calculated underwater area applies only to the hull; for the determination of the total area to be protected $A_{U,\,\mathrm{flat}}$, the appendices, propellers, and shafts must be calculated separately according to the building documents and value for A_U can be added together.

Total protective current

The required total protective current is:

$$I_G = A_{U,flat} \cdot I_S \tag{3.48}$$

with:

I_G	Total protective current (A),
$A_{U,\,\mathrm{flat}}$	Total area to be protective (m^2),
I_S	Protective current density (A/m^2)
I_G	Total protective current (A),
$A_{U,\,\mathrm{flat}}$	Total area to be protective (m^2),
I_S	Protective current density (A/m^2)

Protective current density

For single propeller ships made of shipbuilding steel, a protective current density of 0.015 A/m^2 is taken as a basis. For multi-propeller ships and ships of the German Federal Armed Forces (each made of shipbuilding steel), a protective current density of 0.02 A/m^2 is to be applied for design. When used in predominantly tropical waters, higher protective current densities may be required.

For ships made of shipbuilding steel, which are used for voyages in ice, considerably higher protective current densities are required due to the expected coating damage. Depending on the area of operation, 0.06 A/m^2 must be calculated.

Determination of the total anode weight

$$m_G = \frac{A_{U,\text{flat}} \cdot I_S \cdot t_S}{Q_G} = \frac{I_G \cdot t_S}{Q_G} \tag{3.49}$$

with:

m_G Required total anode weight (kg),
I_G Total protective current (A),
I_S Protective current density (A/m^2),
t_S Term of protection (h),
Q_G Current content (Ah/kg)

The theoretical current content of the anode material can be taken, for example, from Section 7 B of the GL Construction Regulations "VI Supplementary Regulations and Guidelines—10 Corrosion Protection." For zinc anodes, a value from 780 Ah/kg can be found there.

Now, the entry of zinc into the sea is not unproblematic for environmental protection aspects. In this respect, cathodic corrosion protection by means of external current is also common for this reason.

Cathodic corrosion protection by external current
With external current protection, monitoring of the corrosion protection on the hull is easier. The automatically operating, potential-controlled electrical corrosion protection system adapts to the respective condition of the underwater hull to be protected. The protective current strength depends on the electrolytic properties of the seawater, the condition of the outer skin, and the conditions of the berth (stray currents through quay facilities).

Function
The measuring and control electrodes are built into the outer skin in an insulated manner and are connected to the control device of the protective current device. They measure the actual potential. The signal of the control electrode (= difference to the set potential) is used to control the required protective current. The control electrodes are loaded by a current of different magnitude depending on the type of protection current device.

On ships, the protective current devices must be particularly robust and resistant to vibrations. The control is carried out with magnetic amplifiers, via variable transformers with servomotor or via phase angle control with thyristors. The supply systems also contain current and potential measuring devices for the individual external current anodes

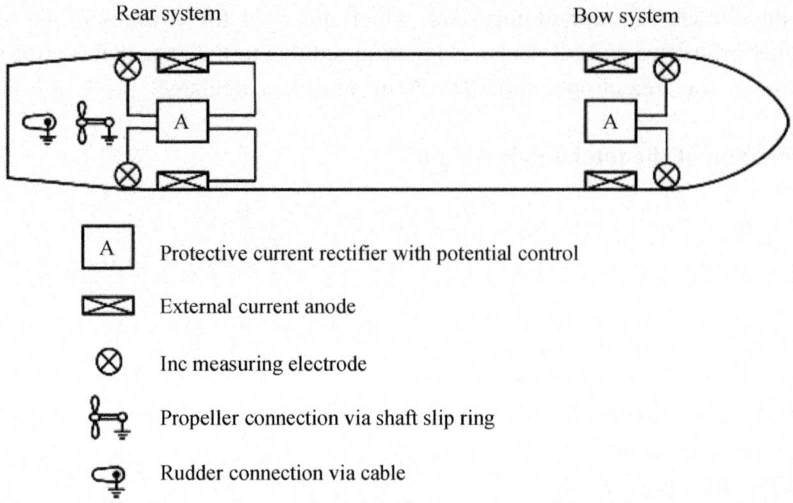

Fig. 3.39 Schematic arrangement of an external power system. *Source* Classification and Construction Regulations GL, VI Supplementary Rules and Guidelines, 10 Corrosion Protection

and measuring electrodes. For larger systems, the most important data are also plotted. Silicon rectifiers are preferably used because of their relatively high power. To protect against overloads in case of low resistance contact to large-area earthed systems, for example, in the harbor, a current limitation or current cut-off must be provided. In the latter case, optical or acoustic warning signals must be used to indicate when the system must be switched on again after the contact is broken. Accordingly, a voltage limitation can also be provided if the external current anodes require this.

Depending on the size of the ship, two external power supply systems can also be installed, which then cathodically protect the stern/midship areas and the forecastle independently of each other.

The anodes are applied to the hull of the ship or embedded in the outer skin. Attachments such as rudders or stabilizers are connected to the hull of the ship by flexible cables so that the surfaces wetted by seawater are integrated into the cathodic protection. Propellers and shafts are connected to the hull of the ship by slip rings in an electrically conductive manner and are thus included in the cathodic protection (Fig. 3.39).

Protective current density for external current protection systems

When determining the protective current densities for the underwater area, the speed of the ship is also decisive. For coated underwater surfaces, the minimum values given in Table 3.6 (protective current densities according to VG standard 81259 Part 1) must be applied.

Table 3.6 Protective current density as a function of ship speed

Ship speed (kn)	Protective current density (A/m^2)
$\leq$20	0.015
>20 to $\leq$ 25	0.030
>25	0.040

For ships used for voyages in ice, considerably higher protective current densities are required due to the expected coating damage. Depending on the sailing area, a minimum of 0.06 A/m^2 must be calculated.

Calculation of the total protective current requirement I_S

$$I_S = \sum I_{Si} \cdot A_i \tag{3.50}$$

with:

I_{Si} Protection current requirement per square meters (A/m^2),
A_i Area (m^2)
I_{Si} Protection current requirement per square meters (A/m^2),
A_i Area (m^2)

3.6.1.2 Hot-Dip Galvanizing

Hot-dip galvanizing is the application of a metallic zinc coating to iron or steel by dipping it into molten zinc (at about 450 °C). A resistant alloy layer of iron and zinc is formed on the contact surface and a firmly adhering pure zinc layer is formed on top [31].

Individual steel components are also hot-dip galvanized for corrosion protection. This protects them in two ways: On the one hand, through the effect of the active cathodic corrosion protection, and on the other hand, the complete coating provides a shield of the component against water and oxygen.

3.6.1.3 Galvanic Separation of Two Different Metals

In order to prevent electrochemical corrosion of a steel hull by direct contact with the lead ballast brought in by sailing ships—for example, Gorch Fock (lead is "nobler" than iron—see Appendix 10), this area of the hull must be coated, for example, with a layer of PA or epoxy resin.

3.6.1.4 Coatings/Paintings

The underwater hull as well as the surface hull are painted not only for optical reasons. The paint or coating prevents the electrolyte from coming into contact with water and oxygen. In many cases, metallic zinc is found in the synthetic resin-based coatings. Thus, both protection principles are combined: active protection by cathodic corrosion protection and passive corrosion protection by barrier for water and oxygen.

Note: With regard to maintenance work to be carried out on paints and coating materials, the manufacturer's instructions must always be observed. This applies in particular to the instructions for occupational health and safety (paints can be flammable liquids, and under certain conditions, they can also form explosive atmospheres; skin contact must be avoided, as must inhalation of paint vapors). In general, the surface must be clean, dry, and free of dust and grease when applying paints. Even though some manufacturers state that the coating can be applied directly to loose rust, it is still advisable to remove rust and loose coatings; rust is the "nobler element" in the electrochemical series of voltages compared to iron or steel. Solid layers of paint must be roughened.

In particular, "smaller" components are *powder-coated* instead of painted.[31] The powder paints used for powder coating generally consist of dry, granular particles that are between 1 and 100 μm in size. Chemically, these are mostly based on epoxy or PA resins. Hybrid systems containing both epoxy and PA resins as binders are also common. These are applied electrostatically to the component to be coated using a spray gun.

Several standards exist for powder coating. In particular, DIN 55633 and EN 15773 should be mentioned. DIN 55633 refers to the corrosion protection and evaluation of coated steel structures. EN 15773 refers to the powder coating of hot-dip galvanized and sherardized[32] steel objects.

The coating of the underwater hull is of major importance. In addition to corrosion protection, the avoidance of animal and plant growth (e.g., barnacles, algae), that is fouling, is also of interest. This fouling increases the ship's resistance considerably, which leads to increased fuel consumption (see also Sect. 4.1).

While some years ago, coatings based on tin butyl (tributyl tin hydride/TBT) were still being used, these substances are now banned by EU Regulation 782/2003 because the substances escape into the water, accumulate in the sediment, and lead to considerable heavy metal pollution of the oceans. There is also a ban on the use of paints containing copper. Various other, allegedly more environmentally friendly biocides are now used as substitutes for TBT. The use of biocides is combined with *self-polishing copolymer* (SPC) systems, which cause a constant biocide release at a permanent polishing rate. The latest developments in the field of antifouling systems are based on silicones, the so-called *silicone fouling release coatings* (silicone FRC). These systems are characterized by an extremely smooth surface, highly flexible, cold resistant, non-eroding, and seawater-resistant properties with a long service life. A still quite new method to combat fouling comes from materials research. For example, the skin of sharks has been studied because, unlike the skin of whales, it is not affected by parasites. An attempt is being made to emulate the biological model with a silicone-like ship paint that forms certain small structures when hardening. The advantage of this method is that no toxic substances are used.

[31]For in-depth powder coating, see [12].

[32]Sherardizing is a zinc diffusion process to form zinc–iron layers on iron-containing workpieces.

Research has shown that tiny nanoparticles of vanadium (V) oxide (vanadium pentoxide) prevent the growth at interfaces. Vanadium (V) oxide acts as a catalyst, which forms highly toxic compounds for microorganisms.

A new type of coating is based on nanocomposite lacquers that have different electrical conductivities and are applied to the ship's hull in a multilayer system. Weak currents in the range of 0.1 mA/cm^2 are allowed to flow through these layers. The currents are then reversed at certain time intervals. As a result of electrolytic processes, the pH value of the water changes at the boundary layer, which counteracts the growth of mussels, algae, and barnacles. The effectiveness was tested in 2012 with a boat belonging to the fisheries control authority in Mecklenburg-Vorpommern [18].[33]

To protect the **ballast water tanks** against corrosion, one or two layers of a two-component epoxy resin coating are used.

The coating of the **cargo holds** depends on the cargo to be loaded; the coating materials must be correspondingly resistant to the load. Here, either a chemical resistance (chemical tankers or also dangerous goods carriers) or a resistance to mechanical stress is superficial (e.g., bulk carriers).[34]

Example for the determination of sacrificial anodes

A tug of the German Navy is to be equipped with sacrificial zinc anodes on the underwater hull to protect it against corrosion. The term of protection is set at 2 years (= 17,520 h). The total area of the underwater hull to be protected is 490 m^2. The weight of an anode should be 10 kg. How many sacrificial anodes are required?

The determination of the required anode weight is determined according to Eq. 3.49:

$$m_G = \frac{A_{U\,flat} \cdot I_S \cdot t_S}{Q_G} = \frac{I_G \cdot t_S}{Q_G}$$

with:

m_G Required total anode weight (kg),
I_G Total protective current (A),
t_S Term of protection = 17,520 h,
Q_G Current content = 780 Ah/kg

The required total protective current results from analogous application of Eq. 3.50 with

$$I_G = A_{U\,flat} \cdot I_S \tag{3.51}$$

[33]See also various product information, for example, under [22].

[34]For more information on coatings in the ballast water and cargo area, see also [15, p. 56 f.].

with

I_G Total protective current (A),
$A_{U,\,\mathrm{flat}}$ Total area to be protected = 490 m^2,
I_S Protective current density 0.02 A/m^2

to

$$I_G = 490\ \mathrm{m}^2 \cdot 0.02\ \mathrm{A/m}^2 = 9.8\ \mathrm{A}.$$

This results in a required anode weight of

$$m_G = 9.8\ \mathrm{A} \cdot 17{,}520\ \mathrm{h}/780\ \mathrm{Ah/kg},$$
$$m_G = 220\ \mathrm{kg}.$$

With a unit weight of 10 kg per anode, 22 sacrificial anodes (11 on each side of the hull) must be installed and replaced every 2 years. ◀

3.6.2 Deck Coverings/Floors

Deck coverings should not only be visually appealing (e.g., the teak deck of a motor yacht, Fig. 3.40).[35] Above all, they must be able to resist any mechanical stresses that may occur and, for reasons of occupational safety, they must be sufficiently slip-resistant.

Slip resistance or perhaps *surefootedness* summarizes the properties of a floor covering in relation to slip-promoting substances, such as water. The slip resistance of persons is particularly at risk on wet and slippery floors because an aquaplaning effect can occur when walking on them.

Deck coverings also serve as corrosion and noise protection as well as thermal insulation. The materials and substances to be used are therefore based on the requirements to be fulfilled.

With the GISA TEX Antislide deck covering, for example, both slip resistance and sealing against water and moisture are achieved. The surface of GISA TEX Antislide consists of PVC and is therefore resistant to temperature, seawater, and UV. The pyramidal structure of its surface gives the flooring slip resistance and at the same time softness of step [19].

A measure of the degree of slip resistance is the so-called R-group. Floor coverings are divided into five R-groups (from R 9 to R 13) depending on their slip resistance, whereby floor coverings with the R-group R 9 meet the lowest slip resistance requirements and those with the R-group R 13 the highest. However, adequate slip resistance is not only required for the area of the outer decks; in principle, all floors must be designed

[35] PU-based teak-look deck coverings are now also available (see e.g. [20]).

Fig. 3.40 Teak deck of a motor yacht

to be sufficiently slip-resistant. The requirements for individual rooms depend on the *Technical Rule for Workplaces—Floors* (*ASR A1.5/1.2*),[36] which specifies the relevant basic requirements of the Workplace Ordinance. Accordingly, tiles laid in the kitchen and galley area must be with *slip resistance index* of at least *R 12* (Fig. 3.41).

Especially in working areas where greasy, pasty, or fibrous-tough materials can get on the floor, tiles may need to have a "displacement space" in addition to the required slip resistance. This is the hollow space below the lifting level, open towards the lifting level, for the absorption or discharge of sliding materials. This is divided into four V-classes. A V-value indicates the quantity of liquid in cubic centimeter that the soil is at least capable of absorbing per square decimeter (see Table 3.7).[37]

According to ASR A1.5/1.2, for example, a tile with the code letters R 12/V4 would have to be chosen for the kitchen or galley area of a ship.

Nonslip deck designs (R 11/R 12) can be produced, for example, using checker plates (also called tear or caterpillar plates) made of steel, stainless steel, or aluminum (see Fig. 3.42).[38]

Such steel plates are standardized in DIN 59220. According to this standard, checker plates are designated with the code letter "R" in the order details, tear plates with the

[36]For the Technical Regulations for Workplace Flooring (ASR 1.5/1.2), see [16]; a selection of slip resistance values according to these ASR can be found in Appendix 12.

[37]In accordance with Workplace Directive "Floors" (ASR A1.5/1.2).

[38]See for example [23].

Fig. 3.41 Tiles in the galley area

Table 3.7 V-class for tiles

Group	Minimum volume (cm^3/dm^2)
V4	4
V6	6
V8	8
V10	10

Fig. 3.42 Deck covering made of "checker plate"

Fig. 3.43 Slip-resistant two-component coating in the traffic area on deck

code letter "T." Steel sheets are standardized in the thickness range of 3–10 mm. Stainless steel sheets are standardized in DIN 5220. Aluminum sheets with rolled patterns are standardized in the European EN 1386. Here, we distinguish between the pattern types "duet," "quintet," "diamond," "barleycorn," and "almond." Plate thicknesses are standardized here in the range of 1.2–20 mm.

In order to achieve or increase the anti-slip properties, the deck, floors, and stairs in the hazardous area can also be covered with particularly anti-slip foils (R 11-R 13) or tapes.[39]

It is also common practice in the deck area to apply anti-slip paints and coatings, usually on a two-component polyurethane or epoxy resin basis (see Fig. 3.43).[40]

The TBS covering consists of polyurethane in combination with polyurethane granulate. A resistant combination is elastic and provides a safe working surface, both in dry and wet conditions. TBS offers maximum comfort when walking and sitting due to its soft, skin-friendly surface. The TBS sheet material lies like a carpet and is particularly suitable for the sports boat sector. PVC coverings with nap structure and cork coverings are also widely used in the sports boat and yacht sector.[41]

Mineral components can also be added to the coating to achieve slip resistance (called sanding). Fire-dried quartz sand with a grain size of 0.1/0.3 mm is suitable (Fig. 3.44).

[39]See for example [14, p. 286].

[40]See for example [20, 21].

[41]See for example [14, p. 286].

Fig. 3.44 Slip resistance through "sanding" in yacht building

Examples of floor design

A customer asks for advice on the floor design of the first aid room of his container ship newbuilding. What advice would you give him in principle?

The first aid room should be easy to maintain and must be sufficiently slip-resistant. For the realization of these two aspects, a synthetic or porcelain stoneware floor covering with slip resistance class R 9 is the best choice.

On a cruise ship, the deck covering in the area of the boat deck is to be renewed and given a teak look. What are the slip resistance requirements in this area?

According to No. 30.1 of Appendix 2 of ASR A1.5/1.2, the slip resistance class R 11 must be fulfilled. Although R 10 would be sufficient for sidewalks, the higher skid resistance should be taken into account here, as a high number of people can be expected in an emergency. Adverse weather conditions (rain, snowfall) favor slipping. In such situations, people who have fallen increase the risk of other people stumbling over them, which ultimately includes the risk of panic. ◀

References

Print media

1. Bernhardt, F., Meier-Peter, H. (eds.): Handbuch Schiffsbetriebstechnik. Seehafen Verlag, Hamburg (2008)
2. Berufsgenossenschaft Verkehrswirtschaft Post-Logistik Telekommunikation (BG Verkehr) (Hrsg.): Richtlinien für zulässige mechanische Schwingungen auf Seeschiffen (2003)

3. Bornemann, S., Harbrecht, J.-P., Kaps, H.: Umweltforschungsplan des Bundesministers für Umwelt, Naturschutz und Reaktorsicherheit, Forschungsbericht UBA FuE-Vorhaben: FKZ 102 04 416, Entwicklung eines Kriterienkatalogs für die Vergabe des Prädikats „Umweltfreundliches Schiff", GAUSS, gem. Gesellschaft für Angewandten Umweltschutz und Sicherheit im Seeverkehr mbH in Kooperation mit der Hochschule Bremen, Fachbereich Nautik, im Auftrag des UBA, Juli (1999)
4. Deutsche Gesetzliche Unfallversicherung (Hrsg.): Handlungsanleitung für die arbeitsmedizinische Vorsorge nach dem Berufsgenossenschaftlichen Grundsatz G 46 „Belastungen des Muskel- und Skelettsystems einschließlich Vibrationen". DGUV Information 240–460, vormals BGI 504-46/GUV-I 504–46 (2009)
5. Deutsche Gesetzliche Unfallversicherung: DGUV Information 208–032 „Auswahl und Benutzung von Steigleitern", vormals BGI/GUV-I 5189" (2013)
6. DNV · GL: RULES FOR CLASSIFICATION, *Ships*, Part 3 Hull. Hamburg, Edition October 2015, Amended January (2017)
7. Dubbel: Taschenbuch für den Maschinenbau, Bd. 1. Springer, Berlin (1974)
8 Germanischer Lloyd: Klassifikations- und Bauvorschriften – I Schiffstechnik – 1 Seeschiffe. Germanischer Lloyd, Hamburg (2012)
9. Koch, St.-W. (Hrsg. der deutschen Übersetzung) von Halliday, D., Resnick, R., Walker, J.: Physik – 16 Schwingungen. Wiley-VCH, Universität Marburg (2009)
10. Lexikographisches Institut: Lexikon der Technik, Bd. 1. 2. Lexikographisches Institut, München (1986)
11. Maschinenbau- und Metall-Berufsgenossenschaft: Unfallverhütungsvorschrift UVV „Leitern und Tritte" – BGV D36 (2006)
12. Pietschmann, J.: Industrielle Pulverbeschichtung. Springer, Wiesbaden (2013)
13. Schröter, W., Lautenschläger, K.-H., Bibrack, H.: Chemie – Fakten und Gesetze. VEB Fachbuchverlag, Leipzig (1977)
14. SVB Spezialversand für Yacht- & Bootszubehör, Katalog 2012, Bremen
15. Verband für Schiffbau und Meerestechnik e. V.: Schiffstechnik und Schiffbautechnologie. Seehafen Verlag, Hamburg (2006)

Internet

16. www.baua.de. Accessed: 13 March 2017
17. http://www.deutsche-flagge.de/de/download/bau-und-ausruestung/neu-und-umbau/zusaetzliche-Informationen/maschine. Accessed: 4 Sept. 2016
18. https://www.fraunhofer.de/de/presse/presseinformationen/2012/dezember/schiffsruempfe-bewuchsfrei-halten.html. Accessed: 11 Nov 2016
19. http://www.gisatex.de/antislide-decksbelag.html. Accessed: 4 Sept. 2016)
20. http://www.gtf-freese.de/de/schiffsdecksbelaege. Accessed: 4 Sept. 2016
21. www.lackundfarbe24.de
22. http://www.lanex.cz/de/cruiser-plus. Accessed: 4 Aug. 2016
23. http://www.mutanox.de/produkt_traenen_triplo.htm. Accessed: 21 Aug. 2016
24. https://rules.dnvgl.com/docs/pdf/dnvgl/ru-ship/2017-01/DNVGL-RU-SHIP-Pt3Ch11.pdf. Accessed: 9 May 2017
25. http://www.schiffslexikon.com/ankerkette-kette-2.html. Accessed: 20 Aug. 2016
26. www.sd-dresden.de
27. http://www.tm-mathe.de/Themen/html/gewdglerzschwing.html. Accessed: 20 Aug. 2016

28. http://www.ttm.tugraz.at/arno/laboruebungen-unterlage-uebung2.pdf. Accessed: 20 Aug. 2016
29. http://de.wikipedia.org/wiki/Drahtseil. Accessed: 20 Aug. 2016
30. http://de.wikipedia.org/wiki/Elektrochemische_Spannungsreihe. Accessed: 4 Sept. 2016
31. http://de.wikipedia.org/wiki/Feuerverzinken. Accessed: 20 Aug. 2016
32. http://de.wikipedia.org/wiki/Schiffsrumpf. Accessed: 20 Aug. 2016
33. http://de.wikipedia.org/wiki/Tauwerk. Accessed: 20 Aug. 2016
34. http://www.wissen.de/lexikon/voelligkeitsgrad. Accessed: 18 July 2016

Further reading (Internet)

35. http://www.din.de. Accessed: 22 Dec. 2017
36. http://www.svb.de. Katalog „Technisches Wassersportzubehör"
37. https://de.wikipedia.org/wiki/Schiffsmaße. Accessed: 20 Aug. 2016

Propulsion Systems

4

When sailing through the water, the ship has to work against wind and waves. The water on the underwater hull and the air on the surface hull cause frictional resistance on the hull due to their flow behavior, which ultimately has to be bridged by the propulsion system. The dimensioning and design of this system and especially of the power generator depend on the ship's resistance.

The resistances acting on the ship are described in more detail below, before the common concepts for power generation are described.

The total resistance to be bridged by the propulsion system, which air and water oppose the direction of motion of the ship, is complex and consists of various individual resistances. The significance of ship resistance from both ecological and economic point of view as well as the individual resistance components is examined in more detail below.

About 90% of the world's goods trade is carried by the sea. Today, the German merchant fleet alone consists of about 3350 ships [34]!

4.1 Introduction

When sailing through the water, the ship has to work against wind and waves. The water on the underwater hull and the air on the surface hull cause frictional resistance on the hull due to their flow behavior, which ultimately has to be bridged by the propulsion system. The dimensioning and design of this system and especially of the power generator (see also Fig. 4.1) depend on the ship's resistance.

The resistances acting on the ship are described in more detail below, before the common concepts for power generation are described.

M. Pfaff, *Ship Operation Technology*, https://doi.org/10.1007/978-3-658-32729-3_4

Fig. 4.1 MaK propulsion diesel

4.2 Ship Resistance[1]

The total resistance to be bridged by the propulsion system, which air and water oppose the direction of motion of the ship, is complex and consists of various individual resistances. The significance of ship resistance from both ecological and economic point of view as well as the individual resistance components is examined in more detail below.

4.2.1 Economic Aspects of Shipping

About 90% of the world's goods trade is carried by sea.[2] Today, the German merchant fleet alone consists of about 3350 ships [33]!

A large part of German foreign trade takes place over the oceans. The cruise industry is also booming. Well-known shipping companies here are MSC, COSTA, or the German AIDA Group. Against the background of the increasing demand for cargo space and passenger seats, competition between shipping companies is emerging in the market. In order to remain competitive, it is important, among other things, to reduce the operating costs[3] of a ship. Two aspects are of particular importance. First: "Time is money." This

[1] See also the following: [27, p. 38 ff.].

[2] Frigate Captain Sauerborn during a lecture at the Armed Forces Command and Staff College on January 14, 2010 in Hamburg.

[3] In addition to fuel costs, operating costs include crew wages, loan repayments, repair and maintenance costs, contractual penalties for late delivery, and much more [54].

Fig. 4.2 Bulbous bow

means that a cargo must reach its destination as quickly as possible. The aim must therefore be to enable ships with the largest possible cargo volume[4] to give the highest possible speed. Second, the fuel costs for a crossing are a considerable financial factor. As a rule, the ships need diesel or heavy oil for their propulsion engine. If energy savings can be achieved here, this will have a positive impact on the marketability of the shipowner. Last but not least, fuel savings also contribute to the reduction of CO_2 emissions and thus to climate protection.[5]

In this respect, it must be the task of shipbuilding designers to pay particular attention to the two aspects mentioned above. They must try to develop the ships in a way that optimizes speed and energy consumption. The bulbous bow (see Fig. 4.2), the design, and coating of the hull contribute significantly to meeting these requirements.

4.2.2 Examples From Nature

When it comes to innovative technical solutions, nature provides interesting models. In the meantime, a real science has developed that examines nature's models more closely in order to use them for technical projects: the **bionics**. The word linguistically combines the two disciplines **biology** and **technology**.

[4] What is possible in individual cases depends on many factors, such as the intended area of operation (for example, a passage through the Panama Canal alone limits the size of a ship), but also ultimately the willingness of the shipowner to invest.

[5] Every combustion process produces, among other things, the climate-relevant gas CO_2.

The guiding idea of bionics can be summarized under the motto "learning from nature." The interesting thing is that nature achieves its goals economically with a minimum of energy.

The following examples are intended to show how nature has implemented the above *economic principle* with similarities to the bulbous bow not to be denied.

The Sperm Whale

Big sperm whale bulls reach lengths of 18 m and a weight of 50 t. On their migrations, they cover long distances. The whale would tire quickly if it could not move through the water to save energy. If one observes a sperm whale while swimming, it seems to move weightlessly with light flukes movements through the sea. Its striking, long head is conspicuous! This gives the sperm whale its streamlined body shape.

The Dolphin

The dolphin also has a snout that resembles the bulbous bow. If we take the dolphin as a model for a ship's construction, an elongated, forward hemispherical "nose" must be attached to the bow of the underwater hull.

Sharkskin

The skin of sharks is covered with many small teeth. These make it difficult for barnacles and mussels to find a hold on the skin. Thus, nature prevents the formation of turbulence on the skin surface and thus reduces the frictional resistance between water and shark skin. This phenomenon is important for the aspect of the *frictional resistance*[6] on the hull of the ship and can thus serve as a model for *surface* quality or rather *surface coating* of the underwater hull.

Mussels and barnacles on the hull make the ship slower—the frictional resistance in the water can increase by up to 15%. As a result, silicone underwater coatings are widely used today [35], particularly in the large shipping industry, against the background of the ban on tin butyl coatings as antifouling paints [71]

4.2.3 Fluid Mechanical Considerations on the Hull

The ship experiences a resistance force R as it moves through the water, which counteracts its direction of movement.[7] This is the force that the water on the hull and the air on the superstructure and possibly the cargo opposes the forward movement of the watercraft. The resistance[8] of bodies within a flowing fluid is generally described by following equation:

---

[6] See Sect. 4.2.3.

[7] From the English "resistance."

[8] Generally also called *flow resistance*.

$$R = C \cdot A \cdot \frac{\rho}{2} \cdot v^2 \tag{4.1}$$

with

- R Resistance,
- C Resistance coefficient,
- A A body cross section opposing the flow, also called reference surface, which is definition-dependent. Usually, it is equal to the frontal surface of body flow,
- ρ Density of flowing fluids (water at hull, air to structures and possibly cargo),
- v Relative speed between ship and water or between ship and surrounding air

The resistance coefficient C is difficult to determine because the resistance R acting on the ship is composed of several individual resistances. The individual resistance components are described in more detail below.

4.2.3.1 The Individual Components of the Resistance

According to Meier-Peter,[9] the total resistance R is composed of the *wave resistance* R_W, the *frictional resistance* R_F, the *form or pressure resistance* R_{PV}, the *air resistance* R_L, and other *additional resistance* R_Z, which cannot be defined more precisely and which can be bypassed by the ship's propulsion:

$$R = R_L + R_Z + R_W + R_F + R_{PV}. \tag{4.2}$$

Air resistance R_L, acting on the superstructure and possibly the cargo is of minor importance in relation to the other resistance values. As can be seen from Eq. 4.1, density is one of the determining factors for resistance. However, as the density of air is about three powers of ten lower than that of water and the wind speed from zero to hurricane force—sometimes acting in the direction of travel, sometimes against the direction of travel—is vectorially attributable to the ship's speed, air resistance is often only recorded with a percentage addition of 2–4% to the total resistance.[10]

Additional resistance components R_Z, which are considered on a percentage basis, are[11]:

- *Steering resistance* (due to course corrections on the rudder blade),
- *Resistance by propeller, bearing blocks, rudder blade, and vegetation* (mussels and barnacles),

along with about 10–20%.

[9] Meier-Peter in [18, p. 354]; in addition, there are often also simplified approaches, which will not be discussed in detail here.

[10] Meier-Peter in [18, p. 356].

[11] Meier-Peter in [18, p. 356].

4.2.3.2 The Resistances on the Hull

The prediction of the resistances on the underwater hull is determined experimentally on a model hull by the so-called tow test in the flow channel, whereby the force required for tow is measured and designated as **tow resistance** R_T. Model calculations are used to transfer the results obtained to the original ship to be constructed.[12] These model calculations are based on the realization that a rigid body and a model of this body, which cause waves by moving through a liquid or are exposed to the waves of a liquid, are comparable on earth exactly when the ratio of the lengths to the square of the speed is identical. This relationship is described by the *Froude number* (see later in this chapter).

According to *Froude*, **tow resistance** R_T is divided into **wave** (R_W), **friction** (R_F), and **pressure or form resistance** (R_{PV}):

$$R_T = R_W + R_F + R_{PV}. \tag{4.3}$$

The sum of *friction* and *pressure or form resistance* is also known as **toughness resistance** R_V[13]:

$$R_F + R_{PV} = R_V. \tag{4.4}$$

It is caused by the movement of the hull in the water and depends on the speed of the ship, the size and roughness of the underwater hull, the shape of the hull, and the toughness of the water. At 75–80%, it represents the largest proportion of the total resistance for the usual merchant ships [6, Volume I, p. 326].

The component of ***frictional resistance*** R_F contained in R_V depends only on the size and roughness of the wetted surface of the hull. It is subject to the fundamental law according to Eq. 4.1 but is provided with indices:

$$R_F = C_F \cdot A_S \cdot \frac{\rho}{2} \cdot v^2 \cdot 10^{-3} \quad [\text{kN}]. \tag{4.5}$$

Herein C_F is the dimensionless resistance coefficient, which in this equation describes the roughness of the hull, ρ the density of water[14] in kg/m^3, and v the ship speed in m/s. A_S in this formula is the surface of the part of the ship in the water and is often only determined by approximate formulas. These formulas give good values for standard ship shapes, while for special shapes, a laborious determination of the surface is necessary [32].

A common approximate equation for A_S is:

$$A_S = 2.6 \cdot \sqrt{V \cdot \text{LWL}}, \tag{4.6}$$

where V the displacement of the vessel in t or m^3 and LWL is the length in waterline in meters.

[12] Meier-Peter in [18, p. 356 f.].

[13] Meier-Peter in [18, p. 354].

[14] Fresh water at 20 °C 1000 kg/m^3, seawater at 20 °C about 1026 kg/m^3.

Representation of the total resistance on a ship

Fig. 4.3 Laminar flow

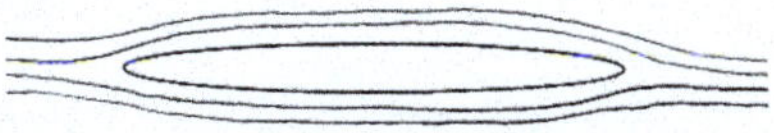

Fig. 4.4 Flow separation with turbulence formation

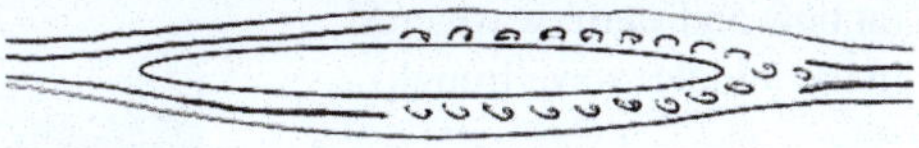

Directly at the "point of contact" water/ship's hull, the water molecules are retained by adhesion. The transition from the speed "zero" to the full value of the water particles flowing past takes place in the narrow area of the so-called boundary layer. This may be *laminar* (Fig. 4.3) or *turbulent* (Fig. 4.4).[15] When *laminar* flow is applied, the resistance is only composed of the shear stresses transmitted in the boundary layer, whereas when the flow is interrupted and the associated turbulence is created (e.g., due to the aforementioned fouling), the resulting negative pressure increases the resistance [6, Volume I, pp. 303, 323].[16]

In case of a **smooth** plate with parallel flow (see Figs. 4.3 and 4.4), the resistance coefficient C_F depends only on the dimensionless *Reynolds number*[17]:

$$C_F = f(Re), \tag{4.7}$$

where

$$Re = \frac{w \cdot l}{\nu} \tag{4.8}$$

where w is the characteristic flow speed, l a typical length dimension (here the plate length), and ν the kinematic toughness (here of water[18]).

The resistance coefficient C_F is described for the usual ship shapes as follows [6, Volume I, p. 326]:

$$C_F = \frac{0.075}{(\lg Re - 2)^2}. \tag{4.9}$$

[15] Turbulent flow = swirl flow; laminar flow = the streamlines run parallel to the main movement.

[16] For large ships, the Reynolds numbers range up to 10^9.

[17] Deepening to corner in [6, Volume I, p. 322, 300 ff.].

[18] At 10 °C about $1.3 \cdot 10^{-6}\,m^2/s$.

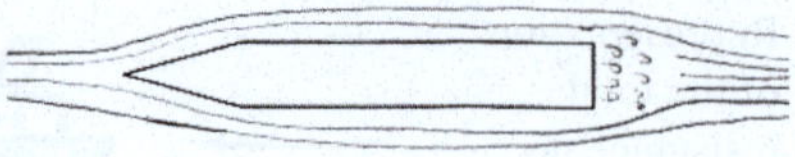

Fig. 4.5 Stall at the stern

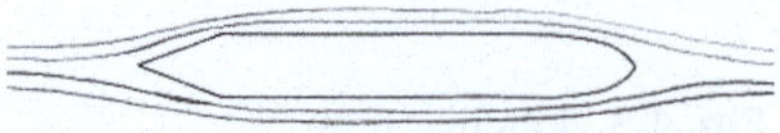

Fig. 4.6 Laminar flow

Fig. 4.7 Typical wave pattern of a boat in displacement mode (at bow and stern: wave crest, in the middle: wave trough)

The ***form resistance*** R_{PV} also owes its origin to the toughness of the water, but is essentially dependent on the hull **shape**. It can be kept low by a good shaping of the underwater hull. The design of the bow and stern is of particular importance: At the stern, a stall, which leads to turbulence and thus to a "braking" negative pressure, must be avoided (Figs. 4.5 and 4.6).

R_{PV} is in principle subject to the same laws as frictional resistance:

$$R_{PV} = C_{PV} \cdot A \cdot \frac{\rho}{2} \cdot v^2 \cdot 10^{-3} \quad [\text{kN}]. \tag{4.10}$$

With C_{PV}, the pressure resistance coefficient of the fuselage is described here. This is difficult to determine, because in the flow channel during a tow test, the friction, shape, and wave resistance forces always act together. In Eq. 4.10, A is the projected "shadow area" of the underwater hull seen from the front. The expression $\frac{\rho}{2} \cdot v^2$ is called "dynamic pressure," which acts on the "shadow area" due to the ship's speed.

The ***wave resistance*** R_W is caused by the wave systems running diagonally to both sides of the ship, which are caused by bow and stern waves (Fig. 4.7). When sailing through the water, the streamlines widen at the bow and stern and constrict in the first third to midships, which, according to Bernoulli's equation,[19] results in a *speed reduction*

[19] Set of the conservation of energy. The B.-G. states that in a stationary flow, the sum of static and dynamic pressure is constant and corresponds to the total pressure of the resting liquid; more [20, p. 108].

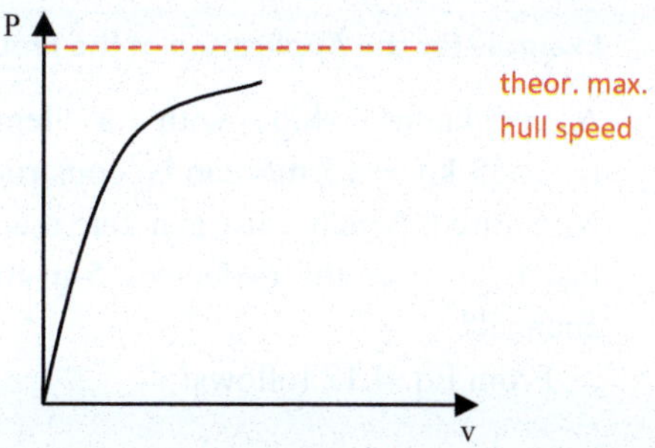

Fig. 4.8 Dependence of the speed on the propulsion power

of the relative water flow at bow and stern and thus a *pressure increase*, while in the middle of the ship, the speed increases and *pressure decreases* accordingly. At the surface of the water, a wave trough is created by the reduction in pressure, while the increase in pressure causes a wave crest [32].

The associated wave elevations move with the ship and to the side, creating a wave field in which the ship moves. The wavelength λ of the inclined systems is proportional to the square of the ship's speed:

$$\lambda = 0.64 \cdot v^2. \tag{4.11}$$

That means fast ships make longer waves. The waves can overlap in such a way that wave crests fall on wave crests, which increases the resistance (*resonance*). If, on the other hand, a wave trough of one system falls on a wave crest of the other, the resistance (*interference*) decreases.[20] This effect is achieved by a bulbous bow (see Sect. 4.2.4). On the basis of these observations, it can be concluded that a ship that produces few waves has a lower wave resistance.

In addition, the ship generates a strong bow wave during pitching movements by the immersion of the bow when passing over the approaching wave, which also exerts considerable resistance against the direction of movement. The bulbous bow reduces the heavy submersion during pitching (and thus a high bow wave), as the bow bead gives the bow area an additional buoyancy component, which makes the submersion smoother.

The ***wave resistance*** in the tow test is determined as the difference between the measured *tow resistance* R_T and the calculated *toughness resistance* R_V. In today's ships, the wave resistance is about 20–25% of the total impedance [6, Volume I, p. 326].

Froude's number is used as the characteristic quantity for the wave resistance[21]:

$$Fr = \frac{v}{\sqrt{g \cdot \mathrm{LWL}}}. \tag{4.12}$$

with $g = 9.81\ \mathrm{m/s^2}$ and LWL = length in waterline. LWL is of particular importance for the consideration of wave resistance and the associated maximum achievable ship speed in displacement mode, as described in Sect. 4.2.4.

---

[20] Meier-Peter in [18, p. 355].

[21] More details on this corner in [6, Volume I, p. 326]; it is generally between 0.20 and 0.35.

Example for the Application of the Froude's Number

A container ship with a length of $LWL_O = 321$ m and a speed of $v_O = 25$ kn $= 13$ m/s can be constructed. To determine the resistance forces, a model is required, which must generate a comparable wave image to the later original. How fast (v_M) must the (selected) 5-m-long model (LWL_M) be pulled through the flow channel?

From Eq. 4.12 follows:

$$Fr = \frac{v_O}{\sqrt{g \cdot LWL_O}} = 0.23 = \frac{v_M}{\sqrt{g \cdot LWL_M}} \Rightarrow v_M = 1.62 \text{ m/s}.$$

◀

4.2.4 The Effect of Bulbous Bow on Tow Resistance

In order to minimize the propulsion energy, the resistors to be bridged must be reduced. For given ship's main dimensions such as length, width, draft, and speed, the only thing that remains is, as can be seen from Eq. 4.1, to minimize the resistance coefficient *C*.[22] In the area of the underwater hull, the wave and form resistance coefficient can be significantly reduced by a bulbous bow (models from nature!), which ultimately reduces the tow resistance and thus also the total resistance of the ship.

The wave pattern described above is strongly dependent on the hull shape. The bead shifts the wave pattern of a moving ship through *interference*. It virtually generates a second wave system, which reduces the bow and stern waves by superimposing them, in the ideal case almost canceling them out [13, p. 96 ff.].

In addition, this "drop shape" has a streamlined shape that guides the flow threads around the hull in a more idealized way, thus reducing the form resistance.

It is interesting to note that not all boats and ships are equipped with a bulbous bow, although this can reduce the total resistance. This is due to the two fundamentally different types of ships and boats: There are *displacer*[23] and *glider*. Whether a watercraft is a displacer or a glider depends on its hull shape. Due to a special bow and hull construction, a glider can "overtake" its wave pattern when propulsion power is strong enough, and then "ride" on its own bow wave, so to speak (which allows very high speeds and is therefore common in sports boat construction). A displacer displaces through the hull just as much water as its mass allows. With increasing speed, only the resistance from its own bow and stern waves increases; however, it cannot leave its wave system. It follows that a displacer—no matter how powerful its engine—can never exceed a maximum possible speed. This is the so-called hull speed. When it is reached, bow and stern lie on a wave crest (see Fig. 4.7). This maximum attainable speed of a displacer is calculated according to Eq. 4.13.

[22] In the literature, a C-value between 0.03 and 0.05 is given as the **total resistance coefficient** for today's ships (see also [27]).

[23] All usual ships of the large shipping like passenger ships, freighters, etc.

$$v_{\text{theor, max}} = 2.42 \cdot \sqrt{\text{LWL}} \text{ in kn } (1 \text{ kn} = 1.852 \text{ km/h}). \tag{4.13}$$

A small container ship with LWL of 110 m, assuming **no** bulbous bow, serves as an example. According to Eq. 4.13, its maximum speed is then 25.4 kn. However, if the ship is given a bulbous bow of approximately 7-m length, its waterline length will then be 117 m, resulting in a maximum hull speed of 26.2 kn. On the Hamburg to Taiwan sea route (approximately 5300 nm[24]), this results in a time saving of approximately 7 h according to Eq. 4.14:

$$v = s/t \tag{4.14}$$

with s the distance and t the time.

Considering how often this route is used per year, this is in any case a plus point for the ship owner's calculations that should not be underestimated.

4.2.5 Required Propulsion Power

The machine power required to maintain forward motion is directly related to the resistance:

$$P_M = \frac{R \cdot v}{\eta_P} \quad \text{[kW]} \tag{4.15}$$

where η_P = degree of propulsion, the total resistance R in kN (see Eq. 4.2), ship speed v in m/s, and $R \cdot v = P_P$ (the propeller power required to bridge the total resistance).

If only Eq. 4.15 is considered, it could be concluded that an increase in power must lead to a virtually unlimited increase in speed. But this is only the case up to the maximum theoretical hull speed (see Sect. 4.2.4). Consequently, the attainable speed will initially approach the hull speed limit value with increasing engine power, and then—with a further increase in power—asymptotically approach it (see Fig. 4.8).

The **quality of the propulsion,** also known as **propulsion efficiency,** is the ratio between power at the ship's propeller (P_p) and the engine power[25]:

$$\eta_P = P_P/P_M. \tag{4.16}$$

This takes into account the fact that various losses occur between the engine and the propeller, for example, in bearings.

From Eq. 4.15, it can be seen that the resistance influences the propulsion power, which in turn influences the fuel costs and these in turn, just like the speed, influence the economy of the ship. Therefore, the bulbous bow also makes a positive contribution in this respect.

---

[24] 1 nm = 1 nautical mile = 1.852 km.

[25] Meier-Peter in [18, p. 354]; this lies between 0.65 and 0.75; see also [41].

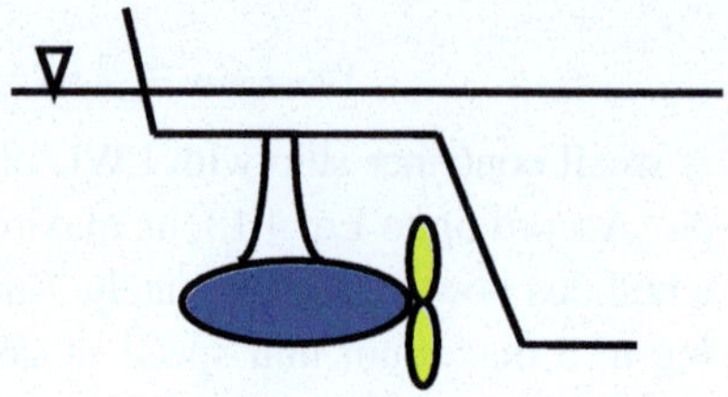

Fig. 4.9 Principle of a POD propulsion

4.2.6 Summary

The ship's speed and the required propulsion power are directly dependent on the (complex) resistance of air and water against the vessel. With regard to the resistance components *form and wave resistance,* the bulbous bow is of particular importance. Optimally designed (by flow experiments on the model and by computer simulations), it can minimize the resistances on the hull: On the one hand, a displacement of the bow wave and thus a reduction of the wave resistance is achieved. Another effect can be observed through the bead: This results in a better flow around the underwater hull and thus a reduction of the form resistance. Both effects result in a reduction of the underwater resistance of up to 10% [43], whereby the wave resistance alone can be minimized by up to 50% [37].

An increase in speed can be achieved by this and probably also by extending the length of the waterline (LWL) by approximately 1 kn.[26]

Example of Ship Resistance

Let us recalculate the required propulsion power for the Queen Mary 2.

Data[27]: 345-m long (LWL approximately 330 m with bulbous bow), width $W = 41$ m, draft $D = 10.3$ m, maximum speed 26.5 kn = 13.6 m/s, propulsion power 86 MW [14, p. 237], kinematic toughness for seawater at 10 °C = $1.356 \cdot 10^{-6}$ m²/s, density of seawater = 1026 kg/m³.

Solution: The total resistance, which is decisive for the required propulsion power, is calculated from Eq. 4.2.

$$R = R_L + R_Z + R_W + R_F + R_{PV}$$

[26] Meier-Peter in [18, p. 357].

[27] See also [48].

under the assumption[28]: $R_W = 15\ \%$, $R_Z = 1\%$, and $R_L = 1\%$ of total resistance $\boldsymbol{R}$.

Determination of the individual sizes:

a) Reynolds number (Eq. 4.8):

$$Re = (w \cdot l)/\nu,$$

$$Re = \frac{13.6\ \text{m/s} \cdot 330\ \text{m}}{1.356 \cdot 10^{-6}\text{m}^2/\text{s}},$$

$$Re = 3.31 \cdot 10^9.$$

b) Coefficient of friction (Eq. 4.9):

$$C_F = 0.075/(\lg Re - 2)^2,$$

$$C_F = \frac{0.075}{(\lg 3.309.734.513 - 2)^2},$$

$$C_F = 0.0013.$$

This corresponds to the diagram according to ITTC for determining the C_F value in shipbuilding (see [32]).

c) Wetted surface (Eq. 4.6):

$$A_S = 2.6 \cdot \sqrt{V \cdot \text{LWL}}$$

with the displacement $V \approx B \cdot T =$

$$A_S = 2.6 \cdot \sqrt{41 \cdot 10.3\ \text{m}^3 \cdot 330\ \text{m}} = 971\ \text{m}^2.$$

d) Frictional resistance $\boldsymbol{R_F}$ from these data (Eq. 4.5):

$$R_F = C_F \cdot A_S \cdot \frac{\rho}{2} \cdot v^2 \cdot 10^{-3},$$

$$R_F = 0.0013 \cdot 971\ \text{m}^2 \cdot 513\ \text{kg/m}^3 \cdot 13.6^2\ \text{m}^2/\text{s}^2 \cdot 10^{-3} = 120\ \text{kN}.$$

e) Pressure or form resistance (Eq. 4.10):

$$R_{PV} = C_{PV} \cdot A \cdot \frac{\rho}{2} \cdot v^2 \cdot 10^{-3}$$

[28] The low approach of R_Z and R_L is based on the one hand on the fact that this ship has no rudder blades for steering movements, but rather nacelles equipped with propellers that can rotate 360°, and on the other hand, the low air resistance might be justified by the rather aerodynamically optimal design of the superstructure; further corner in [6, Volume I, p. 326].

with $A = B \cdot T = 41\ \text{m} \cdot 10.3\ \text{m} = 422.3\ \text{m}^2$ and an assumed value[29] for C_{PV} of 0.11, R_{PV} is calculated:

$$R_{PV} = 0.11 \cdot 422.3\ \text{m}^2 \cdot 513\ \text{kg/m}^3 \cdot 13.6^2 \text{m}^2/\text{s}^2 \cdot 10^{-3},$$
$$R_{PV} = 4408\ \text{kN}.$$

Thus, from Eq. 4.2, taking into account the abovementioned percentages of the other resistance components, the total resistance (R) is (Eq. 4.2):

$$R = R_L + R_Z + R_W + R_F + R_{PV} = 17\% + 120\ \text{kN} + 4408\ \text{kN},$$
$$R = 5455\text{kN}.$$

From Eq. 4.1

$$R = C \cdot A \cdot \frac{\rho}{2} \cdot v^2 \cdot 10^{-3}$$

with $A = 10.3\ \text{m} \cdot 41\ \text{m} = 422.3\ \text{m}^2$ results in **total** resistance coefficient C for the Queen Mary 2 of $C = 0.14$.[30] With a propulsion efficiency of 85%,[31] Eq. 4.15 results in a required machine power P_M

$$P_M = \frac{R \cdot v}{\eta_P} = \frac{5.455\ \text{MN} \cdot 13.6 \frac{\text{m}}{\text{s}}}{0.85} = 87\ \text{MW}.$$

This calculation shows that the assumptions made are real and reflect a very good agreement with the installed propulsion power of the Queen Mary 2. ◀

4.3 Power Generation

The main component of the entire propulsion system of a ship is the power generation unit (*main engine*). Usually, these are diesel[32] or electric motors, gas or steam turbines, gas engines, or *dual fuel engines,* engines that can be supplied with both conventional diesel fuel and liquefied natural gas (LNG) for operation. If the power is often transmitted directly from the propulsion motor to the shaft and the propeller, combined systems are often used, especially in cruise and naval shipbuilding. The combined systems include:

[29] From [6, Volume I, p. 324] for re-sensitive body shape.

[30] This is a realistic value—the Titanic had a resistance coefficient of $C = 0.3$; in many cases, the projected underwater cross-sectional area is only used for rough calculations according to Eq. 4.1, without differentiating between the individual resistance components, which gives sufficiently accurate results for practical use [32, 38].

[31] Assumption due to the particularly efficient "nacelle propulsion."

[32] Rarely—rather in the sports boat sector—also petrol engines.

- **CODAD propulsion** (*Combined Diesel **and** Diesel*): This propulsion concept is used when a lot of power is required but no gas turbines are used. Two—also different—diesel engines are switched to the propulsion shaft via clutches and a collective gearbox. The advantage of this type of propulsion is the low fuel consumption, and the disadvantage is the complicated collective gearbox.
- **CODOG propulsion** (*Combined Diesel **or** Gas*): Propulsion concept in which a diesel engine is used for cruising speed **or** a gas turbine can be connected to the propulsion shaft for maximum speed. The advantage of the CODOG propulsion is the relatively simple design of the main gearbox. Disadvantages are both the additional weight of the propulsion component not in operation and the reduction gearbox required for the gas turbine. The frigates of the F122 class of the German Navy, for example, have such a propulsion system. Two General Electric LM2500 gas turbines, each with approximately 15 MW, and two propulsion diesel engines (AnDiMot), each with 3820 kW, are installed on two shafts with controllable pitch propellers [52].
- **CODAG propulsion** (*Combined Diesel **and** Gas*): Propulsion concept in which two diesel engines **and** a gas turbine can be switched together on the propulsion shaft(s) with controllable pitch propeller. Interconnection is achieved by multistage gearbox and a so-called cross-connect gearbox via clutch systems. A reduction gearbox is required to connect the gas turbine. The first ships built with this system were the frigates of the F120 class of the German Navy. The latest frigate class F124 (Sachsen class) currently in service is also equipped with a CODAG propulsion system. The advantage of this propulsion concept is the lower fuel consumption due to the propulsion diesel in combination with gas turbines that can be switched on at short notice for maximum speed. A disadvantage is the complicated design of the collective gearbox, as very different outputs have to be processed simultaneously (in the order of magnitude: diesel engines several 1000 kW, gas turbines several 10,000 kW, in total up to 80 MW).
- **CODLAG propulsion** (*Combined Diesel-Electric **and** Gas*): This is a further development of the CODAG system [16, p. 616].

  Here, diesel generators supply electricity for the electric propulsion motors. In order to reach maximum speed, a gas turbine is switched on via gearbox and clutches. The advantage of this system is that only one type of diesel generator is needed for **entire** power supply to the ship is required, which minimizes maintenance and repair work. Furthermore, this propulsion concept has the another advantage that the diesel generator sets (electric diesel engines—EDiEng) are not directly connected to the propulsion shaft and can therefore be installed in the most suitable places inside the ship [16, p. 616].

  The new F125 class frigates of the German Navy are equipped with CODLAG propulsion. Up to approximately 20 kn, each of the two electric motors drives a shaft directly without gear reduction. The gas turbine propulsion is switched on to reach maximum speed [1, p. 8 ff.].

- **COGLAG propulsion** (*Combined Gasturbine and Gasturbine*): This propulsion is installed, for example, in the new Japanese destroyer class 25DD. Gas turbines are used. When the energy requirement is low (slow speed), electricity is generated, which is used to drive the propulsion shaft via the electric propulsion motors. At "full speed," the gas turbine is shifted directly to the propulsion shaft(s) via a gearbox and clutch system. This type of propulsion is intended to save fuel at low speeds and reduce the signature[33] [16, p. 616].
- **COGOG propulsion** (*Combined Gas or Gas*): Propulsion, where **two different gas turbines** can be switched to the propulsion shafts by means of reduction gearboxes. A lower powerful turbine is used for cruising speed (cruise gas turbine), and a higher powerful turbine is used for high speeds (high-speed gas turbine). The advantage is reduced fuel consumption because a small turbine designed for cruising speed uses less fuel at 50% power than a turbine twice as powerful (for high speed), which is then only driven at 50% power when at cruising speed. A disadvantage is the complex gearbox.
- **COGAG propulsion** (*Combined Gasturbine and Gasturbine*): In contrast to the COGOG propulsion, this type of propulsion involves two turbines of equal power being connected to the propulsion shaft. These propulsion systems provide high power. Up to 80 MW power can be generated. However, the system is very complex due to the gearbox and consumes a lot of fuel [16, p. 615].
- **IEP propulsion** In contrast to the CODLAG propulsion, a system of diesel engines and/or gas turbines, which only generate electricity for the propulsion motors and have no mechanical connection with the screw shafts, is called an integrated electrical drive ("**i**ntegrated **e**lectric **p**ropulsion" IEP or "**i**ntegrated **f**ull **e**lectric **p**ropulsion" IFEP). The AIDAluna, for example, is equipped with such a propulsion (two electric propulsion motors, each with 12.5 MW; [14, p. 254]). The Hapag's newbuilding Orizaba (4354 BRT) was the first cargo ship with a turboelectric propulsion system (May 17, 1939; [34]).

****

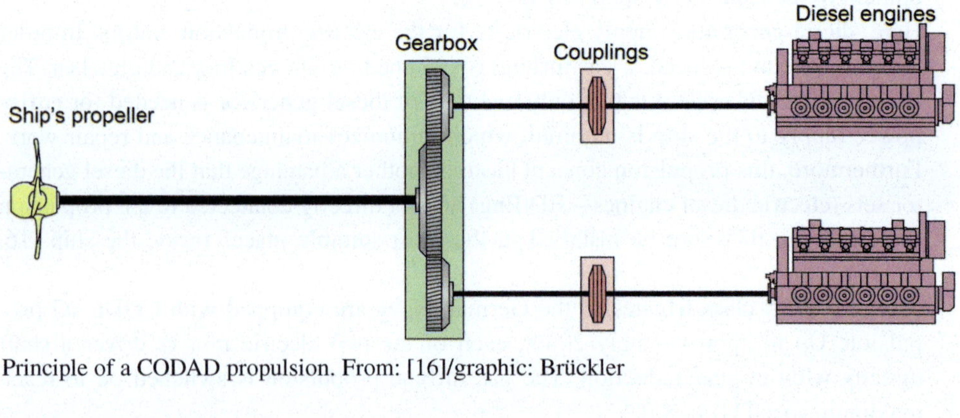

Principle of a CODAD propulsion. From: [16]/graphic: Brückler

[33] Detectability for hostile detection systems (e.g., infrared signature by heat radiation, noise signature by acoustic coupling).

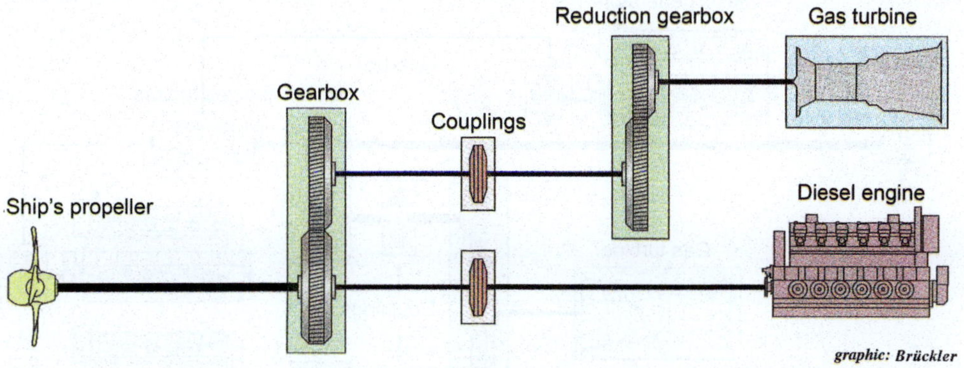

Principle of a CODOG propulsion. From: [16]/graphic: Brückler

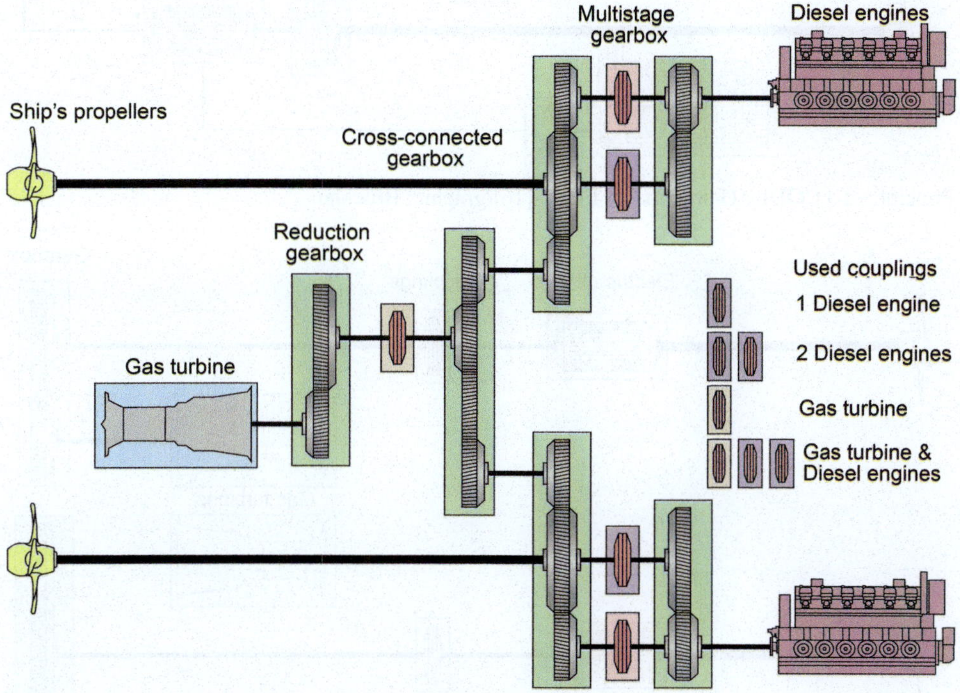

Principle of a CODAG propulsion. From: [16]/graphic: Brückler

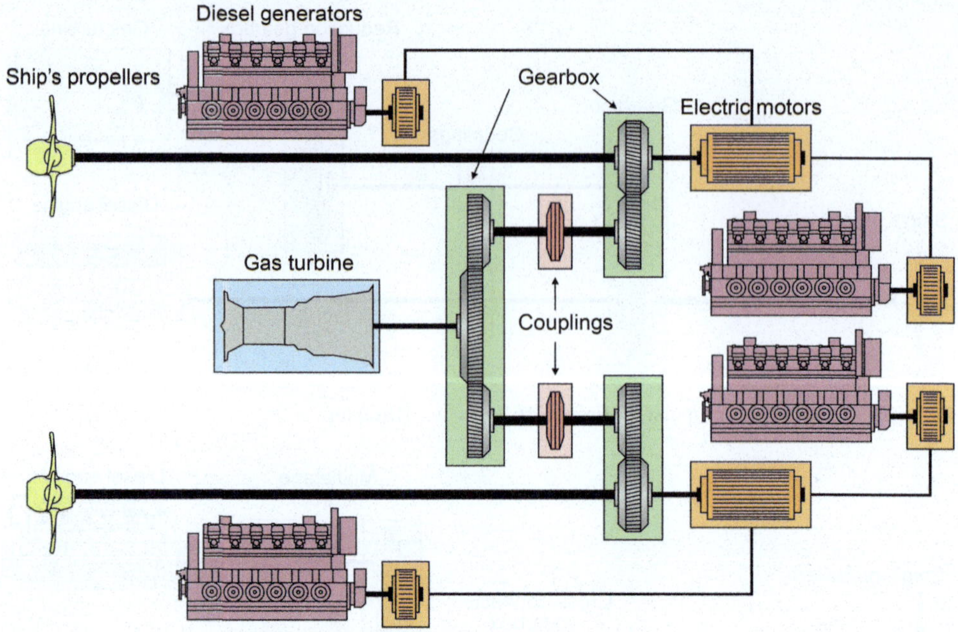

Principle of a CODLAG propulsion. From: [16]/graphic: Brückler

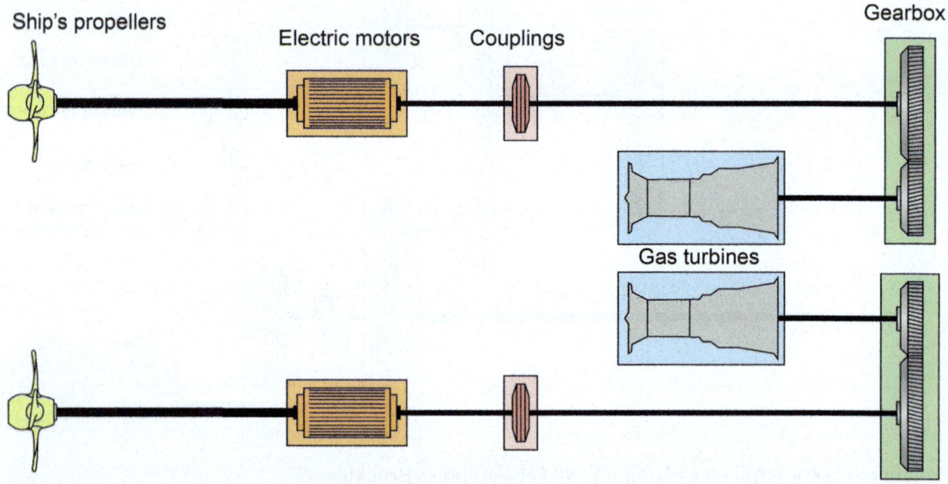

COGLAG propulsion. From: [16]/graphic: Brückler

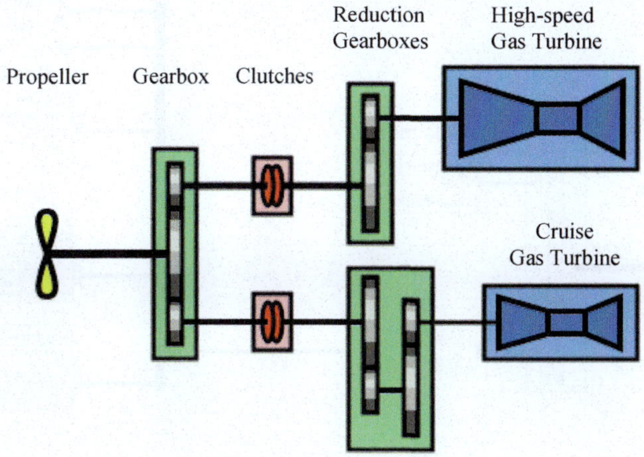

Principle of a COGOG propulsion. *Photo* Alureiter

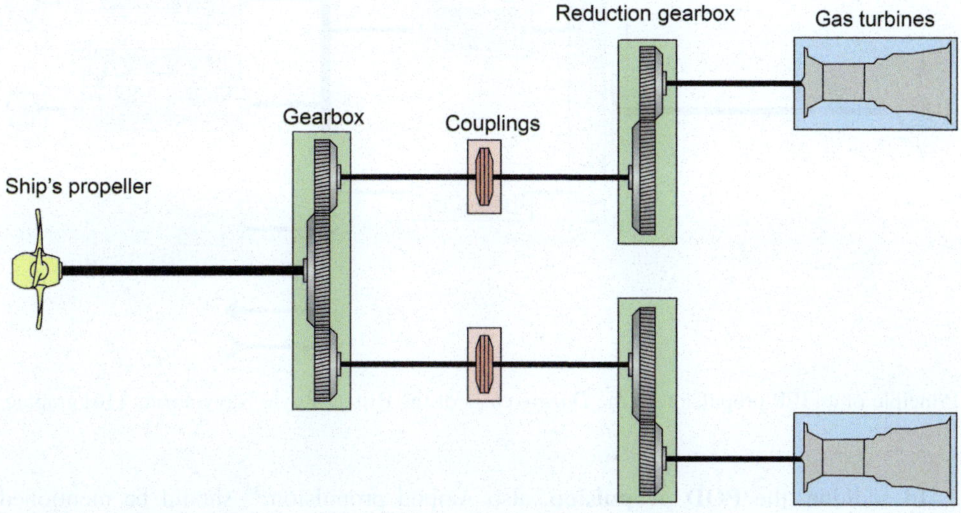

Principle of the COGAG propulsion. From: [16]/graphic: Brückler

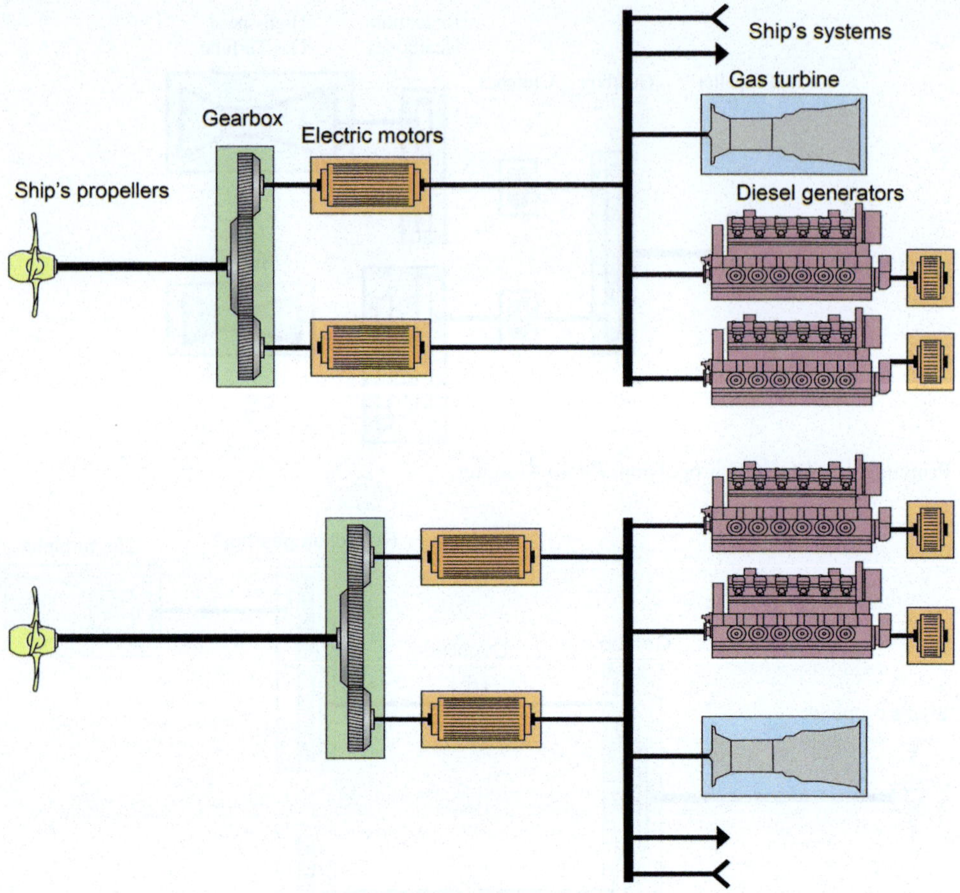

Principle of an IEP propulsion of the *Daring class* of the British Royal Navy. From: [16]/graphic: Brückler

In addition, the **POD propulsion**, also Azipod propulsion,[34] should be mentioned here (see Figs. 4.9 and 4.10). A generator supplies electric current to the electric propulsion motor, which is driven in a 360° rotating nacelle is installed under the hull of the ship. Advantages: With this propulsion system, a rudder blade is not necessary, as the change of direction of the ship is achieved by rotating the nacelles. They can also be mounted at any suitable position under the ship, have a better efficiency than fixed shaft diesel engines, are quieter, and generate less vibration.

Example for Dimensioning a Propulsion System

The rated power of a ship with two diesel propulsion engines rotating in the same direction (two-stroke engines, speed $n = 960\ \text{min}^{-1}$), which combine their power

[34] From the English "pod" = housing; "azi" is a borrowing from Arabic and is supposed to express that the nacelle can be rotated to all angles (360°).

Fig. 4.10 POD nacelles with propeller hub

via a corresponding two-stage collective gearbox on the propulsion shaft (heat-treated steel C45, material no. 1.0503, EN 10,083-1) (CODAD propulsion), must be 12,600 kW. The operating factor[35] f_B of the gearbox is 1.2.

a) How large must the rated power of each motor be, if an efficiency of $\eta = 0.95$ can be accepted for each gear stage?
b) What is the speed of the propeller shaft when the total gear ratio of the gearbox is given as $i_{tot} = 4.0$?
c) What are the nominal and operating torques on the propeller shaft?
d) Roughly determine the diameter of the propeller shaft with a required double safety factor.

To a):

$$P_N = 2 \cdot P_{N,\text{Motor}} \cdot \eta^2 \tag{4.17}$$

(The efficiencies of each stage are multiplied!)

It follows:

$$P_{N,\text{Motor}} = P_N / (2 \cdot \eta^2),$$
$$P_{N,\text{Motor}} = 12{,}600 \text{ kW} / (2 \cdot 0.95^2),$$
$$P_{N,\text{Motor}} = 6981 \text{ kW}.$$

[35] In order to achieve a uniform service life of gearbox and motor, the required torques M must be increased by the respective operating factor f_B at the various operating loads in order not to exceed the maximum permissible gearbox torque.

To b):

$$i_{\text{tot}} = n_{\text{Drive}}/n_{\text{Drive}} \tag{4.18}$$

$$n_{\text{Drive}} = n_{\text{Drive}}/i_{\text{tot}}$$

$$n_{\text{Drive}} = 960\,\text{min}^{-1}/4.0$$

$$n_{\text{Drive}} = 240\ \text{min}^{-1}$$

To c):The torque is dependent on the speed: $M = f(n)$.

In general, the torque is calculated according to the following equations[36]:

a) Two-stroke engine

$$M = \frac{V_h \cdot p_{me}}{2\pi} \tag{4.19}$$

b) Four-stroke engine

$$M = \frac{V_h \cdot p_{me}}{4\pi} \tag{4.20}$$

with V_h the displacement of the engine and p_{me} the effective mean pressure in the cylinders.

The **medium pressure** is a calculated variable for the assessment of the efficiency and the charge exchange of reciprocating engines. The effective mean pressure p_{me} (which can be up to 25 bar for diesel engines) is calculated from the torque output M:

$$\begin{gathered} p_{me} = Z_n \cdot 2 \cdot \pi \cdot M/V_h, \\ (Z_n = 2 \text{ for four - stroke engine; } Z_n = 1 \text{ for two - stroke engine).} \end{gathered} \tag{4.21}$$

When transmitting a power P via a rotating shaft, the dependency of the torque M on the speed n (torque curve) is of interest. For this purpose, the state of constant speed must be established. Power and speed are measured. The evaluation takes place (roughly in certain speed ranges) with the formula

$$M = \frac{P}{2 \cdot \pi \cdot n}. \tag{4.22}$$

The power is measured by means of a so-called dynamometer. Equation 4.22 shows that the torque now decreases with increasing speed for a given power. In fact, this relationship is much more complicated[37]: From the engine characteristic curve shown below (Fig. 4.11), it is clear that the mathematical relationship shown in the above formulas is only valid in part of the speed range. The highest torque is present when at a certain speed n, the optimum speed, the filling level of the cylinder is at its

---

[36] See also Sect. 4.3.1 below.

[37] See below [57].

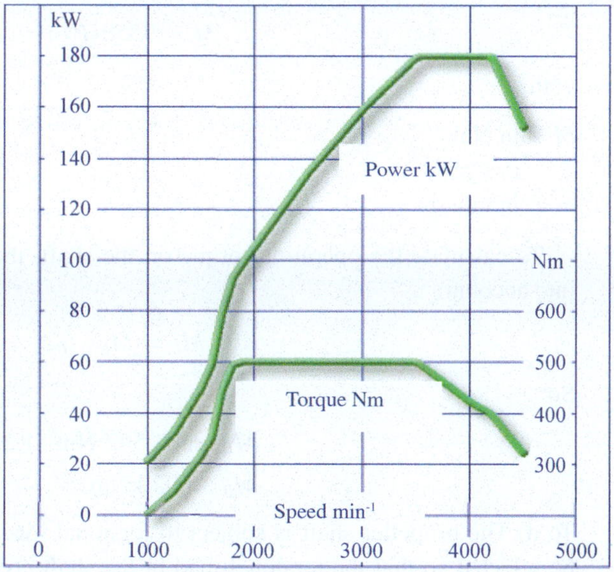

Fig. 4.11 Exemplary motor characteristic curve. *Source* www.kfztech.de

highest and thus the maximum mean working pressure and maximum efficiency are also achieved. In addition, the torque decreases again due to the poorer gas exchange (caused by the higher speed).

The piston speed has a decisive influence on the gas speed. As an internal combustion engine is referred to as an oscillating gas column, the process can be optimized (valve overlap, length of the intake tract, etc.) so that at a certain speed range, the degree of filling is higher than the total volume of the cylinder, thus achieving a slight "supercharging." Also in this case, the motor would have its highest efficiency here. The torque is determined by a number of factors when the above equations are considered closely and is thus proportionally dependent.

- to the mean internal piston pressure,
- to the total cubic capacity,
- on fuel consumption,
- on the specific calorific value of the mixture,
- on the efficiency of use.

So here:

$$M_N = P_N/(2 \cdot \pi \cdot n_{\text{Drive}}),$$
$$M_N = 12{,}600{,}000 \text{ W}/(2 \cdot \pi \cdot 240 \text{ min}^{-1}),$$
$$M_N = 501{,}592 \text{ Nm}.$$

The following numerical value equation is also common, which avoids tedious conversions when the following units are used for calculations:

$$M = 9550(P/n)$$

with

M in Nm,
P in kW,
n in min^{-1}.

To calculate the operating torque on the shaft, the operating factor must be taken into account:

$$M_B = M_N \cdot f_B. \quad (4.23)$$

So:

$$M_B = 501{,}592 \text{ Nm} \cdot 1.2,$$
$$M_B \approx 602 \text{ kNm}.$$

To d):The propeller shaft is subject to torsional load. The diameter of the shaft must be selected so that the torsional load in the shaft cross section caused by this load is less than the permissible load. The torsional load at operating load is calculated using the following equation:

$$\tau_B = M_B/W_p \leq \tau_{B,perm}. \quad (4.24)$$

W_p is the polar moment of resistance of the shaft, that is, a circular area:

$$W_p \approx d^3/5, \quad \text{also} \quad \tau_B = 5M_B/d^3. \quad (4.25)$$

The permissible torsional stress for C45 is 170 N/mm^2. However, as double safety is required, the permissible operating torsional stress may only be half the normally permissible torsional stress:

$$\tau_{B,\text{perm}} = \tau_{\text{perm}}/2,$$
$$\tau_{B,\text{perm}} = 85 \text{ N/mm}^2. \quad \text{Note: } W_p = d^3/5 = M_B/\tau_{B,\text{perm}}.$$

Thus, inserting Eq. 4.24 and changing Eq. 4.25 to "*d*" *results in*:

$$d_{\text{req}} = \sqrt[3]{M_B/(0.2 \cdot \tau_{B,\text{perm}})},$$
$$d_{\text{req}} = 328 \text{ mm}, \quad \text{selected } d = 330 \text{ mm (according to DIN EN 10250).}$$

◀

4.3.1 Internal Combustion Engines

Diesel engines are the most common type of propulsion in the shipping industry, ranging from auxiliary engines on sailing ships up to aggregates with several 1000 kW [10, 62]. They will continue to maintain their place in shipbuilding in the foreseeable future, as they are relatively efficient, especially in comparison with turbine engines [16, p. 618].

Even today, the fuel used in large engines is still mostly inexpensive, unpurified diesel oil, or heavy fuel oil. However, due to IMO resolutions on reducing emissions from propulsion engines, low-sulfur diesel fuels ("marine diesel") are increasingly being used in order to allow passage through so-called Emission Control Areas (ECAs), in which only a sulfur content of 1% in fuel is permitted; in EU ports, even stricter requirements apply: 0.1% sulfur content in fuel.[38]

Especially the larger marine diesel engines with efficiencies up to 50% are designed for operation at low speeds. Four-stroke engines are used for small and medium outputs (up to 24 MW at 400 min^{-1} to maximum 2000 min^{-1}—so-called medium-speed engines), and for large and highest outputs, two-stroke engines (up to 100 MW per engine, rotation 60–250 min^{-1}—so-called low-speed engines) are used. *High-speed engines* at revolutions > 2000 min^{-1} are mainly used in the area of sport and leisure navigation, but are also used in the *fast rescue boats* (fast lifeboats) are used on board professional and naval vessels, although jet propulsion is already becoming more and more common for these boats.

However, not all details of internal combustion engine technology will be discussed in the context of this book; for this purpose, reference is made to the relevant literature on propulsion technology.[39] Rather, at this point, the one or other thing that is relevant for practice should be repeated or even supplemented [10].[40]

The term *diesel* is derived from the working process that takes place in the machine—not from the fuel. The engine developed by Rudolf Diesel (1893–1898) is the most efficient thermal engine ever developed. The engine owes this to the high compression achieved during compression in the working chamber, which is made possible by the fact that, unlike the petrol engine, it compresses pure air and no air–fuel mixture; in the diesel process, the combustible mixture is only produced at the end of compression by injecting the fuel into the cylinder. The injected fuel normally burns automatically under the effect of the high compression temperature without any special ignition device.[41] The diesel engine is therefore also referred to as a compression ignition engine. In contrast to this, the ignition of the fuel–air mixture in the petrol engine takes place via an externally controlled ignition device. In general, electrically operating spark plugs are used, which ignite the intake mixture at the set point in time.

The *type of ignition* of the mixture (petrol or diesel process) therefore determines the type of fuel in such a way that fuels with good self-ignition properties are required for the diesel engine, but fuels with a lower tendency to self-ignition are required for the petrol engine to avoid self-ignition before the intended ignition point.

[38] For more details see also [49].

[39] See also an old, but interesting compilation in [5].

[40] For details on ship diesel engines, see also Behrens and Boy "ship diesel engines" in [23, p. 22 ff.].

[41] Kraemer "diesel engines" in [6, Volume II, p. 141 f.].

Fig. 4.12 "Ideal" diesel or equal pressure process

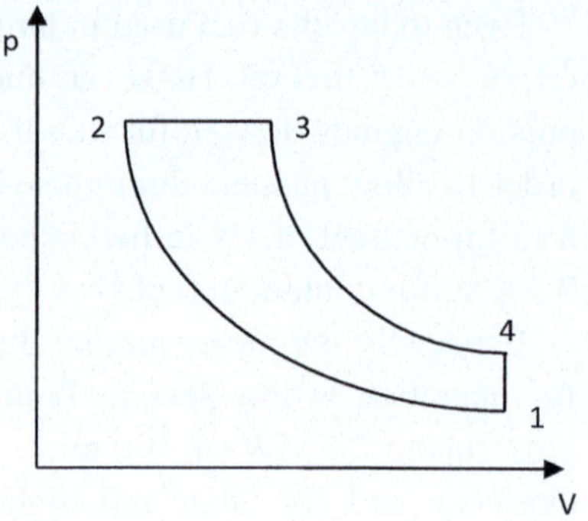

As *working methods*, the two-stroke and the four-stroke processes are known. In the four-stroke process, one working period or working cycle comprises four piston strokes or two crankshaft revolutions. In the two-stroke engine, one working cycle takes place with just one crankshaft revolution or consists of two piston strokes.

In the following, the diesel process will be examined in more detail, since in large ships, among the internal combustion engines, only the diesel engine is used.

The ***diesel process*** is (simplified as ideal circular process—Fig. 4.12) characterized by four steps.

1. **Intake of combustion air and its isentropic compression (1–2)**
   Energy balance:

$$w_{12} = -mc_v \cdot (T_2 - T_1) < 0, \quad q_{12} = 0 \tag{4.26}$$

   In this case, compression work is put into the system. The heat input is "0" as this is an isentropic process.
2. **Isobaric combustion—self-ignition after fuel injection (2–3)**
   Energy balance:

$$w_{23} = 0, \quad q_{23} = mc_v \cdot (T_3 - T_2) > 0 \tag{4.27}$$

3. **Isentropic expansion**
   This step is an isentropic expansion: the piston now moves downward and there is an expansion of volume with simultaneous reduction of pressure in the piston chamber. Energy balance:

$$w_{34} = -mc_v \cdot (T_4 - T_3) > 0, \quad q_{34} = 0 \tag{4.28}$$

4. **Emission of combustion gases and new intake (4–1)**
   Energy balance:

$$w_{41} = 0, \quad q_{41} = mc_v \cdot (T_1 - T_4) < 0 \tag{4.29}$$

   An isochoric recirculation to the initial state takes place, that is, burnt mixture is ejected to suck in fresh combustion air again.

Fig. 4.13 MaK ship diesel engine. MaK = Maschinenbau Kiel, taken over by Caterpillar Motoren GmbH & Co. KG

Diesel engines (Fig. 4.13) are characterized by their operational reliability and reliability. Compared to *petrol engines*[42] running at much higher speed, they have a much longer service life. Another advantage is their ability to run at full load over long distances. Their specific fuel consumption is about 180 g/kWh (four-stroke engine), and for two-stroke engines even only about 170 g/kWh. A low purchase price compared to other propulsion systems and the possibility of a high degree of automation ("wake-free operation": Alarms and other events are transmitted by electronic data processing directly to the bridge or the control center from where the machine is driven) are further positive aspects of the diesel engine.

Mixture Formation [10, 21, p. 261]

A good efficiency of the engine process can be achieved mainly by a complete utilization of the fuel energy, that is, by a complete combustion. The complete combustion of the fuel in the engine in turn requires good mixture formation with regard to the quantity and distribution of fuel and air in the combustion chamber. Furthermore, it must be possible to adjust the mixture formation to various operating conditions determined by the engine process.

The carburetor technology for mixture formation, which is mainly used in petrol engines, will not be considered further here, as it is not used in diesel engines used in large ships. In this respect, the mixture formation in diesel engines will be examined in more detail here; here, mixture formation takes place by injection.

[42] For example, used as an outboard motor in small sports boats.

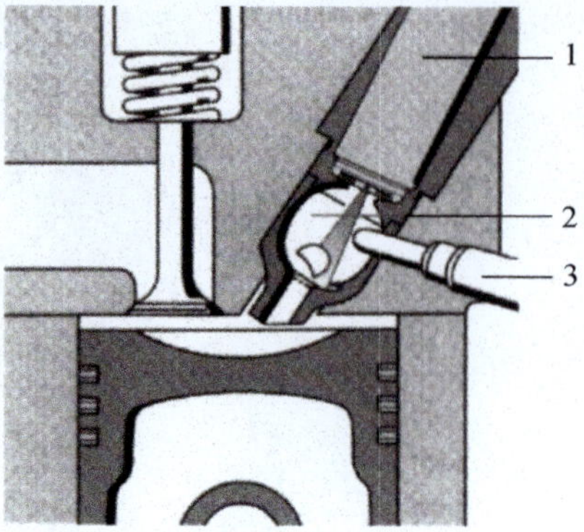

Fig. 4.14 Pre-chamber process. *1* Fuel injector, *2* antechamber, *3* glow plug. *Source* kfztech.de

Here the fuel is injected into the intake air or directly into the combustion chamber. The injection quantity can be precisely dosed. In this way, an optimum fuel–air mixture adapted to the respective operating condition is achieved, which in turn leads to better utilization of the fuel energy. In addition, fuel combustion that is as complete as possible also ensures so-called "clean" combustion.

The injection quantity can be controlled electrically or mechanically. For good mixture formation in the cylinder, which is a prerequisite for good fuel utilization, an intensive swirling of the air and the injected fuel must be achieved. This is achieved by suitable design of the combustion chamber. The most common construction designs are briefly described below:

- In ***pre-chamber method*** (Fig. 4.14), a widely used injection method for diesel engines until the 1990s, a small part of the combustion chamber (the so-called pre-chamber) is separated from the main chamber by a constriction (the firing channel). The fuel is injected at moderate pressure (approximately 100 bar) into the antechamber, ignites and partially burns there; the momentary excess pressure created blows the burning mixture through the firing channel into the air-rich main combustion chamber under a violent atomizing and swirling effect. Today, this process has largely been replaced by *direct injection* and is only used in smaller diesel generators.
- In ***swirl chamber engine*** (Fig. 4.15; [10, p. 298]), a good mixture formation is achieved by the guiding the air in the swirl chamber. The fuel is injected tangentially in the swirl direction. The timing of fuel supply, ignition, and combustion process sequence is analogous to that of the prechamber engine.
- In ***air accumulator engine*** (Fig. 4.16; [10, p. 298]), injection takes place again shortly before top dead center (TDC). The fuel jet is directed toward the inlet cone of the air nozzle of the accumulator so that the injected fuel is carried along by the air flowing into the antechamber. The ignition starts in the prechamber, partial combustion, and overflow with strong turbulence into the cylinder chamber takes place.
  In modern designs, the air accumulator volume is about 20–25 % of the total compression chamber, and the injection pressure is about 10–13 MPa.

Fig. 4.15 Swirl chamber injection

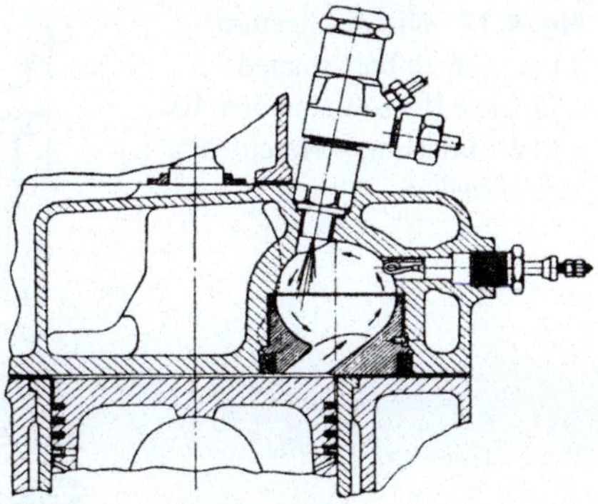

Fig. 4.16 Air accumulator engine

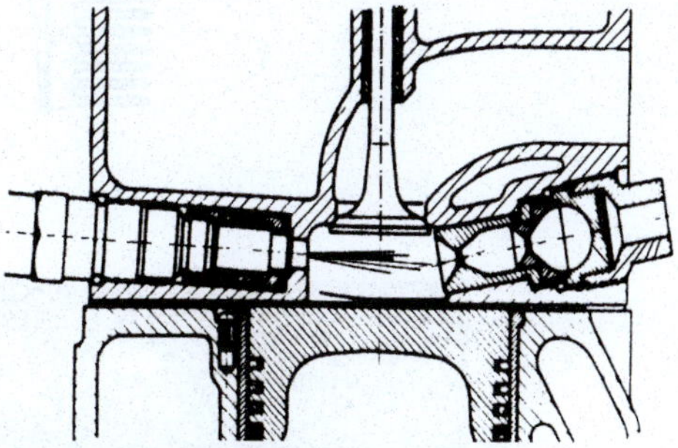

- In the ***direct injection***, where the combustion chamber may be located partly or entirely in the piston crown (Fig. 4.17), the fuel is injected directly into the compression chamber through an injection nozzle containing one or more small holes (Ø 0.1–0.3 mm). The injection pressures are at 400 bar and above to achieve the finest distribution and atomization of the fuel in the combustion air.[43] A suitable shaping of the piston supports the effect of the nozzle arrangement [10, p. 300].
- The ***MAN-M method*** (“M (German: Mittelkugel)” stands for central sphere) is characterized by the typical design of the piston with the recessed combustion chamber in the form of a sphere (Fig. 4.18). Due to the special design of the inlet channel, an air swirl is created in the sphere by injecting the fuel tangentially through a single-hole nozzle into the spherical depression in the piston. This distributes about 95% of the fuel as a film on the wall surface; it only evaporates during combustion and is also removed by the air swirl. Ignition takes place with the small residual proportion of fuel, which has already formed a mixture directly with the air. Now the fuel sprayed onto the hot combustion chamber surface evaporates and continuously mixes with the air swirl. A very soft and relatively complete combustion takes place [10, p. 299].

[43] On the above methods, see also Kraemer in [6, Volume II, p. 142].

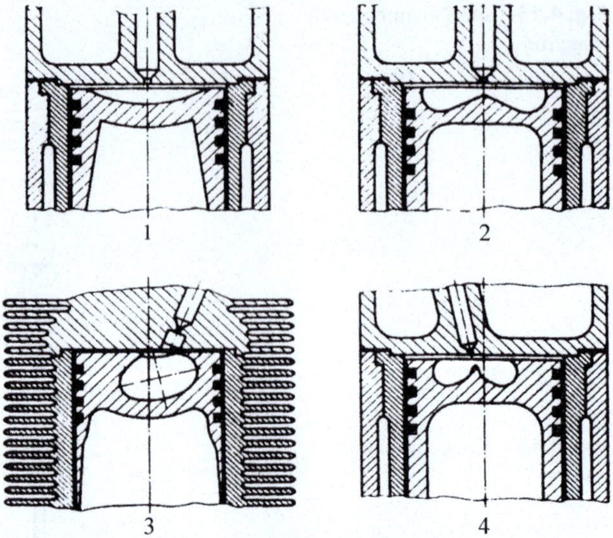

Fig. 4.17 Direct injection. *1* Piston with ball-shaped scraper, *2* Hesselmann pistons, *3* MWM design, *4* Hercules or Saurer design

Fig. 4.18 Four-cylinder diesel engine of the ZT-300 series, M process. *Photo* Sauerlaender, CC BY-SA 3.0

This process is used in the so-called multi-fuel engines.[44]

In the processes mentioned above, different pump systems are used to build up the injection pressures:

- For engines with prechamber or swirl chamber injection:
  - Single injection pump,
  - Distributor injection pump,
  - In-line injection pump,
- For engines with direct injection:
  - Single injection pump, among others in the design as single-stamp pump, colloquially often called *plug-in pump*,
  - In-line injection pump,
  - Distributor injection pump,
  - Pump-nozzle injection systems,
  - Common rail injection.

The Pump-Nozzle System

A characteristic feature for the *pump-nozzle system* is the separate injection pump for each cylinder, which is connected to the injection nozzle via very short pressure lines. The pumps (so-called plunger pumps—piston pumps) are each actuated by a separate cam on the camshaft and a rocker arm (Fig. 4.19).

In order to obtain a pressure curve that is favorable for the injection process, a steep pressure increase over time is required. For this purpose, the kinematics of the actuating path of the working cam is designed so that the piston moves at high speed after strong acceleration. This is achieved purely mechanically by an oval cam shape.

The pressure build-up in the so-called plunger chamber can be controlled by opening and closing a solenoid valve or a valve actuated by a piezo actuator. When the valve is closed, the piston builds up pressure and the fuel is injected through the injection valve. By opening the control valve, the injection process is stopped, whereby optimum combustion is achieved by stopping the injection process as abruptly as possible with a rapid drop in pressure. Piezo actuators work up to three times faster than solenoid valves [76].

Common Rail Injection

Common rail injection (literally: common line; Fig. 4.20), which is also *accumulator injection,* is an injection system for internal combustion engines in which a high-pressure pump brings the fuel to a high pressure level. The pressurized fuel fills a piping system that is constantly under pressure while the engine is running. Here, injection quantity and duration are electronically controllable independently of the crank angle, thus allowing pre-, main, and postinjection; up to eight separate partial injections can be implemented

[44] Kraemer in [6, Volume II, p. 145].

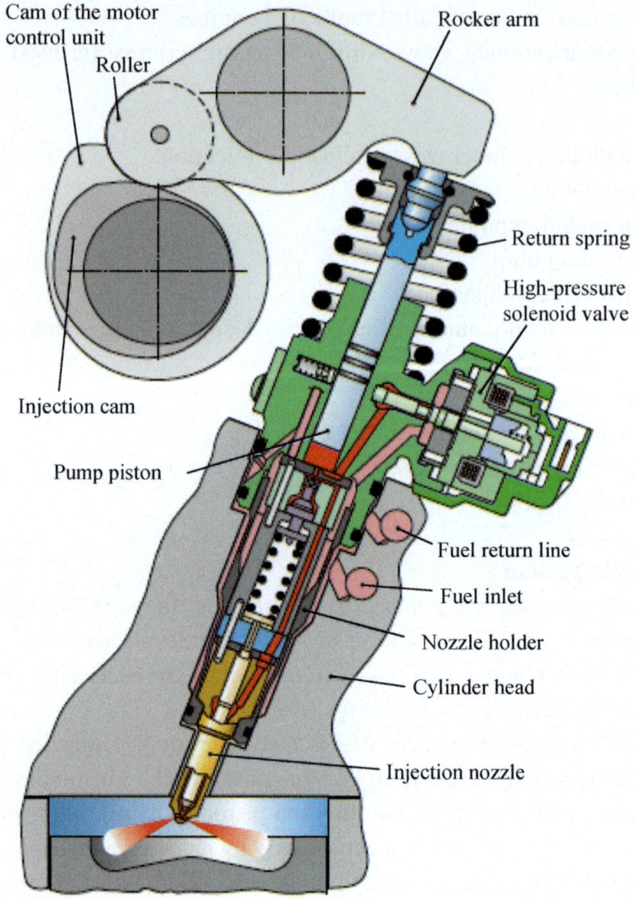

Fig. 4.19 Pump-nozzle element. *Graphic* deckermedia GbR, Rostock

per engine working cycle. Preinjection is primarily used to reduce combustion noise, while postinjection is used for internal engine particulate reduction or to increase exhaust gas temperatures in the free combustion cycles when the pressure loss of the fine particle filters (soot particle filters) in the exhaust system is too high [74].[45]

Ignition Process and Ignition Timing

The ignition initiating the combustion process takes place—as already mentioned above—in the diesel engine by self-ignition, whereby the *injection* timing is decisive for the ignition behavior [10]. In the spark ignition petrol engine, on the other hand, the *ignition* timing is essential for the entire combustion process. The combustion of the fuel

[45] Also in-depth [56].

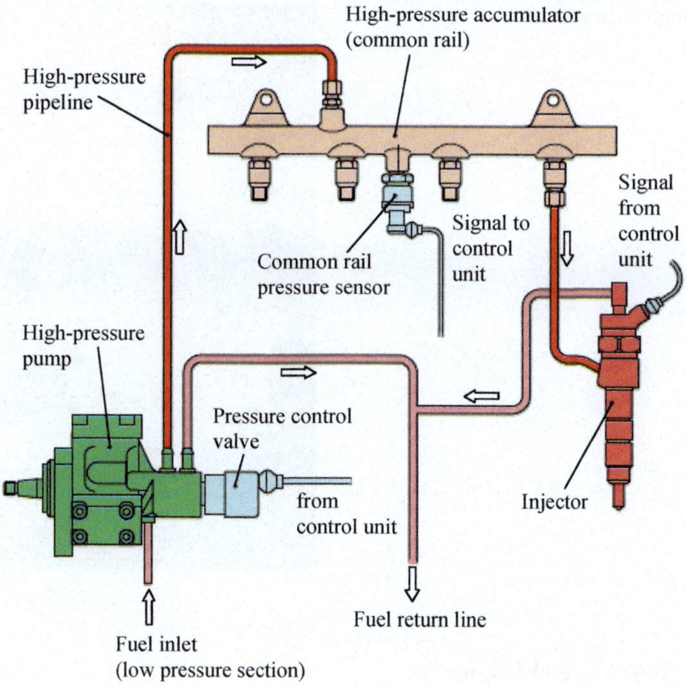

Fig. 4.20 Common rail injection system. *Graphic* deckermedia GbR, Rostock

should ideally take place at TDC of the piston position. However, as the combustion speed is finite (about 15–30 m/s) and depends essentially on the engine speed and air ratio, ignition must take place before the TDC is reached. Depending on engine design and load point, the ignition starts at a crank angle of approximately 0–40° before TDC.

Unintentional and uncontrolled self-ignition places unnecessary mechanical stress on the engine and reduces its efficiency. If the ignition timing is incorrect, the efficiency is generally reduced because the piston stroke is slowed down in the direction of TDC if ignition is premature. If ignition is delayed, the pressure in the combustion chamber no longer reaches its optimum level due to an increase in volume. As the force F on the pistons is the quotient of combustion chamber pressure and piston area, it is immediately apparent that this reduces the engine's power.

In the diesel engine, a defined start of combustion is brought about by the time of injection. The fuel must be injected shortly before the intended ignition, because a certain delay time elapses between the time of injection and ignition. This so-called ignition delay is mainly dependent on the fuel, especially its density and volatility, and on the engine speed; it is approximately 0.0007–0.003s.

Fig. 4.21 Compressor system for starting air

Starting the Ship's Diesel Engine[46]

Large ship diesel engines have such a high compression ratio and such a large displacement that they cannot be started with electric starter motors. Here, oversized electric motors with huge starter gear rims would be necessary. Such diesel engines are therefore started using starting air; this is done using compressed air. This is pumped by a compressor into large compressed air bottles (Fig. 4.21).

If the diesel engine is to be started, the automated starting process of the main engine is initiated from the control station. For this purpose, the air line from the cylinders to the engine is released by means of controlled valves. Each cylinder is pressurized with starting air according to its position (shortly after TDC) and the ignition sequence. The corresponding pistons are pressed down one after the other due to the air expanding in the cylinder chamber and the engine speed is raised to ignition speed. A governor pulls the injection pumps to "filling," fuel is injected, and the first self-ignition occurs. A strong starting compressed air system (usually 30 bar nominal pressure) is required for this process.

This process of starting the engine is also called "blowing on."

In addition, even when the engine is not in operation, the high-temperature (HT) cooling water system and a preheating pump usually continue to keep the engine at a constant lower operating temperature in order to facilitate or accelerate the starting process.

[46] See also [73].

Fig. 4.22 Cylinder and piston system

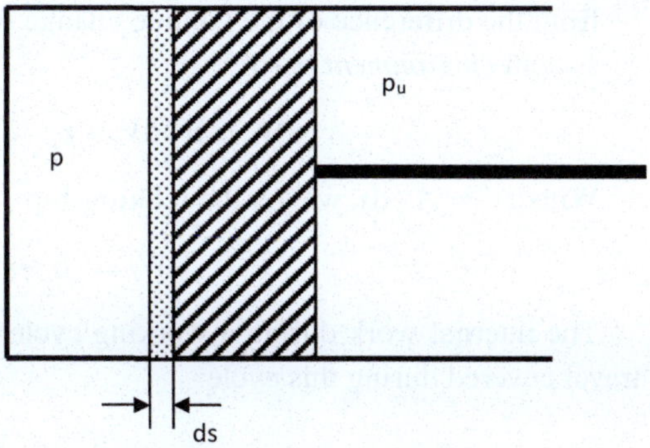

On ships with a controllable pitch propeller, the engine is slowly increased to nominal speed. In this condition, the engine is left idling for a few minutes to stabilize temperatures and pressures. After increasing to constant speed, the control of the engine is transferred to the bridge (remote control) and from there it is accepted by pressing a button.

Ships without controllable pitch propellers are accelerated only very slowly. The reason is the angle of attack of the propeller, which is designed for maximum speed. The acceleration process is therefore comparable to starting a car in top gear: accelerating too fast with a high gear ratio would stall the engine.

Net Power and Some Characteristics of the Diesel Engine[47]

a) Basic considerations

High power output is achieved either by means of high-speed engines combining several rows of small short-stroke crossheadless working cylinders in V or W form or by means of slow-speed long-stroke (two-stroke) large in-line engines with a few large-diameter cylinders.

The multicylinder high-speed engines, which combine several in-line engines on a single crankshaft, have the advantage of the highest liter output and the least input of space and weight. They are therefore preferred for speedboats. For example, the propulsion system of the German Navy's A 143 class speedboats consisted of four high-speed four-stroke 16-cylinder in-line diesel engines with exhaust gas turbocharging (MTU 16 V 956), each with a continuous output of 4000 hp (2942 kW; [45]).

The large engine design has the advantage of low noise, long service life, suitability for use with heavy fuel oil, crude oil, and tar oil. Lower requirements for ongoing operational monitoring and maintenance (due to the manageable number of valves, nozzles, pumps, etc.) are another aspects that speak in favor of these engines in large ships.

b) Calculation of the net power or the effective power P_e [10]

The power of a cylinder of an internal combustion engine is derived from the piston work dW_i. The piston work of a small finite piston movement ds (Fig. 4.22) results

[47] For further details see again Kraemer in [6, Volume II, p. 147 ff.], also [36].

from the difference of the volume change work in the system *cylinder* $p \cdot dV$ and in the system *environment* $p_u \cdot dV$:

$$dW_i = p \cdot dV - p_u \cdot dV = (p - p_u) \cdot dV. \tag{4.30}$$

With $dV = A \cdot ds$, we get the relationship

$$dW_i = (p - p_u) \cdot A \cdot ds. \tag{4.31}$$

The internal work during a working cycle is obtained by integration along the piston travel covered during this cycle:

$$W_i = -\oint (p - p_u) \cdot A \cdot ds = -\Delta p_{mi} \cdot A \cdot s. \tag{4.32}$$

For further calculation, instead of the actual pressure curve during a working cycle, a mean indexed internal pressure is used Δp_{mi} for simplification, the curve of which is assumed to be constant over the piston stroke. The magnitude of pressure is selected so that the actual work done over a complete working period is the product of Δp_{mi} and the *displacement* $V_H = A \cdot s$ results.

The power that results from the combustion of the fuel and the resulting gas pressure on the pistons in the cylinders is the so-called *indexed power* or *internal power* P_i; this is then calculated as

$$P_i = W_i \cdot \frac{N}{Z_N} \tag{4.33}$$

with Z_N=revolution per working cycle of two-stroke engine $Z_N = 1$, and Z_N= revolution per working cycle of four-stroke engine $Z_N = 2$.

In other words:

$$P_i = -\frac{\Delta p_{mi} \cdot A \cdot s \cdot N}{Z_N} = -\frac{\Delta p_{mi} \cdot V_H \cdot N}{Z_N}. \tag{4.34}$$

The *net power* or the *effective power* of the *engine* P_e (which can be taken from the crankshaft) is determined by *internal power* P_i per cylinder and the *mechanical efficiency* η_m of the engine:

$$P_e = P_i \cdot \eta_m. \tag{4.35}$$

The mechanical efficiency η_m takes into account the friction losses P_r which occur in the engine, and is the quotient of the effective power P_e to be tapped at the crankshaft and the internal power P_i:

$$\eta_m = P_e/(P_e + P_r) = P_e/P_i. \tag{4.36}$$

Calculation of the power via piston travel or speed:

General applies: **Power = Force times distance by time ($P = F \cdot s/t$)**

or otherwise expressed: **Power = Force times speed ($P = F \cdot v$)**

Thus, the power would be the product of the force exerted on the piston by the explosion pressure and the piston speed. But an average *piston speed* v_m should be used because this is not constant in the cylinder: The linear movement of the piston is coupled to the rotating movement of the crankshaft. This results in an approximately sinusoidal curve of the piston speed at each stroke. It is $v=0$ at top and bottom dead center and reaches a maximum speed approximately in the middle of the stroke. In practical calculations, the approximately sinusoidal curve is converted into the average piston speed v_m; thus, the engine power can be defined as follows:

$$P = F \cdot v_m. \tag{4.37}$$

But power is also **mechanical work done per time unit ($P = W/t$)**.

In case of a **circular motion**, such as the *crankshaft,* the **speed of a point** on a circular disc is its circumferential speed:

$$v = d \cdot \pi \cdot n. \tag{4.38}$$

If now in Eq. 4.38, d is replaced by $2r$ (with r in meters) and for the expression $F \cdot r$ the torque (work) M written in Nm, the power is in watts:

$$P = M \cdot 2 \cdot \pi \cdot n. \tag{4.39}$$

Often the term "$2 \cdot \pi \cdot n$" is replaced by the circular frequency ω:

$$P = M \cdot \omega. \tag{4.40}$$

Calculation of the indexed or internal power P_i:

As stated above, the general rule is that power is the determining factor:

$$P = F \cdot s/t. \tag{4.41}$$

In addition, the pressure p acting on a piston surface A produces a force:

$$F = A \cdot p. \tag{4.42}$$

The force that is then absorbed by the (mean) internal cylinder pressure p_{mi} on the piston surface A is then:

$$F = A \cdot p_{mi}. \tag{4.43}$$

The indexed internal power or internal capacity P_i of an engine corresponds to the energy released by the combustion gas. As it cannot be measured directly on the crankshaft, the *indexed mean pressure* p_{mi} is first determined by means of pressure measurement in the cylinder and the recording of a p–V diagram from the resulting diagram.

Furthermore, it must be taken into account that not all of the power provided by the combustion of the fuel is available to form the indexed internal power—it is subject to losses described by the indexed (internal) efficiency:

$$\eta_i = P_i/(\dot{m}_B \cdot h_u). \tag{4.44}$$

The following consideration is also made: A working cycle in a four-stroke engine is four strokes, which are performed within two revolutions; Note: The fuel–air mixture is only ignited every second revolution! In a two-stroke engine, two working cycles are performed within one crankshaft revolution. In this respect: Z_N = revolutions per working cycle with $Z_N = 1$ for two-stroke engines and $Z_N = 2$ for four-stroke engines (see above). The distance s corresponds to the piston stroke. By inserting into the above general equation for power, we then obtain the internal power of a cylinder P_i :

$$P_i = (A \cdot p_{m,i} \cdot s \cdot n)/Z_N. \tag{4.45}$$

In multicylinder engines, the total internal power is the product of P_i multiplied by the number of cylinders z:

$$P_i = (A \cdot p_{m,i} \cdot s \cdot n) \cdot z/Z_N. \tag{4.46}$$

Thus, the *effective net power* P_e of the engine is calculated taking into account the mechanical efficiency (see above):

$$P_e = P_i \cdot \eta_m. \tag{4.47}$$

Typical values for η_m are between 0.8 and 0.85 for four-stroke engines and between 0.75 and 0.9 for two-stroke engines [62].

From the knowledge of the effective power P_e, the quantity of fuel supplied, and the net calorific value of the fuel, the *effective efficiency* η_e of the engine can also be determined:

$$\eta_e = \frac{|P_e|}{\dot{m}_B \cdot H_u} \tag{4.48}$$

with $\dot{m}_B$ the quantity of fuel per unit of time, and H_u the lower calorific value of the fuel (for diesel fuel: 42.5 MJ/kg or 11.8 kWh/kg or also 35 MJ/l; [75]).

Analogous to the formula for P_i, the *net power* or rather *effective power* P_e can also be calculated directly if the *effective mean pressure* p_{me} is known:

$$P_e = (p_{me} \cdot V_h \cdot n \cdot z)/Z_N, \tag{4.49}$$

where V_h is the swept volume $A \cdot s$ (with A the piston or cylinder area and s the piston stroke).

Indicative values for effective mean pressures are shown in Table 4.1[48] but the pressures given here must be converted into N/m^2 to be used in the above equation: $1\ \text{bar} = 1 \cdot 10^5\,N/m^2$.

[48] See also [69].

Table 4.1 Effective mean pressures of various engines

Engine	Pressure (bar)
Larger high-speed diesel engines	6–28
Medium-speed diesel engines	15–25
Crosshead engines (two-stroke diesel)	9–15.4

Table 4.2 Average piston speeds

Type of engines	Average piston speed (m/s)
Larger high-speed diesel engines	7–16
Medium-speed engines (diesel)	5.3–9.5
Crosshead engines (two-stroke diesel)	5.7–7

The *effective mean pressure* p_{me} can also be calculated directly from the output torque M, which can be measured at the crankshaft:

$$p_{me} = Z_N \cdot 2 \cdot \pi \cdot M / V_h \tag{4.50}$$

(with $Z_N = 2$ for four-stroke engines; $Z_N = 1$ for two-stroke engines).

Another interesting characteristic of a combustion engine is its *specific fuel consumption* b_e. It is defined as the quotient of the quantity of fuel supplied and the effective net power of the engine:

$$b_e = \dot{m}_B / P_e. \tag{4.51}$$

Furthermore, the *average piston speed* v_m to the most important parameters of an engine:

$$v_m = 2 \cdot s \cdot n \tag{4.52}$$

with s the piston stroke and n of the engine speed; usual values are shown in Table 4.2 [36].

Example for Dimensioning a Propulsion Motor

What is the total stroke volume of the engine "MTU 16 V 956 TB91," which was installed on the German Navy's class A 143 speedboats? The following data are known:

Manufacturer	MTU Friedrichshafen
Type	Diesel engine
Arrangement	2×8 cylinder 60°
Cylinder bore d	220 mm
Piston stroke s	230 mm

The displacement of a cylinder V_h is the difference between the largest and smallest combustion chamber of a working cylinder. It results from the diameter of the cylinder bore and the piston stroke:

$$V_h = (d^2 \cdot \pi/4) \cdot s. \tag{4.53}$$

The sum of the stroke volumes of all working cylinders of the engine is the total stroke volume V_H:

$$V_H = (d^2 \cdot \pi/4) \cdot s \cdot z,$$
$$V_H = (2.2^2 \text{dm}^2 \cdot \pi/4) \cdot 2.3 \text{ dm} \cdot 16,$$
$$V_H = 139.9 \text{ L}.$$

What is the stroke bore ratio of this engine? Is it a "long stroke," "square stroke," or a "short stroke"?

The ratio of "piston stroke to cylinder bore" is called stroke bore ratio "α":

$$\alpha = s/d,$$
$$\alpha = 230 \text{ mm}/220 \text{ mm}, \tag{4.54}$$
$$\alpha = 1.05.$$

This engine is a "long stoke":

Stroke bore ratio "α":

Long stroke $\alpha > 1$,

Square stroke $\alpha = 1$,

Short stroke $\alpha < 1$

The compression ratio of this engine is 13: 1; how large is the compression chamber of this engine?

The compression chamber V_c is the smallest volume of the combustion chamber. The sum of the compression chamber and the displacement of a cylinder is the largest volume of the combustion chamber or the "cylinder volume." The ratio of "largest combustion chamber to smallest combustion chamber" is called compression ratio "ε":

$$\varepsilon = (V_h + V_c)/V_c. \tag{4.55}$$

The compression ratio indicates to how many parts the sucked-in fuel–air mixture or the sucked-in air is compressed. It influences the thermal efficiency and power of the engine.

By switching over, you get

$$V_c = V_h/(\varepsilon - 1) = \frac{V_H}{z}/(\varepsilon - 1),$$
$$V_c = 8.7\ \mathrm{L}/(13 - 1),$$
$$V_c = 0.73\ \mathrm{L}.$$

What is the total cylinder volume of the engine $V_{\mathrm{Cyl,tot}}$?

The cylinder volume of a cylinder V_{Cyl} *is the sum of the compression volume* V_c *and that of its stroke volume* V_h*. The total cylinder volume of the engine is determined by multiplying the cylinder volume by the number of cylinders:*

$$V_{\mathrm{Cyl,tot}} = (V_h + V_c) \cdot z,$$
$$V_{\mathrm{Cyl,tot}} = (8.74\ \mathrm{L} + 0.73\ \mathrm{L}) \cdot 16, \tag{4.56}$$
$$V_{\mathrm{Cyl,tot}} = 152\ \mathrm{L}.$$

The effective mean pressure of the engine of the speedboat type A143 should be determined. The following data must be used.

Type	Four-stroke diesel engine
Cylinder	16
Continuous output	4000 hp (2942 kW) at 1515 min^{-1}
Cylinder bore	220 mm
Piston stroke	230 mm

From Eq. 4.49, the effective engine power of a four-stroke engine (with $Z_N = 2$)

$$P_e = (p_{me} \cdot V_h \cdot n \cdot z)/Z_N$$

is obtained by switching to p_{me}:

$$p_{me} = Z_N \cdot P_e/(V_h \cdot n \cdot z)$$

with

$$P_e = 2{,}942{,}000\ \text{W},$$
$$V_h = 0.0087\ \text{m}^3\ \text{(see above for calculation)},$$
$$n = 1515\text{L/min} : 60\ \text{s/min} = 25.25\ \text{s}^{-1},$$
$$z = 16,$$
$$p_{me} = 2 \cdot 2{,}942{,}000\ \text{Nm/s}/(0.0087\ \text{m}^3 \cdot 25.25\ \text{s}^{-1} \cdot 16),$$
$$p_{me} = 16.7\ \text{bar}.$$

◀

4.3.2 Gas Engines

The gas engine or the dual-fuel engine can be used for LNG.[49] As a reciprocating piston engine, this engine has a design comparable to that of a petrol or diesel engine. The fuel–air mixture is ignited by spark ignition using a spark plug in units based on petrol engines, and by self-ignition of small injected quantities of ignition oil (diesel fuel) in units based on diesel engines. Wärtsilä works with high-pressure gas injection using an engine based on the diesel principle, that is, with self-ignition of the fuel gas injected directly into the combustion chamber at high pressure. Engines operating according to this method achieve the highest mean pressures of all internal combustion engines.

The company MAN has developed the "PGI process." This method works according to the same procedure as the ignition jet principle, but avoids the disadvantage of the additionally required fuel, as it uses the available gas as pilot injection quantity for the initiation of the ignition. Components are a "pilot gas high-pressure injection valve" and a starting aid (glow element), which are integrated in a cooled prechamber. The directly controlled injection valve sprays a small amount of pilot gas into the prechamber shortly before TDC[50] at an inlet pressure of 230 bar generated by a compressor. There, an almost stoichiometric mixture is formed, which ignites on the hot surface of the glow element, which is only electrically heated during the starting phase. In this process, the ignition timing is thus directly controlled by the injection timing of the pilot gas [70].

External mixture formation can be achieved by gas mixing units before or after a possible turbocharger, and internal mixture formation is possible by separate gas inlet valves or injection needles.

---

[49] LNG, "liquefied natural gas.".

[50] TDC = top dead center of the piston.

Fig. 4.23 Marine gas turbine on board the American frigate USS FORD. *Photo* US Navy

4.3.3 Turbines

The use of turbines for power generation in shipbuilding is not new. Whereas in the past, it was usually steam turbines,[51] today, gas turbines are mainly used for the main propulsion (see Fig. 4.23). These are similar to those used in aircraft construction.

The turbine plant is a turbomachine with a rotating propulsion part (*rotors*) to convert the energy of a flowing medium into mechanical energy. The fluid flows through it in the axial direction, whereby the fluid transfers its energy to the rotor. The speeds are in the range of about 3500 min^{-1}.

Gas turbines are also often used in conjunction with diesel engines as a combined propulsion system (see Sect. 4.3) or as a power generator to drive a generator that supplies the necessary voltage for an electric motor (e.g., on the Celebrity Constellation, a cruise ship owned by the shipping company Celebrity Cruises).

4.3.3.1 Function of the Gas Turbine

In principle, the marine turbine plant is a series connection of two fluid energy machines, namely the compressor and the turbine (see Fig. 4.24): the air drawn in from the open air is pre-compressed in the compressor section, preheated, and fed to the combustion chamber, where the injected fuel is burned. For this purpose, the combustion air must be forced into the combustion chamber at a pressure higher than the combustion pressure. This is achieved by a multistage compressor driven by the turbine itself. The combustion chamber itself is not arranged radially around the rotor between the compressor and

[51] As in the case of the guided missile destroyers of the German Navy LÜTJENS, MÖLDERS, and ROMMEL, which have already been decommissioned today.

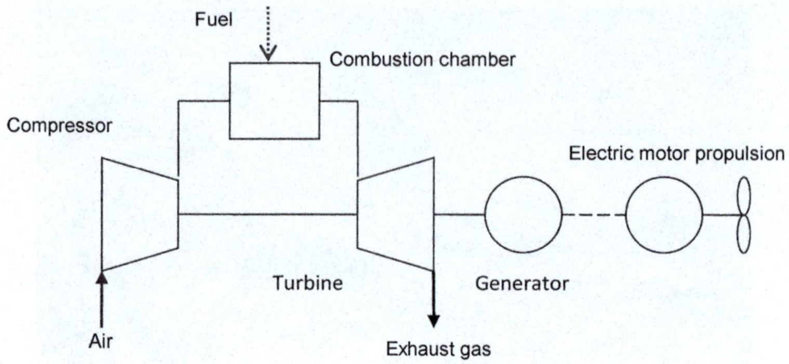

Fig. 4.24 Gas turbine plant

the expansion section (as in jet engines in aircraft construction); in marine gas turbines, there is an external combustion chamber with gas pipes from the compressor and to the expansion section (the actual turbine). Here the exhaust gas is expanded and energy is transferred to the rotor. In the generator, the useful work of the process is converted into electrical energy for the propulsion motor.

To achieve high efficiencies, the heat of the exhaust gas is transferred via a recuperator to the colder air coming from the compressor before it is fed into the combustion chamber (not shown in Fig. 4.24).

This process can be described with good approximation as an open Joule process in a temperature (t)—entropy (s)—diagram (see Fig. 4.25).

4.3.3.2 Fundamentals of Gas Turbine Calculation

Fluid Speed in the Machine Stage

Both the compression and the energy release of the fluid is achieved by a blade system within the compressor and the turbine. We are talking about stationary blades (*guide vanes* or *stators—St)* and the rotating blades (*impellers—Im;* see Fig. 4.26). In this blade system, both the exchange of work between the rotating machine parts and the fluid and an energy conversion between kinetic and potential energy within the fluid takes place. A single *stage* consists of an impeller and a guide wheel. In marine turbine plants, both the compressor and the actual turbine have several *stages,* in these *axial machines* (because the fluid flows axially through the machine) are arranged one behind the other in the direction of flow.

This blade system forms flow channels, also called blade row. These blades are curved like the wing of an airplane. For the description of the flow processes in front of and behind an impeller, the relationship between the absolute speed $\vec{c}$ and the relative speed $\vec{w}$, when the origin of the coordinates of the absolute system and the relative system rotating at constant angular speed coincide, is obtained by the circumferential speed $\vec{u}$ of the relative system.

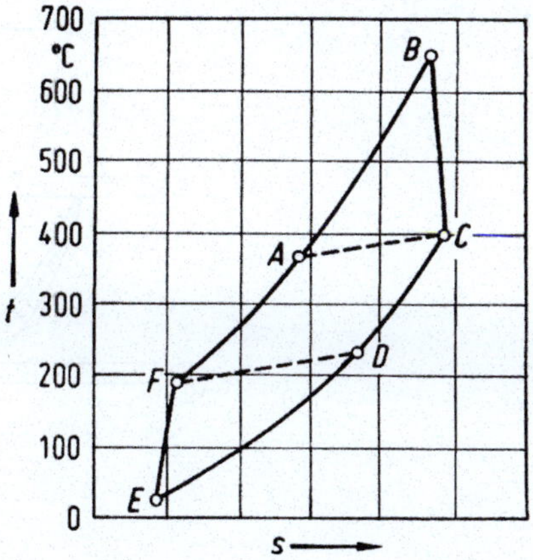

Fig. 4.25 *t–s* diagram of the Joule process "open gas turbine." $A \rightarrow B$ isobaric internal combustion; $B \rightarrow C$ expansion in the turbine, inclination of the line against the vertical indicates the increase in entropy due to losses; $C \rightarrow D$ extraction of heat in the recuperator; $D \rightarrow E$ isobaric heat extraction at the gas outlet from the turbine; E suction of fresh air; $E \rightarrow F$ compression of the intake fresh air, where the inclination of the line represents the increase in entropy due to losses in the compressor; $F \rightarrow A$ supply of heat extracted from the exhaust gas along *C–D*. As the heat exchanger requires a temperature gradient, the points of *CD* are at higher temperatures than those corresponding to the curve *AF*. *Source* [30]

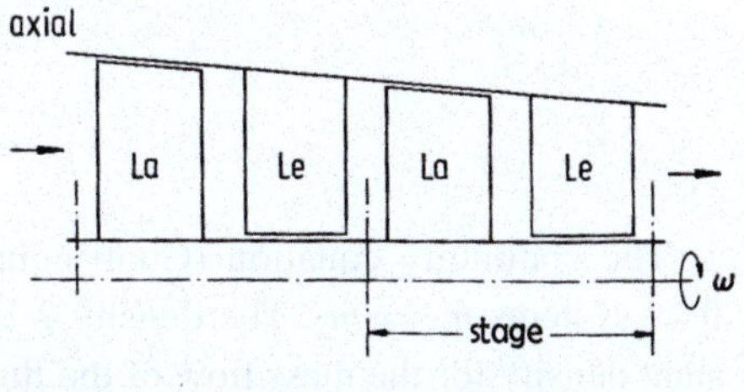

Fig. 4.26 Multistage compressor. *Im*/*St*: impeller/stator vane. *Source* [9]

$$\vec{c} = \vec{w} + \vec{u}. \tag{4.57}$$

If these speed vectors are each added to the representative flow filament at the inlet and outlet of the relative system "impeller," so-called *speed triangles* result at these points (see Fig. 4.27).

Continuity Equation for the One-Dimensional Flow (Flow Filament Theory)

Both the compressor and the turbine section can each be regarded as separate systems with system boundaries (see Fig. 4.28), through which mass, heat, and momentum transfer takes place.

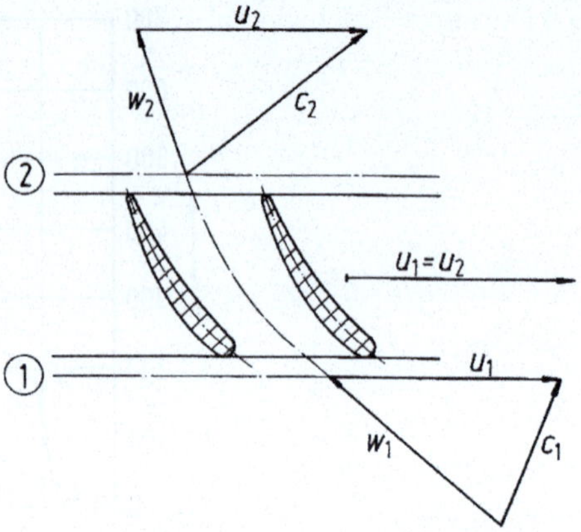

Fig. 4.27 Speed triangles at the inlet and outlet of an axial compressor blade. [9, p. 313]

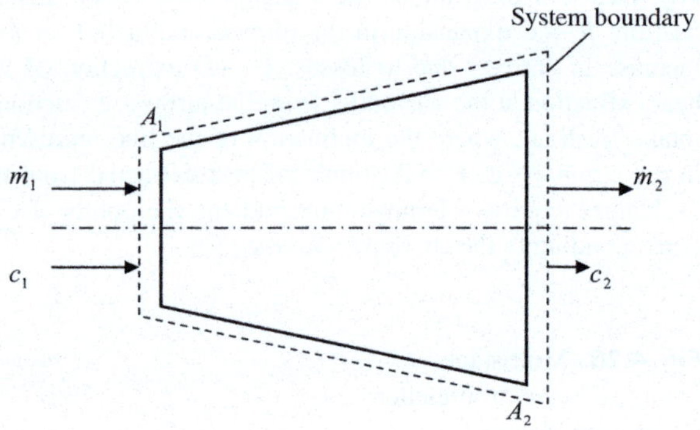

Fig. 4.28 System "turbine" with system boundaries. $\dot{m}_i$ = mass flow per time unit (kg/s), c_i = speed of the fluid, A_i = circular area of the system boundary

The continuity equation (Conti equation) states that the following applies for open flow systems $\dot{m}_1 = \dot{m}_2$. The density ρ of the fluid and its volume flow $\dot{V}$ follow at constant density for the mass flow of the fluid:

$$\dot{m}_1 = \dot{m}_2 = \rho \cdot \dot{V}_1 = \rho \cdot \dot{V}_2 = \rho \cdot c_1 \cdot A_1 = \rho \cdot c_2 \cdot A_2. \tag{4.58}$$

From Eq. 4.58, it can be seen that this also applies: $\dot{V}_1 = \dot{V}_2$.

Consequently, the flow velocities are inversely proportional to the flow cross sections at the inlet and outlet of the system:

$$c_1/c_2 = A_2/A_1. \tag{4.59}$$

Turbine Output

To estimate the turbine power, the following procedure using the "Euler's turbine equation" is recommended.[52]

[52] More in-depth [9, p. 336 ff.].

During the exchange of work between the fluid and the blade row, pressure and thrust forces occur on the wetted surfaces of the blades (F_p and F_τ). The circumferential components of these forces generate the torque at the rotor, and the forces acting in axial direction must be absorbed by the thrust bearing of the machine. By integrating the forces acting on the turbine wheel at a blade surface element dO via the wetted rotor surface O, as shown in Fig. 4.29, the effective forces result and from their circumferential components, the rotor power is obtained as internal power P_i by multiplication with the respective local circumferential speed. In general, however, a determination of the rotor power in this way is hardly of practical importance; as a rule, a correlation between the internal power and the speeds in front of and behind the impeller is derived (see Fig. 4.29), which describes the work conversion without detailed knowledge of the local flow processes. Here, the impulse momentum set (swirl set) is assumed.

This is based on the consideration that a stationary flow process takes place in the turbine (or compressor) system, in which the state variables and speeds at any point in the system are independent of time. The angular momentum can be described as follows:

$$D = m \cdot c \cdot r \tag{4.60}$$

with

D = angular momentum in Nms or kg · m^2 · s^{-1},
m = mass of fluids in kg,
c = speed of fluids in m/s,
r = average distance of flow filament from the center of the rotation axis in m

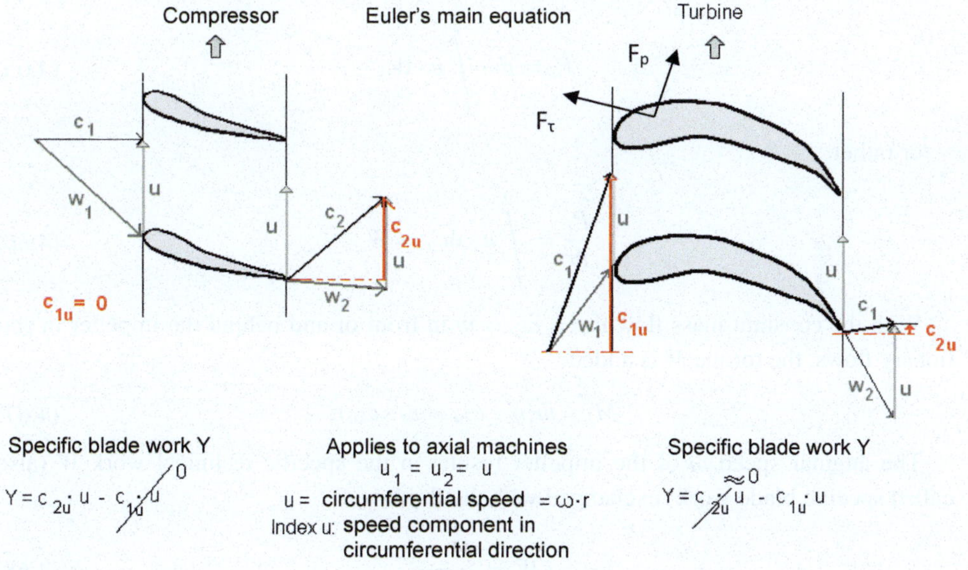

Fig. 4.29 Speed components of the blade systems compressor and turbine. *Graphic* VK, CC BY-SA 3.0

The change of impulse within a subsystem (here: the turbine blades) produces a torque around the center of the turbine:

$$M = \frac{\mathrm{d}D}{\mathrm{d}t} = \frac{\mathrm{d}c}{\mathrm{d}t} \cdot m \cdot r \quad \text{in Nm.} \tag{4.61}$$

Sensibly, only parts of flow speed of fluid can contribute to torque, which are perpendicular to turbine turning point by law of lever arms. Such proportions are marked with the index u.

An integration of Eq. 4.61 yields the following result:

$$\int_{t_1}^{t_2} M \cdot \mathrm{d}t = m \cdot r \cdot \int_{1}^{2} \mathrm{d}c_u. \tag{4.62}$$

From the relationship between torque M and the speed n, the internal power P_i is calculated:

$$P_i = M \cdot 2 \cdot \pi \cdot n = M \cdot \omega \quad \text{in Nms}^{-1}. \tag{4.63}$$

(The term "$2 \cdot \pi \cdot n$" is the angular speed ω.)

So:

$$P_i = \frac{m \cdot r \cdot \omega \cdot \mathrm{d}c_u}{\mathrm{d}t} = \frac{m \cdot u \cdot \mathrm{d}c_u}{\mathrm{d}t} \tag{4.64}$$

where u is the maximum possible circumferential speed in a considered cross section. A renewed integration provides

$$P_i = \dot{m} \cdot \int_{1}^{2} u \cdot \mathrm{d}c_u \tag{4.65}$$

or rather

$$\frac{P_i}{\dot{m}} = \int_{1}^{2} u \cdot \mathrm{d}c_u = W. \tag{4.66}$$

With the constant mass flow $\dot{m}_1 = \dot{m}_2 = \dot{m}$ in front of and behind the impeller in stationary flows, the torque M is added:

$$M = \dot{m}(r_2 \cdot c_{2u} - r_1 \cdot c_{1u}). \tag{4.67}$$

The angular speed ω of the impeller results in the specific technical work W (also called specific blade work) exchanged with the fluid:

$$W = \frac{M \cdot \omega}{\dot{m}} \tag{4.68}$$

and therefore

$$W = \omega \cdot (r_2 \cdot c_{2u} - r_1 \cdot c_{1u}), \tag{4.69}$$

$$W = u_2 \cdot c_{2u} - u_1 \cdot c_{1u}. \tag{4.70}$$

According to sign agreement for the compressors $W>0$
and for the turbine $W<0$

The last equation (Eq. 4.70) is called **Euler's turbine equation**, with u the circumferential speed of the rotating blade tip at the inlet (index 1) and outlet (index 2), and likewise the usable fluid speed c_u at the inlet and outlet.

The circumferential speeds u are referred to the *mean radius* of the blade row r_m (see Fig. 4.30:

$$u = \omega \cdot r_m.$$

The axial flow speed in the considered cross section of the turbine stage c_{ax} is calculated as follows:

$$c_{ax} = \dot{m}/(2 \cdot \pi \cdot r_m \cdot h \cdot \rho). \tag{4.71}$$

In this equation

$\dot{m}$ = Mass flow of exhaust gas,
ρ = Density of exhaust gas,
h = Blade height or rather height of flow channel in the considered cross section,
$2\pi\, r_m\, h$ = Approximation for the annular area of the flow cross section.

The exhaust gas density ρ can be determined with sufficient accuracy from the equation of state for ideal gases:

$$\rho = p/(R \cdot T) \tag{4.72}$$

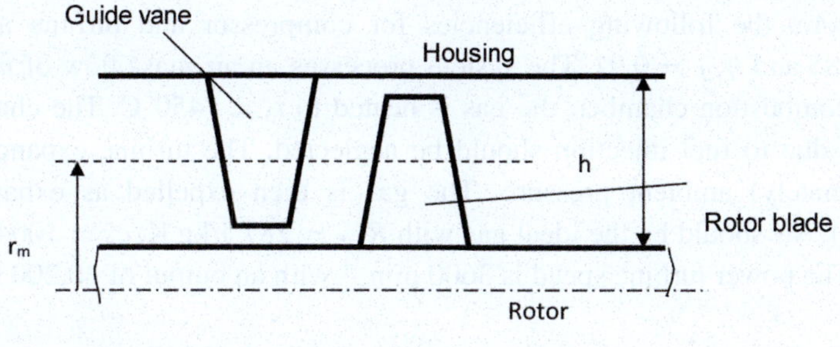

Fig. 4.30 Mean radius of a turbine rotor

with

R = Specific gas constant [J/kg K]; for air: 287.09 J/kg K,
T = Exhaust gas temperature (K)

It is also possible to estimate, with sufficient accuracy, the specific gas constant assumed for air. Strictly speaking, the specific gas constant of the exhaust gas would have to be used. This is the quotient of the general gas constant and the average molar mass of the exhaust gas. However, the average molar mass can only be calculated if either the exact exhaust gas composition is known or the exact composition of the fuel and the excess air λ are given.

The working power of the turbine rotor is calculated using Euler's turbine equation:

$$P_i = \dot{W} = \dot{m} \cdot \omega \cdot r_m \cdot (c_{u1} - c_{u2}) \quad \text{or:} \tag{4.73}$$

$$P_i = \dot{m} \cdot (c_{u1} - c_{u2}) \cdot u. \tag{4.74}$$

From the Conti equation follows for the mass flow $\dot{m}$:

$$\dot{m} = A \cdot \rho \cdot c_{ax}. \tag{4.75}$$

4.3.3.3 Gas Turbine Malfunction Matrix

Like all mechanically and thermally stressed parts, such as plant and machine components, gas turbines are also subject to wear and tear, which can lead to operational disruptions. In order to keep the risk of operating malfunctions as low as possible, the maintenance and inspection instructions in the operating manuals of the manufacturers and suppliers must be observed; this also applies not least to the protection of warranty claims.

Possible malfunctions and their causes are shown in Table 4.3

Examples for the Design of Gas Turbines

A GE LM2500+marine turbine operates between the pressures $p_1 = p_4 = 1\,\text{bar}$ and $p_2 = p_{2'} = p_3 = 23.1\,\text{bar}$. The initial temperature is $t_1 = 20^\circ\text{C} = 293.15\,\text{K}$. Furthermore, the following efficiencies for compressor and turbine are assumed: $\eta_{S,V} = 0.85$ and $\eta_{S,T} = 0.91$. The system processes an air mass flow of $\dot{m} = 85.9\text{kg/s}$. In the combustion chamber, the gas is heated to $t_3 = 1450^\circ\text{C}$. The change in mass flow rate due to fuel injection should be neglected. The turbine expands the gas to (approximately) ambient pressure. The gas is then expelled as exhaust gas (see Fig. 4.31). Air should be the ideal gas with $R_{Air} = 287$ J/kg K, $c_p = 1004$ J/kg K, and $\kappa = 1.4$. The power turbine speed is 3600 min^{-1} with an output of 30,200 kW.

a) The missing temperatures and pressures shall be calculated.
b) Which power P_{tot} can the system deliver, and what is its efficiency?

Table 4.3 Gas turbines malfunction matrix

Malfunction/damage	Possible Cause	Measure
Power loss of the system	Contamination of supply air filter	Pressure loss at supply air filter too high? Clean or replace filter
Power loss of the system	Deposits at the blade system	Optical/endoscopic review of blade system. Cleaning the accessible blades; if malfunction is not eliminated by this: stay in shipyard stay to clean the entire blade system; replace blades if necessary
Power loss of the system	Cavitation[53]/erosion damage or crack formation at blade system; other mechanical damage	Optical/endoscopic review of blade system. Supply air filter defective, resulting in foreign matter entry in system? Check pressure loss of filter systems: Pressure loss to zero! Stay in shipyard—replace blades
Power loss of the system	Leaks at labyrinth seal system; gap distance at blade system to large	Stay in shipyard: replacement of labyrinth seal and damaged blades
Power loss of the system	Fuel supply to combustion disturbed	Fuel supply check; clean the fuel nozzles and replace if necessary
Power loss of the system	Suspend the ignition voltage at combustion chamber system	Coked contacts; clean or rather replace
Crack formation on blade system	Overheating	Check and repair cooling system check
"Abnormal" running noises	Imbalance of rotors due to contamination/erosion of the rotor blades	Check the blade system; replace blades if necessary
"Abnormal" running noises	Bearing damage due to high bearing temperatures or wear	Lubrication check of bearing, replace defective bearing

Preliminary remark on the solution: The real gas turbine process differs from the theoretical Joule process in the irreversibility of the state changes (1–2, 3–4). In addition, pressure losses occur in the combustion chamber (2–3). Nowadays, the pressure change due to heat losses in the combustion chamber can be minimized by suitable measures (high-temperature-resistant ceramic). The differences mentioned are clearly

[53] For *cavitation,* see also Sect. 4.4.2.7.

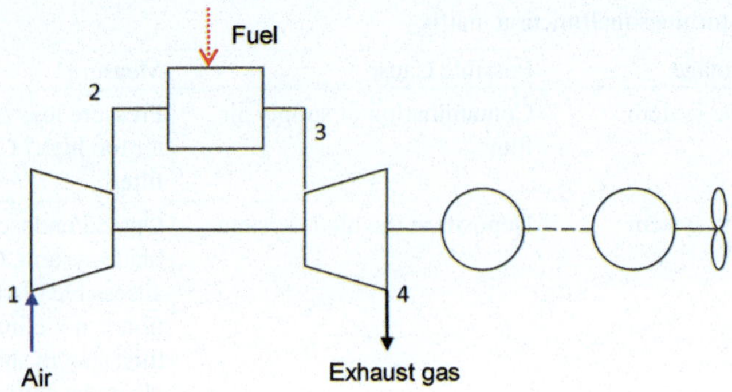

Fig. 4.31 Gas turbine system

Fig. 4.32 *p–V* diagram of gas turbine

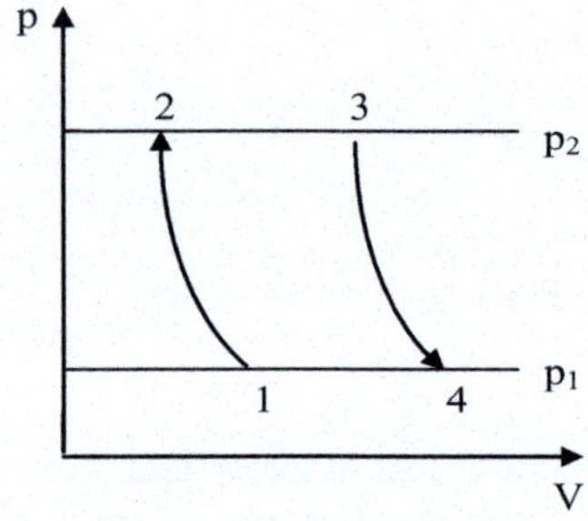

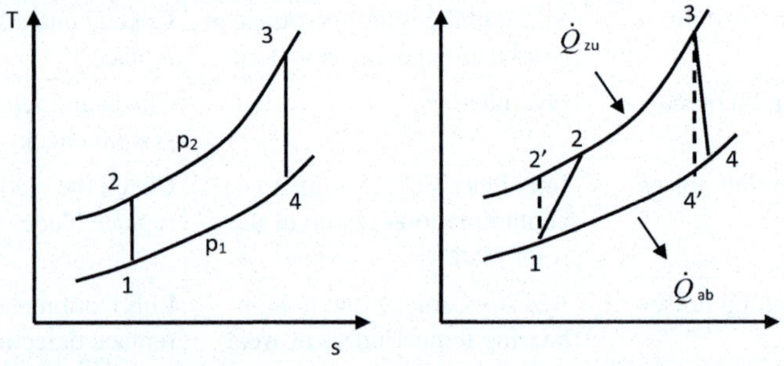

Fig. 4.33 *T–s* diagrams of ideal and real gas turbine process

illustrated in the *p–V* diagram (*p*—pressure, *V*—volume; Fig. 4.32) and in *T–s* diagram (*T*—temperature, *s*—specific entropy; Fig. 4.33).

To a):First, the temperature $T_{2'}$ at the end of the (almost) isentropic compression of p_1 to p_2 is calculated:

$$\begin{aligned} T_1/T_{2'} &= (p_1/p_{2'})^{((\kappa-1)/\kappa)}, \\ T_{2'} &= T_1/(p_1/p_{2'})^{((\kappa-1)/\kappa)}, \\ T_{2'} &= 745.58\ \text{K}. \end{aligned} \tag{4.76}$$

According to the definition of the isentropic compressor efficiency, the following applies $\eta_{S,V} = 0.85$:

$$\eta_{S,V} = (T_{2'} - T_1)/(T_2 - T_1). \tag{4.77}$$

By changing over you get T_2:

$$T_2 = (T_{2'} - T_1 + \eta_{S,V} \cdot T_1)/\eta_{S,V},$$
$$T_2 = 825.42\,\text{K}.$$

At this temperature, the compressed air enters the combustion chamber where it is (approximately) isobarically heated to a temperature $T_3 = 1{,}723.15$ K.

The temperature $T_{4'}$ is now calculated accordingly after an (approximately) isentropic expansion of the hot exhaust gases to ambient pressure:

$$\begin{aligned} T_{4'}/T_3 &= (p_{4'}/p_3)^{((\kappa-1)/\kappa)}, \\ T_{4'} &= T_3 \cdot (p_{4'}/p_3)^{((\kappa-1)/\kappa)} = 693.23\,\text{K}. \end{aligned} \tag{4.78}$$

According to the definition of isentropic turbine efficiency, the following applies

$$\eta_{S,T} = (T_3 - T_4)/(T_3 - T_{4'}). \tag{4.79}$$

By changing over you get T_4:

$$T_4 = T_3 - \eta_{S,T}(T_3 - T_{4'}),$$
$$T_4 = 1{,}723.15\text{ K} - 0.91(1{,}723.15\text{ K} - 693.23\text{ K}) = 785.92\,\text{K} = 513\text{ °C}.$$

In the company brochure, the exhaust gas temperature for this gas turbine is given as 518 °C [53]—insofar a good agreement.

To b):It applies to the power of the system:

$$P = \dot{W}_{t,\text{tot}} = \dot{m} \cdot (w_{t,34} + w_{t,12}) = \dot{m} \cdot c_p \cdot (T_4 - T_3 + T_2 - T_1), \tag{4.80}$$

$$P = \dot{W}_{t,\text{tot}} = 85.9\text{ kg/s} \cdot 1004\text{ J/kg K}$$

$$\cdot (785.92\text{ K} - 1{,}723.15\text{ K} + 825.42\text{ K} - 293.15\text{ K}),$$

$$|P| = |\dot{W}_{t,\text{tot}}| = 34{,}925\text{ kW}.$$

In the literature, the specifications of about 30,200 kW [53] are found for this turbine.

In this respect, the calculation is also in good agreement with the assumptions made. The approximately 10% deviation can be explained by the assumptions made in the conceptual formation, which are certainly different from reality. It is, therefore, questionable whether the assumptions made for the efficiency of the turbine and compressor are exactly correct. Furthermore, neglecting the change in mass flow rate due to fuel injection leads to deviations in the result. Also, the assumption that the turbine expands the gas to ambient pressure leads to further inaccuracies; this also applies to air, which is assumed to be the ideal gas, its real gas behavior is not taken into account.

The (thermal) efficiency of the plant is the quotient of the amount of heat supplied and the power output:

$$\eta = P/Q_{heat} = P/Q_{2-3} \quad \text{(see Fig. 4.40)}. \tag{4.81}$$

The amount of heat supplied $Q_{heat} = Q_{2-3}$ in the combustion chamber is calculated as follows:

$$\dot{Q}_{heat} = \dot{m}_L \cdot c_{pL} \cdot (T_3 - T_2) = \dot{m}_B \cdot H_u, \tag{4.82}$$

$$Q_{heat} = \dot{m} \cdot c_p \cdot (T_3 - T_2), \tag{4.83}$$

$$Q_{heat} = 85.9 \text{ kg/s} \cdot 1{,}004 \text{ J/kg K} \cdot (1{,}723.15 \text{ K} - 825.42 \text{ K}) = 77{,}424 \text{ kW}.$$

So this turbine has an efficiency of $\eta = 34{,}925 \text{ kW}/77{,}424 \text{ kW} = 45\%$.

In a company brochure for this turbine, the thermal efficiency is stated as > 39%, but not 45%. However, a net power of only 30,200 kW is also taken into account [53].

The differences between the calculations and the information in the prospectus also result from the assumptions made in the conceptual formation: They are realistic, but not 100% real.

A company brochure states: Gas turbine LM2500, 22.4 MW power, exhaust gas flow 69.8 kg/s, exhaust gas temperature 524 °C, speed 3000 min^{-1}. What torque can be tapped at the turbine shaft?

The torque is calculated from Eq. 4.63 by changing over:

$$P = M \cdot 2 \cdot \pi \cdot n,$$

$$M = P/(2 \cdot \pi \cdot n).$$

With the given numerical values, the torque is thus obtained:

$$M = 22{,}400{,}000 \text{ W}/(2 \cdot \pi \cdot 50 \text{ s}^{-1}) = 71{,}301 \text{ Nm}.$$

◀

4.3.4 Electric Propulsion

Electric motors are also frequently used as propulsion units (propulsion motors), as for example in the AIDAmar (two Siemens DTMSZ 3352-16YS electric motors with 12.5 MW at 130 min^{-1}).[54]

[54] For the areas of application and special designs, see also [72].

Fig. 4.34 Diesel-electric propulsion

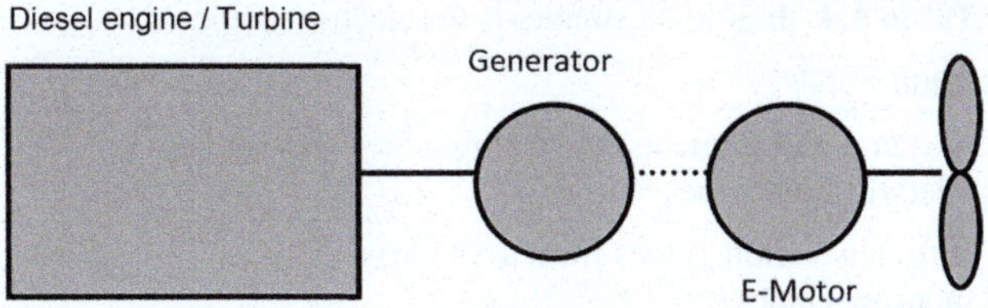

The required electric current is generated by a generator set (diesel engine and generator or turbine and generator—as in a power plant) (Fig. 4.34).

Although the losses between the fuel energy supplied and the net power delivered to the ship's propeller are about 8% higher than with direct mechanical propulsion, this system of power generation also has several advantages: The power generation can be distributed to several generator sets, which increases operational reliability. The individual combustion engines are smaller, which has advantages for maintenance work and possible replacement. Electric motors are easily controllable—the speed is controlled directly via the speed of the motors. It is also easy to switch from "full ahead" to "full aft."

It is possible to react to changing power requirements simply by switching generator sets on or off, so that they can operate almost always in the optimum speed range.

In order to compensate for the somewhat poorer overall efficiency, the electric propulsion motors are often also installed in "nacelles" under the hull (see Sect. 4.3)—so-called Azipod or POD propulsion.

The following motor types are used:

- Three-phase asynchronous motor,
- Synchronous motor, and
- Direct current motor.

The most important standards in connection with the manufacture, commissioning, and operation of electric motors are listed below (Table 4.4).

4.3.4.1 Three-Phase Asynchronous Motor

The asynchronous motor with squirrel cage rotor is one of the most widely used motor types (see also Fig. 4.35) [63]. It is easy to manufacture, robust, and practically maintenance-free.

Function of the Asynchronous Motor

In the stator of the asynchronous motor, three windings are arranged at 120° to each other. The rotor consists of a slotted laminated core. A cage made of aluminum is inserted into the grooves by die-casting, and in newer motors, the cage is made of copper. From an electrical point of view, this cage forms a system of short-circuited electrical conductors.

Table 4.4 Important standards for electric motors

Title	EN/DIN VDE	IEC
General requirements for rotating electrical machines	EN 60,034-1	IEC 60,034-1 IEC 60,085
Efficiency limit values from asynchronous machines		IEC 60,034-30
Rotating electrical machines, determination of losses and of efficiency	DIN EN 60,034-2-1	IEC 60,034-2-1
Three-phase asynchronous motors for general use with standardized dimensions and power ratings Sizes 56–315	EN 50,347	IEC 60,072
Connection designations and direction of rotation for rotating electrical machines	EN 60,034-8	IEC 60,034-8
Rotating electrical machines, designations for construction and installation	EN 60,034-7	IEC 60,034-7
Built-in thermal protection	DIN EN 60,079-14	IEC 60,034-11
Rotating electrical machines, cooling method	EN 60,034-6	IEC 60,034-6
Rotating electrical machines, protection classes	EN 60,034-5	IEC 60,034-5
Rotating electrical machines, mechanical vibrations	EN 60,034-14	IEC 60,034-14
Rotating electrical machines, noise limits	EN 60,034-9	IEC 60,034-9
Rotating electrical machines, start-up behavior of squirrel cage motors at 50 Hz to 660 V	EN 60,034-12	IEC 60,034-12
IEC standard voltages	Verband Deutscher Elektrotechniker (VDE, Association of German Electrical Engineers (AGEE)) 0175	IEC 60,038

If a sinusoidal electric current flows in the windings of the stator and there is a phase shift of 120° between the currents, a rotating magnetic field is formed in the stator of the motor. This magnetic field also permeates the rotor. The rotating magnetic field induces an electrical voltage in the conductors of the rotor (*law of induction*). As the conductors are short-circuited due to their cage design, the induced voltage causes a current flow in the rotor.

The rotor current builds up its own magnetic field, which interacts with the rotating magnetic field of the stator. As a result, a torque acts on the rotor. The rotor reacts, performs a rotary motion, and follows the rotation of the stator field.

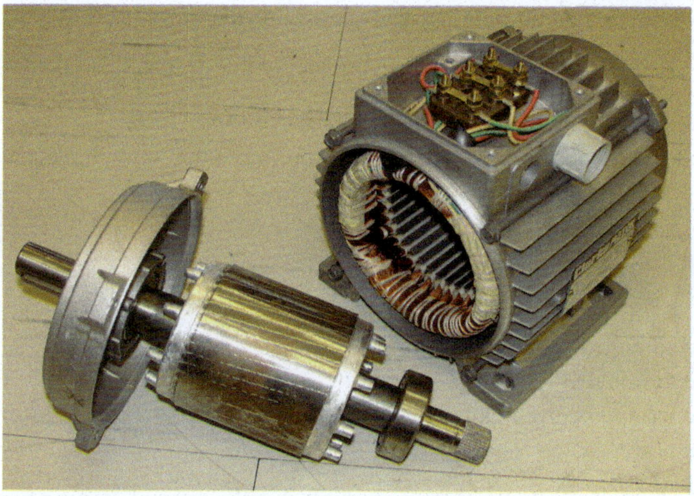

Fig. 4.35 Squirrel cage rotor and stator of a small asynchronous motor. *Photo* Zureks, CC BY-SA 3.0

However, the rotor does not follow the stator field synchronously, but rotates at a lower speed. This is necessary because only under this condition a current flow in the rotor can take place and the rotor can build up its own magnetic field. In this respect, the rotor rotates "asynchronously" to the stator field.

A *slip* occurs between the frequency of the stator field and the rotational frequency of the rotor. The size of the *slip* is load-dependent. In idle speed, the slip is only very small.

Mechanical Speed and Pole Pairs

If currents flow through the three-phase winding system, a stator field with a north and a south pole is formed in the motor. The motor has a so-called pole pair and has the pole pair number 1, which means that the pole pair number is a value determined by the motor design.

By multiple arrangement of the three-phase winding system and series connection of the corresponding phases, motors with more than one pole pair are created. If, for example, current flows through the windings in an arrangement with two pairs of poles, two north and two south poles are created, distributed over the circumference of the stator. The motor has the pole pair number 2.

If the current in the stator windings flows through a full period of time, the magnetic field of the stator continues to rotate by a full pole pitch (one north and one south pole). With two pairs of poles in the stator, this corresponds to a rotation of 180°. The speed of rotation of the stator field has been reduced to half compared to that in a motor with one pole pair, although the frequency of the feeding current has not changed.

Consequently, the number of pole pairs of the motor has an influence on the rotational frequency of the stator field and thus on the speed or rotational frequency of the rotor, which follows the magnetic field asynchronously. It therefore decreases with increasing number of pole pairs. Motors with one to four pole pairs are common.

Electrical and Mechanical Power of the Asynchronous Motor

The absorbed electrical power is calculated according to the following equation:

$$P_{\mathrm{el}} = \sqrt{3} \cdot U \cdot I \cdot \cos \phi \qquad (4.84)$$

with

$\sqrt{3}$	the so-called chaining factor,
U	the terminal voltage,
I	the terminal current, and
$\cos\varphi$	the power factor

The power factor (also called active power factor) is the ratio of active power P at apparent power S. It is equal to the cosine of the phase shift angle φ (see Fig. 4.36) and lies between 0 and 1 (usually around 0.8). The active power P is the product of terminal voltage U and terminal current I, and the apparent power S is composed of the actually converted active power P and an additional reactive power Q and is calculated according to the Pythagoras theorem (see Fig. 4.36):

$$S = \sqrt{P^2 + Q^2}. \qquad (4.85)$$

For active, apparent, and reactive power [50]:

If an inductive or capacitive resistor is connected to an AC voltage, a reactive component appears in addition to the already existing active component, analogous to the resistors.

The reactive component is caused by the phase shift between the current and voltage of the inductor or capacitance. With a pure ohmic resistor, current and voltage are in the same phase, and therefore a pure ohmic resistor has no reactive component.

The unit of reactive power is var. The active power P has the unit "watt" (W). The total power in the AC circuit is the apparent power S and it has the unit VA.

Fig. 4.36 Phase shift angle

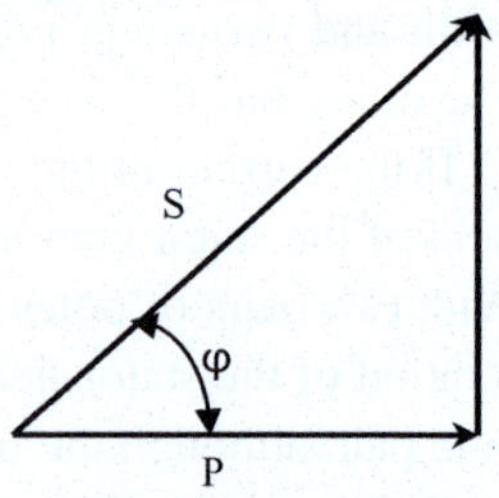

The mechanical power output corresponds to the electrical power consumed minus the power loss component P_V, which consists of copper and iron losses and friction losses:

$$P_{\text{mech}} = \omega \cdot M \tag{4.86}$$

with

ω the angular frequency ($= 2 \cdot \pi \cdot n$), where *n* the speed per second, and
M the required torque

The ratio of mechanical power output to electrical power input is the efficiency of the machine:

$$\eta = P_{\text{mech}}/P_{\text{el}}. \tag{4.87}$$

Start and Speed Control of the Asynchronous Motor
As the inrush current of large motors, including propulsion motors, is very high, the inrush current would lead to a severe dip in the vehicle electrical system. This can be avoided by connecting the stator windings in the star connection during starting process and in delta connection after starting (star-delta starting; Fig. 4.37). Among other methods for starting electric motors,[55] this is the most common method, especially for large engines [7].

This method reduces the starting current to one third of that of direct starting. The disadvantage is that the starting torque is also reduced to one third [25].

Today, the **speed control** and **change of direction of rotation** of electric propulsion motors are mainly carried out by *frequency converter* (*pulse inverter*).[56] *Soft starting* is also possible via the frequency converter.

Speed control is based on the fact that the rotating field speed n_d depends on the frequency *f* of the mains and the number of pole pairs *p*:

$$n_d = 60 \cdot \frac{f}{p} \tag{4.88}$$

with

n_d rotating field speed,
f frequency,
p number of pole pairs

If the frequency *f* is changed, then the rotating field speed n_d also changes and thus (with constant slip) the rotor speed *n*.

[55] See for example [29].

[56] The following from [42]; for the pulse inverter, also see [40, 44].

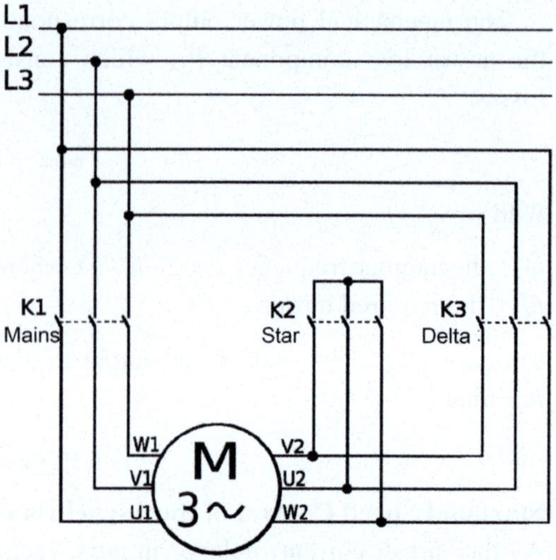

Fig. 4.37 Star-delta connection

A frequency converter is an electronic circuit in which a voltage U_1 at a frequency f_1 is fed to the input side and converted into a voltage U_2 with the same amplitude but a different frequency f_2. The frequency f_2 can be either smaller or larger than f_1.

The speed can be controlled over a wide range by frequency adjustment.[57] As only small losses occur, this method can also be used for large power ratings. Another advantages are:

- Relatively low losses,
- Large control range,
- Speeds higher than the rotating field speed of the mains are possible,
- Suitable for squirrel cage motors.

The only disadvantage is the high circuitry complexity for the implementation of the frequency converter.

The direction of rotation can also be reversed by means of a reversing contactor circuit (Fig. 4.38).

The abovementioned methods eliminate the need for complex mechanical reversing units such as reversing reduction gears or controllable pitch propeller systems.

4.3.4.2 Synchronous Motor

The **synchronous motor** bears its name because of its operating characteristics: the rotor rotates exactly synchronously with the stator rotating field specified by the mains frequency. This distinguishes synchronous machines from asynchronous machines, whose

[57] For details on the function and design of frequency inverters, see [46].

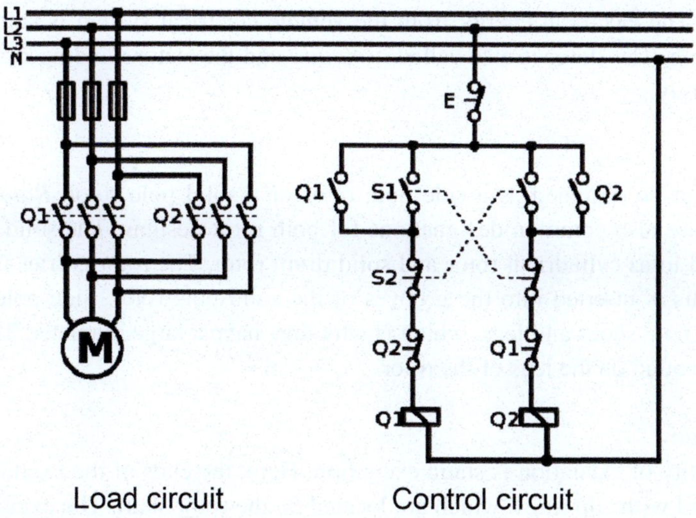

Fig. 4.38 Reversing contactor circuit. *Photo* wdwd, CC BY-SA 3.0

rotors are ahead of the rotating field in motor operation and behind it in generator operation. Another distinguishing feature is that, in contrast to asynchronous machines, an exciter field is required for the operation of synchronous machines [8, p. 291 ff.].

Before a synchronous machine is connected to the mains, it must be synchronized with the mains. The speed change and reversal is carried out by means of the so-called power electronics with frequency converters [8, p. 324 ff.].

A rotary encoder constantly records the change in rotor position during operation. From this, the control electronics determine the actual speed. When loaded, the rotor of the synchronous motor runs at an angle to the rotating field, the so-called pole wheel angle.[58]

Design and Function of the Internal Pole Synchronous Motor

Synchronous motors are manufactured as external or internal pole machines, whereby internal pole machines can be used in shipbuilding for high output. Both machine types have in common that, like all three-phase machines, they have a *rotors* and a *stator*. In any case, an excitation device is required for the operation of the machines. Because of the special importance of the internal pole machine for shipbuilding, this type of motor will be discussed in more detail below.

(a) Stators

The stator winding consists of three windings phases offset by $120°/p$ (p = number of pole pairs), which are designated U, V, and W. They are connected in star or delta

[58] For the *pole wheel*, see below.

connection. The electrical energy from the vehicle electrical system is supplied via the stator winding. The stator is also called *armature* and the stator winding *armature winding* accordingly.

(b) Rotor

The rotor can be designed as a side pole rotor or a solid pole rotor. *Rotor* or perhaps *pole wheel* are also common designations for both rotor designs. The solid pole rotor is also referred to as cylindrical rotor and solid drum rotor. The rotor carries the excitation winding. This is inserted into the grooves of the solid pole rotor. Side pole rotors have pronounced pole shoes and legs, which is why they have a large diameter. The excitation winding is wound on the legs of the rotor.

(c) Excitation

One possibility of excitation is static excitation. Here, the ends of the excitation winding are connected with slip rings, which are located on the rotor shaft. The excitation voltage is applied to the excitation winding via carbon brushes (Figs. 4.39 and 4.40).

If it is a permanent magnet-excited synchronous machine (PSM), the rotor carries permanent magnets for excitation. Permanent magnet excitation is becoming increasingly important [8, p. 287 ff., 11, p. 331].

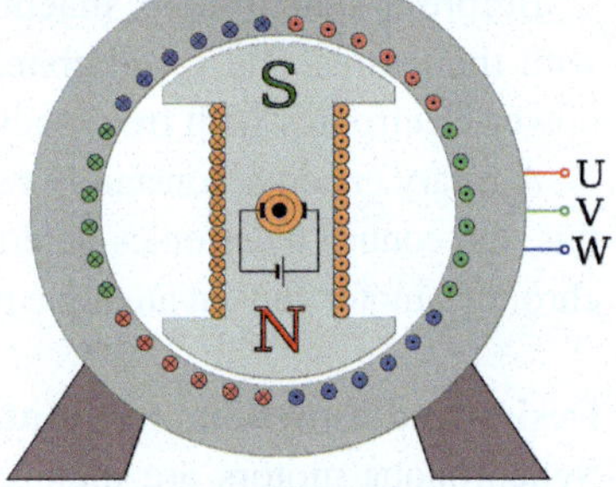

Fig. 4.39 Side pole motor. *Graphic* Biezl

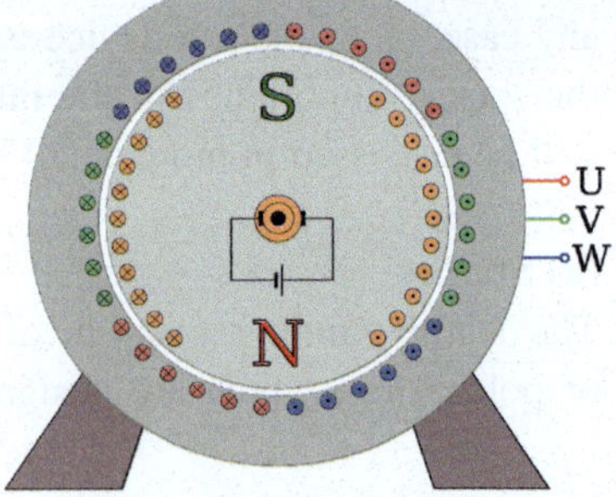

Fig. 4.40 Solid pole motor. *Graphic* Biezl

Further possibilities of excitation can be found in the relevant literature [4].[59]

Characteristics of the Synchronous Motor

As explained above, an excitation winding or a permanent magnet is required for operation to generate an exciter field. In addition, electrical energy must be supplied via the stator windings so that the motor can deliver torque to the shaft.

The absorbed electrical power is calculated as follows:

$$P_{\mathrm{el}} = \sqrt{3} \cdot U_S \cdot I_S \cdot \cos\phi \tag{4.89}$$

with

$\sqrt{3}$	the so-called chaining factor,
U_S	the stator voltage,
I_S	the stator current, and
$\cos\varphi$	the power factor

The mechanical power output P_{mech} corresponds to the consumed electrical power minus the power loss portion P_V, which consists of copper and iron losses and friction losses and is calculated according to Eq. 4.90:

$$P_{\mathrm{mech}} = \omega \cdot M \tag{4.90}$$

with ω the angular frequency ($=2 \cdot \pi \cdot n$), where n is the speed per second,
and M the required torque

The ratio of mechanical power output to electrical power input expresses the efficiency of the machine:

$$\eta = P_{\mathrm{mech}}/P_{\mathrm{el}}. \tag{4.91}$$

Speed Control and Change of Direction of Rotation

For synchronous motors, speed control and reversal of direction of rotation are also generally carried out by frequency converters (see Sect. 4.3.4.1).

4.3.4.3 Direct Current Motor

The use of DC motors is of comparatively little importance in large ships and is limited to special applications, for example, for particularly low-noise or battery-powered propulsion (e.g., in submarine boat construction).

[59] Also for example [67].

Fig. 4.41 Direct current electric motor

Structure

The stator has pronounced north and south poles, which can be generated by a magnet (permanently excited motor) or by excitation windings. The winding of the rotor is constructed in such a way that at the front sides of the rotor, conductors located in the area of a magnetic north pole of the stator are always connected to conductors located at the corresponding area of a south pole. At the front sides of the rotor, which consists of a stack of laminations due to the eddy currents, the winding is connected to the laminations of a commutator (commutator); this is a cylindrical body consisting of sector-shaped copper laminations insulated from one another.

Figure 4.41 shows a small low-voltage electric motor with permanent excitation by horseshoe magnets.

Reversal of Direction of Rotation, Speed Change

The reversal takes place in **direct current motors** in such a way that either the direction of the current in the armature winding or the direction of the current in the excitation winding is reversed.

Normally, the direction of the current is reversed in the armature winding, as in this case, it is not necessary to remagnetize the entire stator. In terms of circuitry, this is done by the pole reversal circuit (Fig. 4.42).

This reverses the polarity of the voltage applied to the motor via a relay so that it runs forward or backward.

The **speed change** in the case of the DC motor is effected via the applied terminal voltage.

Fig. 4.42 Simple-pole reversal circuit

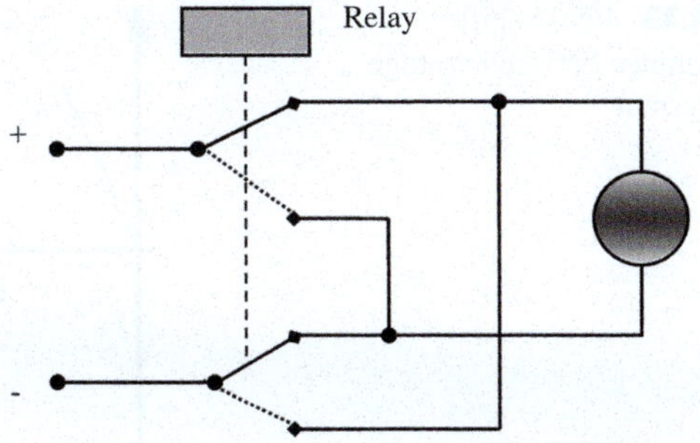

Example Calculation of Electric Propulsion Motor

The propulsion power of a three-phase propulsion motor is 12.5 MW. A terminal voltage of 3610 V is applied. The power factor cosφ of this machine is 0.8, and the efficiency η is 0.85. What is the current I consumed?

The power of a three-phase motor (terminal power or active power) is determined according to Eq. 4.89:

$$P_{\text{el}} = \sqrt{3} \cdot U \cdot I \cdot \cos \phi.$$

"cosφ" is called the efficiency factor or power factor. It is indicated on the type plates of AC and three-phase motors. The power factor is the ratio between active power P and apparent power S and is calculated according to the following equation:

$$\cos \varphi = P/S. \tag{4.92}$$

In this equation, P is the active power and S is the apparent power ($S = U \cdot I$). The power factor indicates which part of the apparent power is converted into active power.

The angle φ describes the so-called phase shift between two sinusoidal oscillations: they are out of phase with each other if their period durations match, but the times of their zero crossings do not. In electrical engineering, the term **phase shift** in an alternating current circuit is used in connection with currents and voltages:

- With an inductance (ideal coil), the current follows the voltage by 90° (the voltage precedes the current by 90°).
- With a capacitance (ideal capacitor), the voltage follows the current by 90° (see Fig. 4.43).
- With ohmic resistance, voltage and current are always in phase.
- With a combination of R, L, and C, the phase shift angle can be any value between -90° and $+90^{\circ}$; it is dependent on the frequency.

In the case of a three-phase motor with a three-phase alternating voltage, this consists of three alternating voltages that oscillate at 120° to each other (Fig. 4.44).

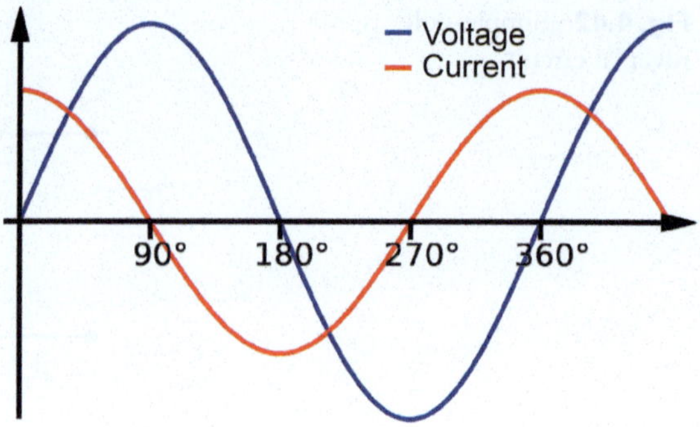

Fig. 4.43 Phase shift capacitance between voltage and current

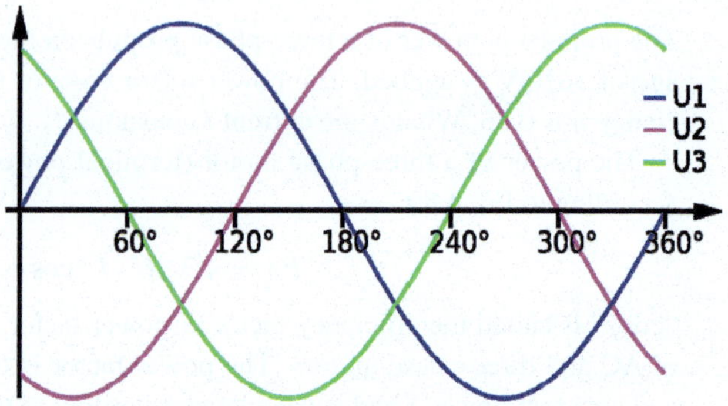

Fig. 4.44 Three-phase alternating voltage

The efficiency is (Eq. 4.91) $\eta = P_{\text{mech}}/P_{\text{el}}$.
Thus, P_{el} is calculated by changing: $P_{\text{el}} = 12.5\ \text{MW}/0.85 = 14.7\ \text{MW}$.
Changing Eq. 4.89 to "*I*" delivers:

$$I = \frac{P_{\text{el}}}{\sqrt{3} \cdot U \cdot \cos\phi} = \frac{14{,}700{,}000}{\sqrt{3} \cdot 3{,}610 \cdot 0.8} = 2.9\ \text{kA}.$$

◀

Example Torque and Rotating Field Speed Synchronous Motor

The AIDAmar has two synchronous motors of the type DTMSZ 3352-16YS. The data sheet states:

Power output:	12,500 kW,
Frequency:	17.46 Hz,
Number of pole pairs:	8

What is the rotating field speed n_d of the engine? The solution is Eq. 4.88:

$$n_d = 60 \cdot \frac{f}{p} = 60 \cdot \frac{17.46}{8} = 131\ \text{min}^{-1}.$$

How high is the tappable mechanical torque? Solution is Eq. 4.90, from:

$$P_{\text{mech}} = \omega \cdot M$$

follows with

$$\omega = 2 \cdot \pi \cdot n.$$

By inserting and rearranging the above equations, the following results are obtained

$$M = \frac{P_{\text{mec}}}{2 \cdot \pi \cdot n} = \frac{12{,}500 \text{ kW}}{2 \cdot \pi \cdot \frac{131 \text{ min}^{-1}}{60 \frac{\text{s}}{\text{min}}}} = 911.2 \text{ kNm}.$$

◄

4.3.5 Fuel Cell Propulsion

For the sake of completeness, the fuel cell propulsion will be discussed here. Although fuel cell propulsion has not yet been able to establish itself in large ships, it is revolutionary for submarine propulsion systems. For example, the German Navy's class 212A submarines and its successor class 214 are equipped with this technology. In addition, a passenger ship on the Alster in Hamburg uses this type of propulsion.

These submarines of class 212A and 214 rely on *hybrid propulsion*.

As before, these submarines, just like all other submarines, are powered by electric motors. The electrical power required for this is generated by diesel generators in conventional submarines or by nuclear fission to generate steam for a turbine in nuclear propulsion.

In the submarines mentioned above, this is done with the help of the *hybrid propulsion*. Hybrid propulsion means that two or more different types of propulsion are used. In these submarines, therefore, the fuel cell is used as a power source in addition to the conventional diesel generator set.

4.3.5.1 Structure and Function of the Fuel Cell

In the fuel cell, the reverse process of electrolysis takes place. The fuel cell reverses this process and generates electrical energy from the reaction of oxygen and hydrogen to water [60].

The actual cell consists of two electrodes in an electrolyte (e.g., potash lye). In the two electrodes (anode and cathode), the so-called reactants hydrogen (on the anode side) and oxygen (on the cathode side) are continuously supplied from storage tanks in the boat as long as the process is running.

Depending on the fuel cell type, methane, methanol, or glucose solution can also be used as fuel. On the cathode side, hydrogen peroxide or potassium thiocyanate can also be used as oxidizing agents.

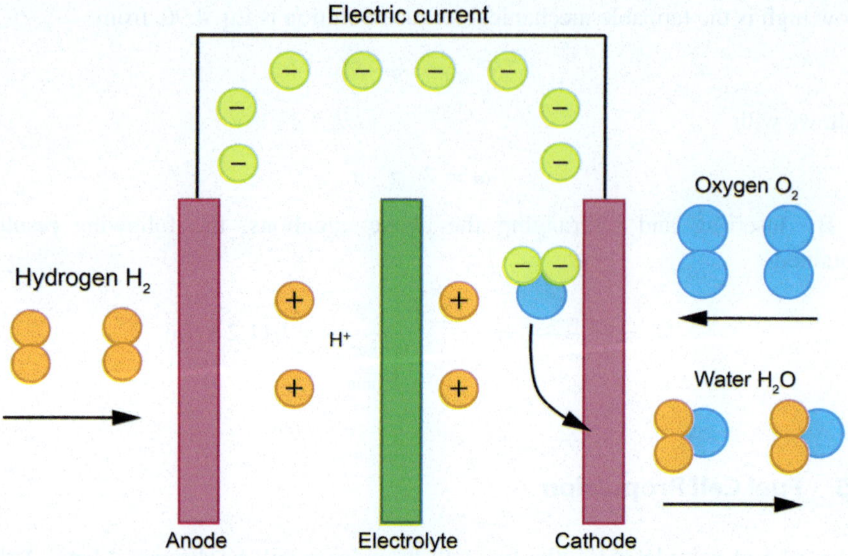

Fig. 4.45 Fuel cell principle. *Graphic* Patrick Schnabel, www.elektronik-kompendium.de

The two electrodes of a cell consist of a substance with catalyst properties, for example, platinum. This means that this substance promotes the reactions to be carried out.

In the electrolyte, the positively charged hydrogen ions migrate to the cathode where they combine with the oxygen to form water. A positive potential is created at the cathode. The electron equalization takes place by the electric circuit via a connected consumer under working power. Figure 4.45 illustrates the process of power generation in the fuel cell.

Sequence of the chemical reaction in the fuel cell with basic electrolyte (e.g., potash lye) in the submarines mentioned above:

	Reaction equation	Types of reaction
Anode (negative terminal)	$2H_2 + 4OH^- \rightarrow 4H_2O + 4e^-$	Oxidation/electron emission
Cathode (positive terminal)	$O_2 + 2H_2O + 4e^- \rightarrow 4OH^-$	Reduction/electron absorption
Overall reaction	$2H_2 + O_2 \rightarrow 2H_2O$	Redox reaction/cell reaction

The fuel, here H_2, is catalytically oxidized at the anode with the release of electrons and converted into ions (H^+). The electrons are discharged from the fuel cell and flow via an electrical consumer (e.g., a battery charger or propulsion motor) to the cathode. At the cathode, the oxidant, in this case O_2, is reduced to form anions (O^{2-}) by accepting the electrons and simultaneously reacts with the protons that have migrated through the electrolyte to the cathode to form water.

In this respect, the chemical energy stored in hydrogen is converted into electrical energy and thermal energy in the fuel cell. The fuel energy is released during combustion of the fuel as heat of reaction; an amount of energy of about 286 kJ is released per mole of hydrogen. This value is called *enthalpy of reaction* ΔH or the *calorific value* at constant pressure and $T = 298.15$ K.

$$2\ H_2 + O_2 \rightarrow 2\ H_2O \qquad \Delta H_0 = 285.8\ \text{kJ/mol}$$

The efficiency of these fuel cells is very high compared to the diesel generator. It amounts to approximately 65%. Good diesel generators have at best an efficiency of about 30%.

The supplied voltage is at the H_2–O_2 cell theoretically at 1.23 V and a power output of 0.1 W/cm^2 electrode area [20, p. 148] at a temperature of 25 °C. In practice, however, only voltages of 0.5–1 V are achieved ("electronic friction losses"): The voltage depends on the fuel, the purity of the reactants, the quality of the cell, and the temperature. To obtain a higher voltage, several cells must therefore be connected in series to form a *stack*.

The theoretically maximum achievable cell voltage $U_{\Delta H}$ is calculated as the quotient of the calorific value of the fuel, the Faraday constant (product of Avogadro number ($N_A = 6.023 \cdot 10^{23}\,1/\text{mol}$) and elementary charge ($e = 1.6 \cdot 10^{-19}\,\text{C}$)) and the exchanged electrons:

$$U_{\Delta H} = -\frac{\Delta H}{n \cdot F}. \tag{4.93}$$

For the fuel cell on class U212A:

Hydrogen: $n = 2$ free electrons,
ΔH_o $= -285.8$ kJ/mol; $\Delta H_u = -241.8$ kJ/mol,
F $= 96{,}485$C/mol

It follows for $U_{\Delta H_o} = 1.48$ V, and for $U_{\Delta H_u} = 1.25$ V.

Differentiation Between Upper and Lower Heating Value (H_o or rather H_u)
The *upper heating value* H_o is a measure of the specific thermal energy contained in a substance per unit of measurement. It indicates the chemically bound energy (reaction enthalpy) that is released during combustion and subsequent cooling of the combustion gases at 25 °C and their condensation.

H_o takes into account both the energy required to heat the combustion air and the exhaust gases and the enthalpy of condensation of the liquids that condense during cooling, especially the water produced during the combustion of hydrogen-containing fuels. In contrast, the *lower heating value* H_u denotes the energy released during combustion and subsequent cooling to the initial temperature of the combustible mixture, the combustion water still being in gaseous form. The calorific value of water-rich fuels is therefore significantly lower than their calorific value by the amount of the condensation enthalpy of the water vapor present.

The current delivered by the fuel cell is calculated from the so-called current density. This is the quotient of the current I and the active electrode area A:

$$j = I/A \quad (\text{A/cm}^2), \tag{4.94}$$

in practice about 0.8 A/cm^2.

4.3.5.2 Advantages of Fuel Cell Propulsion for Submarines

This type of energy generation for propulsion offers great advantages in submarine technology. In comparison to the diesel generator, the noise level and the heat emission is considerably lower, which is otherwise produced by diesel generators. This makes it much more difficult to locate such a submarine. This is also helped by the fact that this type of energy generation produces pure water as the only waste product and no detectable residues (so-called signature reduction).

Furthermore, the hybrid propulsion offers the advantage that the diving times can be considerably longer than with conventional submarine propulsion. Diesel engines need atmospheric oxygen to run, so these submarines often have to come closer to the surface at so-called snorkeling times to absorb it. The class 212A submarines carry pure oxygen and hydrogen for the fuel cells in tanks, making them independent of the outside air supply.

As pure water is emitted during operation of the fuel cell, this propulsion system is particularly environmentally friendly. This applies in particular to nuclear propulsion, which generates radioactive waste that requires a high degree of safety engineering effort in disposal and operation. Added to this are the high space requirements and the considerable noise emission during the cooling of the reactors. Due to the heat and noise radiation and their size due to the space they require, nuclear submarines are also much easier to locate.

To make full use of the power of the new class 212A submarines, they are additionally equipped with a diesel generator for a possible spurt during surface or snorkeling trips, which supply the propulsion battery and recharges it when necessary [60].

4.3.5.3 Future of Fuel Cell Use in Shipping

Long-term shipbuilding research is already intensively concerned with the topic of fuel cells. Companies in the shipbuilding industry and universities are working on the fuel and propulsion technology of the day after tomorrow in the "National Innovation Programmes for Hydrogen and Fuel Cell Technology" (NIP). The joint project "e4ships—Fuel Cells in Maritime Use" prepares the emission-free operation of ships in the future, and also develops practical, modular solutions for onboard power supply and port operation in the present. Building on the success of the demo projects *Pa-X-ell* (passenger ships) and *SchiBZ* (cargo ships), the new project *Rivercell* will now be used in hybrid propulsion systems for river cruise ships, consisting of gas combustion engine, fuel cell, solar cells, and energy storage [22, p. 72 f.].

With the exception of class 212A and 214 submarines, the fuel cell has not yet been able to establish itself—especially in shipping—apart from individual projects: Since 2007, a passenger ship for 100 passengers, powered by electricity (approximately 100 kW) from fuel cells, has been operating on the Alster in Hamburg. The cost of the fuel cells was € 3 million, and the complete ship cost € 5 million.

Hydrogen-powered ocean-going vessels are being tested in various cases. For example, the Norwegian *Viking Lady,* a supply ship, was equipped with a fuel cell in 2009 in addition to the diesel-electric propulsion system [61].

Example of Fuel Cell Application

A single fuel cell delivers a nominal voltage of 0.8 V and a current of 1.0 A. To supply a charger for a battery block, a power of 100 W is required at an operating voltage of about 25 V.

How many cells do you need and how do you have to switch them?

Solution: Number of cells n: For series connection of voltage sources (Fig. 4.46), the following applies

$$U_{\text{tot}} = U_1 + U_2 + \cdots + U_n. \tag{4.95}$$

Thus

$$n = \frac{U_{\text{tot}}}{U_i} = \frac{25\ \text{V}}{0.8\ \text{V}} = 32\ \text{cells}. \tag{4.96}$$

To achieve the charging voltage of 25 V, 32 cells must be connected in series. But these only supply 1 A current. To achieve the required power of 100 W for the charger, the following follows from the relationship between power, current, and voltage

$$P = \frac{U}{I} \tag{4.97}$$

and by switching to I:

$$I = \frac{P}{U} = \frac{100\ \text{VA}}{25\ \text{V}} = 4\ \text{A}.$$

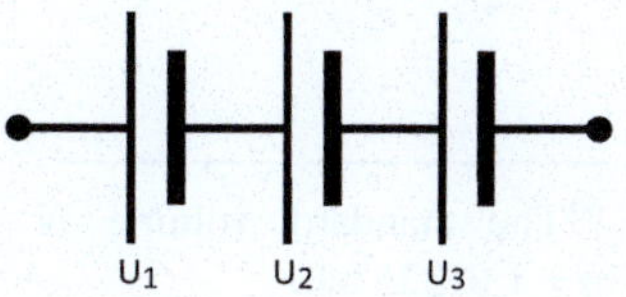

Fig. 4.46 Series connection of direct voltage sources

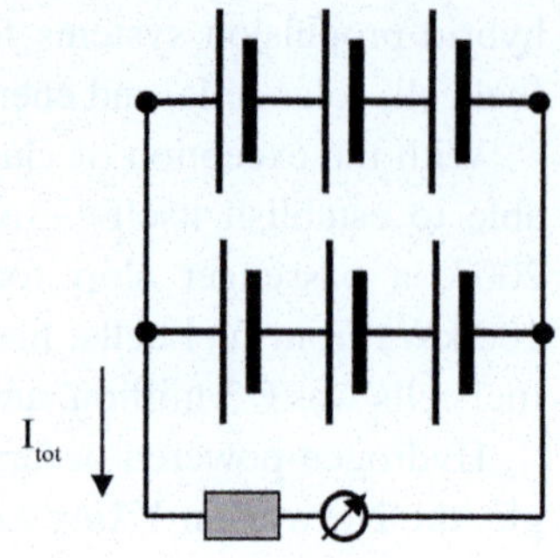

Fig. 4.47 Parallel and series connection of direct voltage sources

As one cell only supplies a current of 1 A, **4 packs** can be connected in parallel to 32 cells connected in series (if voltage sources are connected in parallel, the current intensity is added!—Fig. 4.47).

For the operation of a fuel cell system, 5000 L H_2 at 250 bar is carried in a hydrogen tank. How much energy is stored in the tank?

Solution: Hydrogen has a lower heating value of 10.8 MJ/Nm³.[60] One liter of hydrogen for 1 bar has a calorific value of 10.8 kJ at 0 °C.

One liter H_2 of 250 bar thus has 250 times the amount of energy content:

$$10.8 \text{ kJ} \cdot 250 = 2.70 \text{ MJ}.$$

As the tank contains 5000 L of the compressed hydrogen, the following applies to the total amount of energy stored in the tank and thus usable energy:

$$\Delta E = 5000 \cdot 2.70 \text{ MJ} = 13.5 \text{ GJ}.$$

◄

4.3.6 Sail Propulsion

For the sake of completeness, wind as a propulsion system will also be discussed here. Apart from sailing with yachts and dinghies, tall ships are still in service today, often as traditional sailing vessels, and also especially with the navies of the world for training and representation purposes (e.g., the Gorch Fock in the German Navy).

The fact is that tailwind pushes is a well-known fact that everyone can feel when cycling. According to this principle, a sailing ship or even a sailboat is propelled when the wind comes directly from astern. This effect is also the predominant on space sheet courses, that is, when the wind comes in at an angle from astern (see also Fig. 4.48).

[60] The standard volume is related to the standard physical state: $273.15\ \text{K} = 0^\circ\text{C}$ and $p = 1.01325$ bar.

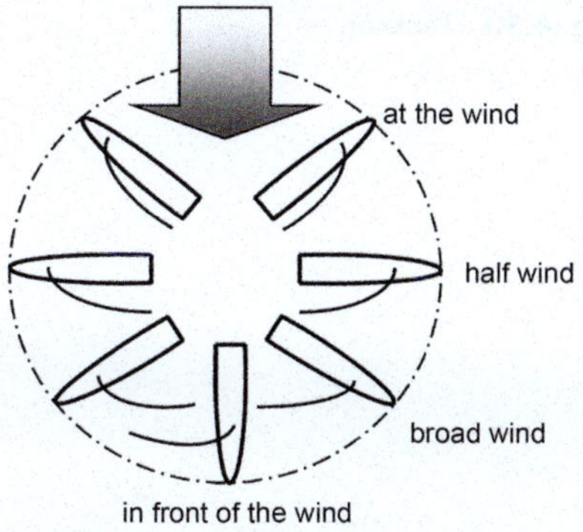

Fig. 4.48 Courses to the wind

Fig. 4.49 *Pinta*—ship of Christopher Columbus

The sails offer resistance to the wind. The air flow is slowed down and interrupted. The larger the surface of resistance—that is, the sail surface—the more air mass is decelerated, the greater the thrust for propulsion. The optimum shape for building up a corresponding dynamic pressure is a hollow hemisphere. Therefore, special downwind sails, such as *square sail* for tall ships (Fig. 4.49) or *spinnaker* on dinghies sewn with a very "bulbous" profile.

However, tall ships as well as yachts and dinghies do not sail exclusively on space sheet courses (in front of the wind and broad-reaching wind), and also on courses where the wind is more or less at an angle of 90° to the longitudinal axis of the ship (half wind) or even falls obliquely from the front (at the wind). Square sails are not suitable for all courses from half wind and forward. This requires sails based on the principle of "propulsion by buoyancy." These sails are cut in profile like the wing of an airplane (see Fig. 4.50) and are called *tacksails*.

Fig. 4.50 Tacksail

Such sails are regarded as a profile with airflow around them on the said courses. The sail generates a buoyancy force according to Bernoulli's principle. This buoyancy force is created by a negative pressure on the upper side or, in the case of a sail, on the lee side of the profile and an overpressure on the underside or windward side. The pressure difference is based on the higher air speed on the lee side than on the windward side due to the curvature. The following applies to incompressible media, according to Bernoulli:

$$\frac{v_{\mathrm{Lee}}^2}{2} + \frac{p_{\mathrm{Lee}}}{\rho} = \text{constant} = \frac{v_{\mathrm{windward}}^2}{2} + \frac{p_{\mathrm{windward}}}{\rho} \quad (4.98)$$

with

v = Air flow speed at the profile,[61]
p = Pressure on the profile sides,
ρ = Density the air

If $v_{\mathrm{Lee}} > v_{\mathrm{windward}}$, p_{windward} must therefore be $> p_{\mathrm{Lee}}$, which generates the buoyancy F_A. In this respect, the pressure difference $\Delta p = p_{\mathrm{windward}} - p_{\mathrm{Lee}}$—which arises due to the different flow speeds on the two sides of the sail—causes this buoyancy force.

By vector decomposition, the buoyancy force F_A results in a force that displaces the sail craft laterally—drift (F_{Drift})—and the propulsive force F_V (see Fig. 4.51).

The drift counteracts a force on the underwater hull, which reduces the lateral displacement of the ship to about 4°.

The propulsive force F_V is counteracted the frictional force F_R on hull and rigging (see Sect. 4.2). The following conditions are to be considered on the sailing ship:

[61] Wind speeds according to the Beaufort Scale see Annex 13.

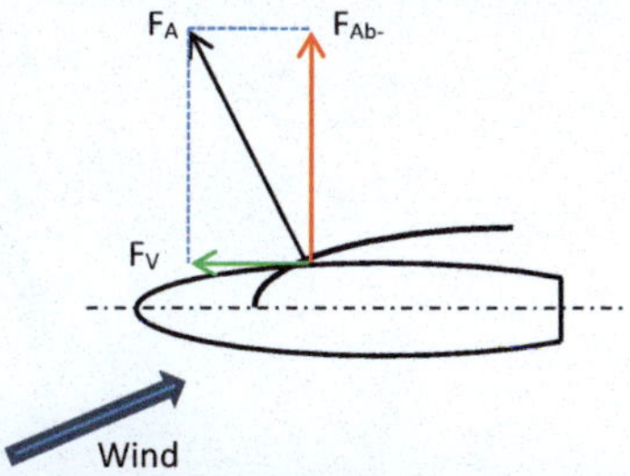

Fig. 4.51 Forces on the sail

a) Sailing ship is still just at rest, has the speed $v = 0$; the sailing force (propulsive force) is maximum:

b) Sailing ship is accelerated, light frictional resistance F_R:

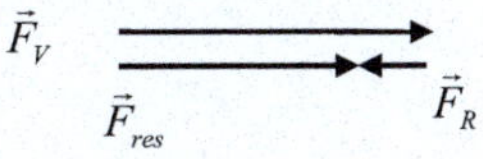

c) Acceleration decreases, ship approaches its maximum speed, frictional force increases further:

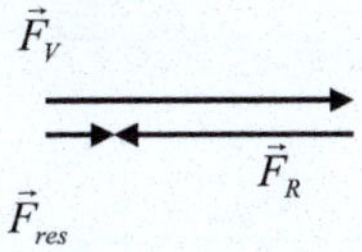

d) Propulsive and frictional forces are equal, sailor has reached its maximum speed (stationary state):

$F_{res} = F_V - F_R = 0$

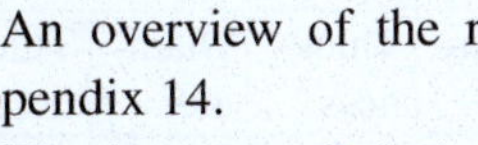

Wind power is generated via the **rig** to the hull of the ship. The rigging of a sailor is considered to be the entire *rigging* with masts, yards, sails, running rigging (sheets and halyards, etc.) and standing rigging (shrouds, guy ropes, stays), etc.

An overview of the rigging types for tall ships, yachts, and dinghies is given in Appendix 14.

The layout and design of a rig with material selection and load assumptions for tall ships can be carried out, unless the customer has other requirements, for example, according to the construction regulations of the former GL in "Classification and Construction Regulations—I Ship Technology, 4 Rig Technology, 1 tall ship rig." The following example task is intended to demonstrate the application of this construction regulation to the rigging details of the *yardarm* and *hanger chain*.

Fig. 4.52 Suspension yards on mast

Example Task "Rig"

On a three-mast barque, the main beam and its hangar chain must be replaced. The main beam has a length of 24 m and is fixed (i.e., cannot be raised and lowered) to the mast. What are the dimensions of the yardarm and which chain should be chosen for the hanger chain?

Solution: The dimensions of the yardarms can be found in Table 1.4 of the abovementioned construction regulations—extract:

Length	in the middle		At the first quarter		At the second quarter		At the third quarter		To the cams	
	Diameter	Thickness	Diameter	Thickness	Diameter	Thickness	Diameter	Thickness	Diameter	Thickness
m	mm	mm	mm	mm	Mm	mm	mm	mm	mm	mm
24	480	8.5	470	8.0	430	7.5	360	6.5	240	5.0

The attachment of a fixed yardarm is shown in Figs. 4.52 and 1.12 of the abovementioned construction regulation. ◀

The hanger chain as well as the hanger rod are used to absorb the vertical force (dead weight and sail wind pressure component), which acts on the yard. It transmits this force into the mast or stalk. The load to be assumed for the hanger chain can be taken from Table 1.5 of the construction regulations—extract:

Length of the yardarm (m)	Load on the hanger chain (kN)
…	…
24	65.8
…	…

Therefore, a chain with a minimum breaking load of 65.8 kN must be selected for the hanger chain, whereby a steel chain with a tensile strength of 400–490 N/mm^2 is to be taken as a basis (see no. 1.1 of the abovementioned construction regulation). For example, a round steel chain according to German Institute for Standardization (GIS, Deutsches Institut für Normung (DIN)) 766 with a nominal thickness of > 13 would be used.[62]

Example Task "propulsive force"

The Gorch Fock sails on a downwind course at maximum speed. What is the sailing force or propulsive force in horizontal direction that the ship experiences?

Data:

Length waterline (LWL):	70.4 m,
Width:	12 m,
Draft:	5.35 m,
Maximum speed:	18 kn = 33 km/h (under sails),
Number of sails:	23,
Total sail area:	2037 m^2

Solution: The Gorch Fock is, according to the conceptual formation, in the stationary state described above, in which sail force or propulsive force and frictional force have the same amount and are directly opposed:

$$F_V = F_R.$$

To determine the resistance, very simplified assumptions are made here; for a detailed calculation of the total resistance to the ship system, please refer to Sect. 4.2. It should be noted at this point, however, that in terms of resistance, the surface vessel is generally negligible, as the density of the air is roughly three powers of ten lower

[62] See for example company Ketten-Fuchs [78].

than that of seawater. However, due to the unfavorable flow dynamic properties of the surface vessel, the rigging of a sailing ship has a higher air resistance than passenger ships, which is taken into account in the approximate assumption of the c_w value. Furthermore, the Gorch Fock does not have a bulbous bow, which would have a positive influence on the flow characteristics of the underwater hull.

The frictional force is calculated according to Eq. 4.1 (see Sect. 4.2):

$$F_R = \frac{1}{2} \cdot \rho \cdot v^2 \cdot A \cdot c_w.$$

with

ρ = 1.025 kg/L(approximate density of seawater at 25 °C),
v = v_{max} at sailing = 22 km/h,
A = 12 m · 5.35 m = 64.2 m²(projected cross-sectional area or rather main clamping surface),
c_w = 0.4 (roughly assumed)

follows in this respect

$$F_R = 0.5 \cdot 1025 \frac{\text{kg}}{\text{m}^3} \cdot 64.2\text{m}^2 \cdot 0.4 \cdot \left(\frac{22,000 \text{ m}}{\text{h}} : \left(3600 \frac{\text{s}}{\text{h}} \right) \right)^2,$$

$$F_R = 491,507 \text{kg/m s}^2 = 491.507 \text{ kN}.$$

As the propulsive force in the stationary state is equal to the resisting force, that is, $F_V = F_R$, this results in a propulsive force of $F_V = 491.507$ kN, which is transmitted to the ship via the rig. To determine the horizontal propulsive force for the **respective** sails, they must be taken into account with their respective percentages of the total sail area.

Assumption: The sail area of the mainsail of the Gorch Fock is 6% of the total sail area. In this respect, 6% of the total propulsive force acts on the mainsail, that is, 30 kN.

However, this is again only a simplifying assumption, as shading of the front sails by the aft ones is not taken into account. ◄

4.4 Power Transmission

The energy generated by the internal combustion engine or turbine or electric motor is fed into the system either *directly* via a *shaft* to the *ship propeller* as the actual driving element or via a *gearbox*. The shaft is guided by the *stern tube* of the hull of the ship to the outside. The *stern tube seal* (see also Fig. 4.66 and Sect. 4.4.4) prevents water from entering the hull at this shaft passage.

The **waterjet propulsion** (Fig. 4.53) will not be considered here in detail. It is used for fast vehicles (sport boats, jet skis, official and pilot vehicles). At this point only that much: It is a propulsion where the water is led through a suction opening in the bottom of the hull to a pump with axial flow. That pump accelerates water to speed several times faster than inlet speed. The water is ejected through a nozzle, thus achieving the feed rate. The propulsion principle is similar to the jet nozzle of a rocket engine. Turning the nozzle controls the vehicle and also changes the direction of travel from forward to aft (see also Fig. 4.54).

The **Voith-Schneider propeller** consists of a round disc, which is installed horizontally in the bottom of the ship and on which four to six vertical spade blades are mounted so that they can rotate (Fig. 4.55). By means of an eccentric control, the angles of attack of these wing-shaped spade blades are permanently adjusted when the disc rotates so that the buoyancy forces acting on these blades move the vehicle in the desired direction. This allows maneuvering in all directions with the same direction of rotation of the circular disc (application: towboats, ferries, working ships, and similar).

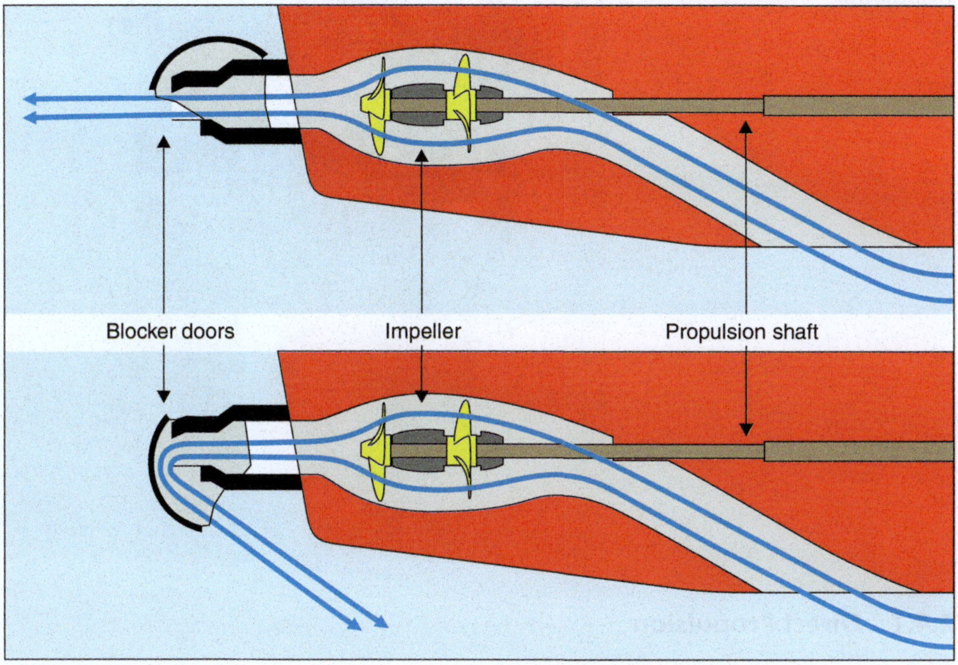

Fig. 4.53 Water-jet propulsion principle. From: [16]/graphic: Brückler

Fig. 4.54 Water-jet propulsion with baffles (deflector caps). *Photo* Doclecter; CC BY-SA 3.0

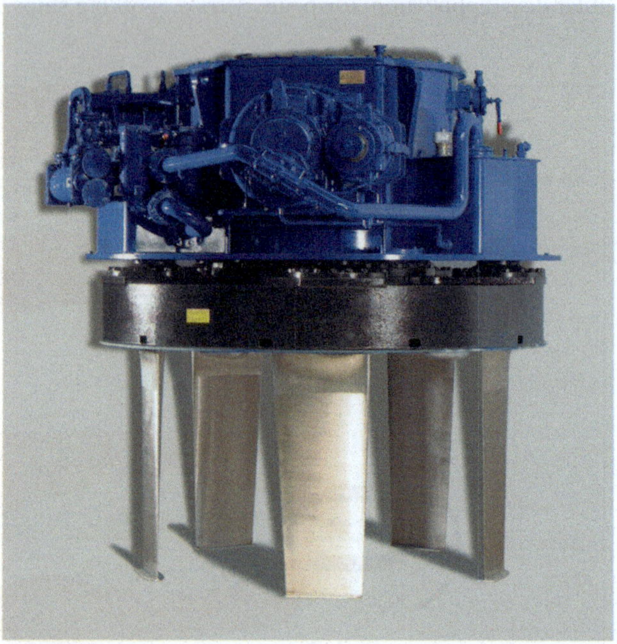

Fig. 4.55 Voith-Schneider propeller. *Photo* Voith AG, Heidenheim

4.4.1 Direct Propulsion

With direct propulsion, the propulsion power of the propulsion engine is transmitted directly to the propeller via a rigid shaft, which is guided outboard through the so-called stern tube. This type of propulsion is usually found in so-called "low-speed engines"—diesel engines with speeds between 60 and 250 min^{-1}—application.

The direction of rotation of the propeller, for example, for reverse, can only be changed here by reversing the engine. The engine must then be stopped from the forward motion, reversed by moving the camshaft and restarted for reverse motion.

Another possibility for changing speed and direction of motion is the use of a *controllable pitch propellers*: To change the ship's speed as well as for forward and reverse motion, the pitch of the individual propeller blades is changed. The engine rotates at constant speed.

The shaft is usually flanged directly to the engine crankshaft by means of a flexible clutch.

So-called medium-speed four-stroke diesel engines with a speed range up to 1200 min^{-1} are primarily used on small to medium-sized cargo ships, passenger ships, and naval vessels.

High-speed engines with speeds > 2000 min^{-1} can be found in the area of inland navigation and in sport and leisure navigation. These speed ranges require a gear reduction, often also in combination with controllable pitch propellers.

4.4.2 Propeller

4.4.2.1 General Principles

The rotational energy generated by the propulsion engine is transmitted via the shaft to the ship's propeller, which converts most of this energy into thrust [20, p. 851 f.]. Today, a ship's propeller has between three and—with propellers that are particularly low in cavitation and therefore quiet—up to seven blades. The shape of the blade is designed to be streamlined.

In the following, fundamental aspects of propeller theory are presented, which are essentially valid for all common types of propellers. As the fixed pitch propeller (see Fig. 4.56) is still the most common propeller, it will be used for the following explanations.

Fig. 4.56 Clockwise rotating propeller

With regard to the peculiarities of other types of propellers (e.g., Voith-Schneider, controllable pitch propellers, jet propulsion), which deviate from the generally valid approaches of propeller theories, reference is made to further literature.

In the *propeller theory*,[63] it is proved that a ship's propeller at the front side sucks in the water and pushes it away at the rear side. According to the laws of hydrodynamics (*Bernoulli's equation*), a negative pressure is created at the front side and an overpressure at the back side. The negative pressure "sucks" the propeller and ship forward, and the positive pressure pushes it in the same direction. The suction side performs about 60% of the propulsion work, the pressure side about 40%.

The thrust T depends on the speed. In ships with direct propulsion, the change in propeller speed is achieved by changing the engine speed. Also the change from forward to aft can only be achieved by reversing the direction of rotation of the engine, provided that the propeller shaft is directly flanged to the engine crankshaft and no reversing gear is used. Because of the large masses moving in the engine, all this is only possible with a time delay after control via the machine telegraph.

This disadvantage can be avoided with the already mentioned *controllable pitch propeller*.[64] The propeller blades are mounted on the propeller hub in such a way that the pitch of the blades can be changed by servomotors (Figs. 4.57 and 4.58); this changes the propulsive force. The engine and propeller thus always maintain the same optimum speed. For sailing astern, the blades are adjusted so that their pitch is reversed [20, p. 850 ff.].

4.4.2.2 The Wheel Effect

Due to the direction of rotation of the propeller, the symmetrical design of the ship's stern, especially in single-screw ships, results in an asymmetrical flow pattern in the area of the propeller due to the so-called wheel effect. This means that the stern has a tendency to migrate in the direction of rotation of the propeller (Fig. 4.59). While this effect is still relatively insignificant and hardly noticeable when moving forward, it is noticeable when sailing astern, as the rudder blade is hardly exposed to the flow of the propeller stream, which is, however, beneficial for the rudder effect.

However, this effect is exploited during berthing maneuvers: For example, if the propeller turns to the right to stop the ship, that is, clockwise when viewed from astern, the stern is moved to starboard. In this respect, the starboard side of this ship would be the "chocolate side" for mooring (see Fig. 4.59).

Experimentally, to avoid this wheel effect, the stern of ships has been designed asymmetrically in order to achieve a symmetrical inflow of water, taking into account the direction of rotation of the propeller. As already mentioned, there have been

[63] This is discussed in more detail [19, 77].

[64] So, for example, also with the frigates of class F 124 of the German Navy.

Fig. 4.57 Controllable pitch propeller

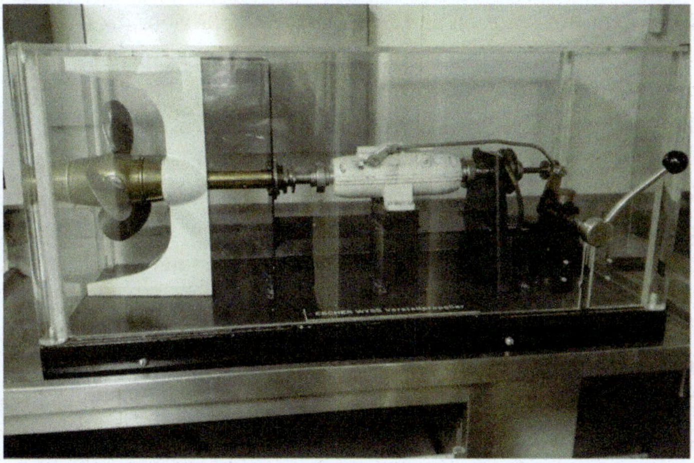

Fig. 4.58 Controllable pitch propeller system model

experiments; the asymmetric stern has generally not been accepted, as the effect is negligible when sailing ahead and can even be used skillfully when mooring.

Furthermore, the wheel effect is compensated by double propeller systems. The port and starboard propellers run in opposite directions.

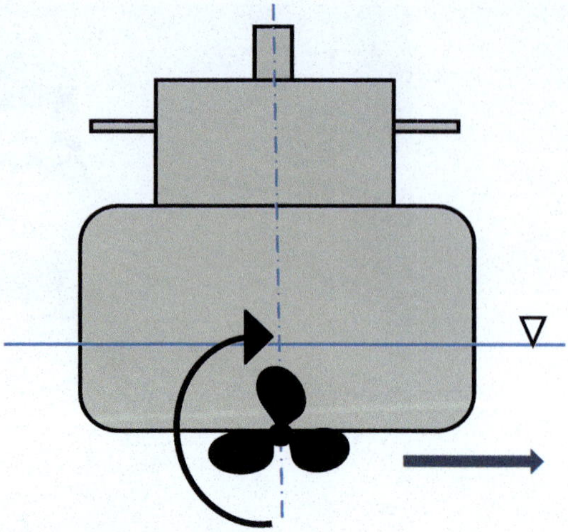

Fig. 4.59 Wheel effect at stern exit

4.4.2.3 Propeller Operation

The mode of operation of the ship's propeller, whose task is to convert rotary power into thrust, is based on various solution and consideration approaches[65]:

- Airfoil consideration: The mode of operation of a propeller is the same as that of an airfoil. The buoyancy generated corresponds to the thrust. The blade model explains the propeller effect at the individual blade sections [68].
- Froude–Rankine propeller theory[66] or jet theory [68]: The propeller generates a pressure jump in its plane of rotation, which causes the flow speed behind the propeller to be higher than in front of it. The resulting change in momentum produces the thrust.

4.4.2.4 Characteristics

An important parameter in connection with observations on the propeller is the area ratio. It describes the actual area described by the blades, in relation to the propeller circular surface A_0:

$$A_0 = \frac{\pi}{4} \cdot D^2 \qquad (4.99)$$

with D = Diameter of the propeller.

In connection with the propeller geometry, both the projected area ratio A_P/A_0 is used, where A_P is the *projected blade area*, as well as the *stretched or unrolled propeller area* A_E (E for "expanded" = unrolled, stretched), that is, A_E/A_0 (usually 0.3–1.5).

[65] For further details on the following see also Meier-Peter in [23, p. 260 ff.].

[66] Lehmann in [23, p. 892 ff.].

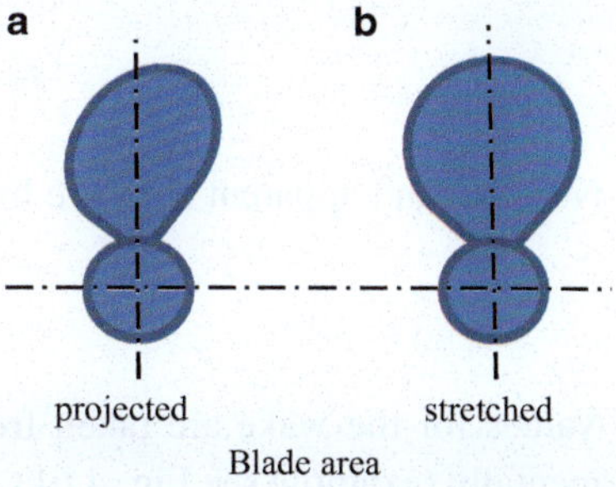

Fig. 4.60 Projected stretched blade area

The *projected area* is the area of all blades (without hub) projected onto a plane perpendicular to the propeller axis, that is, the shadow of the blades on a surface perpendicular to it. The *stretched blade* area is the propeller surface without pitch projected onto a propeller surface perpendicular to the propeller axis. The blade is practically laid flat on a plane (see Fig. 4.60).

Another important characteristic of the propeller is its pitch P. The pitch indicates the distance that a fixed point on a propeller blade, for example, at the outer edge of a blade, would travel in the axial direction during one revolution in a solid medium. The following relationship exists between the pitch P, the radius r, and the corresponding blade angle or angle of attack of the blade δ:

$$P = 2 \cdot \pi \cdot r \cdot \tan \delta. \tag{4.100}$$

However, in water, the actual propeller travel is smaller than the theoretically possible. The ratio of this difference to the theoretically possible path is called *slip* and it is about 20%. The *slip* establishes a relationship between the rotational speed of the propeller shaft and thus the propeller and the ship's speed. This is based on the aforementioned consideration that the propeller moves in water like a screw with thread in a solid material (e.g., wood screw in wood). Thus, the *theoretical speed of advance* v_{th} can be given as the product of the rotational speed and the propeller pitch:

$$v_{th} = n \cdot P. \tag{4.101}$$

However, the water actually flows to the propeller at a lower speed v_A, also called *actual speed of advance*. Putting these two speeds in relation to each other, one obtains the nominal slip s_R:

$$s_R = \frac{v_{th} - v_A}{v_{th}} = 1 - \frac{v_A}{n \cdot P}. \tag{4.102}$$

This indicates by what percentage of the theoretical speed the propeller moves relative to the water.

If one sets the theoretical speed of advance to the speed of the ship v_S in proportion, you get the apparent slip s_A:

$$s_A = \frac{v_{th} - v_s}{v_{th}} = 1 - \frac{v_s}{n \cdot P}. \tag{4.103}$$

Nominal and apparent slips are linked by wake w:

$$s_A = 1 - \frac{1 - s_R}{1 - w}. \tag{4.104}$$

Values for the wake are taken from the propeller slip diagrams and determined experimentally (example see Fig. 4.61).

As the *pitch* is changed over the radius, the reference value for the propeller pitch is usually the value at $0.7r$ and related to the diameter (pitch ratio P/D). Therefore P/D designates the *pitch ratio* at $0.7r$. Usually for freighters, tankers, vehicles, etc., pitch ratios of 0.5–1.4 are selected.

Furthermore, the speed of rotation is important for the propeller geometry. A point on a blade of a propeller moves at the circumferential speed v_u, which is dependent on the radius r and the speed n:

$$v_u = f(r, n) = 2 \cdot r \cdot \pi \cdot n. \tag{4.105}$$

In general, it can be said for profiles similar to airfoils that an increased slope is chosen at low approach speeds than at higher speeds. For this reason, the pitch of a ship propeller blade changes from a larger angle of attack at the hub to a smaller angle of attack at the outer diameter of the propeller.

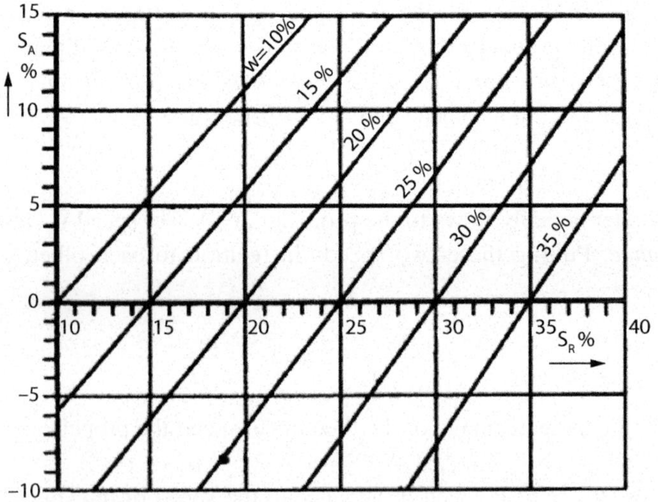

Fig. 4.61 Example of a propeller slip diagram. *Source* [18]

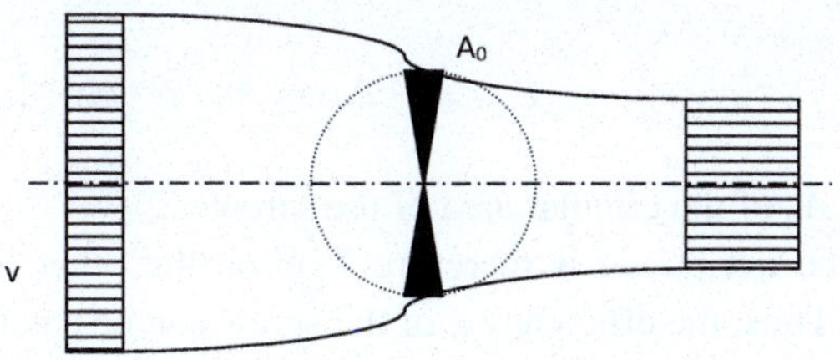

Fig. 4.62 Acceleration of the water jet

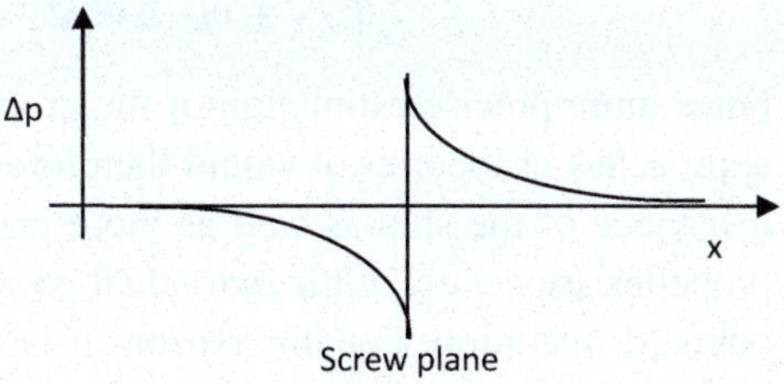

Fig. 4.63 Pressure jump in screw plane

With regard to the airfoil-like profile of the propeller blade, the following applies: The higher the incident flow speed, the slimmer the profile. Therefore, the profile of the propeller blade is more pronounced at the hub and becomes slimmer toward the outside.

The number of blades z of the propeller depends on the diameter, the cavitation properties,[67] and the vibration behavior of the entire propulsion system and is between $z=2\ldots7$ blades (e.g., the low-noise *skew-back propeller* for modern submarines).

The classical beam theory is still frequently used to estimate the thrust generated. The thrust is generated by an impulse change ΔI of the moving water mass m:

$$\Delta I = m \cdot \Delta v. \tag{4.106}$$

A driven propeller moving freely in the water at the speed v accelerates the part of the flow it detects by the amount Δv (Fig. 4.62). According to the impulse set, the screw presses with the force (the thrust T)

$$T = \dot{m} \cdot \Delta v \tag{4.107}$$

(with $\dot{m}$ is the accelerated mass of water in kg/s) and causes a contraction of the detected beam. In the plane of the screw, there is a mean speed of $v + \frac{\Delta v}{2}$.

The screw also generates a static pressure jump (Fig. 4.63) of the size

$$\Delta p = \rho \cdot \Delta v \cdot \left(v + \frac{\Delta v}{2} \right) \quad \text{in N/mm}^2. \tag{4.108}$$

The thrust T can now be written as follows:

[67] See Sect. 4.4.2.7.

$$T = A_0 \cdot \Delta p = A_0 \cdot \rho \cdot \Delta v \cdot \left(v + \frac{\Delta v}{2}\right) \quad \text{in N} \tag{4.109}$$

with A_0 of the circular area of the screw.

The net power is therefore $T \cdot v$; on the other hand, the kinetic energy $\dot{m} \cdot \Delta v^2/2$ is lost. Thus, the efficiency η of the screw can be calculated:

$$\eta = \frac{T \cdot v}{T \cdot v + \dot{m} \cdot \Delta v^2/2} = \frac{T \cdot v}{T \cdot v + T \cdot \frac{\Delta v}{2}} = \frac{v}{v + \Delta v/2}. \tag{4.110}$$

For a more precise estimation of the propeller characteristics using the abovementioned approaches or theories, it would therefore be important to have precise knowledge of the resistance of the ship as well as more precise information on the flow conditions at the propeller, especially at the individual propeller blades, which are not available in normal onboard operation. For this reason, it is sufficient to estimate the propeller thrust T in practice with various rules of thumb, for example:

$$T = 1{,}450 \cdot P \cdot \eta_D / v_A. \tag{4.111}$$

T is the thrust in Newton, P is the engine power in hp, η_D is the propulsion efficiency, and v_A is the propeller's advance speed in knots.

This formula is only suitable for calculating the thrust at rated engine power and the corresponding speed of the ship. The static thrust, which results, for example, from the so-called bollard pull test, cannot be calculated with this formula, because here it would have to be divided by zero (because the ship does not make any movement and v_A insofar as it is zero). For these cases, the following formula exists, with which the thrust of the propeller can be **approximately** determined:

$$T = 279 \cdot (P \cdot D/12) \cdot 0.67 \tag{4.112}$$

with D the propeller diameter in inches!

It is clear from both equations that the thrust at a given engine power is determined exclusively by the propeller diameter.

Furthermore, the following rule of thumb is commonly used to estimate the thrust:

$$T = k \cdot \frac{P}{v_S}. \tag{4.113}$$

This formula contains the factor k between 1.5 and 2.0 and takes into account the effective wake[68] and the efficiency of the free-running propeller, and the conversion of the ship's speed from m/s to knots (1 kn = 0.5144 m/s). The power P is in kW, and the speed of the vessel v_S to be inserted in kn.

[68] The wake is given as a percentage of the ship speed.

For the propulsion projection, the effective wake is reduced to the effective **wake figure** w. To determine the effective wake coefficient, the ratio of the difference between the ship speed v_S and the speed of advance v_A at the same ship speed is formed [39]:

$$w = \frac{v_S - v_A}{v_S} = 1 - \frac{v_A}{v_S}. \tag{4.114}$$

Furthermore, the following figures are commonly used to describe the ship propeller (with T in N, torque M_d in Nm transmitted to the screw, and density of water ρ in kg/m³):

Load Index

$$c_s = \frac{T}{\frac{\rho}{2} \cdot v^2 \cdot A}, \tag{4.115}$$

Torque Number

$$k_d = \frac{M_d}{\frac{\rho}{2} \cdot u^2 \cdot A \cdot \frac{d}{2}} \tag{4.116}$$

or also written as

$$k_d = \frac{M_d}{\rho \cdot n^2 \cdot d^5} \tag{4.117}$$

where u is the circumferential speed of the propeller with the diameter d.

Thrust Value

$$k_s = \frac{T}{\frac{\rho}{2} \cdot u^2 \cdot A_0}. \tag{4.118}$$

The following general relationships between the figures are also common:

Efficiency

$$\eta = \frac{2}{1 + \sqrt{1 + c_s}} = J \cdot \frac{k_s}{k_d}, \tag{4.119}$$

Thrust Value

$$k_s = c_s \cdot J^2. \tag{4.120}$$

In Eq. 4.120, J is the so-called degree of progress and indicates the ratio between the speed of the freely moving propeller in the water and its circumferential speed with n as speed in $1/s$ and D as propeller diameter:

$$J = \frac{v}{n \cdot D} \quad \text{or} \quad J = \frac{v}{u}. \tag{4.121}$$

4.4.2.5 Selection Criteria for Propellers

(a) Number of Propellers

The following reasons may speak for the use of two—or perhaps three or four—propellers [19, p. 218 ff.]:

- Draft restriction/high width to draft ratio at the stern,
- A better thrust load ratio and a better propeller efficiency,
- The improvement of maneuverability,
- Greater operational reliability (failure of a propeller or propulsion train does not lead to the ship's maneuverability, important especially in warship building; see, e.g., frigates of the German Navy class F 124 with double-shaft system),
- Greater cavitation safety,
- Avoidance of vibration excitation.

However, the following aspects can speak against the arrangement of several propellers:

- A higher power requirement,
- Higher production costs,
- Higher personnel costs for operation and maintenance.

(b) Number of Blades

Usual blade numbers are shown in Table 4.5

(c) Area Ratio

The propellers are mounted with the area ratio $A_E/A_0 = 0.40$–1.50. Typical values of the ratio are shown in Table 4.6

The increase in the ratio A_E/A_0 reduces the risk of cavitation, but at the same time, the efficiency of the propeller is reduced. For example, if the ratio is increased by 10%, it decreases by about 1.5–2%.

(d) Pitch Ratio

The following relationship can be used in an initial design phase for a propeller based on the pitch ratio *P/D*:

$$\frac{u}{v_A} \cdot \frac{P}{D} \approx 5. \tag{4.122}$$

In Eq. 4.122, u is the circumferential speed of the propeller at the reference radius or mean radius $= r/R = 0.7$ and v_A is the speed of advance.

Table 4.5 Usual number of propeller blades

Number of blades	Ship
2	Sail, small motorboats, fishing boat
3	Usually twin screwdriver, coaster
4	Screwdriver for the normal case
5	Screwdriver, for vibration safety reasons, reduction of the cavitation load (see Sect. 4.4.2.7)
6	Screwdriver, for vibration safety reasons, reduction of the cavitation load
7	Screwdriver, for vibration safety reasons, reduction of the cavitation load, reduction of the noise signature (e.g., U 212A of the Germans Navy)

Table 4.6 Usual propeller area ratios

Area ratio	Ship
0.55…0.65	Cargo ships
0.6…0.7	Towboats
0.9…1.4	Fast ships

4.4.2.6 Adaptation of the Engine and the Propulsion

When using fixed pitch propellers, the ship, engine, and propeller must be carefully matched. This means that the speed of the propeller must be selected so that the intended engine power can be used effectively for the intended ship speed. Therefore, when the engine power is transmitted directly to the propeller, the speed of the engine shaft = speed of the propeller. Furthermore, it must be ensured that the torque M_d to be delivered by the machine is equal to the hydrodynamic torque Q_d acting on the propeller, taking into account the friction losses at the shaft: $M_d = Q_d$.

The hydrodynamic torque is calculated from the definition of the torque coefficient K_Q:

$$Q_d = K_Q \cdot \rho \cdot n^2 \cdot D^5, \tag{4.123}$$

where K_Q depends on the degree of progress J and on the propeller pitch P/D. The torque to be delivered by the machine can be represented in the same form:

$$M_d = K_Q' \cdot \rho \cdot n^2 \cdot D^5, \tag{4.124}$$

where the coefficient K_Q' is only dependent on the speed. For cargo ships, the dependency of the speed on the ship speed is usually linear in certain areas (see Fig. 4.64):

$$v_S = c_1 \cdot n \tag{4.125}$$

where c_1 = Slope of the straight line.

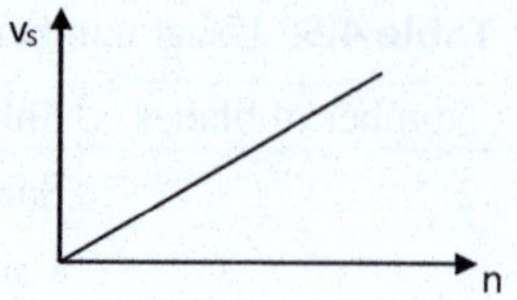

Fig. 4.64 Dependence of ship speed on the propeller speed

Therefore, the degree of progress $J = v_S/(n \cdot D)$ and the torque coefficient K_Q hardly change at constant gradient P/D. As a result, the power P_D absorbed by the propeller is proportional to the speed n with the third power:

$$P_D = 2 \cdot \pi \cdot n \cdot Q_d = 2 \cdot \pi \cdot K_Q \cdot \rho \cdot n^3 \cdot D^5 = c_2 \cdot n^3. \qquad (4.126)$$

In Eq. 4.126, ρ is the density of water in kg/m^3; the term $2 \cdot \pi \cdot n \cdot K_Q \cdot \rho \cdot D^5$ is designated with c_2.

4.4.2.7 Cavitation

Each substance can occur in all three states of aggregation depending on pressure and temperature [9, p. 215 ff.]. For the *fluid energy machine* "ship propellers" only the liquid and vapor states are of importance. In this respect, the vapor pressure curve, which describes the relationship between vapor pressure and temperature that is peculiar to the substance, is of interest here. Vapor pressure is the pressure that occurs when a vapor and its liquid phase are in thermodynamic equilibrium in a closed system. The vapor pressure increases exponentially with temperature for all fluids (Fig. 4.65).

If the static pressure in a liquid flow, as here on the ship's propeller, is locally reduced below the vapor pressure corresponding to the fluid temperature, cavities filled with vapor are formed there, which, transported by the flow into areas of higher pressure, collapse abruptly within milliseconds. This process is called cavitation.

This formation of steam bubbles affects both the operating behavior and the material of the ship propeller:

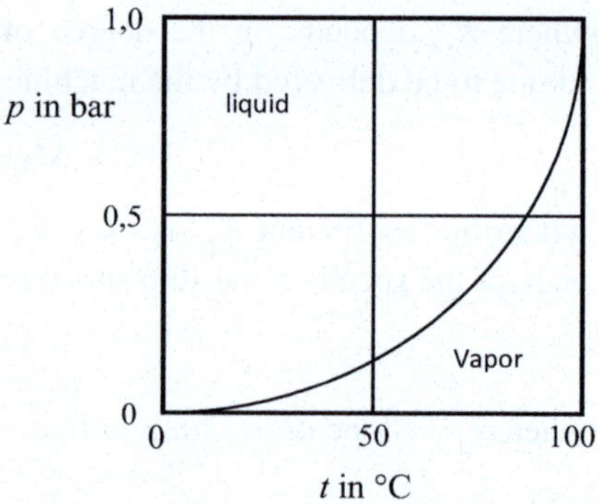

Fig. 4.65 Vapor pressure curve for water

- Near the wall of the propeller blades, the boundary layer is detached. The cavitation areas constrict the free flow cross section, which is also reflected in the buckling of the machine characteristics.
- Due to the implosion of the steam bubbles near the wall, the former gas space is filled by a microfine liquid jet in the range of the sonic speed of water, which in the long run leads to erosion of the propeller by removing particles from its material matrix. The effect of the individual microjet is in the range of a few milliseconds.
- As a result of the high pressure when a steam bubble collapses, intensive contact between O_2 and the propeller material can occur in oxygen-containing liquids and thus lead to increased oxidative processes.

A high proportion of free and dissolved gases in water generally favors the onset of cavitation and bubble growth. In addition, rough surfaces, especially on the suction side of the propeller and in the area of the blade edges, considerably favor the cavitation tendency. As a result of incipient cavitation, flow separation and reduction of buoyancy are detected on the profile of the propeller blade. In this respect, special care must be taken in the manufacture of the propellers to ensure that the surface treatment and design of the blade edges are particularly careful.

4.4.2.8 Causes of Loss of Thrust During Travel

In onboard practice, it can lead to noticeable thrust losses, which are not attributable to the propulsion system, that is, the ship's engine; these losses are then to be sought in the propeller. Possible causes are shown in Table 4.7

4.4.3 Propulsion Shaft Assembly

4.4.3.1 General

The **shafting** (marine shaft device) serves to transmit the rotary motion/torque power of the propulsion engine to the propeller and to absorb the propeller thrust and transfer it to the hull. The shaft device may consist of one or more parallel propulsion trains (multi-propeller vessels; see Fig. 4.66).

If the propulsion engine is not located very far aft, the shaft system is usually a system divided into several segments. The shaft assembly includes the propeller shaft, also known as the tail shaft (1), which passes through the stern tube, the line or propulsion shaft (2), the pressure shaft (3), the shaft clutches, the bearings, and the stern tube with the stern tube seal.

The ship's propeller sits on the propeller shaft (1), which is guided in the stern tube by means of a water-lubricated plain bearing. At least to the inside of the ship, the stern tube is sealed to prevent water from entering the ship's interior. The propulsion or line shaft (2), mounted on bearing blocks in plain, ball, or roller bearings, is flanged to the pressure shaft (3). This is equipped with a pressure flange and guides the propeller feed

Table 4.7 Causes for thrust loss

Cause	Remedial action
Pulling a fishing net or cordage into the propeller	Remove net or cordage by divers; preventive measure: install swap cutter behind the propeller on the propeller shaft
Cavitation damage	Propeller not optimally designed; replace if necessary during the next stay in the shipyard
Vegetation	Remove vegetation during the next stay in the shipyard
Mechanical damage, e.g., through ground contact	Rework propeller blades if necessary during the next stay in the shipyard. However, this may lead to unacceptable imbalances on the propeller. If the propeller is too badly damaged, it must be replaced
Propeller hub partly slips on the cone of the propeller shaft (slip)	Propeller nut has loosened—retighten; hydraulic clamping bush defective—check
Sudden thrust loss, accompanying with sudden increase in speed of engine	Shear pin sheared off due to ground contact; key or keyway connection between propeller hub and shaft defective; loss of propeller. Ship is unable to maneuver with a single propeller system: Stay at the shipyard

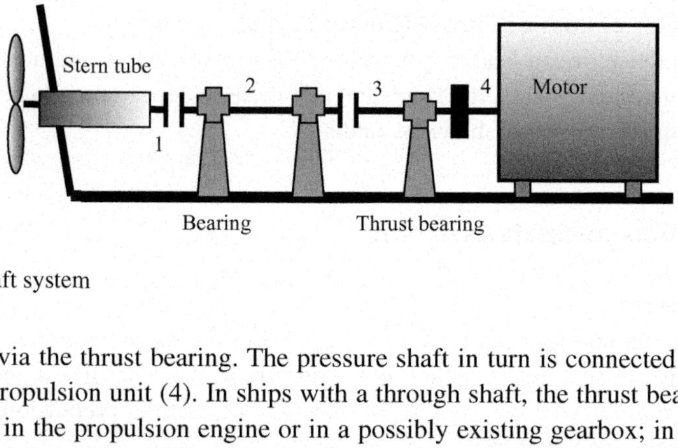

Fig. 4.66 Shaft system

into the hull via the thrust bearing. The pressure shaft in turn is connected to the crankshaft of the propulsion unit (4). In ships with a through shaft, the thrust bearing can also be integrated in the propulsion engine or in a possibly existing gearbox; in this case, the thrust bearing shown here is not required.

For maintenance and assembly reasons, long shaft systems are designed with several partial shafts. Clean alignment of the propulsion shaft is important to ensure optimum operating conditions for the ship's propulsion. An inaccurate alignment of the propulsion shaft can cause vibrations of the ship and damage to the shaft bearings.

In the case of very long shaft installations, a walk-through shaft tunnel surrounds the shaft installation when passing through holds or other ship spaces.

Gearboxes are often part of the shaft system. They are used to reduce the engine speed, as speeds below 200 rpm, if possible even below 100 rpm, are aimed for in order to achieve favorable propulsion efficiency in large ships. In addition, the gearboxes often have other outputs for operating auxiliary machines such as generators or pumps.

In complex propulsion systems, the gearboxes of the shaft system play a central role, as several motors and possibly turbines with very different speeds have to be connected to one or more shafts (see also Sect. 4.3).

4.4.3.2 Advantages and Disadvantages of Shaft Systems

The majority of all ships—especially cargo ships—are still equipped with shaft systems today. This is due to the fact that direct transmission of the rotary power causes only very small power losses: only about 1% of the transmitted power is lost due to friction losses in the bearings. For single gear units, the losses are still very low at 2–3%. In comparison, a Z-gearbox like the one in the Schottel Rudderpropeller already has 5% losses. The losses of an electric propulsion are even higher.

The shaft system normally does not include maneuvering devices, such as rudder. Other propulsion systems without a shaft system have these integrated (see POD propulsion, Voith-Schneider propulsion).

Disadvantages of shaft systems are the changeover from fore to aft, which can only be done slowly and in a controlled manner: For reverse driving, the so-called reversing, either the engine itself or the gearbox must be reversible or the propeller itself must be designed as a controllable pitch propeller. Reversing the engines is usually associated with high loads on the engine (in combustion engines; electric motors are insensitive to reversing) and a time delay (certainly of several minutes). Controllable pitch propellers and gearboxes for reversing are expensive.

Another disadvantage: Especially in shaft systems with only one propeller, starting astern leads to a strong lateral offset of the stern due to the so-called wheel effect of the propeller. However, this can only be counteracted to a very limited extent by rudder setting, as the rudder effect is only very slight when sailing astern (see Sect. 4.4.2.2)!

In shaft systems with two propellers, the direction of rotation of both propellers is in opposite directions, so that almost no wheel effect occurs and maneuverability can be significantly improved.

4.4.3.3 Shaft Assembly Design Guidelines

As a rule, simple heat-treatable steels (Ck 35, Ck 45 with tensile strengths of $R_m > 700\ \text{N/mm}^2$) or heat-treatable steels of higher strength are used.[69]

With regard to production, the shaft should be forged for reasons of strength. This avoids interruptions in the fiber flow of the material, especially when the diameter changes, which would be the case with turned parts. The surface is re-turned, ground, polished, or lapped; at the bearing points of the shaft, roughness depths of approximately 2 μm should be aimed for (grinding, polishing).

Hollow shafts are also frequently used, not only when controllable pitch propellers are used. Hollow shafts with $d_i = 0.5\,d$ have only 75% of the weight, and 94% of the moment of resistance of solid shafts!

[69] For detailed shaft calculations, see also Böge et al. in [3].

If a ship is to be classified by DNV GL, the requirements for the shafting system can be found in its classification and construction regulations "I Ship's machinery, 1 seagoing vessel, 2 machinery, (GLRP I-1–2), Sect. 4."

Basically, diameter changes should be carried out by means of a cone or large fillet. The radii should at least correspond to the change in diameter. This means that if the shaft diameter is changed from 500 to 550 mm, the radius of the fillet at the diameter transition should be carried out with $r = 50$ mm.

For intermediate and thrust bearing shafts, the fillet radius of forged flanges should be at least 8% of the calculated minimum diameter for a solid shaft at the relevant location.

At the rear propeller shaft flange, the radius of curvature should be at least 12.5% of the calculated minimum diameter for a solid shaft at that point.

With regard to the design of shafts, reference is first made to the following standards:

DIN 743-1 Calculation of load capacity from shaft s and Axles—Fundamentals, Introduction,

DIN 743-2 Calculation of load capacity from shaft s and axles—Notch effect and shape numbers,

DIN 743-3 Calculation of load capacity from shaft s and axles—Material strength values,

DIN 743-4 Calculation of load capacity from shaft s and axles—Fatigue limit, endurance limit,

DIN 748-1 Cylindrical shaft ends—Dimensions, nominal torques,

DIN 1448 Tapered shaft ends with long cone (1: 10) and threaded pin and with short cone and threaded pin,

VDI 3840 Vibration engineering calculations

Concrete calculation bases for shaft systems of ships can be found in the abovementioned construction regulations of DNV GL. A shipyard may therefore also depend on the customer's requirements with regard to the dimensioning of the shaft system. In the following, a propeller shaft is first calculated according to DIN 748, and then according to the GL construction regulation, in order to see what differences occur here.

Example for Dimensioning a Propulsion Shaft

For a container ship with an engine power of 20,000 kW, shaft speed $= 120\ \text{min}^{-1}$, the propeller shaft must be dimensioned.

The propeller shaft is the component from which the propulsion power $\boldsymbol{P}_{an}$ introduced directly via the propulsion motor or the propulsion power introduced via the gearbox is tapped at its shaft end in the form of the propeller power $\boldsymbol{P}_{ab}$:

$$P_{ab} = 2 \cdot \pi \cdot n \cdot M_{ab}. \tag{4.127}$$

As both the power $\boldsymbol{P}_{ab}$ as well as the speed $\boldsymbol{n}$ is known in Eq. 4.127, the underlying torque $\boldsymbol{M}_{ab}$ is calculated and the diameter of the shaft end can be deduced using the **DIN 748**, in which the transmittable torque is compared to the respective diameter $\boldsymbol{d}$ of the shaft end.

Extract from DIN 748 page 1 on the relationship between torque and shaft diameter:

Column a	Column b	Column c
Transmission of a pure torque $M = \frac{9.8065 \cdot \pi}{4} \cdot 10^{-3} \cdot d^3$ in Nm (d in mm)	Simultaneous transmission of a torque and a corresponding known bending moment: $M = 58.8399 \cdot 10^{-5} \cdot d^{3,5}$ in Nm (d in mm)	Simultaneous transmission of a torque and a unknown bending moment: $M = 27.45862 \cdot 10^{-5} \cdot d^{3,5}$ in Nm (d in mm)

The above DIN 748 takes into account that a propulsion shaft is not only subject to torsional stress due to the torque to be transmitted; due to its dead weight, it has a deflection between the bearings, which leads to an additional bending stress of the shaft to be considered.

Solution: By rearranging the equation from column c (diameter and thus deflection of the shaft are still to be determined—thus the bending moment is not known) to $\boldsymbol{d}$**,** the following results are obtained

$$d = \sqrt[3.5]{\frac{M}{27.45862 \cdot 10^{-5}}}.$$

$$M = \frac{9550 \cdot P}{n}$$

M**,** in turn, is obtained by changing Eq. 4.39 for the net power of a diesel engine[70] with

M in Nm,
P in kW,
n in min^{-1},

$M = \frac{9550 \cdot 20{,}000}{120} = 1{,}591{,}667$ Nm.

Inserting and solving the above equation for $\boldsymbol{d}$ results: $d = 616$ mm.

Design in accordance with "GL I Ship Technology, 1 seagoing vessel, 2 machinery installations, Sect. 4 C.2":

The calculation is based on the following equation:

[70] See Sect. 4.3 and also "Net power and some diesel engine characteristics," Sect. 4.3.1.

$$d_a \geq d \geq F \cdot k \cdot \sqrt[3]{\frac{P_W}{n \cdot \left[1 - \left(\frac{d_i}{d_a}\right)^4\right]}} C_W, \tag{4.128}$$

d Minimal required outside diameter the shaft (mm),

d_a Executed outside diameter the shaft (mm),

d_i Executed diameter the shaft bore (mm); if $d_i \leq d_a$, $1 - \left(\frac{d_i}{d_a}\right)^4 = 1$ will be set

P_W Nominal power (kW) of propulsion motor without bearing and gearbox losses,

n Shaft speed (min^{-1}) at nominal power,

F Factor for the type of propulsion
a) Propeller shafts 100 (for all systems),
b) Intermediate shafts and thrust bearing shafts 95 (for turbine systems, diesel engine systems with hydraulic slip clutches, propulsion by electric motors), 100 (for all remaining propulsion systems),

C_W Material factor, to calculate from

$$C_W = \frac{560}{R_m + 160}, \tag{4.129}$$

R_m Specified minimum tensile strength of shaft material (N/mm^2; see Sect. B.1 of this construction regulations),

k Factor for the type of the shaft; according to construction regulation, values between 1.00 and 1.20—depending on the shaft design—are specified for **intermediate shafts**. For **pressure shafts**, “k” is 1.10 and for **propeller shafts**—depending on the design and bearings—between 1.15 and 1.40

Here $k = 1.26$ (propeller shaft on which the propeller is fixed by means of a cone and a feather key).

By inserting it in Eq. 4.128 follows:

$$d_a \geq d \geq 100 \cdot 1.26 \cdot \sqrt[3]{\frac{20{,}000}{120 \cdot 1}} \cdot C_W,$$

where C_W with $R_m = 600$ N/mm^2 is also calculated to:

$$C_W = \frac{560}{600 + 160} = \underline{0.74}.$$

Note: For R_m, the following values are relevant according to B.1 of this GL Regulation:

Fig. 4.67 Thrust bearing (oil lubrication) of the AIDAmar. *Photo* AIDA

- 600 N/mm^2 for propeller shafts (exceptions require the express approval of GL),
- 760 N/mm^2 for shafts made of carbon or carbon–manganese steel, except propeller shafts,
- 800 N/mm^2 for alloy steel shafts except propeller shafts.

Thus, the following is calculated: $d_a \geq d \geq 627\,\text{mm}$.

A comparison of both calculations shows a slight deviation in the determined diameter. The DNV-GL construction regulation provides for better consideration of specific stresses on the shaft parts due to their design and installation or bearing types by factorization in the diameter calculation. ◀

4.4.3.4 Shaft Bearing

The bearing system of the propulsion shaft consists of a *locating bearing* and one or more *floating bearings*. The locating bearing (so-called bivalent bearing) absorbs forces in both radial and axial directions. The thrust bearing (see Fig. 4.67) usually fulfills this task.

For the locating bearing, however, a pure axial bearing (pure thrust bearing) and a radial bearing arranged in the immediate vicinity thereof can also fulfill this task. The other bearings, viewed from the thrust bearing in the direction of the propeller, are designed as non-locating bearings to compensate for axial length variations of the shaft, particularly due to temperature fluctuations.

Example of Length Variation of Propulsion Shaft

A propulsion shaft of 10-m length is mounted at 15 °C. During the operation of the ship, the temperature of the shaft increases to 45 °C. By how many centimeters does the shaft expand in axial direction?

Solution: The linear expansion Δl of a body is calculated according to the equation

$$\Delta l = l_0 \cdot \alpha \cdot \Delta t \tag{4.130}$$

with

- l_0 The initial length,
- α Coefficients of linear expansion specific to the material (for steel $11.7 \cdot 10^{-6}$ 1/K; further Coefficients of linear expansion can be found in Annex 15),
- Δt The temperature difference

This results in the linear expansion of this shaft:

$$\begin{aligned}\Delta l =& 10{,}000 \text{ mm} \cdot 11.7 \cdot 10^{-6}/\text{K} \cdot 30 \text{ K},\\ \Delta l =& 3.51 \text{ mm}.\end{aligned}$$

◀

Experience has shown that the bearing distance l_a for long shafts should be approximately[71]

$$l_a \approx 300\sqrt{d} \tag{4.131}$$

with d as shaft diameter. When selecting the bearing spacing, however, it is important that all bearings are loaded as evenly as possible. For this purpose, a compact design with small bearing distances should be aimed for. As a result, the shaft has a lower bending moment, which in turn leads to a smaller shaft diameter and thus also to smaller bearings. These can basically be designed as *plain* or *rolling bearings*.

Regarding the lubrication of the bearings, a distinction is made between *oil and grease lubrication*. Where plain bearings are fitted, oil lubrication is mainly used, whereas rolling bearings can also be grease-lubricated in the case of propulsion shaft bearings. Maintenance-free *rolling bearings* are filled with rolling bearing grease, the bearing cage is sealed on both sides with sealing washers. This prevents grease from escaping from the bearing and dirt from entering. Figure 4.68 shows the bearing arrangement of a propulsion shaft using rolling bearings and oil lubrication. The bearing housing is filled with oil, the oil level can be checked visually through the inspection window at the top. A brass thermometer can be seen on the right-hand side of the bearing housing, with which the oil temperature can be monitored visually on site—in addition to a recording in the control room.

[71] See also Böge et al. in [3, p. I 119 ff.].

Fig. 4.68 Bearing of propulsion shaft in roller bearing. *Photo* AIDA

The oil lubrication of the *plain bearing* serves on the one hand to make the shaft "float" in the bearing, and on the other hand, the oil serves to dissipate the heat generated by the bearing friction. The resulting oil loss must be compensated. For small bearings, lid oilers, wick oilers, drip oilers, or similar are therefore used. However, circulation lubrication systems are commonly used for large shaft systems with plain bearings. By means of a pump (a separate lubrication system for each bearing or as central lubrication for all plain bearings), the lubricating oil is fed through the upper bearing shell to the plain bearing(s), drained and cleaned (filtered) at the bottom of bearing and pumped in a circle. Losses are compensated for by the storage tank.

In this way, a uniform required oil supply is achieved.

First of all, the bearing of the *propeller shaft* at *stern tube* is considered. Through this, the propeller shaft leads into the interior of the ship. As a rule, a water- or oil-lubricated plain bearing is used as a bearing arrangement, in the case of nacelle propulsion (POD propulsion) also roller bearings.

In *plain bearings*[72] generally used here, the bearing material should always be softer than the shaft material so that the shaft is not attacked and seizes into the bearing material. Therefore, plastics, special rubbers, bronze (copper–tin casting alloys), tinplate (lead–tin casting alloys), or occasionally cast iron (EN-GJL-250 and EN-GJL-300) are used for the lining of the bearing bush. The bearing shells have cleverly designed grooves, which ensure optimum wetting of the bearing surfaces with water or oil.

[72] For further information on plain bearings, see Böge et al. in [3, p. I 177 ff.].

Table 4.8 Reference values for bearing clearance

Shaft diameter (mm)	Bearing clearance (mm)	
	Rubber	White metal/bronze
100–200	0.3–0.8	0.2–0.3
200–300	0.8–1.0	0.3–0.4
300–400	1.0–1.2	0.4–0.5
400–500	1.2–1.5	0.5–0.6
500–600	1.5–1.7	0.6–0.7
600–700	1.7–1.9	0.7–0.8
700–800	~ 2.0	0.8–0.9
800–900		0.9–1.0
900–1000		1.0–1.1
1000–1100		1.1–1.2

With a water-lubricated bearing, the stern tube usually has only an inner stern tube seal. If the bearing is oil lubricated, the stern tube must also be sealed to the outside. On the one hand, for environmental reasons, and on the other hand, it must be avoided that the entering water is suspended with the oil, which reduces the lubricating properties of the oil.

Modern water-lubricated stern tube plain bearings also have a seal to the outside, as they are lubricated with filtered cold seawater from the engine's sea-cooling water system.

Although plastics or special rubbers have low wear rates and favorable sliding properties, they only have low emergency running properties because they react sensitively to high temperatures (through friction).

In contrast, bronze and lead–tin alloys have much better emergency running properties.

In order that a sufficient lubricant film can form in the bearing, a minimum amount of bearing clearance is required; Table 4.8 contains reference values for the bearing clearance depending on the bearing material.

The significantly greater bearing clearance with rubber is due to the fact that this material tends to swell—on the one hand due to water or oil absorption, and on the other hand also due to the thermal loads present during operation (volume expansion). This means that if the bearing gaps are too narrow, the friction within the bearing would become too high. In addition to the rubber bearings, bearing shells made of *Vesconite* [81] are widely used today. This thermoplastic material has good dry-running properties (due to embedded molybdenum sulfide as lubricant), absorbs little water, and therefore swells only slightly. It is similarly resilient as white metal. With regard to the required bearing clearance, values comparable to those for the metallic plain bearings can therefore be assumed.

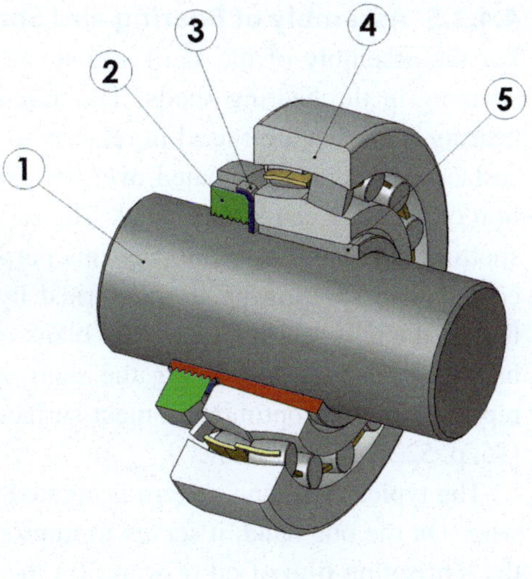

Fig. 4.69 Spherical roller bearings. *Photo* Silberwolf, CC-BY-SA-2.5

However, the manufacturer's instructions must always be observed with regard to the required bearing clearance.

The *propulsion shafts*—and, if necessary, for long shaft systems, also the *intermediate shafts*—can also be guided in plain bearings. In addition to plain bearings, rolling bearings,[73] for example, spherical roller or deep groove ball bearings (see above, Fig. 4.68), are used (see Fig. 4.69) with the assemblies:

- One shaft,
- Two locknut according to DIN 981,
- Three lock washer according to DIN 5406,
- Four outer and inner ring, rollers or balls, and roller or ball cage,
- Five adapter sleeve according to DIN 5415.

The bearings are mounted on bearing blocks, which are connected to the ship structure by so-called integrated foundations (see Fig. 4.67).

The *pressure shaft* finally transmits the propeller thrust to the ship via the *thrust bearing* during forward and aft sailing. A roller bearing is used (e.g., double-direction thrust ball bearing or tilting pad thrust bearing[74]), which absorbs the axial forces of the thrust. If a gearbox is installed between motor and shaft, the thrust bearing is integrated in the gearbox. In direct propulsion, the thrust bearing with its bearing block is directly connected to the ship's structure via an integrated foundation; in some cases, it is mounted on the motor base plate.

[73] For further information on rolling bearings, see Böge et al. in [3, p. I 156 ff.].

[74] See for example [79].

4.4.3.5 Assembly of Bearing and Shaft

For the assembly of the shaft system into its *plain bearing*, this is optimally fitted by scraping in the bearing shells. The aim of the process is to remove those parts of the bearing shell that are raised in relation to the ideal plane. As all load-bearing points of a surface are evenly distributed over the entire surface, the best possible uniform distribution of the load can be achieved. The respective quality of the surface is determined by spotting with spotting paint. For this purpose, the shaft is inserted into the bearing shell coated with spotting paint and turned by hand. The paint remains in the depressions, that is, the non-bearing areas. The blank areas, that is, the steles where the spotting paint has been removed by turning the shaft, are reworked with the scraper. This process is repeated until an optimum contact surface is achieved. Flatness tolerances of 0.001 mm [15, p. 522] are achievable.

The typical scraping pattern is created by changing the direction of the scraper crosswise. On the one hand, it serves to improve the surface optically, and on the other hand, the lubricating film of oil or water (in the stern tube) is better maintained in the recesses in hydrodynamically lubricated plain bearings.

The slow disappearance of the pattern allows the wear to be estimated.

For mounting and dismounting of *rolling bearings*, the following must always be observed[75]:

Before starting (dis)assembly, a diagram of the individual operations should be drawn up. Information on the required heating temperatures, the forces required for mounting and dismounting the bearings and the required quantity of grease must be available.[76]

Rolling bearings are preserved with an anti-corrosion oil. It does not need to be washed out when mounting the bearings. It combines with the lubricant during operation and ensures sufficient lubrication for a short time during start-up. The anti-corrosion oil is wiped off the seat and contact surfaces before assembly. From tapered bearing bores, on the other hand, the corrosion protection should be washed out before fitting to ensure a secure, tight fit on the shaft or sleeve. After washing out with cold cleaner, the bore is thinly moistened with a machine oil of medium viscosity. Used and dirty bearings must be washed out in kerosene or cold cleaner before installation and then immediately re-oiled or greased. Rolling bearings must be protected against dirt and moisture under all circumstances, as even the smallest particles penetrating the bearing can damage the running surfaces. Therefore, the installation location must be dust-free and dry. Attention must also be paid to the cleanliness of the shaft and housing. Carefully remove anti-rust coatings and paint residues at the bearing seating points on the shaft and in the housing. Burrs and sharp edges on shafts and housings must be removed and broken.

All parts belonging to the bearing arrangement must be checked for dimensional and geometrical accuracy before assembly. In principle, both rolling bearing rings should be

[75] For further details see: "FAG—Mounting of Rolling Bearings" [80].

[76] The following from [47].

well supported by the seating surface and therefore fitted as tightly as possible. However, this is not always possible because this makes mounting and dismounting more difficult or because a ring must be easily displaceable in non-locating bearings. The interference in tight fits leads to an expansion of the inner ring or to a contraction of the outer ring and thus to a reduction of the radial internal clearance. Therefore, the radial internal clearance must be matched to the fits.

For the *installation* of the rolling bearings, please refer to the following:

Due to different designs and sizes, rolling bearings cannot all be mounted using the same method. A distinction is made between *mechanical*, *hydraulic,* and *thermal* assembly procedure.

As the hardened bearing rings are sensitive to impact stress, the rings must not be hit directly with a hammer during *mechanical assembly*. When fitting non-demountable bearings, the mounting forces must always act on the tightly fitted ring. This ring is also mounted first. Forces acting on the loosely fitted ring would be transmitted by the rolling elements, which could damage raceways and rolling elements. Bearings up to a bore diameter of approximately 80 mm can be cold pressed onto the shaft during the usual tightening process. A mechanical or hydraulic press should be used for this purpose. If no press is available, the bearing can be driven onto the shaft with light hammer blows if necessary, and if the fits are not too tight, by means of a hammer sleeve or a tube.

If tight fits on the shaft are specified for cylindrical seats, the *thermal assembly method* is selected. In this process, the bearings are heated for mounting in order to temporarily increase the internal diameter of the bearing by utilizing the thermal expansion. Sufficient expansion is achieved at temperatures around 80–100 °C. However, it must not exceed 120 °C, as there is then a risk that the metal structure of the bearing parts will change. This can result in a decrease in hardness and changes in the bearing dimensions. The same temperature limits apply to bearings with solid cages made from glass fiber-reinforced polyamide as to purely metallic rolling bearings. Bearings with shields and sealing washers are already filled with grease. They may be heated up to a maximum of 80 °C during installation, but not in an oil bath. Rolling bearings can be temporarily heated on a temperature-controlled heating plate. During this process, the bearing should be turned several times to ensure that it is evenly heated.

As a rule, the bearings are heated in a temperature-controlled oil bath. This method ensures uniform heating and the installation temperature of 80–100 °C can be safely maintained. After heating, the oil must drip off well. Wipe all fitting and contact surfaces carefully!

Then push the bearing quickly and without tilting in one go until it stops at the fitting position. A slight screwing rotation when placing the shaft facilitates rapid mounting. Wear suitable protective gloves to prevent burns to the hands during assembly!

If the outer ring of the bearing must be tightly fitted in the housing, the housing is heated. With bulky and large housings, this sometimes causes difficulties; in this case, the bearing must be cooled in a mixture of dry ice and alcohol. A temperature must not fall below −50° C. The condensation water that occurs during temperature equalization must be completely flushed out of the bearings with oil, otherwise there is a risk of corrosion.

At the (more complex) *hydraulic method,* machine oil or oil with rust-dissolving additives is pressed between the mating surfaces. The oil film largely eliminates contact between the mating parts, so that they can be moved against each other with little effort and without risk of surface damage.

When *dismounting* rolling bearings, the following must be observed:

If the bearings are to be reused, care must be taken when dismounting; in particular, the extraction tool must be applied to the ring to be extracted, otherwise the rolling elements will press into the raceways and damage the bearing. In addition, there is a risk of breakage with thin-walled outer rings. In non-separable bearings, the ring fitted with a sliding fit is first removed from its seating surface. Then the tight-fitted ring is pressed off. The force required for pressing off is usually considerably greater than the press-on force, as the ring becomes stuck over time. Even with loosely fitted rings, removal can be difficult if fretting corrosion has formed after long periods of operation. Alternatively, the shaft can be cooled (in front of and behind the bearing seat with lobes soaked in refrigerant—see above) and—especially with smaller bearings—the rolling bearing can be heated with a hot air dryer via the outer ring.

4.4.3.6 Bearing Calculation

Once the required shaft diameter has been determined, the appropriate bearings must be selected. The catalogs of the bearing manufacturers are to be referred to.

Decisive criterion for the selection of the "correct" *rolling bearings* is its expected life, which is determined by the life factor f_L. The life factor should be 3…4 for ship thrust bearings and 4…6 for ship shaft bearings. It is determined according to the following equation:

$$f_L = \frac{C}{P} f_n \tag{4.132}$$

with

- C The basic dynamic load rating (see further below),
- P The equivalent dynamic bearing load (see below),
- f_n The speed factor (from manufacturer's catalog)

The following calculation steps according to ISO 281 are to be carried out:

a) Determination of the equivalent dynamic load

The equivalent dynamic load is the purely radial, and in axial bearings the purely axial load (caused by the propeller thrust), which the bearing will achieve under the actual operating conditions. If a radial bearing is loaded by a radial force alone, the equivalent dynamic bearing load is

$$P = F_r. \tag{4.133}$$

To calculate the radial forces acting on the individual bearings, a shaft system such as a two- or multi-bearing beam (see Fig. 4.74) can be calculated according to the laws of statics. Here the dead weight of the shaft is included in the calculation as line load, and the propeller also generates an additional load on a cantilever arm (see Fig. 4.75). An example for determining the bearing forces is given in Sect. 4.4.3.7d.

For radial bearings loaded radially with a radial force F_r and axially with an axial force F_a, the equivalent dynamic bearing load is

$$P = XF_r + YF_a. \tag{4.134}$$

X is the radial factor, which takes into account the ratio of radial to axial force, and Y is the axial factor for converting the axial force into an equivalent radial force. These factors can be taken from the manufacturers' bearing catalogs.

b) Basic dynamic load rating and service life

The life of a rolling bearing is the number of revolutions or hours before the first signs of wear on the rolling element or raceway become apparent. The basic dynamic load rating C is the load that can support a nominal life of $L = 10^6$ revolutions or $L_h = 500$ h at $n = 331/3\ \text{min}^{-1}$ ...to be expected. C can be found in the manufacturer's catalogs.

Calculation of the service life in millions of revolutions at 10 % probability of failure:

$$L_{10} = \left(\frac{C}{P}\right)^p \tag{4.135}$$

where p is the service life exponent ($p = 3$ for ball bearings; $p = 10/3$ for all other rolling bearings).

Calculation of the service life in hours at 10 % probability of failure:

$$L_{10\text{h}} = \frac{10^6}{60 \cdot n} \cdot L_{10} = \left(\frac{16,666}{n}\right) \cdot \left(\frac{C}{P}\right)^p. \tag{4.136}$$

Example of Rolling Bearing Selection

A radial force of 12 kN acts on the floating bearing of a propulsion shaft. The shaft rotates during operation at 160 min^{-1}. This bearing shall be a cylindrical roller bearing. It should reach service life of at least 10,000 h.

Solution: To select a bearing that can carry this load, the necessary dynamic load rating must first be determined:

For the bearing type "cylindrical roller bearing," the bearing catalog of the selected manufacturer contains the values $X = 1$ and $Y = 0$.

From Eq. 4.134 follows:

$$P = X\,F_r + Y\,F_a = 1 \cdot 12\ \text{kN} + 0 \cdot F_a = 12\ \text{kN}.$$

By changing the equation for $L_{10\text{h}}$ (Eq. 4.136) to C now follows:

$$C = \left(\frac{L_{10\text{h}} \cdot 60 \cdot n}{10^6}\right)^{\frac{1}{p}} \cdot P = \left(\frac{10{,}000 \cdot 60 \cdot 160}{10^6}\right)^{\frac{3}{10}} \cdot 12 = 47.2\ \text{kN}.$$

The dynamic load rating should therefore be **at least 47.2 kN**. This allows a suitable bearing to be selected from the manufacturer's catalog.

Checking the service life: A value of 0.625 for f_n can be found in the manufacturer's catalog for the cylindrical roller bearing at shaft speed from 160 min^{-1}. This gives f_L according to Eq. 4.132:

$$f_L = \frac{C}{P} f_n = \frac{47.2\ \text{kN}}{12\ \text{kN}} \cdot 0.625 = 2.458.$$

However, as stated at the beginning, f_L should reach values between 4…6. That is not the case here. In this respect, a roller bearing with a higher dynamic load rating than 47.2 kN would have to be selected; by converting Eq. 4.132 to C follows:

$$C = \frac{f_L \cdot P}{f_n} = \frac{5 \cdot 12}{0.625} = 96\ \text{kN}.$$

The selected bearing should therefore have a dynamic load rating of 96 kN. ◄

When *designing plain bearings*, the *permissible specific bearing load* is the decisive factor for dimensioning the bearings.[77] In principle, the *permissible specific bearing load* (surface pressure) must not be exceeded; it applies:

$$p = \frac{F}{b \cdot d} \leq p_{perm} \tag{4.137}$$

with

p = specific bearing load (N/mm^2),
p_{perm} = permissible specific bearing load (N/mm^2),
F = radial force on the bearing shell (N),
d = diameter of the bearing shell (mm),
b = width of the bearing shell (mm)

From the determined shaft diameter, which is entered as d in Eq. 4.137, the required bearing shell width can then be calculated with the permissible specific bearing load by changing the equation.

[77] In detail also again Böge et al. in [3, p. I 181 ff.].

Table 4.9 Approximate values of the permissible bearing load for plain bearings

Bearing material		p_{perm} (N/mm^2)
PbSn alloy		12.5
Bronze	CuSn7Zn4Pb7-C	20
	CuSn12-C	25

Table 4.9 contains approximate values for p_{perm}.

4.4.3.7 Shaft Clutches

Rarely, the propeller shaft is directly connected to the gearbox or crankshaft of the engine. As a rule, the shaft systems are split, particularly for reasons of ease of installation (see Sect. 4.4.3.1). The connection of the individual shafts is made by shaft clutches. Both *fixed (rigid)*—Fig. 4.70—as well as *flexible clutches* can be used. A further distinction is made between shiftable clutches (which allow the propulsion train to be connected and disconnected to the engine) and non-shiftable clutches. Shiftable clutches work on the principle of frictional engagement. There are also hydraulic and electromagnetic clutches. It would be beyond the scope of this paper to go into detail here about all the special features of the variety of clutch systems—reference is made to the relevant literature, for example [2, 24, 26].

In the following, fixed (rigid) and flexible clutches used to connect the individual shaft segments will be considered in more detail.

When using *fixed* or rather *rigid clutches*, attention must be paid to exact alignment of the shaft components!

Flexible clutches, on the other hand, are able to compensate for deviations in the longitudinal direction caused by temperature changes or variable thrust forces. They compensate for shaft movements in radial direction, caused by bending moments (dead weight, torsional vibration). They can also compensate for small angular deviations between two shafts. In addition, they compensate for abrupt torque fluctuations (e.g., due to rapid speed changes); the flexible clutches can reduce such shock loads by temporarily accumulating mechanical work.

For the calculation of clutch, the maximum torque $M_{t,\max}$ to be transmitted is relevant, where

$$M_{t,\max} = k \cdot M_{t,\mathrm{normal}} \tag{4.138}$$

with k the so-called impact coefficient. The impact coefficient takes into account the starting and operating behavior of the propulsion system; with a good approximation, it can generally be assumed to be 1.1…2.5 [28, p. 64 ff.].[78]

[78] Values for k, see also [6, Volume I, p. 739].

Fig. 4.70 Fixed or rigid shaft clutch. *Photo* AIDA

(a) Basic Standards

The essential technical standards for clutches on propulsion shaft systems are.

- DIN 115 shell clutches,
- DIN 116 disc clutches,
- DIN 740 flexible shaft clutches,
- VDI 2240 shaft clutches.

(b) Shell Clutch

Shell clutches belong to the fixed clutches and are used for shaft connections to *long rigid continuous* shaft systems and a torque up to 2500 Nm. The shells are clamped onto the shaft ends so that the torque is transmitted by frictional connection. Additional securing is often provided by feather keys (Fig. 4.71). The advantage of these clutches is that they allow easy mounting and dismounting without the simultaneous removal of shaft parts. For dimensions and number of clamping screws, please refer to the relevant manufacturer and supplier catalogs. Selection criterion is the torque to be transmitted $M_{t,\max}$:

$$M_{t,\max} = F_S \cdot n \cdot \mu \cdot D \tag{4.139}$$

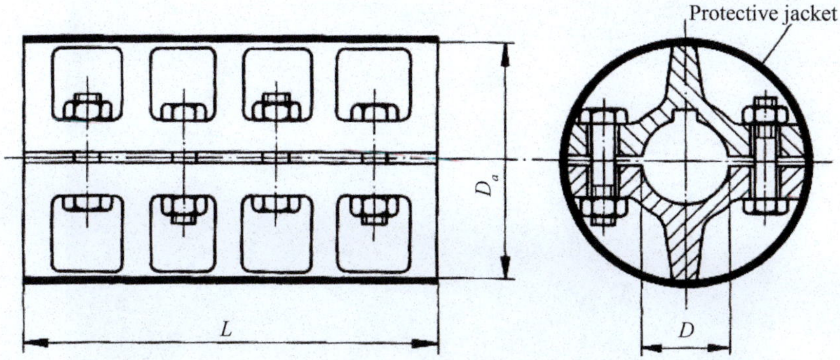

Fig. 4.71 Shell clutch. *Source* [3]

with

F_S Contact pressure of the individual screw,
n Number of the screws,
μ Coefficient of static friction material shaft/clutch,
D Shaft diameter

(c) Disc Clutch[79]

They are designed both as fixed (see Fig. 4.72) and as flexible clutches (see Fig. 4.73).

A flange hub (disc) is fixed (shrunk on with key or welded) on each shaft end. Both elements are screwed together. A centering lug between the two discs ensures the correct fit. The shafts must be absolutely aligned (on clutches with centering collar). The power transmission is based on the static friction between the two clutch surfaces (static friction coefficient $\mu \approx 0.2$ for roughed friction surfaces), and torque transmission through *static friction* and *form fit* when using *fitting bolts.*

On disc clutches with flexible intermediate ring, position tolerances can be compensated. However, if possible, they should be installed at the point on the shaft where their bending moment is zero (see below under d).

The shafts must be supported in the immediate vicinity of flanged joints.

Further details:

- Advantage compared to shell clutch: higher torques can be transmitted with the same shaft diameter; disadvantage: more difficult disassembly,
- Suitable for shock and alternating loads, axial forces, and bending loads (clutch with flexible intermediate ring),
- For highly stressed shafts available up to 1,450,000 Nm,

[79] See also [58].

Fig. 4.72 Fixed disc clutch

Fig. 4.73 Flexible disc clutch. *Source* http://www.reich-kupplungen.com

- For shafts up to 500 mm,
- Speed at $\text{Ø}_{\text{Shaft}} = 25$ mm until $n = 6850\ \text{min}^{-1}$,
- Speed at $\text{Ø}_{\text{Shaft}} = 500$ mm until $n = 750\ \text{min}^{-1}$.

Calculation:

$$M_{t,\max} = 0.5 \cdot F_S \cdot n \cdot \mu \cdot D_1 \tag{4.140}$$

with

F_S Contact pressure of the individual screw,
n Number of the screws,
μ Coefficient of static friction material clutch discs,
D_1 Screw circle diameter

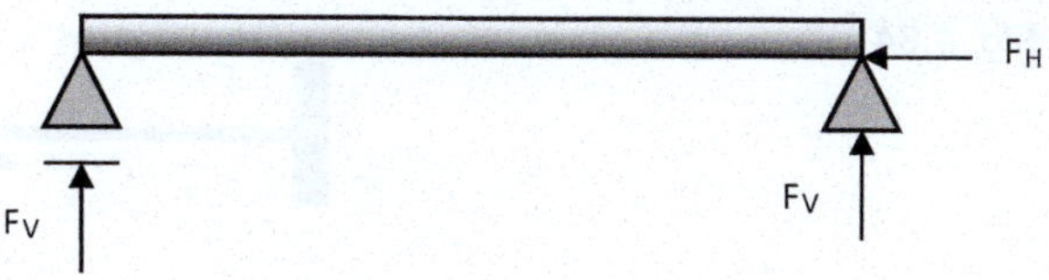

Fig. 4.74 Shaft as "beam on two supports"

(d) Static Aspects for the Selection of the Clutch Type

From a static point of view, the clutch type has to be considered when selecting the clutch[80]: A shaft is statically supported when the sum of the vertical forces, the sum of the horizontal forces, and the sum of the acting moments is zero. A shaft that can be regarded as a continuous beam (Fig. 4.74) is therefore statically determined (externally statically determined) if no more than three unknown support reactions are present.

If a shaft (carrier) is to be statically determined on two bearings (supports), then a fixed (two bearing reactions—in Fig. 4.74, the right bearing—for example, the thrust bearing acting in axial and radial direction at the crankshaft of the engine) and a movable support (one bearing reaction—in Fig. 4.74, the left bearing—for example, the stern tube bearing) must be present, because only then unknown bearing reactions are present in total $2+1=3$.

In contrast, a system is called statically indeterminate if more than three unknown quantities occur. This is often the case with long shaft systems. To prevent inadmissible deflections of the shaft, intermediate bearings must be provided between the stern tube bearing and the end bearing. This leads to static indeterminacy (depending on the number of intermediate bearings, to a corresponding static overdeterminacy: in the case of one intermediate bearings, simply statically overdetermined, and in the case of two intermediate bearings, twice statically overdetermined, etc.). A statically overdetermined system can be more stable in itself than a statically determined system, but it is more complicated to calculate.

The following is important for the selection of the clutches: If *fixed clutch* is used, its installation location must be chosen constructively (aspects of practicality such as ease of installation usually count here). However, *flexible clutches* in statically overdetermined systems must be arranged where the moment line in the load case *dead weight shaft* has a zero passage, as these clutches can hardly absorb radial forces—they act as joints as in a Gerber beam.

With the appropriate number and arrangement of flexible clutches, a shaft with multiple bearings can then become a statically determinate shaft system again, so that no forces resulting from the deflection of the parts are transmitted between the individual parts. The shaft is statically determined if the following rules are followed:

[80] See also [55] for further details.

Fig. 4.84 Shafting

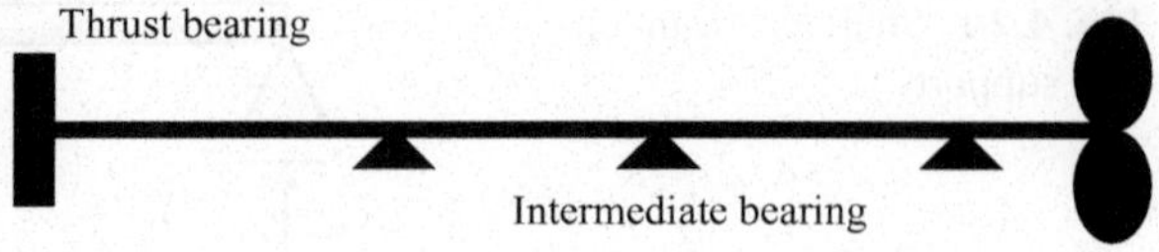

- The number of joints (flexible clutches) is one less than the number of fields.
- A maximum of two joints may be arranged in the center fields, and a maximum of one joint in the end fields.
- Center fields adjacent to center fields with two joints may have a maximum of one joint.
- End fields adjacent to center fields with two joints must not have any joint.

The *static aspect* of the multiple-bearing and subdivided shaft system is examined in more detail below.

Example Task "statics of the shaft system"[81]

A 10-m long shaft is supported on the motor side in a bivalent thrust bearing. In addition, there are two further intermediate bearings between the thrust bearing and the stern tube bearing. The propeller (weight 1.3 t) is mounted on a 0.4-m long shaft end. The entire shaft has a constant diameter and can be considered uniform with a dead weight of 12 kN/m.

How high are the bearing forces? Where are moment zero points located? The static system can be seen in Fig. 4.75.

Solution: Statically speaking, this is a three-span beam **with** cantilever arm of unequal spans. A uniform line load (from the dead weight of the shaft) as well as a single load at the end of the cantilever arm (dead weight of the propeller) can be applied as load.

For the solution of such and similar load cases (e.g., same spans, different line loads due to differences in shaft diameter), the relevant literature on statics[82] contains corresponding solution approaches.

[81] In collaboration with *Pfaff, R.*, B. Eng.

[82] For example [17] or also [31].

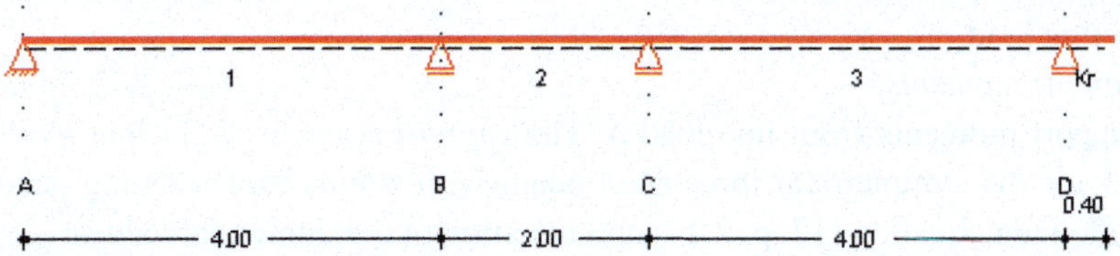

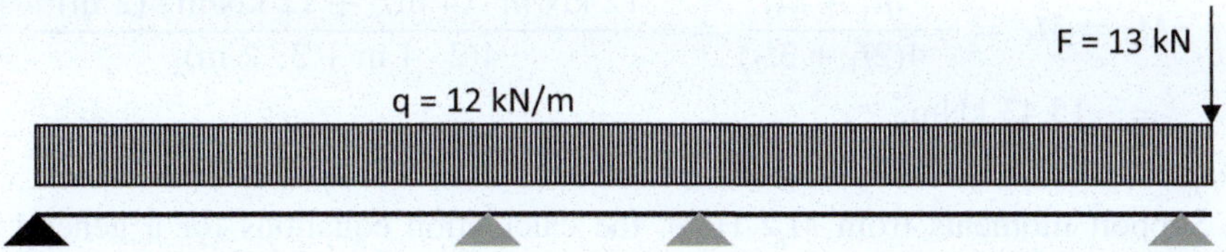

Static system and loads:

System simplification and loads: At first, only the cantilever arm is to be considered. As a result of the specified load, a moment and a vertical force arise at the stern tube bearing.

Moment from line load on cantilever arm:

$$M_{K,q} = -\frac{q \cdot l_K^2}{2} = -\frac{12 \text{ kN/m} \cdot (0.4 \text{ m})^2}{2} = -0.96 \text{ kNm}.$$

Moment from propeller weight on cantilever arm:

$$M_{K,Pr} = -F \cdot l_K = -13 \text{ kN} \cdot 0.4 \text{ m} = -5.20 \text{ kNm}.$$

Total moment M_K:

$$M_K = M_{K,q} + M_{K,Pr} = -6.16 \text{ kNm}.$$

Vertical force from line load on cantilever arm:

$$V_{K,q} = q \cdot l_K = 12 \text{ kN/m} \cdot 0.4 \text{ m} = 4.80 \text{ kN}.$$

Vertical force from propeller weight on cantilever arm:

$$V_{K,Pr} = F = 13 \text{ kN}.$$

Total vertical force V_K:

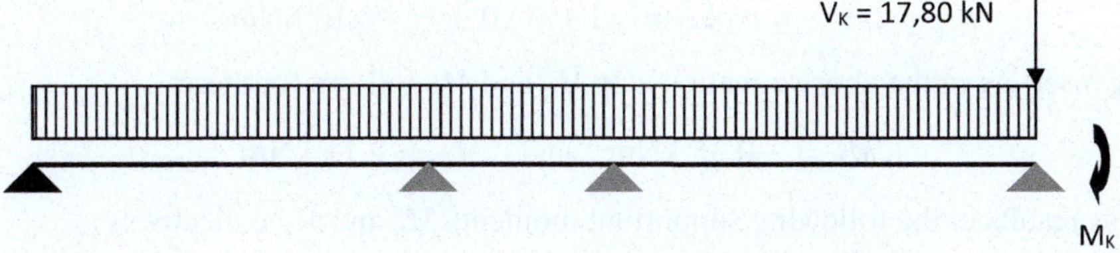

$$V_K = V_{K,q} + V_{K,Pr} = 17.80 \text{ kN}.$$

From a static point of view, it is now a three-span beam **without** cantilever arm with projecting load.

Support moments:

Support moments from line load q: The *support moments from line load* is determined for the symmetrical three-span beam with symmetrical loading according to the following equation [17, p. 2.14] and insertion of the numerical values:

$$M_B = M_C = -\frac{ql_1^3 + ql_2^3}{4(2l_1 + 3l_2)} = -\frac{12 \text{ kN/m} \cdot (4 \text{ m})^3 + 12 \text{ kN/m} \cdot (2 \text{ m})^3}{4(2 \cdot 4 \text{ m} + 3 \cdot 2 \text{ m})}$$

$$= -15.43 \text{ kNm}.$$

Support moments from V_K*:* $M_B = M_C = 0$.

Support moments from M_K: Here, the calculation equations for a general three-span beam with any spans and any load are used; [17, p. 2.13] applies:

$$M_B = \frac{L_3 \cdot l_3 \cdot l_2}{K}$$

and

$$M_C = -\frac{L_3 \cdot 2l_3 \cdot (l_1 + l_2)}{K}$$

with

$$K = 4(l_1 + l_2) \cdot (l_2 + l_3) - l_2^2,$$
$$K = 4(4 \text{ m} + 2 \text{ m}) \cdot (2 \text{ m} + 4 \text{ m}) - (2 \text{ m})^2,$$
$$K = 140\text{m}^2$$

and

$$L = -M_K \cdot (1 - 3 \cdot \beta^2)$$

with

$$\beta = b/l \quad [17, \text{ p. } 2.10].$$

In this equation, l is the length of a considered beam, and b is the distance from the end of the beam to the point where the moment strikes. As in the present case, the moment at the end of the beam is affected, β is therefore zero. Thus, L is calculated:

$$L = -6.16 \text{ kNm} \cdot (1 - 3 \cdot 0^2) = -6.16 \text{ kNm}.$$

By inserting in the above equations for M_B and M_C follows for these:

$$M_B = -0.35 \text{ kNm} \quad \text{and} \quad M_C = 2.11 \text{ kNm}.$$

This results in the following supporting moments M_B and M_C collectively:

$$M_B = -15.43 \text{ kNm} - 0.35 \text{ kNm} = -15.78 \text{ kNm},$$
$$M_C = -15.43 \text{ kNm} + 2.11 \text{ kNm} = -13.32 \text{ kNm}.$$

Calculation of internal forces [17, p. 2.14]:

$$V_a = \frac{q \cdot l_1}{2} + \frac{M_B}{l_1},$$

$$V_a = \frac{12\ \text{kN/m} \cdot 4\ \text{m}}{2} + \frac{-15.78\ \text{kNm}}{4\ \text{m}} = 20.06\ \text{kN},$$

$$V_{bl} = -\frac{q \cdot l_1}{2} + \frac{M_B}{l_1},$$

$$V_{bl} = -\frac{12\ \text{kN/m} \cdot 4\ \text{m}}{2} + \frac{-15.78\ \text{kNm}}{4\text{m}} = -27.95\ \text{kN},$$

$$M_1 = \frac{V_{bl}^2}{2q} + M_B,$$

$$M_1 = \frac{(-27.95\ \text{kN})^2}{2 \cdot 12\ \text{kN/m}} - 15.78\ \text{kNm} = 16.77\ \text{kNm},$$

$$V_{br} = \frac{q \cdot l_2}{2} + \frac{M_c - M_B}{l_2},$$

$$V_{br} = \frac{12\ \text{kN/m} \cdot 2\ \text{m}}{2} + \frac{-13.32\ \text{kNm} + 15.78\ \text{kNm}}{2\ \text{m}} = 13.23\ \text{kN},$$

$$V_{cl} = -\frac{q \cdot l_2}{2} + \frac{M_c - M_B}{l_2},$$

$$V_{cl} = -\frac{12\ \text{kN/m} \cdot 2\text{m}}{2} + \frac{-13.32\ \text{kNm} + 15.78\ \text{kNm}}{2\ \text{m}} = -10.77\ \text{kN},$$

$$M_2 = \frac{V_{br}^2}{2q} + M_B,$$

$$M_2 = \frac{(13.23\ \text{kN})^2}{2 \cdot 12\ \text{kN/m}} - 15.78\ \text{kNm} = -8.49\ \text{kNm},$$

$$V_{cr} = \frac{q \cdot l_3}{2} + \frac{M_K - M_C}{l_3},$$

$$V_{cr} = \frac{12\ \text{kN/m} \cdot 4\ \text{m}}{2} + \frac{-6.16\ \text{kNm} + 13.32\ \text{kNm}}{4\ \text{m}} = 25.79\ \text{kN},$$

$$V_d = -\frac{q \cdot l_3^2}{2} + \frac{M_K - M_C}{l_3},$$

$$V_d = -\frac{12\ \text{kN/m} \cdot 4\ \text{m}}{2} + \frac{-6.16\ \text{kNm} + 13.32\ \text{kNm}}{4\ \text{m}} = -22.21\ \text{kN},$$

$$M_3 = \frac{V_{cr}^2}{2q} + M_C,$$

$$M_3 = \frac{(25.79\ \text{kN})^2}{2 \cdot 12\ \text{kN/m}} - 13.32\ \text{kNm} = 14.39\ \text{kNm}.$$

Vertical forces and field moments:

Moment zero points [17, p. 2.17] (preferred locations for installation of clutches, especially flexible clutches):

Field 1:

$$a_1 = -\frac{2M_B}{q \cdot l_1} = -\frac{2(-15.78\ \text{kNm})}{12\ \text{kN/m} \cdot 4\text{m}} = 0.6575\ \text{m},$$

$$x = l_1 - a_1 = 4\text{m} - 0.6575\ \text{m} = 3.3425\ \text{m}.$$

Field 3:

$$x_{1,2} = \frac{V_{cr}}{q} \pm \sqrt{\frac{2M_3}{q}} = \frac{25.79\ \text{kN}}{12\ \text{kN/m}} \pm \sqrt{\frac{2 \cdot 14.39\ \text{kNm}}{12\ \text{kN/m}}},$$

from this follows $x_1 = 0.60$ m and $x_2 = 3.70$ m.

Representation of the internal forces $\mathbf{M}_i$ *and* $\mathbf{V}_i$:

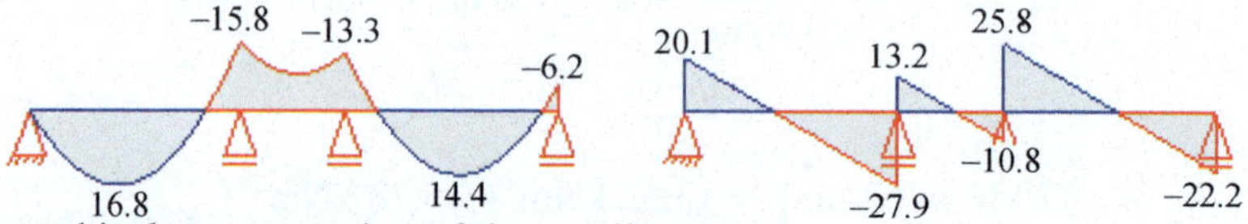

Without graphical representation of the cantilever arm.

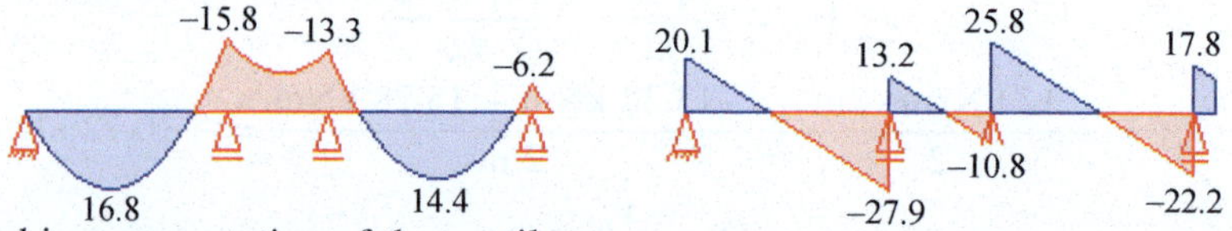

With graphic representation of the cantilever.

From the representation of the internal forces, the *bearing forces* that are decisive for the dimensioning and selection of the plain or rolling bearings cab be determined:

$$\begin{aligned} \text{Thrust bearing A} & & &= 20.1\ \text{kN},\\ \text{Intermediate bearing B} &= 27.9\ \text{kN} + 13.2\ \text{kN} & &= 41.1\ \text{kN},\\ \text{Intermediate bearing C} &= 10.8\ \text{kN} + 25.8\ \text{kN} & &= 36.6\ \text{kN},\\ \text{Stern tube bearing D} &= 22.2\ \text{kN} + 17.8\ \text{kN} & &= 40.0\ \text{kN}. \end{aligned}$$

◀

4.4.4 Stern Tube Seal

The propulsion or propeller shaft is guided out of the hull through the stern tube (see Fig. 4.66). To prevent water from entering the ship through the stern tube, various stern tube sealing systems are used.

There is no absolute tightness in the physical sense. It is necessary to agree on what is meant by "tight" in a specific case (molecules, moisture, drops…). This "tightness" is called technical tightness. If the joint between two parts to be sealed (here shaft and stern tube) is filled with a suitable auxiliary material (seal) and this auxiliary material is pressed so strongly that both its internal "pores" and the micro-gaps between seal and parts to be sealed become so small that the substance to be retained can no longer penetrate, *contacting* seal is achieved. Seals that do not require mechanical contact between the two parts and do not require a solid "intermediate" material are *non-contact* seals (also *gap seals*), but these are not discussed in detail here.

The following systems are used in shipping for stern tube seals:

- Mechanical seal,
- Rotary shaft seal.

The **stuffing box** is found at best still on small boats or in historical vehicles. Therefore, the following only refers to the *mechanical seal* and *rotary shaft seal* systems in more detail.

4.4.4.1 Mechanical Seal

Mechanical seals (GLRD = short form for the German word **Gl**eit**r**ing**d**ichtung) have been able to establish themselves over time as shaft seals in the marine sector, mainly due to their comparatively low leakage and friction.[83]

Main components of the *mechanical seal* system (Fig. 4.76) are two rings (seal ring and a counter ring) sliding on each other, which are pressed together axially by a spring and between which a sealing gap with a gap dimension $< 1\ \mu m$ is formed by hydrostatic and hydrodynamic effects. This sealing gap contains a lubricating film of the substance to be sealed (in this case water or oil), which has a decisive influence on the function of the mechanical seal. The distance between the two surfaces depends on several factors. These include, among others, the nature of the seal faces, the pressure to be sealed, and the seal speed. The lubricating film leads to a reduction in friction, which practically prevents wear of the seal faces [59].

Possible wear of the sliding surfaces is automatically compensated for by the spring pressure; mechanical seals are therefore maintenance-free.

One of the two rings sits rigidly in the stern tube, and the other is fixed to the shaft by means of anti-rotation pins. Sliding materials are usually made of carbon graphite materials, metal, ceramics, plastics, or resin-bonded carbon.

Some mechanical seals are designed in such a way that they can be serviced without dry-docking [51].

While the expected leakage is difficult to calculate due to the many different influences, the frictional power P between the sliding surfaces of standard mechanical seals and thus the moment can be calculated using the relationship

$$P = f \cdot \kappa \cdot \Delta p \cdot v \cdot A \tag{4.141}$$

[83] For further details, see also [12].

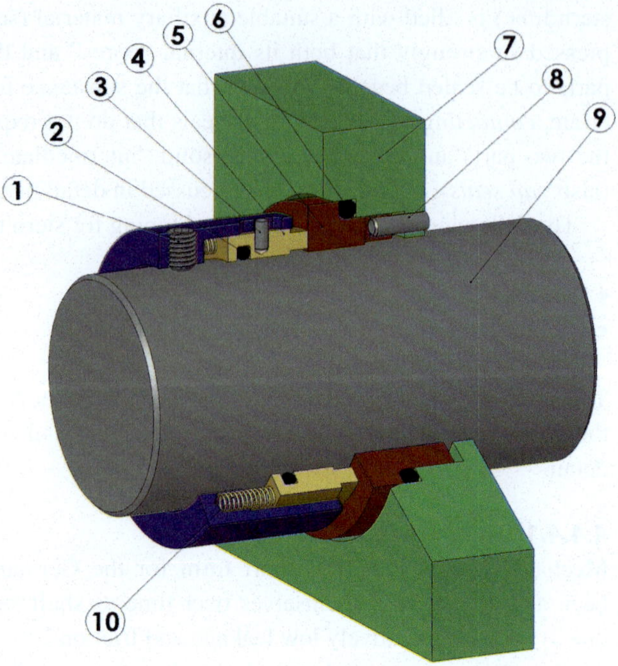

Fig. 4.76 Sectional view of a single-acting mechanical seal, independent of direction of rotation. *1* Grub screw, *2* O-ring (secondary seal), *3* tension pin as anti-rotation lock for the slip ring (*4*), *4* slip ring, *5* counter ring, *6* O-ring (secondary seal), *7* casing wall or stern tube (only indicated), *8* tension pin as anti-rotation lock for the counter ring (*5*), *9* propulsion or propeller shaft, *10* springs. *Source* Silberwolf, CC BY-SA 2.5

and

$$\kappa = k + \frac{p_f}{\Delta p} \tag{4.142}$$

with

f = Friction coefficient,
κ = Stress factor,
Δp = Pressure difference,
v = Sliding speed,
A = Sliding surface,
p_f = Spring compression,
k = Area ratio

(κ, p_f, A, and k are parameters of the mechanical seal) and the diagram (Fig. 4.77).

The above equations illustrate that narrow sliding surfaces (small A) and balanced mechanical seals (area ratio $k < 1$) lead to higher power limits (p–V value). Therefore, for pressures > 1.5 MPa, practically only balanced mechanical seals are used.

4.4.4.2 Radial Shaft Seals (RSS)

Rotary shaft seals [12] are lip seals, where the primary sealing lip is in radial contact with a rotating part. The classic, best-known, and very frequently used element in this

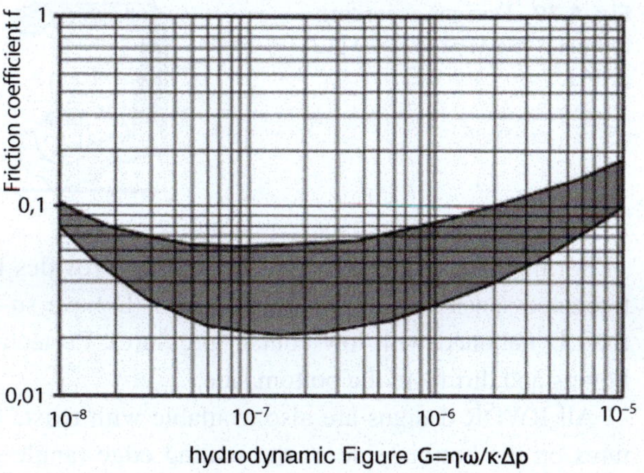

Fig. 4.77 Range of friction coefficients of standard mechanical seals. *Source* [12, p. 22]

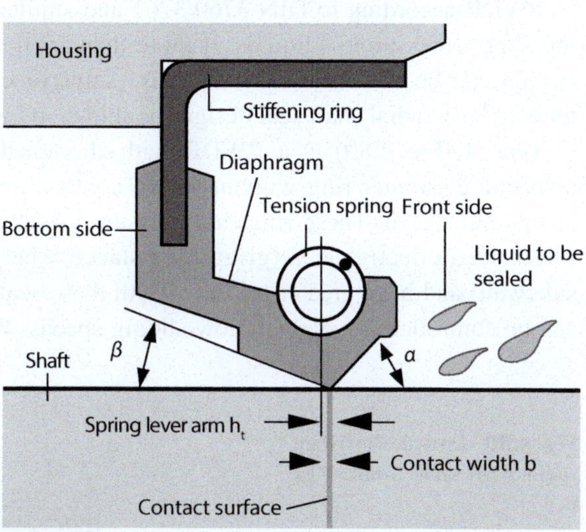

Fig. 4.78 Elastomer radial shaft seal (RWDR) with important designations. *Source* [12, p. 16]

group is the elastomer radial shaft seal RWDR, commonly known as the Simmerring (see Fig. 4.78). With a well-lubricating fluid, optimum flushing, and suitable material, RWDR up to 35 m/s sliding speed can be used. The sealing effect of the RWDR is based on an elastohydrodynamic return flow effect. This is based on the two different sealing edge angles α and β and a characteristic contact surface wear. The liquid to be sealed must always be on the side with the larger sealing edge angle α. If liquid is supplied to the side with the smaller sealing edge angle β, it is pumped to the side with the larger sealing edge angle. Radial shaft seals are therefore active sealing elements and can therefore seal without leakage.

Figure 4.79 shows some designs of RWDR according to DIN.

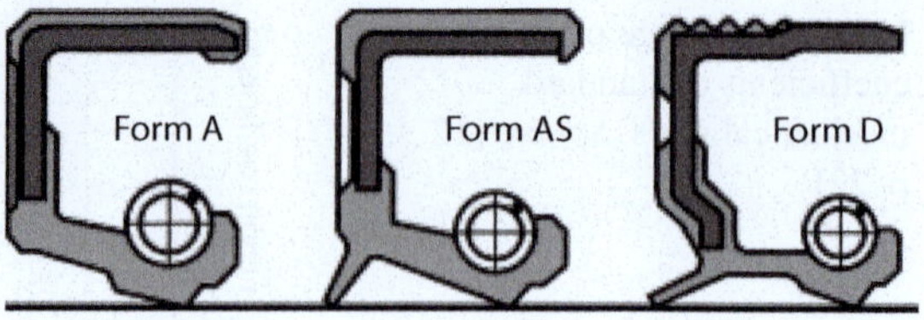

Fig. 4.79 Designs according to DIN 3760/3761. *Source* [12, p. 16]

Form A with rubberized stiffening ring provides better secondary sealing and lower tolerance sensitivity. Form AS and form D have so-called protective lips (AS contact-free, D flattened with low contact pressure). Protective lips reduce the accumulation of liquids and dirt from the bottom side.

All RWDR designs are also available with twist. Twist are macroscopic return structures on the bottom side of the sealing edge (angle β). They considerably improve the sealing capacity of RWDR.

RWDR according to DIN 3760/3761 and similar elements are only suitable for sealing almost pressureless liquids. If more than approximately 0.02 MPa have to be sealed (which can be the case, e.g., on cruise ships or container ships with drafts of sometimes > 7 m), radial shaft seal designs as shown in Fig. 4.80 are used.

Type a (Fig. 4.80) is a RWDR with shortened and reinforced diaphragm, type b a special elastomer ring without spring contact, and type c a sleeve seal with a PTFE compound sleeve. These rings are pressure-loadable because they have a "small" pressure-loaded effective area (given by distance x between sealing edge and low-pressure side wall) and have greater stability. With these sealing elements, pressures up to 1 MPa can be controlled at relatively low sliding speeds. With sealing systems such as d and e

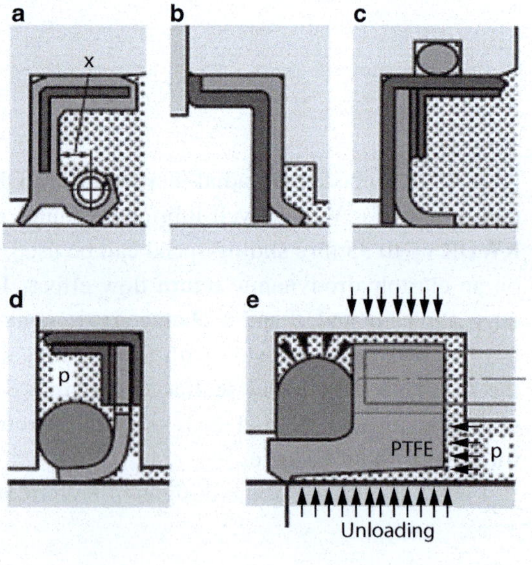

Fig. 4.80 Pressure-resistant radial shaft seals. *Source* [12, p. 17]

with sealing rings made of PTFE, due to their special design (small value *x*), pressures up to 3 MPa at sliding speeds up to 12 m/s (simultaneously) can be sealed.

PTFE sleeve seals are always used when the temperature stability or chemical resistance of elastomer materials is no longer sufficient or when poorly lubricating fluids have to be sealed (including seawater).

The Shaft Running Surface [77]

The *shaft surface* on which the sealing edge of the shaft sealing ring runs is decisive for the tightness of the system. It must be so smooth that excessive wear due to abrasion on the sealing lip is avoided. Usually, surface roughness of $Ra = 0.2–0.8\ \mu m$ ($Rz = 1–5\ \mu m$, $Rmax < 6\ \mu m$) is recommended. Such surface qualities are achieved, for example, by twist-free grinding in the recess (see DIN 3760/3761).

With regard to the *hardness of the shaft surface*, at least 55 hardness Rockwell C (HRC)[84] is recommended.

Frictional Power Loss

The friction of the sealing lip on the shaft causes a frictional power loss, which can be calculated according to the following equation:

$$P = f \cdot p_m \cdot \pi \cdot b \cdot d \cdot v \cdot n \tag{4.143}$$

with

f = Friction coefficient,
p_m = Mean surface pressure,
b = Contact width,
d = Seal diameter,
v = Circumferential speed,
n = Number of the lip seals

The coefficient of friction *f* is dependent on the "Gümbel number" *G*. Figure 4.81 shows the friction coefficient range of RWDR as a function of *G*. Herewith the friction power *P* of an RWDR can be estimated.

For a first estimation, a mean surface pressure $p_m = 1\ N/mm^2$ at a contact width of $b = 0.15$ mm can be assumed.

Example

Shaft system with three lip seals; $d = 350$ mm, $n = 150\ min^{-1}$, $f = 0.4$. How large are the friction torque and friction power?

Solution: From Eq. 4.143, it follows by inserting the given values:

[84] Hardness Rockwell Cone, a hardness testing method using a conical test piece to measure the penetration depth.

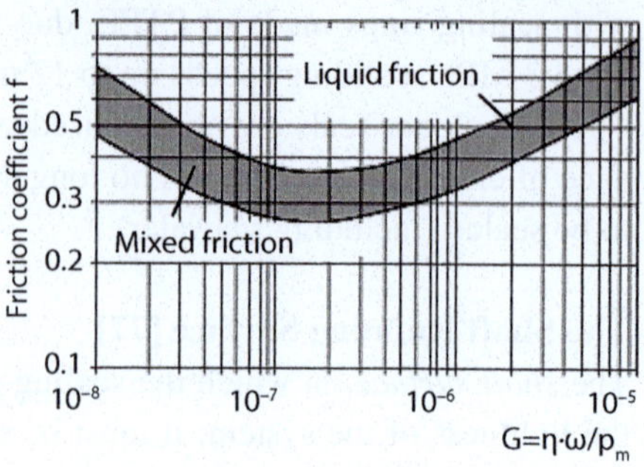

Fig. 4.81 Friction coefficient range of RWDR with ω = angular speed and η = dynamic viscosity in the sealing gap. *Source* [12, p. 18]

$$P = 0.4 \cdot 1\ \text{N/mm}^2 \cdot \pi \cdot 0.15\ \text{mm} \cdot 350\ \text{mm} \cdot 2{,}749\ \text{mm/s} \cdot 3,$$
$$P = 544\ \text{W},$$

where

$$v = d \cdot \pi \cdot n = 350\ \text{mm} \cdot \pi \cdot 150/60\ \text{s}^{-1} = 2749\ \text{mm/s}.$$

◀

Widespread are lip seal systems in which the stern tube is sealed both outward toward the water and inward toward the engine room (Fig. 4.90). The interstitial space is filled with oil via a container located above the waterline. This achieves two things: first, the sealing system is lubricated with oil, and second, a falling oil level in the elevated tank indicates a leak in the sealing system. By placing the elevated tank above the waterline, the pressure in the space between the inner and outer seal is minimally higher than the water pressure from outside. In the event of a leak, oil would thus leak to the outside.

Well-known representatives of this type of installation are stern tube seals according to the SIMPLEX systems [64]. Figure 4.83 shows the automatic monitoring unit of a SIMPLEX-Compact stern tube seal.

The SC2, also called Simplex-Compact 2000, is one of the latest developments in oil-lubricated stern tube seals; it is currently the preferred technology in ocean-going vessels. The main features of SC2 are [65]:

- Circumferential speed of the bushings up to 10 m/s,
- Unrestricted draft of the ships,
- Tolerance of axial and radial displacements,
- Operating hours up to 6000 h/year,
- Classification periods of up to 10 years.

The SC2 seal can also be supplied in a completely split version to allow replacement of the entire seal including the bushing without pulling the shaft. The housing of the rear seal is also axially split. The split sealing components allow easy and fast service and are therefore ideal for saving time and costs in dry dock.

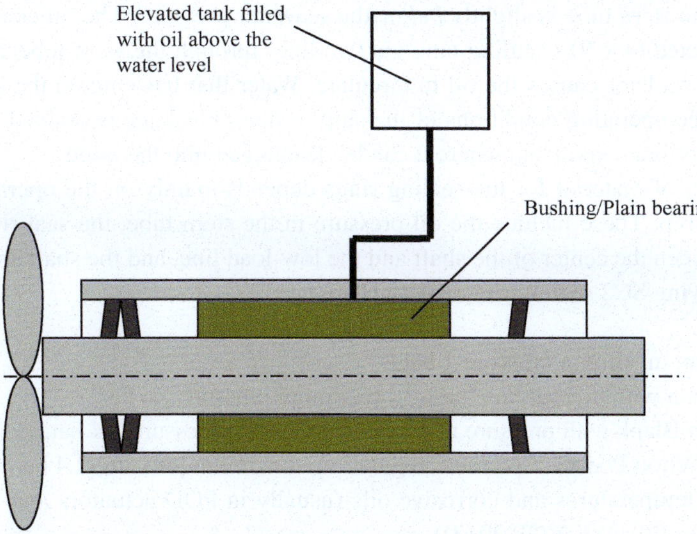

Fig. 4.90 Principle of inner and outer stern tube sealing with leakage monitoring by oil filling

Fig. 4.83 Monitoring unit of the SIMPLEX-Compact stern tube seal. *Photo* AIDA

The SC2 seals are available in various basic designs and can be adapted to meet special requirements.

A modification of the principle of leak monitoring by means of elevated tanks (where oil would leak into the water in case of a leakage of the outer sealing) is the "leakproof" system by SIMPLEX. The stern tube high tank is arranged in such a way that the oil pressure in the stern tube is slightly *below* the seawater pressure. The oil chamber of the seal is connected to a 50-*l* settling tank via two lines through the stern tube. As the shaft rotates, the circulator causes the oil to circulate. Water that has entered the oil chamber due to extreme operating conditions of the ship or due to leakage is drained through the pipes to the settling tank, from where it can be discharged into the bilge.

The choice of material for the sealing rings depends mainly on the operating conditions of the ship. These include the oil pressure in the stern tube, the seal ring size, the distance between the center of the shaft and the low-load line, and the shaft speed.

The following SC2 sealing ring materials are used:

- **P** Perbunan, maximum pressure 1 bar,
- **P+** Perbunan with pressure regulation, maximum pressure 1.6 bar,
- **VB+** Viton Blank with pressure regulation, maximum pressure 2.4 bar,
- **VB-Pod+** Viton Pod with pressure regulation, maximum pressure 2.4 bar, for systems with high temperatures and corrosive oils (usually in POD actuators and when using special bio-oils (EAL-VGP 2013)),
- **VS+** Viton Superlip with pressure regulation, pressure more than 2.4 bar.

In addition, water-lubricated mechanical seals have become established, which have the advantage over the abovementioned oil chamber seal that in case of leakage, no oil will escapes into the water.

Installation of Radial Shaft Seals

To ensure that the sealing rings function properly during operation, expertise, suitable tools, and cleanliness are of particular importance during installation. Shaft and stern tube must be clean and free of chips, dust, etc.

To ensure easy installation of the sealing ring and initial lubrication when the propeller shaft starts up, it is recommended to grease the sealing ring according to the manufacturer's instructions before installation. While the outer diameter of lip seals with metal housing can be easily lubricated for ease of installation, the outer diameter of elastomer jacketed seals should always be lubricated.

For sealing rings with a grinding protective lip, the space between the sealing and protective lip can be filled with a lubricating grease to reduce the frictional torque. This does not apply to sealing rings made of silicone rubber or with hydrodynamic sealing aid, with the exception of SKF-WAVE sealing lip designs.

To press the sealing rings into the housing bore, the use of a hydraulic pressing-in device and suitable pressing-in tools is recommended. It must always be ensured that the installation force is applied as close as possible and evenly distributed on the outer surface of the sealing ring.

Tools should be used when pressing seals against a shoulder or snap ring. When installing sealing rings on stepped shafts whose shoulder cannot be provided with the recommended chamfer or rounding, a mounting sleeve must be used. If the sealing ring has to be mounted via spring grooves, threads or a serration, thin-walled mounting sleeves ensure that the sealing edge is not damaged. The outer surface of the mounting sleeves should be coated with the same lubricant as the sealing ring and the mating surface. Radial shaft seals made of silicone rubber must always be installed with mounting sleeve [66].

Appendix 16 Shows a Matrix for Clarifying the Cause of Faults in the SIMPLEX-Compact Stern Tube Sealing System for Ship Operation.

References

Print media

1. Becker, T.: Fregatte Klasse F 125, Neue Wege für die nächste Fregattengeneration der Marine. MarineForum 11/2005. Deutsches Maritimes Institut (DMI), Mittler-Report-Verlag, Bonn (2005)
2. Binsau, A.: Schaltbare Kupplungen. Hochschule Magdeburg-Stendal, Stendal (2004)
3. Böge, A., Böge, W.: Handbuch Maschinenbau. Springer Vieweg, Wiesbaden (2017)
4. Budig, P.-K.: Stromrichtergespeiste Synchronmaschine – Theorie und Anwendungen. VDE-Verlag, Berlin (2003)
5. Deutsche Verlagswerke Strauss: Deutsche Schiffs-Dieselmotoren. Vetter & Co, Berlin (1935)
6. Dubbel: DUBBEL Band I und II. Springer, Berlin (1974)
7. FANAL: Schaltungspraxis. 7. Aufl., Metzenauer & Jung GmbH, Wuppertal
8. Fischer, R.: Elektrische Maschinen, 14th edn. Hanser, München (2007)
9. Fister, W.: Fluidenergiemaschinen, vol. 1. Springer, Berlin, Heidelberg, New York (1984)
10. Fister, W.: Vorlesungsskript „Grundzüge der Fluidenergiemaschinen – Teil II". Ruhruniversität Bochum, Bochum (1979)
11. Giersch, H.-U.: Elektrische Maschinen, 5th edn. Teubner, Wiesbaden (2003)
12. Haas, W.: Grundlehrgang Dichtungstechnik, Institut für Maschinenelemente, Universität Stuttgart. http://www.ime.uni-stuttgart.de. Accessed 28 Dec 2017
13. Hager, M.: Die Nase macht den Unterschied. In: BOOTE EXCLUSIV 3/06. Delius Klasing, Bielefeld (2006)
14. Heine, F., Lose, F.: Die großen Passagierschiffe der Welt. Koehlers Verlagsgesellschaft, Hamburg (2010)
15. Heisel, U., Klocke, F., Uhlmann, E., Spur, G.: Handbuch Spanen. Hanser, München (2014)
16. Hofbauer, B.-G.: TRUPPENDIENST-Handbuch ‚Moderne Seemacht' Bd. 2. Bundesminister für Landesverteidigung und Sport, Wien (2016)
17. Holschemacher, K.: Entwurfs- und Berechnungstafeln für Bauingenieure, 5th edn. Beuth, Berlin (2012)
18. Illies, K.: Handbuch der Schiffsbetriebstechnik. Vieweg & Sohn, Braunschweig (1972)
19. Kornev, N.: Propellertheorie. Universität Rostock, Fakultät für Maschinenbau und Schiffstechnik, Rostock (2009)
20. Lexikographisches Institut: Lexikon der Technik Bd. 1. 2. Lexikographisches Institut, München (1986)
21. Magg, J.: Die Steuerungen der Verbrennungskraftmaschinen. Springer, Berlin (1914)

22. Marinekommando Rostock: Jahresbericht (2015)
23. Meier-Peter, H., Bernhardt, F. (eds.): Handbuch Schiffsbetriebstechnik. Seehafen Verlag, Hamburg (2008)
24. Micknass, W.: Kupplungen, Getriebe, Antriebswellen. Vogel Business Media, Würzburg (2004)
25. Moeller, F., Vaske, P. (eds.): Elektrische Maschinen und Umformer, Teil 1 Aufbau, Wirkungsweise und Betriebsverhalten, 11th edn. Teubner, Stuttgart (1970)
26. Peeken, H.: Elastische Kupplungen. Springer, Heidelberg (1986)
27. Pfaff, M., Pfaff, R.: Strömungsmechanische Grundlagen zum Widerstand von Schiffen. Zeitschrift Hansa **148**(6), 38 (2011)
28. Pinnekamp, W.: Zur Auslegung von Antriebselementen. Eine Analyse des Betriebsfaktors (Stoßbeiwert) *k*. Konstruktion **18**(2), 1966 (1966)
29. Roseburg, D.: Elektrische Maschinen und Antriebe. Fachbuchverlag Leipzig, Hanser (1999)
30. Schmidt, E.: Technische Thermodynamik, vol. 1. Springer, Berlin (1975)
31. Schneider: Bautabellen für Ingenieure. Werner Verlag, Bundesanzeiger Verlag, Köln (2016)
32. Verband für Schiffbau und Meerestechnik e. V.: Schiffstechnik und Schiffbautechnologie. Seehafen Verlag, Hamburg (2006)
33. Witthöft, H.-J.: „Schifffahrt“. In: Köhlers Flottenkalender. Koehler, Hamburg (2010)
34. Witthöft, H.-J.: Erinnerungsdaten – Köhlers Flottenkalender. Koehler, Hamburg (2014)

Internet

35. Dambeck, H.: Haihaut soll Schiffe schützen, Spiegel Online. http://www.spiegel.de (2010). Accessed 18 July 2005
36. Herzog, K.: https://www.thm.de/me/images/user/herzog-91/Kolbenmaschinen/Kolbenmaschinen%5Cbreak_3_KenngrosenKennfelder.pdf. Accessed 8 Oct 2016
37. Köchert, C.: Untersuchung einer Bugwulstmodifikation. http://www.ssi.tu-harburg.de/doc/webseiten_dokumente/ssi/veroeffentlichungen/sdari02.pdf (2002). Accessed 21 Aug 2016
38. Müller, R.: „Eine Formel in vielen Kontexten: Fermi-Probleme zum Strömungswiderstand“, Universität München. http://www.tu-braunschweig.de/Medien-DB/ifdn-physik/leistung.pdf. Accessed 21 Aug 2016
39. Vorhölter, H.: Vereinfachte Bestimmung des effektiven Nachstroms in Manövrierzuständen, Institut für Entwerfen von Schiffen und Schiffssicherheit, Technische Universität Hamburg-Harburg. http://www.ssi.tu-harburg.de (2017). Accessed 26 July 2017
40. http://institute.unileoben.ac.at/etechnik/Ger/Lehre_G/Skripten_G/Files_G/El_Antriebst/T_Pulswechselrichter.pdf. Accessed 9 Oct 2016
41. Hamburg Harburg, T.U.: Grundlagen der Propulsion. http://www.ssi.tu-harburg.de/doc/webseiten_dokumente/ssi/vorlesungsunterlagen/propulsion.pdf (2004). Accessed 21 Aug 2016
42. http://elektronik-kurs.net/elektrotechnik/drehzahlregelung-und-drehrichtungssteuerung. Accessed 9 Oct 2016
43. http://hansamare.de/newsMedia/archiv/onboard/58.html. Accessed 11 Nov 2016
44. http://link.springer.com/chapter/https://doi.org/10.1007/978-3-8348-2330-4_5. Accessed 9 Oct 2016
45. http://s-boot.net/sboote-bmarine-klasse-143.html. Accessed 17 Nov 2016
46. http://ueba.elkonet.de/static/ueba/kundenauftraege/elkonet.de/home/ausbilder_bc/bc/Kernauftrag_02/fachliche%20grundlagen/fachtheorie/frequenzumrichter.aspx.htm. Accessed 9 Oct 2016

47. http://www.bbs-buxtehude.de/sats/starnet/studentpages/119/Lagermontage.pdf. Accessed 4 Nov 2016
48. http://www.cruiseferry.de/kreuzfahrer.html. Accessed 12 Nov 2016
49. http://www.cruisetricks.de/marinediesel-schweroel-co-schiffstreibstoffe-im-ueberblick/. Accessed 24 Nov 2016
50. http://www.elektrotechnik-fachwissen.de/wechselstrom/leistung-wechselstrom.php
51. http://www.fluiten.de/PDF2008/CAT025_ENG_L.pdf. Accessed 4 Feb 2017
52. http://www.fregatte-klasse122.de/technische-daten/antrieb/. Accessed 13 Nov 2016
53. http://www.geaviation.com/marine/engines/military/lm2500plus/. Accessed 20 Nov 2016
54. http://www.geschlossener-fonds.net/betriebskosten.html. Accessed 21 Aug 2016
55. http://www.goepf.bettschen.org/Statik12.pdf. Accessed 25 July 2017
56. http://www.kfztech.de/kfztechnik/motor/diesel/commonrail.htm. Accessed 17 Nov 2016
57. http://www.kfztech.de/kfztechnik/motor/steuerung/fuellung.htm. Accessed 13 Nov 2016
58. http://www.konstruktionsatlas.de/antriebstechnik/kupplungen-starre.shtml. Accessed 4 Nov 2016
59. https://www.ksb.com/ksb-de/Produktbauarten/Gleitringdichtung/. Accessed 4 Feb 2017
60. http://www.marine.de/portal/a/marine/!ut/p/c4/04_SB8K8xLLM9MSSzPy8xBz9CP3I5EyrpHK93MQivfLEtLTUvNI8vdKk_PySVD0jQ6NEvZLU5Iy8zGz9gmxHRQBPgSDG/. Accessed 31 Oct 2016
61. http://www.mtu-report.com/pdf/1003/2010_03_MTU_Report_Seefest.pdf. Accessed 1 Nov 2016
62. http://www.saiya.de/maschbau/odm_Otto-Diesel-1.3.5.pdf. Accessed 8 Oct 2016
63. http://www.schrittmotoren.de/fachwissen/motoren/f_beitr_00_311.htm. Accessed 8 Oct 2016
64. http://www.sealsystems.de/fileadmin/mediaFiles/Kundentag/2014/Vortraege/05_SKF_BVI_Anwenderbericht.pdf. Accessed 4 Feb 2017
65. http://www.simplexturbulo.com/de/schiffsindustrie/komponenten-antriebswellen/simplex-sterntube-seals-sc2/. Accessed 4 Feb 2017
66. http://www.skf.com/de/products/seals/industrial-seals/power-transmission-seals/radial-shaft-seals/seal-installation-general-industrial-applications/index.html. Accessed 4 Jan 2017
67. https://www.springerprofessional.de/energietechnik/elektrotechnik/der-ideale-elektromotor-fuer-die-elektromobilitaet/6592306. Accessed 20 Nov 2016
68. http://www.ssi.tu-harburg.de/doc/webseiten_dokumente/ssi/vorlesungsunterlagen/propeller.pdf. Accessed 4 Nov 2016
69. http://www.thm.de/me/images/user/herzog-91/Kolbenmaschinen/Kolbenmaschinen_3_KenngrosenKennfelder.pdf. Accessed 8 Oct 2016
70. https://www.tib.eu/de/suchen/id/tema%3ATEMA20061106135/32-40-PGI-Neuer-Otto-Gas-motor-ohne-Z%C3%BCndkerzen/. Accessed 19 July 2017
71. https://www.umweltbundesamt.de/antifouling-mittel. Accessed 21 Aug 2016
72. http://www.vem-group.com/fileadmin/content/pdf/Download/Brosch%C3%BCren/Flyer_VEM-Sachsenwerk/schiffbau_de.pdf. Accessed 20 Nov 2016
73. http://www.wer-weiss-was.de/t/start-eines-schiffsdiesels/2886892. Accessed 17 Nov 2016
74. http://de.wikipedia.org/wiki/Common-Rail-Einspritzung. Accessed 17 Nov 2016
75. http://de.wikipedia.org/wiki/Dieselkraftstoff. Accessed 8 Oct 2016
76. http://de.wikipedia.org/wiki/Pumpe-D%C3%BCse-System. Accessed 28 Dec 2017
77. http://www.fachwissen-dichtungstechnik.de. Accessed 28 Dec 2017
78. http://www.ketten-fuchs.de. Accessed 28 Dec 2017
79. http://www.orion-corp.com. Accessed 28 Dec 2017

80. http://www.schaeffler.com. Accessed 28 Dec 2017
81. http://www.vesconite.com. Accessed 28 Dec 2017

Further reading

82. http://www.ssi.tu-harburg.de. Accessed 28 Dec 2017
83. Illies, K. (ed.): Handbuch der Schiffsbetriebstechnik – Teil 1. Vieweg, Braunschweig, Wiesbaden (1984)
84. Schnabel, P.: http://www.elektronik-kompendium.de. Accessed 28 Dec 2017

5 Ship Operating Systems/Auxiliary Systems

The image of an engine room in a ship is determined not only by the drive motor(s), the drive shaft(s), and if applicable, the gears, but also by piping, valves, pumps, and other aggregates. The main maneuvering device is the steering gear. In addition to this, there are other facilities and installations for operating the ship, some of which are specific to the type of ship and its range of tasks.

The main auxiliary systems and operating equipments are described in the following sections.

5.1 Steering Gear

The rudder enables a ship to maintain its course or change direction. The steering effect on the ship is achieved by the rudder laying in the driving and propeller current (see Fig. 5.1). Annexure 17 contains a selection of constructive rudder designs.

The steering forces acting on an occasional rudder result from the distribution of pressure around the rudder blade, which is now predominantly drop-shaped. The rudder angle is generally designated by the symbol δ (Fig. 5.2). The optimum rudder angle at which the rudder achieves its maximum rudder force is—depending on the rudder blade shape—about 35° to 38°.

5.1.1 Size of the Rudder Area

The rudder surface area A is roughly dependent on the lateral area of the ship (simplified: Ship length LWL multiplied by its draft T; [64]).[1]

[1]The construction regulations of the classification societies may provide for different equations.

M. Pfaff, *Ship Operation Technology*, https://doi.org/10.1007/978-3-658-32729-3_5

Fig. 5.1 Rudder of a freighter, placed directly behind the propeller. *Photo* Brosen, CC BY 2.5

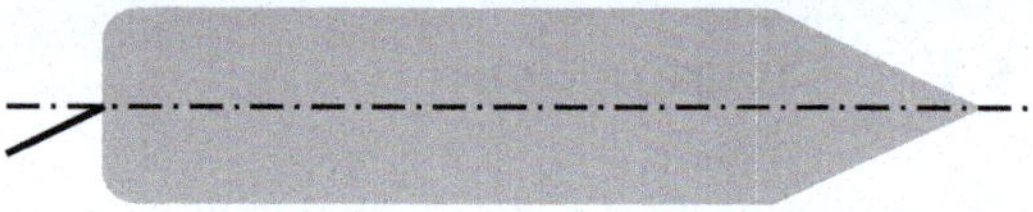

Fig. 5.2 Paddle-ship

The following applies to complete ships

$$A = \frac{1}{30} \quad \text{to} \quad \frac{1}{40} LWL \cdot T, \tag{5.1}$$

for sharp ship sections

$$A = \frac{1}{50} LWL \cdot T. \tag{5.2}$$

5.1.2 Calculation of the Rudder Force and Rudder Moment

The oblique flow of the laid rudder creates a force perpendicular to the rudder surface, known as the **rudder force** F_R.[2]

[2]More details [22].

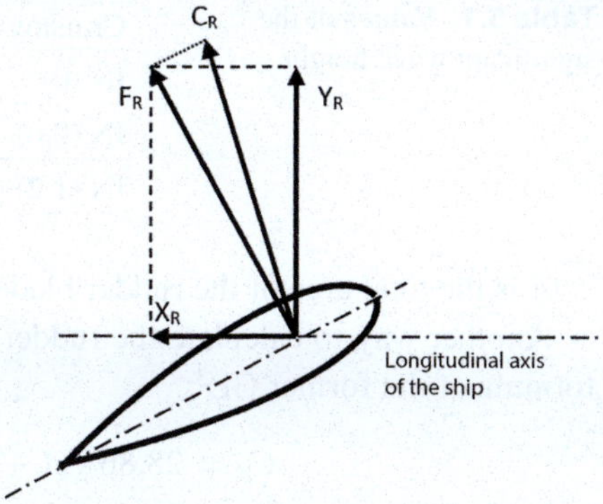

Fig. 5.3 Forces at the rudder blade

The rudder effect is determined by the lateral rudder force Y_R (perpendicular to the ship's longitudinal direction) and a force in the ship's longitudinal direction X_R (see Fig. 5.3). The rudder force F_R approximately matches to the lateral rudder force C_R (the buoyancy force at the profiled rudder blade) (for plate rudders, $F_R = C_R$ can therefore be set).

For the calculation of the rudder force, also rudder pressure, there are various approaches in the literature, which have generally been obtained empirically. The construction regulations of the classification societies sometimes contain more, different formulae for this purpose, leading to some considerable deviations—depending on which approach is followed.

The Joessel equation is widely used [10, p. 248].

However, this equation must be modified, as the *Joessel equation* still determines the rudder force in kg; therefore the result has to be multiplied by the acceleration of gravity g:

$$F_R = \frac{20 \cdot A \cdot v^2 \cdot \sin \delta}{0.2 + 0.3 \cdot \sin \delta} \cdot 9.81 \frac{\text{kg m}}{\text{s}^2} \text{ [N]}. \tag{5.3}$$

The rudder force F_R (or C_R for the profiled rudder) acting on the rudder blade at its center of gravity is therefore dependent on the rudder area A (m^2), the rudder angle δ and the ship speed v (m/s).

Another approach is based on the theory of resistance of bodies in a fluid. Here, both the density of the water (sea or fresh water; see table in Appendix 2) and a constant K, *which depends on the rudder blade design, are used*:

$$F_R = \frac{\rho}{2} \cdot v^2 \cdot \sin 2\delta \cdot A \cdot K \text{ [N]}. \tag{5.4}$$

The constant K is determined as follows:

$$K = 2\pi \cdot \frac{\Lambda \cdot (\Lambda + 0.7)}{(\Lambda + 1.7)^2} \cdot \sin \delta, \tag{5.5}$$

where the aspect ratio of the rudder is $\Lambda = h^2/A$ [22, p. 4].

Table 5.1 Values of the significant wave height

Cruising range	Wave height H (m)
IN (0)	0
IN (0.6)	0.6
IN (1.2) to IN (2.0)	1.2–2.0

A is the total area of the rudder blade, h the rudder height.

Another way to calculate the rudder force C_R (N) for **inland vessels** is according to a formula of the former GL[3]:

$$C_R = 28.86 \cdot (1 + n)^{0.15} \cdot A \cdot v^2 \cdot r_1 \cdot r_2 \cdot r_3 \text{ [N]} \tag{5.6}$$

with

N = navigation coefficient = $0.85 \cdot H$ (H is the significant wave height—see Table 5.1),
V = the ship speed (km/h),
r_1 = form factor, like follows:

$$r_1 = \frac{\Lambda + 2}{3}, \tag{5.7}$$

Λ = factor (not greater than 2), as follows:

$$\Lambda = \frac{h^2}{A}. \tag{5.8}$$

h = average height the rudder surface (m),
A = rudder surface (m^2),
r_2 = coefficient, to be determined from Table 5.2,
r_3 = coefficient, as follows:
= 0.8 for rudder outside of propeller jet (central rudder at twin-screw vessels or similar cases),
= 1.15 for rudder behind a fixed propeller nozzle,
= 1.0 in others cases

A comparison of the two approaches shows that in Eq. 5.3, the rudder angle as a determining variable has an influence on the rudder force, Eqs. 5.4 and 5.6 on the other hand, essentially consider the profile of the rudder.

For **seagoing vessels,** GL Construction Regulation I-1-1 provides for a further calculation equation:

$$C_R = 132 \cdot A \cdot v^2 \cdot \kappa_1 \cdot \kappa_2 \cdot \kappa_3 \text{ [N]} \tag{5.9}$$

[3]See Construction Regulation GL I 2 Ship Technology—inland navigation vessels Chap. 7, Nr. 3.

Table 5.2 Factors r_2 for various rudder profiles

Ruder profile types	r_2 for headway	r_2 for reversing
NACA 00—Göttinger profile 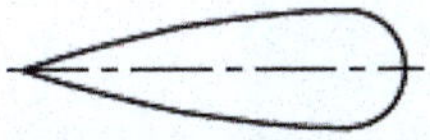	1.10	0.80
Hollow profile 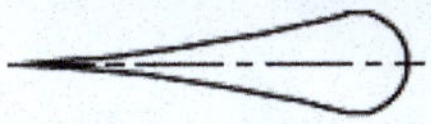	1.35	0.90
Profile without curvature 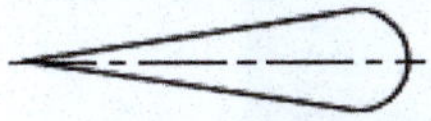	1.10	0.90
High performance rudder 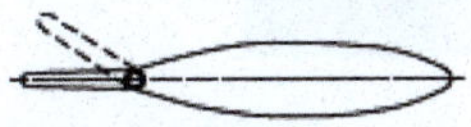	1.70	1.30
Fish tail profile 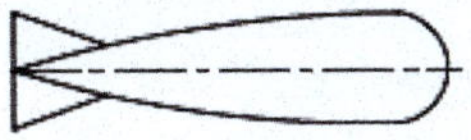	1.40	0.80
Single-plate rudder 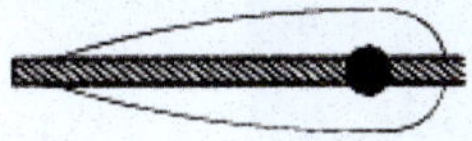	1.00	1.00

with

A = total rudder area (m^2),
v = ship speed (kn),
κ_1 = factor, dependent on the aspect ratio Λ of rudder blade,
κ_2 = rudder profile factor (see Table 5.2),
κ_3 = rudder arrangement factor (see above values for r_3)

Whereas in the past the forces to be applied from the steering wheel to the rudder blade were obtained by chain or cable pulls (today only in small traditional vehicles or in the sports boat sector; see Fig. 5.4), today hydraulic rowing machines are used (Fig. 5.5).

The amount to be generated by the steering gear **rudder moment** M_R results from the lever arm r and the rudder force F_R or rather C_R:

$$M_R = F_R \cdot r \text{ [Nm]}. \qquad (5.10)$$

The lever arm r results from the distance between the center of gravity (pressure point) of the rudder and the axis of rotation (center of rudder shaft) with

Fig. 5.4 Transmission of the rudder rotation to the rudder by means of a chain hoist

Fig. 5.5 Hydraulic steering machine

$$r = b \cdot \left(\alpha - \frac{A_F}{A}\right). \tag{5.11}$$

(Note: The lever arm r is not less than $0.1b$ for headway).

Fig. 5.6 Area proportions of rudder blade

b = mean rudder width (m),
$\alpha = 0.33$ for headway,
$\alpha = 0.66$ for reverse,
A_F = area (m^2) of rudder blade before the centerline of rudder ship (see red area in Fig. 5.6),
A = total area of rudder blade (see red and blue area in Fig. 5.6)

5.1.3 Kort Nozzle

A **kort nozzle** is a tapered ring, profiled like an airfoil, which surrounds the propeller. By using a kort nozzle, the flow conditions at the ends of the propeller blades are optimized, resulting in a higher mass flow. This leads to an increase in efficiency. Both the higher mass flow and the quieter wake ensure an optimum flow to the rudder.

Kort nozzles can be pivotally mounted and thus used directly as rudders (Fig. 5.7). In sports shipping and other small vessels, kort nozzles are often used as a protective device for the propeller against damage from flotsam and/or grounding.

Calculation example rudder

A single-screw motor vessel is specified with a length L_{wl} from 107 m and a draft $T = 5.45$ m; maximum speed $v = 17$ kn (= 31.5 km/h = 8.75 m/s). The rudder can be assumed to be a hollow (see Table 5.2). The effective rudder angle δ is 35°. How large should the rudder area be?

Fig. 5.7 Rotatable kort nozzle

Solution: The ship can be regarded as sharply cut. According to Eq. 5.2, this results in an area of

$$A = \frac{1}{50} L_{wl} \cdot T = A = \frac{1}{50} 107 \text{ m} \cdot 5.45 \text{ m},$$
$$A = 11.6 \text{ m}^2.$$

How great is the rudder force (to be determined according to GL equation for seagoing vessels) in advance?

From Eq. 5.9, Table 5.2 and the above comments on r_3 follows:

$$C_R = 132 \cdot A \cdot v^2 \cdot \kappa_1 \cdot \kappa_2 \cdot \kappa_3 = 132 \cdot 11.6 \text{ m}^2 \cdot 17^2 \text{ kn}^2 \cdot \kappa_1 \cdot 1.35 \cdot 1.$$

The factor κ_1 depends on the aspect ratio Λ of the rudder blade; here assumed as follows (see Eqs. 5.7 and 5.8 for inland vessels)

$$r_1 = \kappa_1 = \frac{\Lambda + 2}{3} = \frac{2 + 2}{3} = 1.3.$$

This results in a rudder force $C_R = 777$ kN.

A calculation according to the modified Joessel Eq. (5.3) gives the following result:

Fig. 5.8 Bilge keel on the model of the battleship Bismarck

$$
\begin{aligned}
F_R =& \frac{20 \cdot A \cdot v^2 \cdot \sin\delta}{0.2 + 0.3 \cdot \sin\delta} \cdot 9.81 \frac{\text{kg m}}{\text{s}^2},\\
=& \frac{20 \cdot 11.6\ \text{m}^2 \cdot 8.75^2\ \text{m}^2/\text{s}^2 \cdot \sin 35°}{0.2 + 0.3 \cdot \sin 35°} \cdot 9.81 \frac{\text{kg m}}{\text{s}^2},\\
F_R =& 269\ \text{kN}.
\end{aligned}
$$

These two calculations illustrate the different results, depending on which approach is followed. ◀

5.2 Stabilization Systems

5.2.1 Introduction

On the one hand, the rolling motions of a ship[4] make it difficult for the ship to continue its course, and on the other hand they can cause seasickness in passengers and crew. Therefore, eespecially on passenger ships, stabilization systems are therefore used to provide greater comfort for passengers and crew. However, such systems are also used on ferries and cargo ships, where slippage of the cargo (so-called spillage) could cause major damage or even capsize the ship. In the case of warships, it is also important that

[4]Rotations around the ship's longitudinal axis.

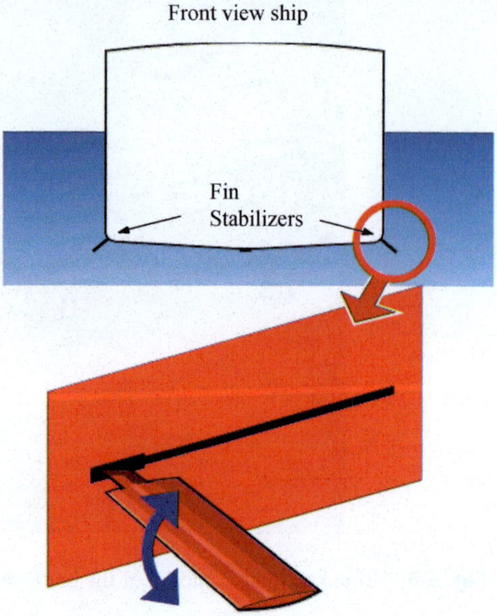

Fig. 5.9 Active principle of the fin stabilizers (Fin Stabilizer). *Photo* Tosaka, CC BY-SA 3.0

stabilizers improve safety when planes or helicopters land and that the less the ship rolls at the moment of launch, the greater the accuracy of the weapon systems.

5.2.2 Bilge Keels

Bilge keels are flat steel profiles permanently welded to both sides of a ship to dampen the rolling movements of the ship around its longitudinal axis (Fig. 5.8). Usually a Holland profile[5] welded perpendicularly to the hull is used as a bilge keel. In the position of the bilge, that is, the transition of the ship's floor, the bilge keels are fitted in the side walls. They usually run only in the area of the parallel central nave, that is. on the greatest width of the hull.

5.2.3 Fin Stabilizers

These are movable streamlined fins, which can be retracted into the hull on both sides, using the principle of lift around a hydrofoil (see Fig. 5.9).

The angle of attack of the fins is adjusted electrohydraulically to the rolling motion of the ship. The fin position is optimized on the basis of the speed, heel angle, angular velocity, and angular acceleration of the ship. The parameters change with the size,

[5]Profile according to EN 10,067.

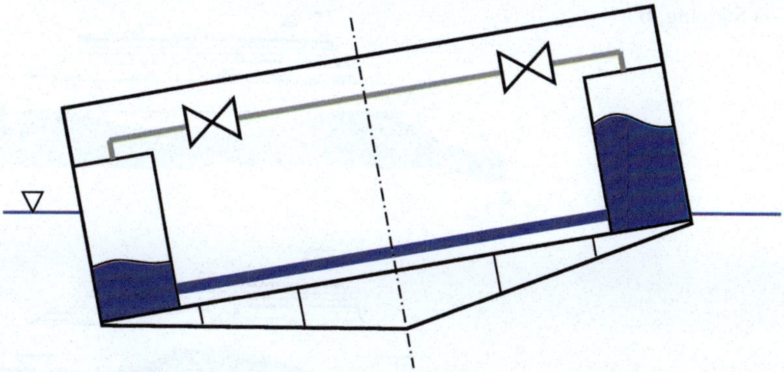

Fig. 5.10 Principle sketch of a roll damping tank

weight, and payload of the ship. This system is very effective, but has the disadvantage that the buoyancy effect of the fins decreases with decreasing ship speed.

5.2.4 Roll Damping Tanks

Rolling is counteracted by an adjustable exchange of water via tanks connected to pipes on the starboard and port side (Fig. 5.10). These tanks are connected at the bottom with an overflow channel and at the top with an air suspension pipe. Due to the rolling movement, the water always flows from one tank to the other. The flow rate of the water in the connecting pipes of the tanks is optimally adjusted depending on the angle of heel, the angular velocity and the angular acceleration. This ensures that the tank whose side is moving straight upward due to rolling always contains a larger mass of water than the other tank (see Fig. 5.10). This counteracts the rolling movement. The optimum control of the water flow rate is achieved by computerized regulation of the overflow valves in the air suspension pipe.

Features of the roll damping tank:

- Less effective than fin stabilizers,
- Also works when the ship is not speeding through the water,
- High space requirement within the ship,
- Also compensates for heeling during curving and by loading.

5.3 Heel Compensation and Ballast Water Systems

When sailing empty or with very little cargo, the stability of the ship's structure, especially in heavy weather, can be jeopardized by high compressive and tensile stresses due to states of *sagging* and *hogging*.

Fig. 5.11 **a** Sagging, **b** hogging

Sagging (Fig. 5.11a) can be translated as "durchsacken". It describes the bending load acting on the structure of a ship when the wavelength is approximately equal to that of the ship. The bow and stern are both on the wave crests at the same time. As a result, the buoyancy force of the water increases locally there, while in the middle of the ship, in the trough, the buoyancy force decreases significantly. This results in bending forces, which lead to high tensile forces in the keel area and compression forces in the deck area (main chord). *Hogging* (Fig. 5.11b) describes the condition in which the stern and bow are in the wave trough and the center of the ship is on the wave crest.

To avoid these conditions **ballast water** is recorded. The ballast tanks are filled and drained via sea valves, which are conveniently located above the sea chests (openings in the ship's side below the waterline). The ballast water pump is located below the waterline and therefore does not need to be self-priming.

Ballast water tanks are mainly located in the double bottom of the ship, but may also be arranged laterally (see Figs. 5.10 and 5.12). By selective filling and draining of the bottom and side high tanks (similar to roll damping tanks—see Sect. 5.2.4), longitudinal trim and heel balancing (cross-trim), required by appropriate cargo conditions, may also be carried out.

For the design of pumps and piping, please refer to Sect. 5.4

5.4 Pumps, Pipelines, and Fittings

5.4.1 Pumps

On board of ships there are a large number of pumps to handle a wide range of liquids: Cooling water, oil in the lubricant circuit, service water, firefighting water, ballast water etc. Different pump systems are used depending on the application and task.

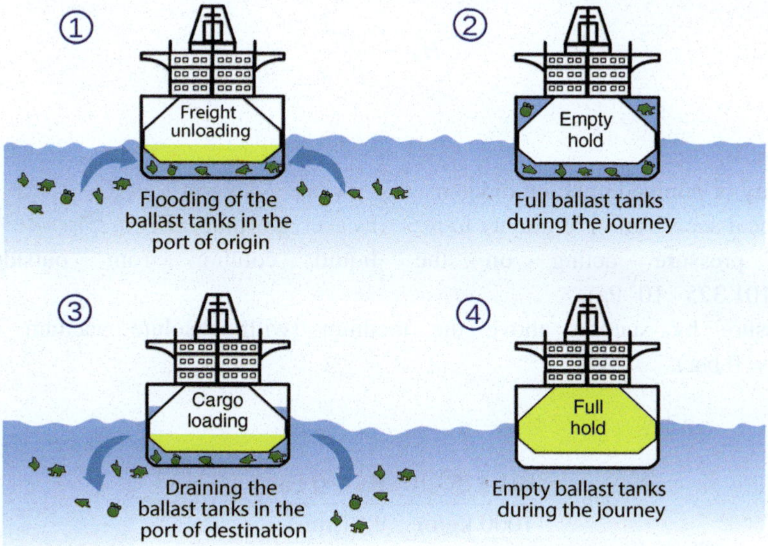

Fig. 5.12 Principle of the ballast water system. *Photo* Mxxl 2, CC BY-SA 3.0

5.4.1.1 General Information About Pumps

The task of pumps is to pump liquids from a lower to a higher level or from a room with lower pressure to a room with higher pressure. In shipping, the low level is given by the liquid level of the water being navigated, a room in the ship (e.g., the bilge) or a container from which the liquid is taken. The higher level is determined by the liquid level of an elevated tank (e.g., side ballast water tanks, cargo tanks) or by the pressure line (e.g., firefighting line) into which the pump delivers.

Pumps are divided into two main groups according to their operating principle: *Positive displacement pumps* and *flow pumps*. These groups are further subdivided according to design; centrifugal, gear, and piston pumps are mainly used in shipbuilding.

5.4.1.2 Theoretical Foundations

The **theoretical maximum (suction lift)** H_{th} (this is the theoretical maximum suction lift) of a pump is about 10 m for water and the suction tank open to the free atmosphere. This results from the following relationship:

The hydrostatic basic equation is changed according to the height h:

$$p_1 = p_0 + \rho \cdot g \cdot h, \tag{5.12}$$

$$h = \frac{p_1 - p_0}{\rho \cdot g}, \tag{5.13}$$

with $h = H_{th}$ follows

$$H_{th} = \frac{\Delta p}{\rho \cdot g} \tag{5.14}$$

with

ρ = density of pumped medium in kg/m³ (e.g., water = 1 kg/m³),
g = the local acceleration of gravity in m/s² (as average value 9.81 m/s²),
p_1 = air pressure acting on the liquid column from outside (e.g., 1 bar = 101.325 · 10³ Pa),
p_0 = pressure by suction above the medium (with absolute vacuum in pump chamber = 0 bar),

so:

$$H_{th} = \frac{101.325 \cdot 10^3 \text{ Pa} - 0 \text{ Pa}}{1000 \text{ kg/m}^3 \cdot 9.81 \text{ m/s}^2} = 10.33 \text{ m}.$$

The theoretical suction lift is therefore dependent on the locally prevailing conditions of air pressure and local acceleration of gravity as well as on the density of the pumped medium; real suction lift for water is approximately 7 m.

Example

A bilge pump is to pump water from the bilge. The density of the bilge water can be checked with 1021 kg/m³ the prevailing air pressure is 1023 mbar. The ship is just at the equator—there the acceleration of gravity[6] is only 9.78 m/s². What is the theoretical suction lift H_{th} ?

Solution: Inserting the given numerical values in Eq. 5.12 results in

$$H_{th} = \frac{1.023 \cdot 10^5 \text{ kg} \cdot \text{m}^3 \cdot \text{s}^2}{\text{m} \cdot \text{s}^2 \cdot 1021 \text{ kg} \cdot 9.78 \text{ m}} = 10.2 \text{ m}.$$

The theoretical suction lift would therefore be 10.2 m. ◀

The **effective head** H_N is the difference in height between the level of the suction and discharge liquid level. The effective head is made up of the suction head H_S and the pressure head H_D together (Fig. 5.13).

$$H_N = H_S + H_D. \tag{5.15}$$

The suction lift at *piston pumps* is the vertical distance between the liquid level in the suction tank or the free liquid surface (e.g., the surrounding seawater) and the sealing

[6]See also [49].

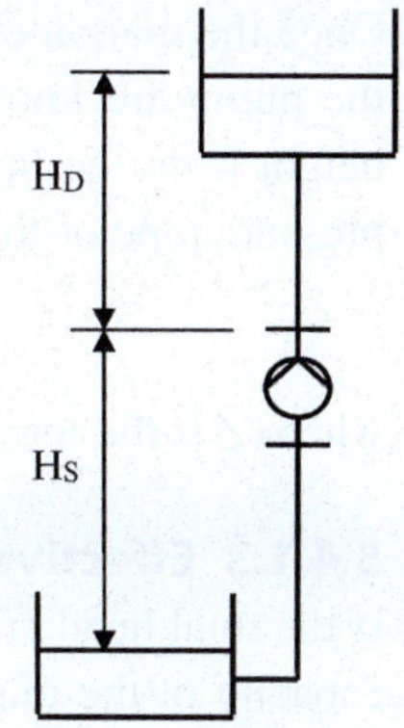

Fig. 5.13 Suction lift and pressure height

surface of the pressure valve. The pressure height is the vertical distance from the pressure gate valve to the liquid level of the pressure vessel, the elevated tank or the free outlet of the pressure line.

At *centrifugal pumps* the suction lift is the vertical distance between the liquid level in the suction tank or the free liquid surface (e.g., the surrounding seawater) and the center of the pump shaft. The pressure head is the vertical distance from the center of the shaft to the liquid level of the pressure vessel, the elevated tank or the free outlet of the pressure line.

Resistances in the pipelines must be overcome, which are noticeable as pressure losses (see also Sect. 5.4.2). They are dependent on the

- Type of pipe,
- Pipe length,
- Pipe diameter or the flow rate of the fluid,
- Number and type of valves and fittings,
- Number of pipe bends,
- Type of strainer.

The sum of the resistances occurring in the pipelines is called **head loss** H_V is called:

$$H_V = H_{VS} + H_{VD} \tag{5.16}$$

with

H_{VS} Resistance in the suction line in m,
H_{VD} Resistance in the pressure line in m

The **manometric delivery height** H_{man} is the sum of the effective head H_N and the resistances of the suction and pressure lines, that is, the head loss H_V:

$$H_{\text{man}} = H_N + H_V. \tag{5.17}$$

Once the manometric delivery height has been calculated and the internal resistances of the pump are known, the total delivery height, that is, the maximum possible delivery height from the liquid surface of the medium to be sucked in to the highest point of the pressure pipe or the liquid level in the elevated tank, can be determined:

$$H = H_{\text{man}} + Z, \tag{5.18}$$

where Z is the internal resistances of the pump expressed as pressure losses in m.

5.4.1.3 Effective Power of the Pump and Power of the Pump Engine

If the total head H, the flow rate Q and the density ρ of the liquid is known, the effective capacity of the pump P_W which is the pump shaft power, can be calculated according to the following equation:

$$P_W = (Q \cdot \Delta p)/\eta = (Q \cdot m \cdot g \cdot H)/\eta \tag{5.19}$$

with

P_W Effective capacity of the pump shaft (W),
Q Delivery volume (m^3/s),
m Mass flow of pumped medium (kg/s),
Δp Total pressure loss (Pa),
G Acceleration due to gravity (9.81 m/s^2),
H Delivery height the pump,
η Efficiency the pump

The required power of the pump motor is then calculated as follows:

$$P_M = P_W/\eta_M \tag{5.20}$$

with η_M the efficiency of the motor.

5.4.1.4 Positive Displacement Pumps

Positive displacement pumps are pumps in which a reciprocating (oscillating) piston (displacer) sucks in and pushes the liquid. For this purpose, the medium is conveyed through self-contained volumes, prevention of backflow is achieved by non-return valves or flaps. Except for the leakages due to design, the fluid cannot flow through the pump against the direction of flow even when it is at a standstill. Positive displacement pumps are generally self-priming, whereby the maximum suction lift (geodetic suction lift) is limited by the attainable vacuum, the local air pressure, the density of the medium and the flow resistance to be overcome. Positive displacement pumps should not be shut off on the discharge side, unless, for example, overpressure or bypass valves prevent an excessively high pressure from building up in the system. On board, they are used, for example, as bilge and feed water pumps.

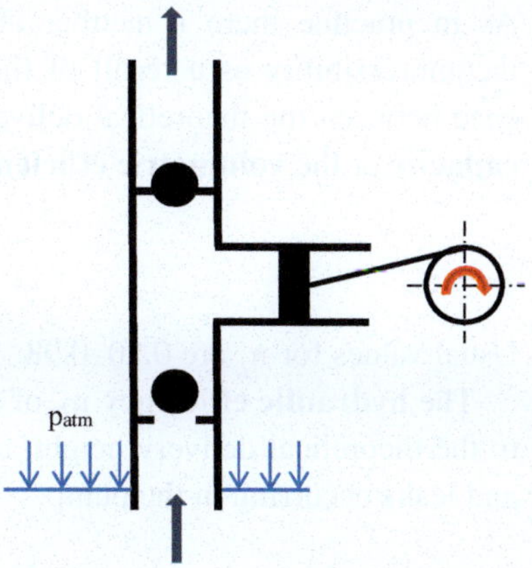

Fig. 5.14 Principle piston pump

Well known representative of this type of pump is the *(stroke) piston pump.*[7]

Its suction effect is based on the driving force of atmospheric air pressure, which forces the water into the pump chamber (Fig. 5.14). Despite different designs, all piston pumps follow the same operating principle:

First stroke: Suction. The piston movement increases the volume of the cylinder space, creating a vacuum in the system. The inlet valve opens and the atmospheric pressure allows the medium to flow in (principle of establishing thermodynamic equilibrium).

Second stroke: The piston moves into the cylindrical chamber and reduces its volume. The suction valve closes, the pressure gate valve opens and the fluid is forced out.

The **theoretical delivery volume** Q_{th} with absolute impermeability, pressure loss-free delivery, and theoretical incompressibility is calculated according to the following numerical value equation:

$$Q_{th} = \frac{d^2 \cdot \pi}{4} \cdot s \cdot \frac{n}{60} \cdot z \ [\mathrm{m^3/s}] \tag{5.21}$$

with

d = cylinder diameter in m,
s = piston stroke length in m,
n = speed of eccentric drive in min^{-1},
z = number of cylinders

[7]For more details on this and on the different types of piston pump, see also [19, p. 213] and Bouché "Kolbenpumpen" in [6, Volume II, p. 247].

As in practice there is neither absolute impermeability (seals, valves) nor complete incompressibility as a result of the gases dissolved in the liquids, the resulting difference between the theoretical delivery volume and the **actual delivery volume** Q by the **capacity** or the **volumetric efficiency** η_L or η_V taken into account:

$$\eta_L = \frac{Q}{Q_{th}}. \tag{5.22}$$

Usual values for η_L are 0.90–0.98.

The **hydraulic efficiency** η_H of a piston pump is the ratio of the actual delivery height to the theoretical delivery height, and takes into account pressure losses, friction losses, and leaks occurring in the pump:

$$\eta_H = \frac{H}{H_{th}}. \tag{5.23}$$

Usual values for η_H are between 0.85–0.98.

The **mechanical efficiency** η_M is the ratio of the indicated pump capacity P_i to the drive power P_M:

$$\eta_M = \frac{P_i}{P_M}. \tag{5.24}$$

The mechanical losses are friction losses in the moving pump parts (piston sealing, bearing friction, etc.). Usual values for η_M are 0.88–0.95.

The **indicated pump capacity** P_i is determined by the pump manufacturer and results from the *indicated pump pressure* $p_{i,}$ which is determined by measurement.

In order to avoid acceleration losses caused by the acceleration of the liquid columns in the suction and pressure line, air receivers are often installed in piston pumps. The pressure and volume fluctuations caused by the oscillating piston are also reduced. The air receiver consists of a reservoir with an air cushion.

Other types of positive displacement pumps are:

- Twin piston pump,
- Diaphragm pump,
- Axial piston pump,
- Radial piston pump,
- Peristaltic pump,
- Gear pump,
- Screw pump,
- Rotary piston pump.

The most frequent representative of the rotating positive displacement pumps is the **gear pump** (Fig. 5.15).

Fig. 5.15 Open gear pump

In these pumps, two meshing gear wheels rotate, with the cogs displacing the pumped medium. Sealing from the pressure to the suction chamber is achieved inside the pump by combing the cogs, in the outer space between the upper edge of the cogs and the housing. Valves and air vessels are not required. They are mainly used in the lubricant circuit.

The theoretical delivery volume $\dot{Q}_{th}$ is calculated according to the following equation:

$$\dot{Q}_{th} = \frac{\pi}{2} \cdot \left(d_K^2 - c^2 - d_g^2 \cdot \frac{\pi^2}{3 \cdot Z^3} \right) \cdot b \cdot n \tag{5.25}$$

with

d_K = outside diameter in m,
C = center distance in m,
d_g = base diameter in m,
Z = number of cogs,
b = width of gear wheel in m,
n = speed in 1/s

The actual volume flow Q is smaller than the theoretical volume flow, as some part of the pumped medium flows back into the suction chamber via the meshing cogs and the circumference of the wheels, and a small part escapes to the outside via the bearings and sealing system. These losses are defined by *efficiency* or perhaps *volumetric efficiency* η_L:

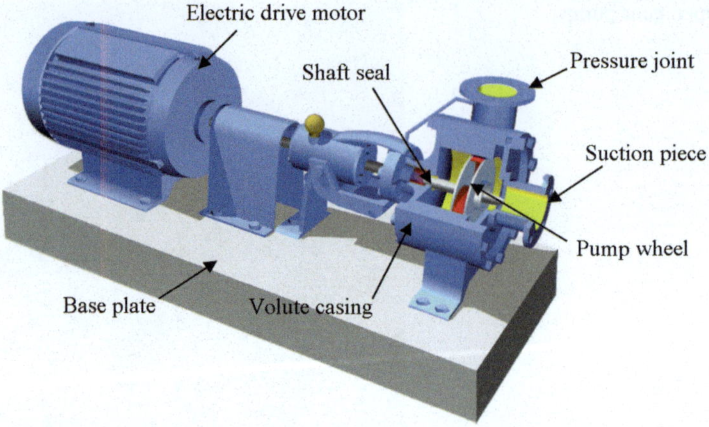

Fig. 5.16 Principle centrifugal pump with drive motor. *Photo* Mokery J., CC BY-SA 3.0

$$\eta_L = \frac{\dot{Q}}{\dot{Q}_{th}}. \tag{5.26}$$

5.4.1.5 Flow Pumps

The best known representative of flow pumps and frequently encountered on board ships is the *centrifugal pump* (Fig. 5.16).

It uses the *centrifugal force* to pump the liquid. For this reason it is called as **centrifugal pump**. The medium to be pumped enters the centrifugal pump (volute casing) via the suction line, it is then collected by the rotating pump wheel and carried outward on a spiral track. The resulting decreasing radial velocity of the liquid toward the outside leads to an outwardly increasing pressure inside the pump, which transports the liquid into the pressure pipe arranged tangentially on the volute casing. If the impeller and casing are designed accordingly, liquids mixed with solids (e.g., waste water) can also be pumped. A measure of the permissible solid size is the so-called ball passage, indicated as the maximum diameter of the ball that could pass through the pump.

Standard centrifugal pumps, unlike positive displacement pumps, are not self-priming. They must therefore be filled with the fluid before start-up. For this purpose, they are installed in the inflow of the medium to be pumped. If this is not possible, the pump and the suction line must be filled before starting, for example, by means of a piston pump. However, a special design of centrifugal pump, the side channel pump (Fig. 5.17), is self-priming; it is not necessary to fill the pump before starting.

However, centrifugal pumps[8] have the advantage over piston pumps that they deliver continuously and the flow rate can be controlled continuously at a constant drive speed (by throttling) or by changing the speed.

[8]More in-depth [2].

Fig. 5.17 Side channel pump with NPSH pre-stage. *Photo* Chris1207/SERO, CC BY-SA 3.0

Despite different designs of the pump casing, arrangement of suction and discharge connections, number and shape of impellers and guiding devices in the pump, they all follow a basic design principle. It generally comprises two blade systems, one of which is connected to the casing (guiding devices) and rests, while the other is connected to the drive shaft and rotates (impeller).

Figure 5.18 shows a single-stage centrifugal pump; single-stage because there is only one impeller to which the fluid flows. A single-stage pump can achieve average delivery heights of approximately 150 m wc [9, p. 8], which corresponds to a pressure of approximately 15 bar.

After the maximum pressure has been reached, one also speaks of

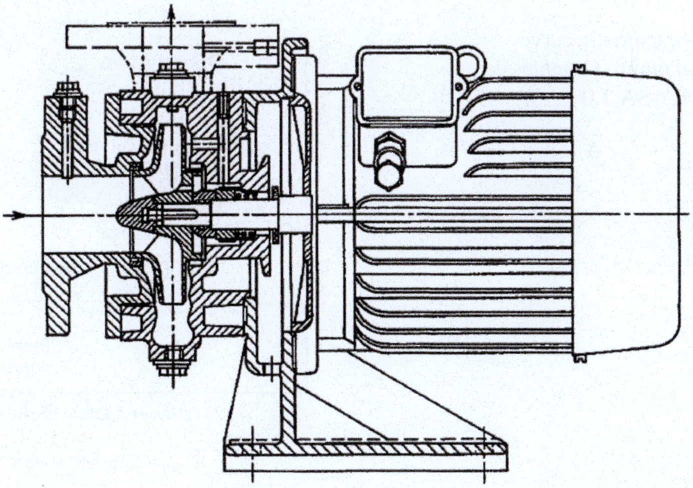

Fig. 5.18 Sectional view of a single-stage centrifugal pump with electric motor. *Source* [2]

- Low-pressure pumps: up to about 1 bar,
- Medium-pressure pumps: up to about 6 bar,
- High-pressure pumps: >6 bar.

The **delivery process of a centrifugal pump** whose casing and suction line must be filled with the pumped medium, as it is not self-priming, is carried out as follows: If the impeller is turned, the centrifugal force causes the fluid in the impeller to be pumped outward and passes through the guiding device into the pressure line. In the center of the impeller vacuum is created compared to the pressure on the surface of the liquid level of the liquid to be sucked in, through which further liquid flows in via the suction line of the pump.

If the required delivery height or required pressure is so high that a single-stage pump is not sufficient, several impellers are connected in series and such pumps are called multistage centrifugal pumps.

The *characteristic curve* of a centrifugal pump describes the relationship between pressure increase or delivery height and flow rate (see Fig. 5.19). The highest pressure in a centrifugal pump is theoretically generated at zero flow rate. This is the case when the pump delivers against a closed gate valve. Combined with the characteristic curve of the connected pipe network, the *operating point* as the intersection of the pump and pipe network characteristic curve. If several centrifugal pumps are connected in series, the delivery pressure is added, if they are connected in parallel, the achievable flow rate is added. Changes in the speed of the pumps change both the flow rate and the pressure and thus the power consumption.

Laws of dependence of flow rate, delivery height, and drive power of a pump in relation to its speed:

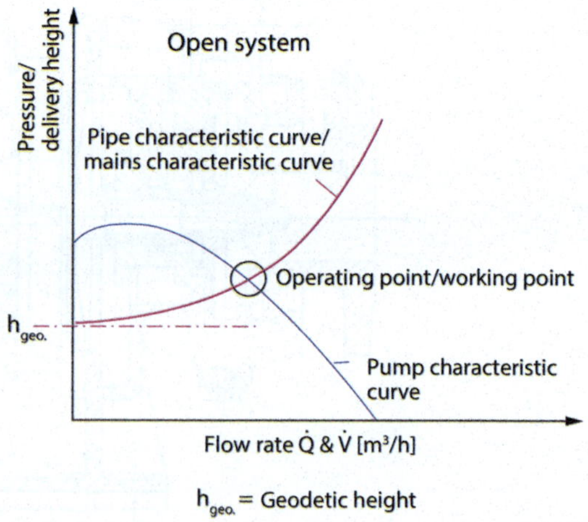

Fig. 5.19 Characteristic curve of a centrifugal pump. *Graphic* Konwiki, CC BY-SA 3.0

- $Q \sim n$,
- $h \sim n^2$,
- $P \sim n^3$.

The delivery height in the *operating point* of a pump results from the point of intersection of the characteristic curve of the pump with the characteristic curve of the pipeline, which is a combination of the static (geodetic) height difference (H_{geo}) and the pure flow losses H_V.

The following parameters characterize the centrifugal pump:

Flow rate	Q	(m^3/h)
Delivery height	H	(m)
Coupling performance	P	(W)
Efficiency	η	
Net positive suction head NPSH at inlet		(m)
Speed	n	(min^{-1})

Furthermore, the so-called NPSH value is a characteristic parameter of a centrifugal pump. According to DIN EN ISO 12723, the term NPSH (m) is the abbreviation for the English term **N**et-**P**ositive-**S**uction-**H**ead (in German: "nettopositive Saughöhe" or also "Gesamthaltedruckhöhe"; [45]).

The NPSH is linked to the term *cavitation*. It is one of the most important operating parameters of a pump, along with delivery height, flow rate, and power requirement. A distinction is made between the NPSH of the system ($NPSH_A$ or $NPSH_{before}$) and the NPSH of the pump ($NPSH_P$ or $NPSH_{erf}$). By simply comparing $NPSH_{before}$ with $NPSH_{erf}$ it is possible to assess whether or not the operational safety of a selected pump is ensured for the system concerned. The following must apply for cavitation-free operation,

$$NPSH_{exist} > NPSH_{required}. \tag{5.27}$$

This requirement must be fulfilled for the entire permissible flow range of a pump system; that is, if $NPSH_{before}$ is greater than the value for $NPSH_{erf}$ by a safety margin—usually 0.5 m.

5.4.1.6 Maintenance and Servicing of Centrifugal Pumps

Generally, the pumped medium should not fall below certain flow rates in pipes and in the pump itself to protect against deposits [2]. This applies in particular to centrifugal pumps, which transfer fluids containing solids:

- Water with normal pollution: 1.0 m/s,
- Water with sand (sand particles < 0.1 mm): 1.5 m/s,
- Water with sand (sand particles < 0.6 mm): 2.5 m/s.

Furthermore, the following must be observed when maintaining and operating pumps:

Suction line and centrifugal pump must be filled before commissioning. As a rule, centrifugal pumps are therefore installed—as far as possible—below the level of the liquid to be pumped. When filling, the gate valves located at the highest point of each pump stage must be opened and kept open until liquid flows out. In the case of self-priming centrifugal pumps and piston pumps, filling is only necessary during initial start-up or after a longer standstill.

When starting up a centrifugal pump, the pressure gate valve must be closed. Only when the operating pressure has been reached (pressure reading!), the gate valve can be opened.

Ensure sufficient lubrication of the bearings (check oil level indicator!).Centrifugal pumps with so-called sealing liquid in the shaft seal must be filled with sealing liquid before commissioning. The separate sealing liquid circuit must be drained and checked for leaks. The correct direction of rotation of the centrifugal pump (usually indicated by an arrow on the pump casing) must be observed. This must be checked in particular after replacing the electric drive motor (polarity and direction of rotation of the motor!).

Furthermore, a pump must not run dry, as there is a risk of the shaft seal or the piston seal of piston pumps becoming leaky. The tightness of the seals is to be checked visually, for example, during the round blank run.

Table 5.3 shows common causes for pump operation malfunctions.

Example of Pump Calculation

A pump with a capacity of 1 kW, pumps 4000 L water ($\rho = 1$ kg/m^3) 5 m up. The overall efficiency of the plant is 80%. How long does the pumping process take?

Table 5.3 Disturbance matrix for pumps

Malfunction	Cause
Pump does not deliver	Not properly filled; speed too low; shaft seal leaking; foot valve (suction valve) of the piston pump leaking
Too low flow rate	Too low speed; shaft seal leaking; wrong direction of rotation; heavy wear the internal parts (cavitation); viscosity the liquid too high; foot- and/or pressure valve of the piston pump leaking
Heating of bearings	Leaky shaft sealing; dry-running bearing; pump poorly aligned with motor; insufficient lubrication
Shaft sealing leaking	Stuffing box as shaft sealing leaking: Packing worn out; shaft or rather shaft sleeve has scratches, gland packing tightened at an angle; imbalance in the drive shaft; at mechanical seal: seal defective, change
Rattling noise in the pump	Air in the pump, flow rate is to high

Solution: Use Eq. 5.19. The volume flow Q is the volume of 4000 L to be pumped in the required time t. In this context, in Eq. 5.19, Q must be replaced by "V/t" and the equation must be changed to "t":

$$t = \frac{1000\ \text{kg} \cdot 9.81\ \text{m} \cdot 5\ \text{m} \cdot 4\ \text{m}^3}{1000\ \text{W} \cdot \text{m}^3 \cdot \text{s}^2 \cdot 0.8} = 245.45\ \text{s} = 4.1\ \text{min.}$$

◀

5.4.2 Pipelines and Fittings

5.4.2.1 Pipelines

a) General

Pipes and pipelines are classified according to their nominal diameters (inside diameter or inner width of the pipe) and according to their permissible operating pressures.

According to EN ISO 6708, the *nominal diameter* is indicated by the designation DN (French "diamètre nominal"), followed by a dimensionless number approximately corresponding to the inside diameter in millimeters. It is a parameter for related parts such as pipes, flanges, fittings, and valves.

Example: According to DIN EN 10357 series 2, DN 125 is a pipe with the outside diameter of 129 mm and a wall thickness of 2 mm (inside diameter thus 125 mm). Or a pipe with the following designation: According to EN ISO 1127 series 1, DN 125 is a pipe with an outside diameter of 139.7 mm and a wall thickness of 2.6 mm (inside diameter thus 134.5 mm).

As the examples show, the actual inside diameter can often deviate from the nominal diameter by several millimeters. In this respect, it can only be assumed with certainty that pipes from different manufacturers can be combined if the nominal diameter is stated with reference to the same standard. The deviations can be explained by the fact that the wall thicknesses of the steel pipes traditionally used for sanitary installations grow inward as the pressure range increases, that is, the free cross section decreases. As the outside diameter remains the same, the same threads can be cut on pipes with the same nominal diameter. This means that the same pipe sockets (fittings) can be used for all pressure ranges.

Only in the case of a very thick-walled pipe under very high pressure, the outside diameter increases. The same applies to pipes made of glass fiber reinforced plastic (usually synthetic resin): These are wound around a mandrel or a plastic pipe (inliner); therefore, the wall thickness here must increase outward.

According to ANSI[9] the nominal diameter is given in NPS (Nominal Pipe Size) in inches. A pipe of NPS 2 corresponds approximately to the nominal diameter DN 50.

In addition to the nominal diameter, the *nominal pressure* is also an indicator describing the pipe. According to DIN, EN, and ISO, this is indicated by the designation *PN* (Pressure Nominal) followed by a dimensionless integer representing the delivery pressure in bar at room temperature (20 °C). The permissible operating pressure at a given temperature is usually given as a percentage of the nominal pressure. At higher and lower temperatures, the permissible pressure is correspondingly lower due to the decrease in the permissible material characteristics (yield strength). PN 10, for example, designates a pipeline with the maximum permissible pressure of 10 bar at a fluid temperature of 20 °C.

According to EN 1333, certain nominal pressure levels are specified: PN 2.5, PN 6, PN 10, PN 16, PN 25, PN 40, PN 63, PN 100, PN 160, PN 250, PN 320, and PN 400. The wall thickness of the pipes and also the dimensions of the flanges within the pipeline depend on the nominal pressure level.

In the American system of standards (ASME B 16.5), and also as an alternative in European standards, the designation *class* can be used instead of *PN*. In this system of standards, the class designation means a load capacity of the pipe by specifying fixed pressure–temperature combinations. In the case of the class specifications, the pressure was originally given in psi, but today they are also given in bar.

Class levels: Class 25, Class 75, Class 150, Class 250, Class 300, Class 600, Class 900, Class 1500, Class 2500, and Class 4500.

b) Marking of Pipelines According to the Fluid Medium

The marking of pipelines on ships is important to ensure safety at all times. Accidents, injuries, and damage to machinery and equipment can be caused by persons who are not aware of the fluid media flowing through the pipes on board.

If, for example, in the chemical industry, pipelines are marked according to their fluid medium in accordance with DIN 2403, the Ship Safety Division of German Employer's Liability Insurance Association for Transport and Traffic (German: Berufsgenossenschaft für Transport und Verkehr, BG Verkehr), prescribes marking in accordance with DIN EN ISO 14726.[10] According to these regulations, pipelines in operating rooms must be clearly and permanently marked if there is a risk of danger due to confusion in connection with contents, temperature or pressure. As the pipelines are uniformly designed in light colors when the ship is put into service, additional color coding is required to prevent confusion. This is intended to enable faster identification of the operating fluids transported in the pipeline and the direction of flow.

[9]ANSI = American National Standards Institute.

[10]D.6 Marking of pipelines on seagoing vessels—Annexure 5, BG Verkehr, Ship Safety Division, Status 02/2012, Reference: [34].

Fig. 5.20 Marking of a pipe according to the fluid medium—here ballast water. *Photo* AIDA

DIN EN ISO 14726-1 specifies the basic colors for the identification of substances in pipes. DIN EN ISO 14726-2 specifies additional color codes, which enable detailed identification of the substances.

The mentioned ISO standard applies to ships with keel laying on or after October 1, 2010. Existing standards of shipyards or shipping companies will still be accepted according to the mentioned BG regulation if the main colors comply with the ISO standard.

The marking can be made by colored adhesive strips (Fig. 5.20) or by painting the entire line in the appropriate color.

If a further distinction beyond the main colors is to be made for the group of fuels, oils, and sludge/waste water, further identification colors in accordance with the appendix to the BG regulation mentioned above are to be used (Fig. 5.20 and Appendix 19).

In Table 5.4, the media are shown with the corresponding main colors.

Notes on pipe marking:

- At least one marker in each room,
- At any point where the pipes run through a wall, ceiling or floor,
- Near each valve,
- Every 3–5 m (sometimes the authorities/classification societies require additional markings)
- Arrows to indicate the flow direction,
- An additional indication of the medium in the form of text makes the contents of the pipe even clearer.

Table 5.4 Main colors and media

Main color	Medium	RAL number
Black	Contaminated media[a]	9005
Blue	Fresh water	5015
Brown	Fuel	8001
Green	Seawater[b]	6018
Gray	Non-combustible gases	7001
Red-brown	Masses (dry and moist)[c]	8015
Orange	Oils, other than fuels	2003
Silver	Steam	9006
Red	Fire extinguishing media	3000
Purple	Acids, alkalis	4001
White	Air in ventilation systems	9010
Yellow ochre	Combustible gases	1021

[a]Black water, gray water, waste oil, exhaust gas
[b]For ships with different trade lane (sea/river ships) all water from outboard
[c]Excluding powder and foam for fire extinguishing

In accordance with the standard DIN ISO 14726:2010-10, the following color scheme is recommended by **ebu**,[11] to be used for color coding of pipelines on board (Table 5.5).

c) Pressure Losses in Piping Systems

The **pressure loss** (also **pressure drop**) in pipeline systems results from the individual losses (caused by wall friction and dissipation) of all pipeline parts, such as pipes, fittings and valves, and also from the influence of the geodetic height and from changes in cross section.

For gases, the volume change due to expansion must also be taken into account. However, it can be neglected if the pressure drop is only a few percent of the absolute pressure. Under this condition, the calculations of pressure losses for liquid and gaseous media can be made using the same approaches.

In general, the pressure loss is calculated according to the following equation:

$$\Delta p = C \cdot (\rho \cdot w^2)/2 \tag{5.28}$$

with

$$C = \lambda \cdot l/d \tag{5.29}$$

[11]ebu = European Barge Union, "Recommendation on color coding of pipelines for tankers" [38].

Table 5.5 Marking of pipelines on aboard tank ships

Contents	Main color/additional color	RAL number
Fire-extinguishing water	Red/Green/Red	3000/6018/3000
Spray water (sprinkling)	Red/Violet/Red	3000/4001/3000
Drinking water	Blue/Green/Blue	5015/6018/5015
Ballast water	Green/Violet/Green	6018/4001/6018
Diesel fuel	Brown/Yellow/Brown	8001/1021/8001
Exhaust steam	Silver/White/Silver	9006/9010/9006
Steam	Silver/Yellow/Silver	9006/1021/9006
Gas discharge	Yellow (only main color)	1021
Nitrogen	Gray/Green/Gray	7001/6018/7001
Compressed air	Gray/Orange/Gray	7001/2003/7001
Breathing air	Gray/White/Gray	7001/9010/7001
Thermal oil	Orange/Blue/Orange	2003/5015/2003
Hydraulic fluid	Orange/Gray/Orange	2003/7001/2003

results for the pressure loss from wall friction for pipes:

$$\Delta p = \lambda \cdot (l/d) \cdot (\rho \cdot w^2/2). \tag{5.30}$$

The equation $C = \lambda \cdot l/d$ is also called as **form**- or **proportionality factor**.

In the above equations,

- Δp Pressure loss in Pa (Pascal); conversion Pa in cash: 1 bar $= 1 \cdot 10^5$ Pa,
- ρ Density of fluids in kg/m^3,
- w Average flow rate of fluids in m/s,
- l Characteristic pipe length in m,
- d Inner pipe diameter in m,
- C Form or proportionality factor, dimensionless,
- λ Pipe friction coefficient, dimensionless (see Fig. 5.21)

The pipe friction diagram (Fig. 5.21) shows k the roughness of pipes (Table 5.6).

The pressure loss Δp in *valves and fittings* is proportional to the dynamic pressure $\rho \cdot v^2/2$. The proportionality factor is the **resistance coefficient** or perhaps **resistance figure** ζ, from which tables can be obtained (see Annexure 19—Table of selected resistance coefficients). The following applies: $C = \zeta$. Thus, the pressure loss in valves and fittings is calculated according to the following equation

$$\Delta p = \zeta \cdot (\rho \cdot w^2)/2. \tag{5.31}$$

The resistance coefficient itself depends on the volume flow, geometry, Reynolds number, etc.

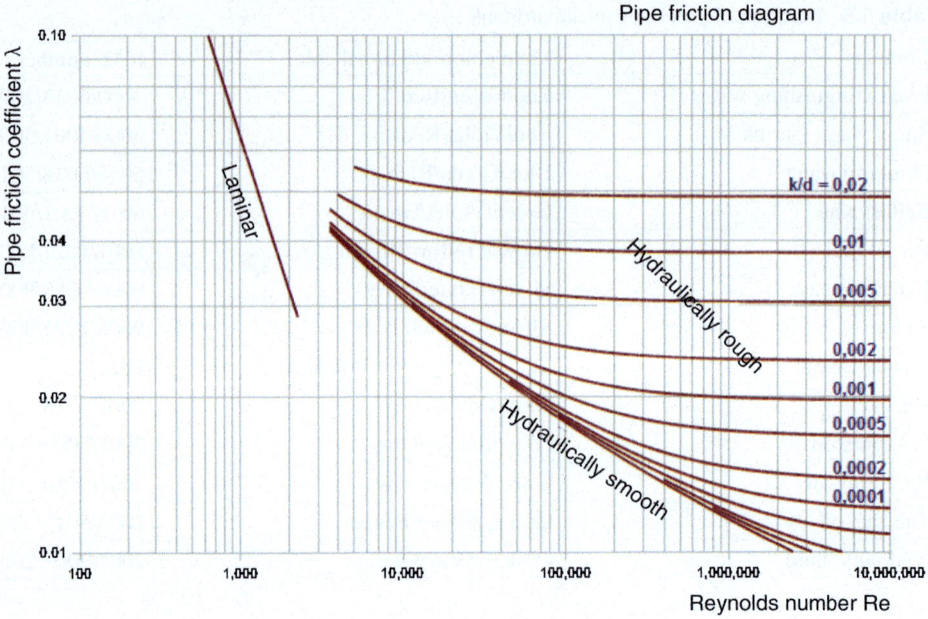

Fig. 5.21 Pipe friction diagram

Table 5.6 Roughness values

Material and Pipe type	Condition of the pipe	*k* (mm)
New pulled and pressed pipes from Cu, Ms, bronze, Al, glass, and plastic	Technically smooth	0.001–0.0015
New seamless steel pipes, rolled or pulled	New Commercially rusted Encrusted	0.25–0.5 1.0–1.5 1.5–3.0
New longitudinally welded steel pipes	With rolling skin	0.04–0.1
New steel pipes with coating	Metal spraying Dip galvanized Commercially galvanized Electroplated	0.08–0.09 0.07–0.1 0.1–0.16 approx. 0.008
Used steel pipes	Uniform rust scars Light encrustation Middle encrustation Strong encrustation	approx. 0.15 0.15–0.4 approx. 1.5 2.0–4.0
Rubber pressure hose	New, not embrittled	0.0016

In another common notation, C is also replaced with $\zeta \cdot a$, with a being the so-called external factor. For *pipes* applies $a = l/d$, for *valves and fittings* is $a = 1$. Thus, Eq. 5.28 gives the equation for pressure loss in pipes (without fittings, etc.),

$$\Delta p = \zeta \cdot l/d \cdot (\rho \cdot w^2/2). \tag{5.32}$$

The **Reynolds number** *Re* is a dimensionless parameter and indicates the ratio of load capacity forces to toughness forces. It provides information about the type of flow present: A flow is laminar at $Re < 2000$, possibly turbulent at $Re \geq 2000$, usually turbulent with $Re > 2320$ in technical pipelines.

$$Re = \frac{\rho \cdot w \cdot d}{\eta} = \frac{w \cdot d}{\nu} = \frac{4 \cdot \rho \cdot \dot{V}}{\pi \cdot \eta \cdot d}. \tag{5.33}$$

Here ρ is the density of the fluid, w is the flow rate of the fluid relative to the body and d is the characteristic length of the body. The characteristic length, also called reference length, is defined or should be defined for the respective problem. For fluid bodies, the length of the body in the direction of flow is usually selected. In the case of resistances, the width or height transverse to the direction of flow is usually taken as the characteristic length, in the case of pipe flows it is usually the internal diameter of the pipe and in the case of flumes, the depth or width at the flume surface. The kinematic viscosity ν of the fluid is different from the dynamic viscosity $\eta = \nu \cdot \rho$ by the factor ρ.

The velocity of the flow can be described by the term

$$w = \frac{\dot{V}}{A} \tag{5.34}$$

where A is the clear cross-sectional area of the pipe.

The pressure loss Δp caused by friction in a pipe is proportional to the **specific pipe length** l/d and proportional to the dynamic pressure of the flow $\rho \cdot w^2/2$. The proportionality factor is the **pipe friction coefficient** λ (see Eqs. 5.29 and 5.30). The pipe friction coefficient in turn is a function of **Reynolds number** *Re*. In case of a *laminar flow*, the pipe friction coefficient is solely dependent on the Reynolds number ($\lambda = f(Re)$) and is committed to a fully developed flow in a circular pipe according to the Hagen-Poiseuille law:

$$\lambda = 64/Re. \tag{5.35}$$

However, if the flow is turbulent, the roughness of the surface is also taken into account—it is affected by *hydraulically rough pipe* spoken. The value for λ is then calculated, taking into account the actual roughness, according to Prandtl and von Karman as follows[12]:

---

[12] In the literature, different calculation approaches are sometimes chosen for different Reynolds ranges (see also [52]).

$$\frac{1}{\lambda} = -2\log\left(\frac{k}{3.71 \cdot d}\right). \tag{5.36}$$

Some values for the roughness k can be found in Table 5.6[13]:

In the case of several **individual resistances** (pipe elbow, shut-off valve, non-return valve, etc.) of the same nominal size, the pressure loss for these components is determined as follows using Eq. 5.31:

$$\Delta p = \sum \zeta_i \cdot \rho \cdot w^2/2. \tag{5.37}$$

The resistance coefficient ζ is determined experimentally and can be found in the relevant tables (see Annexure 19—Table of selected resistance coefficients).

The pressure loss in a pipe element can also be determined without the resistance coefficient by means of the same friction loss of an equivalent straight pipe section. For the so-called equivalent pipe length l then applies:

$$l = (\zeta \cdot d)/\lambda. \tag{5.38}$$

With this equivalent pipe length, plus the actually given pipe length, the pressure loss of the entire pipeline can be calculated according to Eq. 5.39 in one run. For this purpose, the term in above Eq. 5.32 is used. It results according to Darcy-Weisbach:

$$\Delta p = \frac{\rho \cdot w^2}{2} \cdot \left(\lambda \cdot \frac{l}{d} + \sum \zeta_i\right). \tag{5.39}$$

With a vertical pipe routing, the *potential energy* of the medium flowing through changes. According to the law of conservation of energy (Bernoulli's law for frictionless flows[14]), the pressure in the pipeline also changes as a result (influence of the geodetic height on the pressure loss). Example: The **pressure loss from the geodetic height** which a pump has to overcome is to be considered according to the following equation:

$$\Delta p = \rho \cdot g \cdot h \tag{5.40}$$

with

Δp Pressure in Pa,
ρ Density of fluids in kg/m^3,
g Acceleration of gravity in m/s^2,
h Height of the fluid column in m

[13]Based on [31].

[14]In-depth [50].

Pipe friction resistances are sometimes also referred to as resistance heights H_v in m (pressure head loss) (see Sect. 5.4.1.2); based on the above equations, the following applies

$$H_v = C \cdot w^2/(2g) \text{ in m,} \tag{5.41}$$

$$\Delta p = H_v \cdot \rho \cdot g \text{ in Pa.} \tag{5.42}$$

d) Recommended Flow Rate in Pipes

In order to avoid pressure surges due to acceleration or deceleration, the flow rate in pipelines should be kept constant. *Economical speed* is also a decisive factor for dimensioning. It is the optimum of the sum of the investment costs for the pipeline, the investment costs of the machinery (pumps, compressors), and the energy and maintenance costs over the entire operating period. Furthermore, the flow rate of the medium must be selected so high that no deposits occur in the pipe system. On the other hand, a particle-laden flow should not contribute to increased erosion of the inner pipe wall due to excessive flow rates. Furthermore, the flow rate of the fluid depends on whether laminar or turbulent flow in the pipe is desired.

Appendix 20 Contains an Overview of Recommended Flow Rates.

e) Compensation of Length Changes in Piping Systems

Expansion of the pipelines due to temperature changes or displacement of their suspensions as a result of ship deformations can be prevented by pipe bends (e.g., "skillful" arrangement of 90° bends or U-bends) expansion joints (see Figs. 5.22 and 5.23) or flexible pipe connections (e.g., hoses) [63]. In this context, care must be taken to ensure a suitable arrangement of fixed points.[15]

The calculation of the length change due to temperature changes is carried out according to Eq. 5.43:

$$\Delta l = l_0 \cdot \alpha \cdot \Delta T \tag{5.43}$$

with

Δl Change of length in m,
l_0 Output length in m,
α Linear expansion coefficient in 1/K[16],
ΔT Temperature change in K,

[15]Germanischer Lloyd, Construction Regulations I-1 Ship Technology/Sea Vessels, pp. 11–19, Edition 2014.

[16]Please refer to the relevant tables (see Annexure 15).

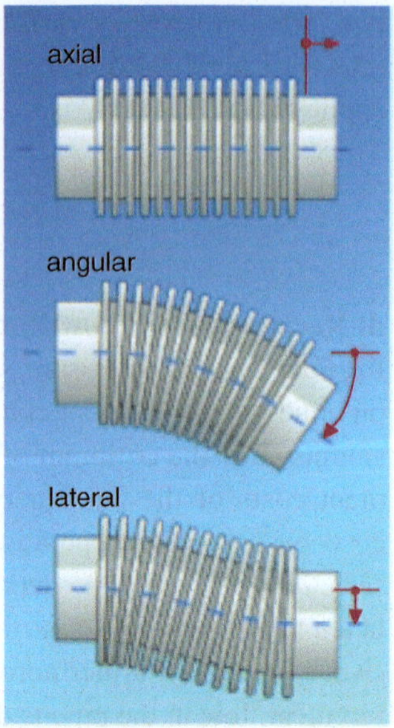

Fig. 5.22 Compensation of pipe position changes. *Photo* Witzenmann GmbH, Pforzheim

Fig. 5.23 Rubber compensator. *Photo* Yeti

f) Pipe Connections

The choice of the pipe connection depends on several boundary conditions, such as:

- Pipe material,
- Ease of assembly,

- Ease of maintenance (should a detachable or non-detachable connection be made),
- Customer request.

The following pipe connections can be selected[17]:

- Fully welded butt welds with/without measures to improve the quality of the weld,
- Socket welds of adequate fillet weld thickness and, if necessary, according to recognized standards,
- Steel flanges can be used under consideration of the permissible pressures and temperatures specified in the relevant standards,
- Mechanical connecting elements (pipe fittings, pipe couplings, compression fittings, cutting ring fittings, etc.) of an approved type.

The "Guidelines for the assembly of flanged joints in process plants" [20] contain valuable information on the assembly of flanged joints (see Fig. 5.24).

Temporary coatings, for example, as protection against corrosion, must be removed from the flange sealing surfaces without leaving any residue before installation (e.g., with cleaning agent, suitable wire brush). When replacing a seal, care must be taken to ensure that the old seal is completely removed from the flange sealing surface without damaging it. The flange sealing surfaces must be clean and free from damage; the same applies to the seal. Special attention must be paid to the thread and the contact surfaces of the flange bolts, which must not be damaged. To minimize the frictional forces and also in view of a flange connection that may have to be loosened again, the bolt threads, nuts and washers should be treated with suitable lubricants before tightening (e.g., Molykote). The seal must be centered accurately. The screws must be tightened evenly crosswise with the specified tightening torques.

Cutting ring fittings (e.g., ERMETO system) are standardized according to EN ISO 8434 or DIN 2353 and are used primarily for connecting hydraulic lines. The components of a cutting ring screw connection are (see Fig. 5.25a,b) union nut, clamping cone and cutting ring. The screw connection has a 24°-sealing cone, the nut has a metric thread.

By tightening the union nut, which is tapered on the inside, the cutting ring is compressed, causing the wedge-shaped inside of the ring to cut into the pipe wall and create a tight form fit.

[17]Germanischer Lloyd, Construction Regulations I-1 Ship Technology/Sea Vessels, pp. 11–19, Edition 2014.

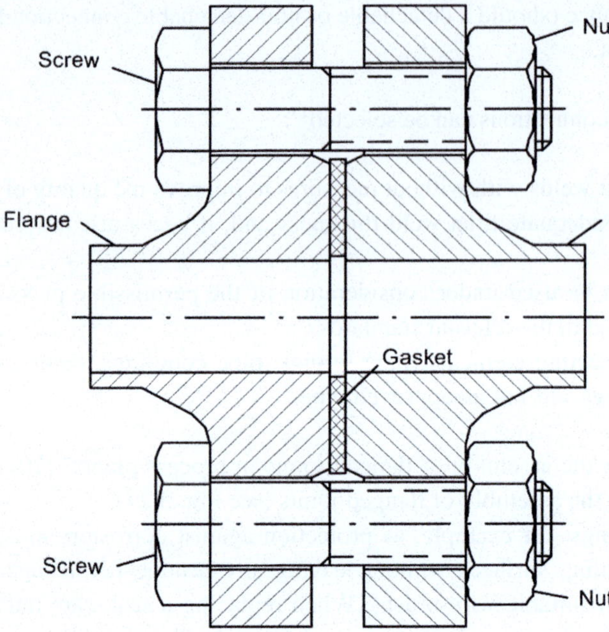

Fig. 5.24 Pipe connection with welding neck flanges DIN/EN 1092-1/ASME. *Source* LINNEMANN GmbH, 72070 Tübingen, Rohrverbindungen und Edelstahlarmaturen

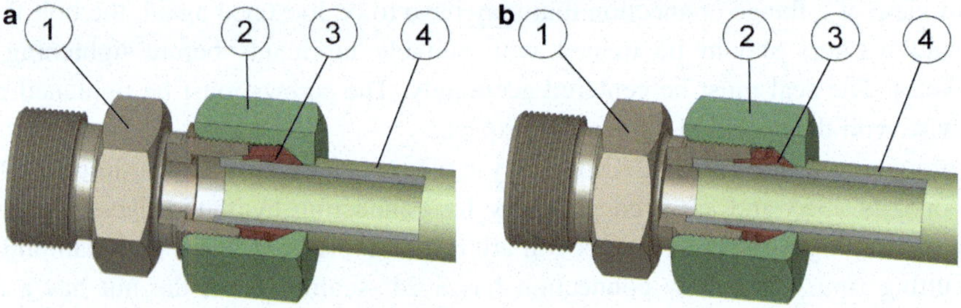

Fig. 5.25 Cutting ring bolt connection. **a** before tightening, **b** tightened. *1* Union nut, *2* clamping cone, *3* cutting ring, *4* pipe wall. *Photos* Lindecke, Chr., CC BY-SA 3.0

Example Pump and Pressure Loss[18]

A fire pump (centrifugal pump) delivers a volume flow $Q = 300$ L/min through an 18 m long pipe with the inner width of 52 mm and a roughness of $k = 0.2$ mm to the tapping point in the form of an angle valve for connecting a C-hose. The pump is

[18]For planning and execution of fire extinguishing systems, see also [56].

located one meter below sea level, the tap 15 m above sea level. The density of the water can be assumed to be 1020 kg/m³. In the course of the pipeline there are four pipe bends of 90° (with a ratio of pipe bending diameter to pipe inside diameter of two each and a resistance number $\zeta_{smooth} = 0.14$; see Annexure 19). Pressure losses in the suction line and in the pump itself can be assumed to be 0.2 bar.

a) What is the inlet speed at the suction port of the centrifugal pump?
b) How large is the pressure loss in the pressure line?
c) What is the required pump shaft power at a pump efficiency of 0.85?

Solution:

a) From the continuity equation for incompressible flows it follows by rearranging the Bernoulli equation:

$$Q = w \cdot A = \text{const},$$

$$w = \frac{Q}{A} = \frac{Q \cdot 4}{d^2 \cdot \pi} = \frac{0.05\ \text{m}^3 \cdot 4}{0.052^2\text{m}^2\text{s} \cdot \pi} = 23.6\ \text{m/s}.$$

b) Eq. 5.39 provides the following approach for the solution.

$$\Delta p = \frac{\rho \cdot w^2}{2} \cdot \left(\lambda \cdot \frac{l}{d} + \sum \zeta_i \right).$$

The sum of the individual resistances is calculated with the data from Annexure 19:

$$\zeta_{\text{Rohrbögen}} = 4 \cdot 0.14 = 0.56,$$

$$\zeta_{\text{Eckventil}} = 2.00,$$

$$\rightarrow \sum \zeta_i = 2.56.$$

The pipe friction losses λ can be determined using the Reynolds number according to Eq. 5.36:

$$\frac{1}{\lambda} = -2 \log \left(\frac{k}{3.71 \cdot d} \right) = -2 \log \left(\frac{0.2\ \text{mm}}{3.71 \cdot 52\ \text{mm}} \right) = 5.97 \rightarrow \lambda = 0.17.$$

This results in the pressure loss due to friction and form resistance:

$$\Delta p = \frac{1020\ \text{kg} \cdot 23.6^2\text{m}^2}{2 \cdot \text{m}^3 \cdot \text{s}^2} \cdot \left(0.17 \cdot \frac{18\ \text{m}}{0.052\ \text{m}} + 2.56 \right) = 284{,}111\ \text{Pa} = 2.84\ \text{bar}.$$

c) From Eq. 5.19, the pump capacity can be determined. In this equation, however, Δp is the pressure loss determined above plus the pressure loss due to the geodetic height loss to be bridged and the pressure losses on the suction side and in the pump itself.

$$P_W = (Q \cdot \Delta p)/\eta$$

with

$$\Delta p = 2.84 \text{ bar} + 1.50 \text{ bar} + 0.20 \text{ bar} = 4.44 \text{ bar},$$
$$P_W = (0.05 \text{m}^3/\text{s} \cdot 444.000 \text{ kg/ms}^2)/0.85 = 26.118 \text{ kW}.$$

Since a fire hose is still connected to the tap and the jet pipe still has to reach a certain throw distance, the pump output will have to be much higher. For firefighting, a powerful water jet is required at the jet pipe. When using both water and foam pipes, a jet pipe pressure of at least 5 bar is required.

For example, a CM jet pipe with a nozzle diameter of 9 mm and a jet pipe pressure of 5 bar has a throwing distance of about 25 m. The flow rate there is approximately 130 L/min [29].

The pressure loss in a C-hose (52 mm) can be assumed to be 0.7 bar for 100 m with a flow rate of 200 L/min.

Taking this information into account, Δp would still increase by 0.7 bar/4 = 0.175 bar if a 25 m C-hose is connected. In addition, 5 bar minimum pressure at the jet pipe must be taken into account:

$$\Delta p = 2.84 \text{ bar} + 1.50 \text{ bar} + 0.20 \text{ bar} + 0.18 \text{ bar} + 5 \text{ bar} = 9.62 \text{ bar}.$$

The required pump shaft power is then

$$P_W = (0.05 \text{ m}^3/\text{s} \cdot 962.000 \text{ kg/ms}^2)/0.85 = 57 \text{ kW}.$$

◀

Example Throwing Distance of Fire Monitor

On board a firefighting boat (operational area Baltic Sea—density of water can be checked with 1023 kg/m^3 are accepted) there is a fire monitor with the following data:

- Nozzle diameter $A = 150$ mm,
- Working pressure in the jet pipe 10 bar,
- Flow rate $Q = 20{,}000$ L/min.

What is the theoretical maximum throwing distance s and the theoretical maximum throwing height h of the fire monitor?

Solution: The throwing range of the fire monitor is based on the law of the "oblique throw", because here the throwing distance of a water particle can be considered. The oblique throw is composed of a uniform translation at the angle α to the horizontal and a free fall; the maximum values for throwing distance and height are calculated according to the following equations [8, p. L8]:

$$s = v_0^2 \cdot \sin 2\alpha / g \tag{5.44}$$

and

$$h = v_0^2 \cdot \sin^2 \alpha / 2\ \mathrm{g} \tag{5.45}$$

with

S = throwing distance,
v_0 = outlet velocity of water particle,
G = acceleration of gravity 9.81 m/s^2

Equation 5.45 shows that the maximum throwing distance is achieved by a throwing angle α of 45°, whereby the water particle describes a parabola, the maximum height is reached at 90° (there sin 90° = 1).

The outlet velocity at the nozzle of the fire monitor is calculated according to Eq. 5.46 [8, p. N7]:

$$v_0 = \phi \cdot \sqrt{2 \cdot \frac{p_{ü}}{\rho}} \tag{5.46}$$

with

φ = fluid friction value—sometimes also called velocity number (for water φ = 0.97),
$p_{ü}$ = overpressure compared to external pressure,
ρ = density (for water depending on the shipping area; see Annexure 2)

The overpressure acting in the monitor or also within a jet pipe of a fire hose is the driving force that gives the water particle the impulse, with the corresponding velocity v_0 to leave the system limit "outlet opening".

The initial velocity v_0 can also be calculated using the continuous equation $Q = A \cdot v = \text{const}$. Then the following applies: $v_0 = Q/A$ with Q. The more sharp-edged the nozzle opening, the more the jet narrows; that is, the extinguishing jet at the nozzle outlet has a smaller cross section A2 than the nozzle cross section $A_1 : A_2 < A_1$. The ratio of A_2 to A_1 is called the contraction number μ:

$$\mu = A_2/A_1. \tag{5.47}$$

The product of the contraction coefficient and the liquid friction value is the discharge coefficient α:

$$\alpha = \mu \cdot \varphi \tag{5.48}$$

with

$$Q = \alpha \cdot \sqrt{2 \cdot g \cdot p_{ü}} \cdot A_1 \tag{5.49}$$

and thus $v_0 = Q/A$ can also be determined by knowing the flow rate and the jet pipe outlet cross section.

Fig. 5.26 Fire monitors in use. *Photo* Arcudaki, CC BY-SA 3.0

By inserting the mentioned numerical values into Eqs. 5.46 and 5.44 and you get $v_0 = 43$ m/s the theoretical maximum throwing distance $s = 188$ m; the theoretical throwing height of the fire monitor is then $h = 94$ m (according to Eq. 5.45).

In reality, the maximum throwing distance in reality is at a much smaller throwing angle. This is due to the fact that the water particle is exposed to frictional forces with the air during its flight. Practice has shown that the maximum throwing distance at a throwing angle of about 32° can be achieved (see also Fig. 5.26). For example, for the aforementioned fire monitor, an achievable throwing distance of 132 m is given [66]. ◀

5.5 Heat Exchanger

5.5.1 Introduction

A **heat exchanger** (colloquially also heat exchanger) is an apparatus that transfers thermal energy from one material flow to another. For example, they are used to cool the lubricating oil in central lubrication circuits (bearings of the drive shaft) in the context of air conditioning technology (water chiller) and many other applications.

In a heat exchanger, the heat is transferred from a warm material quantity $\dot{m}_1$ with its specific heat capacity c_1 along the surface A of a partition wall, to a cold material quantity $\dot{m}_2$ with its specific heat capacity c_2, which flows along the other surface of the partition wall (e.g., material 1 around a pipe, material 2 through the pipe). The temperature

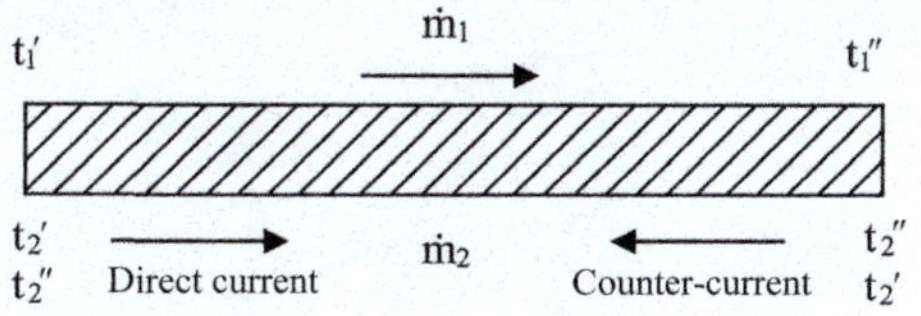

Fig. 5.27 Temperatures at the heat exchange wall

difference between the two materials changes along the partition wall. In the general scheme of Fig. 5.27 we refer to t_1' and t_1'' the temperatures of the warm material and t_2' and t_2'' the temperatures of the cold material at the points on the wall indicated in Fig. 5.27, the amount of heat $\dot{Q}$ transferred along the surface A:

$$\dot{Q} = \dot{m}_1 \cdot c_1 \cdot (t_1' - t_1'') = \dot{m}_2 \cdot c_2 \cdot (t_2'' - t_2') = k \cdot A \cdot \Delta t_m \tag{5.50}$$

with

Q = heat quantity in kJ/s or kW,
$\dot{m}_i$ = mass flow in kg/s,
$t_i{}^i$ = temperatures in K,
c_i = specific heat capacity of Media in kJ/(kg K; Values s. Annexure 22),
Δt_m = average logarithmic temperature difference in K (see explanations below),
K = heat transfer coefficient the partition wall in W/m^{2}K (see explanations in Sect. 5.5.4.2)

5.5.2 Types of Heat Exchangers

Tube bundle and plate heat exchangers are commonly used in shipbuilding.

5.5.2.1 Shell and Tube Heat Exchangers

In principle, a shell and tube heat exchanger is a hollow cylinder (see Figs. 5.28 and 5.29), inside which there is a bundle of tubes.

The tube bundle is an assembly consisting of tubes arranged in parallel. They are rolled, soldered or welded into round plates with holes, the tube sheets. Shell and tube heat exchangers have two separate spaces, the tube space and the shell space. One medium flows through the pipe bundle, the pipe space, while the other flows between the pipes in the shell space. Baffles can be installed and divert the fluid in the shell space to improve heat exchange. In a gas or vapor and liquid exchange, the gas or vapor phase usually flows around the tubes and the liquid flows through the tubes.

There are heat exchangers with two tube sheets (Fig. 5.29) and heat exchangers with U-tubes (Fig. 5.30) and one tube sheet.

Applicable DIN standards:

Fig. 5.28 Shell and tube heat exchanger (see also Sect. 5.5.2.1)

Fig. 5.29 Tube bundle heat exchanger principle

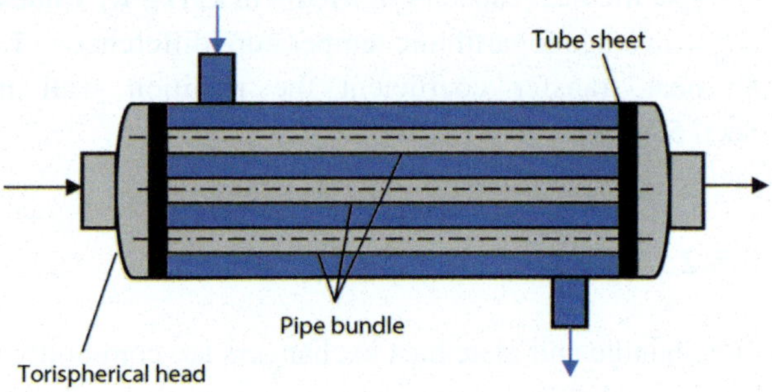

Fig. 5.30 U-tube heat exchanger

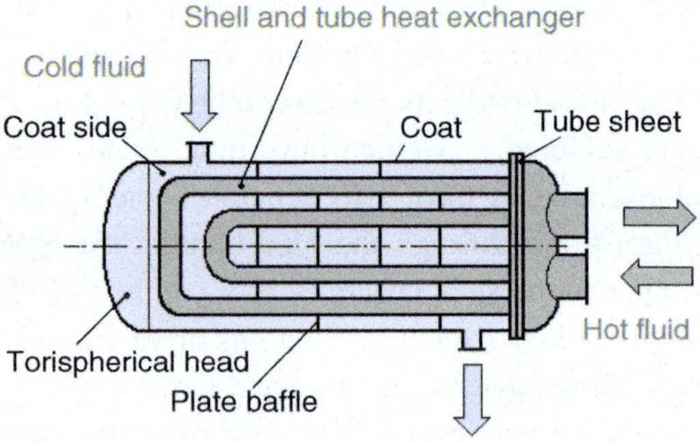

- DIN 28008 Tolerances and limiting dimensions for shell and tube heat exchangers,
- DIN 28183 Shell and tube heat exchangers—designations,
- DIN 28184 Tube bundle heat exchanger with two fixed tube sheets,
- DIN 28185 Fittings for shell and tube heat exchangers,
- DIN 28190 Shell and tube heat exchangers with welded floating head—design examples for tube arrangement and floating head.

5.5.2.2 Care and Maintenance of Shell and Tube Heat Exchangers

Due to the narrow pipe spacing between the individual pipes of the pipe bundle, the flow can easily be impaired by dirt and deposits. This also applies to the inner pipe walls, then the equipment must be rinsed or cleaned in some other way. Acidic solutions are usually used for rinsing.

5.5.2.3 Plate Heat Exchanger

A plate heat exchanger consists of corrugated plates (Fig. 5.31), which are assembled in such a way that the fluid to be heated and then the heat-emitting fluid flows in the consecutive spaces between them.

The plate pack is sealed to the outside and between the media and is held together with screws. Often a soldered construction is also preferred.

Fig. 5.31 Open plate heat exchanger

Due to their special design, these transformers are very compact and easily expandable and flexible with regard to the design of the flow path, which is determined by the position of the seals.

5.5.2.4 Care and Maintenance of Plate Heat Exchangers

Due to the small distances between the individual plates, the flow can be impaired by dirt and deposits. Then the equipment must be rinsed or cleaned in some other way. Here too, acidic solutions can generally be used for rinsing, as in the case of shell and tube heat exchangers. In case of massive contamination the bolted heat exchangers can be disassembled for cleaning.

5.5.3 Construction of Heat Exchangers

Heat exchangers are operated according to the co-current, counter-current, or cross-flow principle.

5.5.3.1 Direct Current Heat Exchangers

Here both fluids are guided in such a way that they flow in the same direction on both sides of the wall. A DC unit is always used when fast, reliable cooling is required.

Figure 5.32A shows the temperature curve (in K) of the warm (red) and cold fluid (blue) over the heat transfer surface of a direct current heat exchanger.

5.5.3.2 Countercurrent Heat Exchanger

Here, the two fluids are guided on both sides of the plate so that they flow in opposite direction. Ideally, the temperatures of the material flows are exchanged, that is, the originally cold medium reaches the temperature of the originally hot medium and vice versa ($t_2 = t'_2$; see Fig. 5.33). The prerequisite for this *ideal case* is that the heat capacity flows are equal on both sides of the heat exchanger. In addition, the heat exchanger would have to have an efficiency of 100%, which is not the case in reality; therefore, in practice, exchanging temperatures is only approximately possible.

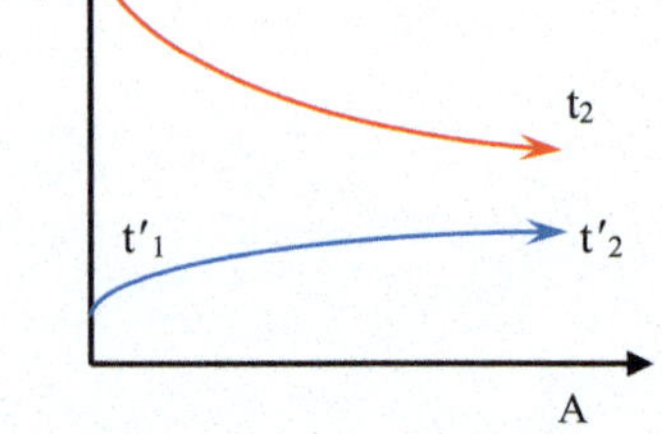

Fig. 5.32 Direct current heat exchanger

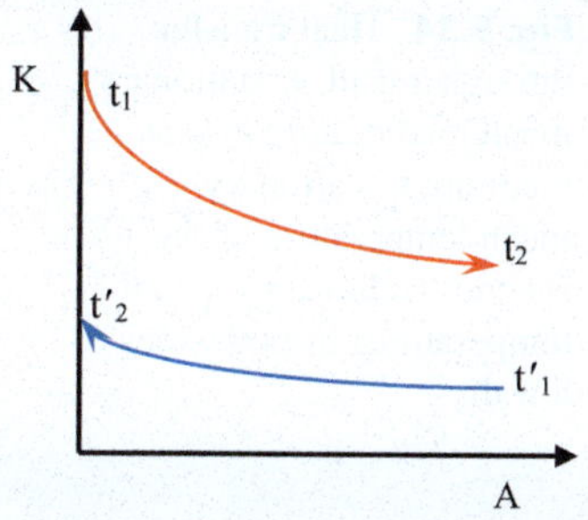

Fig. 5.33 Countercurrent heat exchangers

Figure 5.33 shows the temperature curve (in K) of the warm (red) and cold fluid (blue) above the heat transfer surface A of a countercurrent heat exchanger.

5.5.3.3 Cross-Flow Heat Exchangers

Here the material flows are guided in such a way that their directions cross. As far as inlet and outlet temperatures are concerned, the result is between counter-flow and co-current flow apparatuses.

5.5.4 Characteristics of Heat Exchangers

5.5.4.1 Mean Logarithmic Temperature Difference

The mean temperature difference between the two media along the heat exchange surface is given as the average logarithmic temperature difference:

$$\Delta t_m = (\Delta t_a - \Delta t_e)/\ln(\Delta t_a/\Delta t_e), \tag{5.51}$$

where Δt_a the difference in temperature between the two media at the beginning a and Δt_e the temperature difference at the end e of the heat exchange surface; thus

$$\text{for direct current } \Delta t_a = t_1 - t_1' = \Delta t_{\text{high}}, \tag{5.52}$$

$$\Delta t_e = t_2 - t_2' = \Delta t_{\text{low}}, \tag{5.53}$$

$$\text{for countercurrent } \quad \Delta t_a = t_1 - t_2' = \Delta t_{\text{high}}, \tag{5.54}$$

$$\Delta t_e = t_2 - t_1' = \Delta t_{\text{low}}. \tag{5.55}$$

A look at Eqs. 5.52–5.55 and on Figs. 5.32 and 5.33 shows that it is generally valid for the mean logarithmic temperature difference (in this respect also for the cross-flow heat exchanger):

$$\Delta t_m = (\Delta t_{\text{high}} - \Delta t_{\text{low}})/\ln(\Delta t_{\text{high}}/\Delta t_{\text{low}}). \tag{5.56}$$

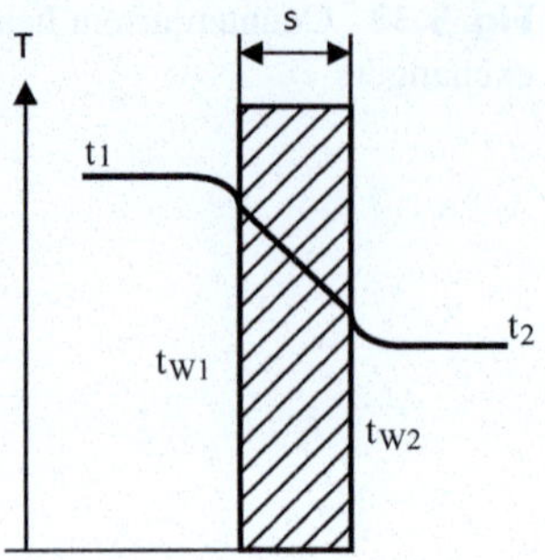

Fig. 5.34 Heat transfer through a wall. t_1 stationary mean temperature of one medium, t_2 stationary mean temperature of the second medium, $t_{W1,2}$ wall temperatures of both sides of a wall

5.5.4.2 Heat Transmission Coefficient *k*

The heat passage (characterized by the heat transfer coefficient k[19]) through a plate or a pipe is understood to be the combination of all individual processes involved in heat transport: heat is transferred from one flowing fluid by convection at the surface A, is carried away through the wall with thickness s and finally is transferred by convection from the other wall surface to another flowing fluid (Fig. 5.34). The heat transfer coefficient is measured in the unit W/m^{2}K.

The *convective transition* is described by heat *transfer coefficients* or the *transmission coefficient* α.

The heat passage through the wall is characterized by *thermal conductivity* λ of the material of the wall. In order to guarantee good heat transfer, the material of the wall (pipe) must have good thermal conductivity.[20] A wall can also consist of several layers, for example, a ship's wall: 1 cm aluminum sheet, 5 cm hard foam insulation, 0.5 cm plywood. The same applies to a tube in a heat exchanger: encrustations can form a layer in the tube and/or on its outside, which is important for the heat transfer. The same applies to coated pipes.

The heat transfer coefficient is given in W/m^{2}K, the thermal conductivity is given in W/mK.

For a single-layer wall (pipe) the heat transfer coefficient k is calculated as follows:

$$1/k = 1/\alpha_i + 1/\Lambda + 1/\alpha_a, \qquad (5.57)$$

$$1/\Lambda = s/\lambda \qquad (5.58)$$

with

[19]In building physics today, the symbol U is used for the heat transfer coefficient; k is common in mechanical and process engineering.

[20]Table with thermal conductivity values, see Annexure 23.

α_i and α_a Heat transfer coefficients to the internal and external wall side,
s Wall thickness

For a multi-layer wall, the heat transfer coefficient k is calculated as follows:

$$1/k = 1/\alpha_i + 1/\Lambda_{\text{tot.}} + 1/\alpha_a, \tag{5.59}$$

$$1/\Lambda_{\text{tot.}} = \sum (s_i/\lambda_i). \tag{5.60}$$

According to the above equations, the heat loss from the ship to the outside environment can also be calculated, which is important, for example, for the design of radiators in ship rooms.

5.5.4.3 Heat Passage Through a Pipe

In the technical application of heat transfer in heat exchangers, preheaters, condensers, etc., one speaks of heat transfer and means the summary of the following processes[21]: Heat transfer from the flowing medium to one pipe wall (e.g., outer pipe wall) by convective processes (described by the heat transfer coefficient α_a), heat conduction through the pipe wall (described by the thermal conductivity λ) and heat transfer from the other pipe wall (e.g., of the inner wall) by convective processes (described by the heat transfer coefficient α_i) to the other flowing medium (see also Sect. 5.5.4.2).

Once a uniform heat flow has been achieved in the apparatus (stationary state) $\dot{Q} = \text{const.}$, it is then obtained by the insertion of Eqs. 5.57, 5.58, or 5.59 and 5.60, Eq. 5.50 is obtained:

$$\dot{Q} = k \cdot A \cdot \Delta t_m.$$

The pipe surface area A is calculated from the average pipe diameter $r_m = r_a - r_i$ (see Fig. 5.35 and Eq. 5.61) and the total length of the pipe or the sum of all individual pipe lengths in the pipe bundle l.

$$A = 2 \cdot (r_a - r_i) \cdot \pi \cdot l = 2 \cdot r_m \cdot \pi \cdot l. \tag{5.61}$$

The temperatures t_{high} and t_{low} are the average temperatures of the media inside and outside the pipe.

Table 5.7; [7][22] contains reference values for *k-values* to *rough* calculation of heat exchangers.

5.5.4.4 Heat Transfer Coefficient Α

If the heat transfer is to be determined exactly, the heat transfer coefficient k must be determined taking into account the heat transfer coefficient α. The heat transfer

[21] For the derivation of the following equations see [32].

[22] Furthermore [53, 58].

Fig. 5.35 Dimensional convention on the tube

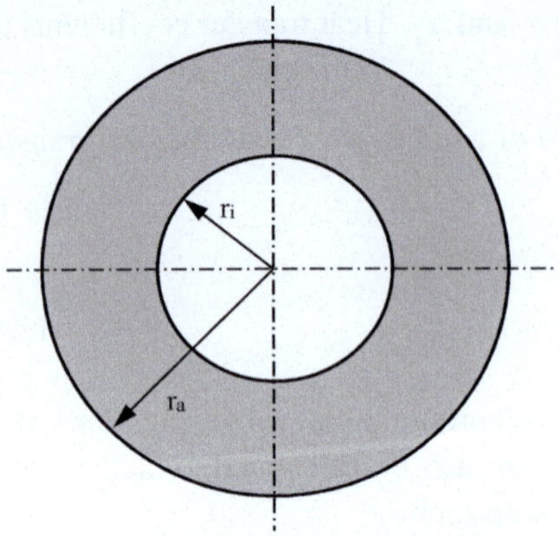

Table 5.7 *k*-values

Heating medium	Material of the pipe wall	Medium to be heated	Heat transfer coefficient *k* W/m^2K
Water	Cast iron	Air (smoke)	8
Water	Copper	Air (smoke)	13
Water	Cast iron	Water	291
Water	Copper	Water	407
Oil	Copper, ferrous metals	Water	110–350
Oil	Copper; ferrous metals	Oil	80–120
Air (flue gas)	Cast iron	Air (flue gas)	6
Air (flue gas)	Copper	Air (flue gas)	10
Steam	Cast iron	Air	12
Steam	Copper	Air	16
Steam	Cast iron	Water	907
Steam	Copper	Water	1163
Steam (condensation)		Water	1000–6000
Gas general 1 bar	Copper; ferrous metals	Gas general 1 bar	5–35

Gas general 200–300 barCopper; ferrous metalsGas general 200–300 bar150–500

coefficient α is, among other things, a function of the flow rate of the media, that is, also of the Reynolds number *Re*, insofar as it depends on whether the flow is laminar or turbulent. In case of a turbulent flow ($Re > 2300$), the heat transfer is greater than with a laminar flow. For the consideration of the two flow types, a complex determination of this characteristic is then necessary. Reference is made to the relevant literature [6, Volume I, p. 470 ff.].

Basically α is dependent on the Nusselt number. It is a dimensionless characteristic figure from the similarity theory of heat transfer, which is used to describe the convective heat transfer between a solid surface and a flowing fluid. It can also be considered as a dimensionless gradient of temperature on a surface.

a) The following applies to forced convection in the tube

$$\alpha = Nu \cdot \frac{\lambda}{d}. \tag{5.62}$$

The Nusselt number now depends on the type of flow.

For the laminar flow ($Re < 2300$) applies:

$$Nu = \left[3.65 + \frac{0.0668\left(Re \cdot Pr \cdot \frac{d}{L}\right)}{1 + 0.045\left(Re \cdot Pr \cdot \frac{d}{L}\right)^{2/3}}\right] \cdot \left(\frac{\eta_{Fl}}{\eta_W}\right)^{0.14}. \tag{5.63}$$

For the turbulent flow ($Re > 2300$) applies:

$$Nu = 0.116\left(Re^{2/3} - 125\right)Pr^{1/3} \cdot \left[1 + \left(\frac{d}{L}\right)^{2/3}\right] \cdot \left(\frac{\eta_{Fl}}{\eta_W}\right)^{0.14}. \tag{5.64}$$

They are in the above equations:

- d Inner diameter of pipes,
- L Pipe length,
- Pr Prandtl number $= (\eta \cdot c_p)/\lambda$,
- Re Reynolds number $= (\rho \cdot w \cdot d)/\eta$,
- ρ Density of fluids at its mean temperature,
- w Flow rate,
- η Dynamic toughness (from relevant tables, e.g., [8]),
- η_{Fl} Dynamic toughness at the mean fluid temperature,
- η_W Dynamic toughness at the wall temperature,
- λ Thermal conductivity of fluids

b) For free convection on a horizontal pipe

$$\alpha = Nu \cdot \frac{\lambda}{D}, \tag{5.65}$$

$$Nu = 0.41 \cdot \sqrt[4]{Gr \cdot Pr}, \tag{5.66}$$

$$Gr = \frac{g \cdot \beta \cdot \Delta t \cdot \rho^2 \cdot D^3}{\eta^2}. \quad (5.67)$$

They are here:

D Outside diameter of pipes,
Gr Grashof number,
G Acceleration of gravity (9.81 m/s^2),
β Volume expansion coefficient (from relevant tables, e.g., [8]),
Δt Absolute temperature difference between wall and fluid in the thermally unaffected area

If λ_{Wall} is known, table values for α can be used to roughly determine *k* (Table 5.8; [7]).

Example

A countercurrent tube bundle heat exchanger is to be used for the central lubrication of the rolling bearings of the drive shaft.

a) How large must the required heat exchanger surface be if 220 L/h lubricating oil ($\rho = 890$ kg/m^3) with a specific heat capacity of $c = 2.1$ kJ/(kgK) is to be cooled from 47 °C to 25 °C? The sea water used for this purpose is supplied to the heat exchanger at 15 °C and should heat upto 22° C.
b) How many tubes must the tube bundle consist of if the tubes are to have an inside diameter of 22.3 mm and a wall thickness of 1.00 mm and the tube bundle must not be longer than 1.00 m due to available space resources?

Solution:

For a)

Table 5.8 Average values for heat transfer coefficient α

Medium	Heat transfer coefficient α W/m^{2}K
Boiling water at vertical wall	3489
Boiling water at horizontal wall	1745
Flue gas	$4.7w^{0.8}$
Overheated steam	$52w^{0.8}$
Highly compressed air for intercoolers	$233w^{0.8}$
Air in air preheaters	$5.8w^{0.8}$
Condenser steam	11,630
Current water in preheat, coolers and so on	$3489w^{0.8}$

Here Eq. 5.50 provides the solution:

$$\dot{Q} = k \cdot A \cdot \Delta t_m.$$

The heat flow $\dot{Q}$ to be dissipated is calculated according to Eq. 5.68:

$$\dot{Q} = \dot{m} \cdot c \cdot \Delta t = \dot{V} \cdot \rho \cdot c \cdot \Delta t, \tag{5.68}$$

$$\dot{Q} = 0.220\ \text{m}^3/\text{h} \cdot 890\ \text{kg/m}^3 \cdot 2.1\ \text{kJ/kgK} \cdot 22\ \text{K} = 9046\ \text{kJ/h} = 2513\ \text{W}.$$

For the mean logarithmic temperature difference Δt_m, Eqs. 5.54–5.56 provide the solution:

$$\begin{aligned}\Delta t_m &= (t_{\text{high}} - \Delta t_{\text{low}}) / \ln(\Delta t_{\text{high}} / \Delta t_{\text{low}}),\\ \Delta t_m &= ((47°\text{C} - 22°\text{C}) - (25°\text{C} - 15°\text{C})) / \ln(25\ \text{K}/10\ \text{K}) = 16\ \text{K}.\end{aligned}$$

By changing Eq. 5.50 to A with a heat transfer coefficient $k = 200\ \text{W/m}^2\text{K}$ (according to Table 5.7) the required heat exchanger area is obtained:

$$A = \frac{2513\ \text{W}}{200\ \text{W/m}^2\text{K} \cdot 16\ \text{K}} = 0.8\ \text{m}^2.$$

For b)

The calculation of the required number of heat exchanger tubes is done by changing Eq. 5.61 to n:

$$\begin{aligned}A &= n \cdot d_m \cdot \pi \cdot l = n \cdot (d_a + d_i)/2 \cdot \pi \cdot 1.00\ \text{m} = 0.8\ \text{m}^2,\\ n &= A/((d_a + d_i)/2 \cdot \pi \cdot 1.00\ \text{m})\\ &= 0.8\ \text{m}^2/((0.0243\ \text{m} + 0.0223\ \text{m})/2 \cdot \pi \cdot 1.00\ \text{m}),\\ n &= 11.\end{aligned}$$

◄

5.6 Refrigeration, Ventilation and Air-Conditioning Technology, Heating Systems

5.6.1 Refrigeration

On ships, refrigeration systems are used to cool the provision loads, possibly the cargo and, if necessary, also for the waste storage rooms.[23] Refrigeration systems are also found in connection with the operation of air conditioning systems.

[23]For example, the task force suppliers of the German Navy.

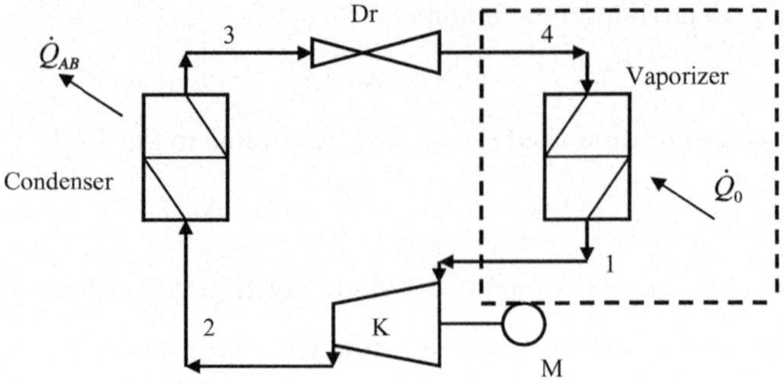

Fig. 5.36 Principle of a single-stage compression refrigeration system. The area within the dotted line is the cold store

5.6.1.1 Refrigeration Systems

a) Structure and Functioning

The single-stage compression refrigeration system is predominantly used (Fig. 5.36), as it is also widely used in domestic refrigerators. In addition, absorption refrigeration systems are also increasingly used in shipping. However, these systems will not be discussed in detail here, as they are still the exception. The basic operation of the absorption chiller is described in Sect. 5.6.2.3.

The basic principle of this refrigeration system is based on the evaporation and condensation of a refrigerant boiling at low temperatures, which is circulated ("cold vapor refrigeration system"—see Fig. 5.36). The compressor *K* driven by the motor *M* brings the refrigerant vapor sucked from the evaporator at a low temperature T_0 and low pressure p_0 to the higher pressure p_1 (state change 1→2), so that its liquefaction heat $\dot{Q}_{AB}$ can be released to the environment (cooling water, air) by the condenser at temperature T (state change 2→3); in this process the refrigerant vapor is liquefied (condensed) in the condenser. The liquefied refrigerant flows through the controlled throttle valve *Dr* with the help of which it is expanded again to pressure p_0 (change of state from 3→4 to the evaporator, in which it evaporates again under heat supply (cooling capacity $\dot{Q}_0$) (change of state 4→1).

b) Calculation of the Refrigeration Machine

For dimensioning the refrigeration system, it is important to know which cooling capacity (cooling requirement) is required.

Whether it is just a refrigerator or a cold room—basically, the required cooling requirement is made up of several individual components including:

- Cooling requirement for cooling the chilled or frozen goods,
- Cooling requirement for cooling the renewal air due to door openings,

- Cooling requirement for heat penetration through walls, ceilings, and floors,
- Cooling requirement for the heat equivalent by persons, machines, and lighting.

The share of the cooling requirement for cooling the chilled goods is usually the largest. However, if frozen goods are already stored in a deep-freeze room, the cooling requirement for cooling the goods can be neglected. A similar approach will be followed in the design of the waste load. The primary task here is to determine the cooling requirement in order to cool the air in the room down to storage temperature and to keep it at this temperature.

However, if food is brought into a cold store which have a temperature of 20 °C and have to be cooled down to 7 °C, energy must be used. The cooling requirement required $\dot{Q}_{AB}$ to cool the goods to be cooled is calculated as [48]:

$$\dot{Q}_{AB} = \frac{m \cdot c \cdot \Delta T}{t} \text{ [kW]} \tag{5.69}$$

with

m = mass of refrigerated goods in kilos,
c = specific heat capacity of chilled goods during cooling in kJ/kgK (see Table 5.9),
ΔT = temperature difference of chilled goods between the insertion temperature and the desired storage temperature in K (in above Example 293 K − 280 K = 13 K),
t = required time for cooling (usually 24 h = 86,400 s)

Table 5.9 gives a selection of some specific heat capacities.

Example

How large is the cooling heat flow for 1000 L (≈ 1000 kg) of various beverages, which are to be cooled down to 8 °C within one day at an input temperature of 20 °C in the provision load?

Solution: The solution is provided by Eq. 5.69:

$$\dot{Q}_{AB} = \frac{m \cdot c \cdot \Delta T}{t} = \frac{1000 \text{ kg} \cdot 4.2 \text{ kJ/kgK} \cdot (293 \text{ K} - 281 \text{ K})}{86{,}400\text{s}} = 0.6 \text{ kW}.$$

◀

Table 5.9 Average specific heat capacities for rough calculation of a cold room cooling capacity 12]

	Storage temperature (°C)	Specific heat capacity before solidification (kJ/kgK)
Fruit and vegetables	+2/+18	3.85
Frozen goods	−18/−22	4.4
Beverages	+8/+10	4.2

Furthermore, the so-called heat of respiration $\dot{Q}_{AT}$ of food (and certainly also of waste) due to organic processes must also be taken into account. This is done using the *specific heat of respiration of the chilled good(s)* q_{AT}[24]:

$$\dot{Q}_{AT} = \frac{m \cdot q_{AT}}{t}. \quad (5.70)$$

Furthermore, the quantity of heat to be dissipated from the packaging material is to be determined in the application of Eq. 5.71, whereby the specific heat content of the packaging material in the temperature window under consideration is to be used here.

In addition, the heat transfer through walls, ceilings, floors, and doors must also be taken into account, as described in Sect. 5.5.4.2.

Although the heat emission of persons entering the cold store is very low, it can be roughly estimated as follows[25]:

$$\dot{Q}_P = \frac{i \cdot q \cdot t}{24} \quad (5.71)$$

with

$\dot{Q}_P$=heat emission of the person in W,
q=heat emission of one person (see Table 5.10),
i=number of persons in the cold room,
t=time spent by a person in hours per day

Furthermore, the heat emission via lighting and, if necessary, other heat-emitting machinery and equipment in the cold rooms must be taken into account. For cold rooms on land, the calculation of the cooling requirement can become very complex.

Once all heat flows is recorded, which in total represent the total heat flow $\dot{Q}$ the further conceptual design of the plant can be carried out.

The evaporator capacity $\dot{Q}_0$ results from the total heat flow plus a safety margin of about 20% [46].

$$\dot{Q}_0 = \dot{Q} + 20\%. \quad (5.72)$$

To describe the refrigeration process (Fig. 5.36), the Carnot cycle is used as a comparison process (see Fig. 5.37).

This diagram (Fig. 5.37) describes the thermal energy $\dot{Q}_{41}$ supplied to the evaporator:

$$\dot{Q}_{41} = T_0 \cdot \dot{m} \cdot (s_1 - s_4) \quad (5.73)$$

and $\dot{Q}_{23}$ the amount of heat to be dissipated in the condenser:

---

[24]Can be taken from relevant tables of the food industry, e.g., in [15].

[25]See [46].

Table 5.10 Heat output from persons in dependence the ambient temperature

Temperature cold room in °C	Heat emission per Person in W	Temperature cold room in °C	Heat emission per Person in W
20	180	−5	300
15	200	−10	330
10	210	−15	360
5	240	−20	390
0	270	−25	420

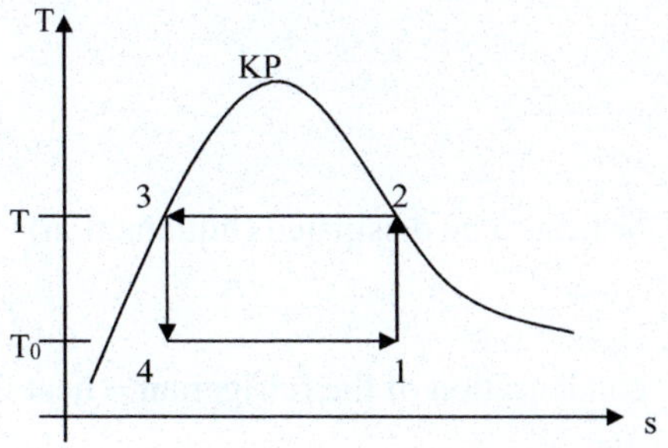

Fig. 5.37 Carnot cycle in the *T–s* diagram (= temperature entropy diagram)

$$\dot{Q}_{23} = T \cdot \dot{m} \cdot (s_3 - s_2) \tag{5.74}$$

with $\dot{m}$ the mass flow of the refrigerant.

The *critical point CP* describes the thermodynamic state of a substance (in this case the refrigerant used), which is characterized by the equalization of the densities of the liquid and gas phases. The differences between the two states of matter cease to exist at this point.

The process can also be shown in in log p–h diagram (Fig. 5.38), which has been created for the individual refrigerants. In this diagram, the pressure of the refrigerant in the system is plotted in logarithmic form above the enthalpy of the refrigerant. Annexure 23 shows such a diagram for the refrigerant ammonia.

This diagram (Fig. 5.38) shows

Required compressor capacity (internal or indexed capacity P_i)

$$P_{12} = P_i = \dot{m} \cdot (h_2 - h_1) \tag{5.75a}$$

or compressor work W_t

$$W_t = h_2 - h_1, \tag{5.75b}$$

heat flow to be dissipated at the condenser

$$\dot{Q}_{23} = \dot{Q}_{AB} = \dot{m} \cdot (h_3 - h_2) \tag{5.76a}$$

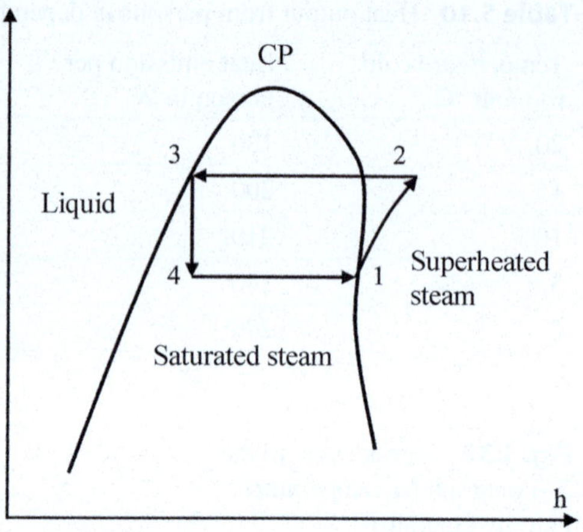

Fig. 5.38 Log p–h diagram for the single-stage refrigeration process (CP = "critical point")

or heat to be dissipated (liquefaction)

$$q_{23} = h_3 - h_2, \tag{5.76b}$$

condensation of the refrigerant = heat flow to be dissipated from the cold room

$$\dot{Q}_{41} = \dot{Q}_0 = \dot{m} \cdot (h_1 - h_4) \tag{5.77a}$$

or heat to be dissipated (evaporation)

$$q_{41} = h_1 - h_4. \tag{5.77b}$$

The expansion of the refrigerant takes place at a constant enthalpy $h_4 = h_3$ (adiabatic state change).

The required mass flow of the refrigerant results from the determined heat dissipation from the cold room and the log p–h diagram to the enthalpies shown in the diagram:

$$\dot{m} = \dot{Q}_0/(h_1 - h_4), \tag{5.78}$$

Compressor drive power (effective power consumption at the compressor shaft)

$$P_e = P_i/\eta_e \tag{5.79}$$

with η_e the overall efficiency of the compressor (effective efficiency). This consists of an internal or indexed efficiency η_i and the mechanical efficiency η_m together:

$$\eta_e = \eta_i \cdot \eta_m. \tag{5.80}$$

Other key figures are common:

To describe the refrigeration process, the Carnot cycle is described with the absolute condensation temperature T and the absolute evaporation temperature T_0 (see above); the so-called degree of cold ε_C of this process is described as follows:

$$\varepsilon_C = T_0/(T - T_0). \tag{5.81}$$

The actual degree of cold ε is below Carnot's degree of cold due to exergy components in the process; the ratio of both degrees of cold is called the *quality of the refrigeration machine*:

$$\varepsilon = \frac{\dot{Q}_0}{P_i} = \frac{\dot{m}(h_1 - h_4)}{\dot{m}(h_{2s} - h_1)}. \tag{5.82}$$

The enthalpy "h_{2s}" indicates that the compaction is isentropic.[26]

$$\eta_G = \varepsilon / \varepsilon_C \tag{5.83}$$

c) Refrigerant

For a long time, ammonia was the refrigerant of choice. Due to its toxicity and flammability, other refrigerants were sought—the so-called safety refrigerants, which are compounds of chlorofluorocarbons (CFC refrigerants), came onto the market. However, these have a strong influence on the depletion of the ozone layer. When the CFC halon prohibition ordinance (FCKW-Halon-Verbotsverordnung 1991) came into force in 1991, Germany gradually phased out the use of these *fully halogenated* refrigerants by 1994. Increased use has been made of partially halogenated hydrocarbons, the hydrofluorocarbons (HFCs). For example, R404A and R410A are often used as refrigerants on ships.

R404A is a Mixture of 44% Trifluoroethane, 52% Pentafluoroethane, and 4% Tetrafluoroethane It is Non-Toxic and Non-Flammable.[27] However, from January 1, 2020 It is Prohibited to Use This Refrigerant in Existing Refrigeration Systems [59].

R410A also HFC-410A or Suva 410A, Consists of 50% R32 (Difluoromethane) and 50% R125 (Pentafluoroethane). However, the Refrigerant Cannot Be Used in the Deep-Freeze Range Because the Final Compression Temperature is Too High. Although R410A Has no Ozone Depletion Potential, It is a Greenhouse Gas and Thus Contributes to Global Warming.

In addition (especially in view of the phasing out of the use of halogenated hydrocarbons as refrigerants), CO_2 (R744) and ammonia (R717) are used. CO_2 is also a climate-relevant gas, but it is not toxic and not flammable. Ammonia (NH_3) has no climate- and ozone layer-damaging properties, but is flammable and toxic.

d) Components of the Refrigeration System

The *compressors* used are piston, rotary, turbo and screw compressors. The *condensers* are mostly air-cooled. They consist of finned tubes, which in some cases are blown on by a fan. In large systems, the refrigerant can also be passed through a heat exchanger in which water is used as the recooling agent for condensation. In shipbuilding, there are essentially two basic types of *evaporators*: the steep tube evaporator and the coil tube evaporator. The *throttle* in smaller systems is a capillary, in larger systems a controlled valve.

[26] *Isentropes* are lines of equal entropy.

[27] See product data sheet of TYCZKA Industrie-Gase.

Example

A compressor chiller should be operated with ammonia at an evaporation temperature of − 10 °C (cold room temperature) and a condensation temperature of 25 °C. Compression is approximately isentropic.[28] What is the cooling degree of the plant? To how much bar is the refrigerant compressed?

Solution: From the log p–h diagram in Annexure 23 and Fig. 5.36 can be found for h_1 at −10° C 1748 kJ/kg. For h_2, one reads at the temperature of 25 °C 1925 kJ/kg. To this end, one goes from h_1 along the isentropes to the temperature 25 °C.

The heat to be dissipated is the difference of $h_2 - h_3$ (see Eqs. 5.76a and 5.76b). When the heat is dissipated in the condenser, the refrigerant is liquefied. The enthalpy h_3 then lies on the condensation curve and can be read with $h_3 = 620$ kJ/kg. In the throttle, the refrigerant is expanded again to point 4, which corresponds to the enthalpy $h_4 = 620$ kJ/kg. Thus the cooling degree ε is calculated according to Eq. 5.82 to

$$\varepsilon = \frac{\dot{Q}_0}{P_i} = \frac{\dot{m}(h_1 - h_4)}{\dot{m}(h_{2s} - h_1)} = \frac{1748 - 620}{1925 - 1748} = 6.4.$$

The pressure to which the refrigerant (here NH_3) is compressed can also be read off from the log p–h diagram,: At the temperature of 25 °C and the isentrope $s = 5.75$ the left ordinate of the diagram shows 1 MPa = 10 bar. ◀

5.6.2 Ventilation and Air-Conditioning Systems

5.6.2.1 Introduction

Regardless of the type of ship, the air conditioning and ventilation conditions on ships have a considerable influence on the well-being of crew and passengers as well as the trouble-free operation of machines, systems and equipment.

Air conditioning create the necessary environmental conditions in ships for climate-sensitive technical equipment and ensure a pleasant indoor climate for the people on board, which is usually achieved at a temperature of about 20–22 °C and a relative humidity of 40–50%.

In addition, the workplace directive ASR A3.5 "Room temperature" [30] prescribes differentiated room temperatures for special working areas (see Table 5.11).

Table 2 of the abovementioned ASRs provides the following indications for the classification of working condition (Table 5.12).

Furthermore, according to the ASR A3.5, the air temperature in working rooms +26° C must not be exceeded.

---

[28] Since entropy and potential temperature are directly related, the term *isentrope* is also used synonymously for lines of equal potential temperature.

Table 5.11 Minimum values of air temperature in working rooms according to Table 1 of ASR A3.5

Predominant posture	Working condition		
	Light	Medium	Heavy
Sitting	+20° C	+19° C	–
Standing, Walking	+19° C	+17° C	+12° C

Table 5.12 Classification of work severity according to Table 2 of ASR A3.5

Work severity	Examples
Light	Light hand/arm work at quiet sitting or rather standing connected with occasional walking
Resources	Medium difficulty hand/arm or leg work at sitting, walking or standing
Heavy	Heavy hand/arm, leg- and torso work at walking or standing

It also states that in break, stand-by, sanitary, canteen and first-aid rooms, the air temperature during the period of use must be at least +21° C; the air temperature in toilet rooms may be temporarily lower than this temperature due to ventilation processes triggered by the users. In washrooms where showers are installed, in the course of usage the air temperature should be at least +24° C.

If the air temperature in a room of +35° C is exceeded (which may be the case in the engine room), the room must be left without

- Technical measures (e.g., air showers),
- Organizational measures (e.g., heat dissipation phases) or
- Personal protective equipment (e.g., heat protective clothing), as for heat work,

not suitable as a workroom.

Furthermore, the "feeling of well-being in a room" depends not only on the room temperature, but also on a corresponding relative air humidity. Table 2 of the workplace directive ASR A3.6 "Ventilation" specifies the following values (Table 5.13).

5.6.2.2 Ventilation Systems

The room temperatures and humidity values mentioned in the previous section can be achieved by a suitable **ventilation.** First of all, *free ventilation* should be mentioned here. The simplest form of free ventilation is window ventilation. It has a high acceptance if the opening of the windows can be determined by the persons on board. Other forms of free ventilation are, for example, shaft, roof-top or chimney ventilation (wind shelter).

The free ventilation of rooms can be described as *shock ventilation* or *continuous ventilation*. In working areas a sufficient *free ventilation* is only guaranteed if the required ventilation cross sections and the maximum permissible room depths according to

Table 5.13 Maximum relative humidity according to Table 2 of ASR A3.6

Air temperature (°C)	Relative Air humidity (%)
+20	80
+22	70
+24	62
+26	55

Table 3 of ASR A3.6 are observed (Table 5.14). Deviations from the required ventilation cross sections specified there are possible if the requirements in Table 5.13 are met even with smaller ventilation cross sections.

Requirements for Free Ventilation

For window ventilation, at least ventilation cross sections according to Table 5.14 are necessary to achieve the requirements for a pleasant room climate. The window openings must be arranged in such a way that a sufficiently even ventilation of the rooms is ensured.

Other forms of free ventilation through natural ventilation (e.g., ventilation chimneys, wind shelters; Fig. 5.39) must be designed accordingly.

The duration and intensity of the air exchange during free ventilation shall be such that drafts are avoided as far as possible.

Table 5.14 Minimum opening area for continuous ventilation and for shock ventilation according to Table 3 of ASR A3.6

System	Maximum permissible room depth relative to the clear room height (h) in m	Opening area to ensure the minimum air exchange	
		For continuous ventilation (m^2/present person)	For shock ventilation ($m^2/10\,m^2$ floor space)
I Single-sided ventilation	Room depth = $2.5 \times h$ (at $h>4$ m: max. room depth = 10 m) (assumed air velocity at cross section = 0.08 m/s)	0.35	1.05
II Cross ventilation	Room depth = $5.0 \times h$ (at $h>4$ m: max. room depth = 20 m) (assumed air velocity at cross section = 0.14 m/s)	0.20	0.60

[a]The specified opening areas are the sum of supply air and extract air areas

Fig. 5.39 Windshield on board the four-masted barque Passat

Ventilation Systems[29]

Free ventilation—natural ventilation—is now only used in exceptional cases on small ships. Today, air supply and disposal is usually carried out as positive pressure ventilation by means of room air conditioning systems (AC systems). For this purpose, the supply air is conveyed to the areas to be ventilated (in particular the engine room, cargo holds on ferries) by means of electromotive fans. The exhaust air is led outside via exhaust chimneys (Fig. 5.40), flaps or louvers.

In the case of air conditioning systems, the supply air (outside air/circulating air) must be cleaned by state-of-the-art air filters before being fed into the areas to be ventilated, in accordance with the requirements for the use of the rooms. The AC system must not itself become a source of danger (e.g., through hazardous substances, bacteria, mold, or noise).

[29]More detailed information on ventilation and air-conditioning technology: FVLR guideline "Natürliche Lüftung großer Räume" (Natural ventilation of large rooms) [42].

Fig. 5.40 Air supply and exhaust stacks of a ferry

The outdoor air volume flow rate must be designed according to the state of the art in such a way that room air loads (material, moisture, heat loads) are reliably discharged.

Exhaust air from rooms with loads (material, humidity, heat loads) may only be used as recirculated air if health hazards and nuisance can be eliminated. Exhaust air from sanitary rooms, smoking rooms (fishing factory ships) and galleys shall not be used as supply air.

The AC system must not generate any unacceptable drafts. Draft is mainly dependent on air temperature, air speed, air turbulence in the room and the type of activity (i.e., heat generation through physical work). At an air temperature of $+20^{\circ}$ C and an average air velocity < 0.15 m/s there is usually no unacceptable drafts in light work. With greater physical activity and other air temperatures, the value for the average air velocity may deviate. Too high air speeds can lead to drafts. For example, the air velocity at office workplaces and comparable rooms (e.g., the bridge) ≤ 0.2 m/s should be included.

Ventilation measures are also aimed at reducing CO_2 value in room within tolerable limits. A too high CO_2 value leads to tiredness, lack of concentration and can cause headaches; this can already be the case with CO_2 concentrations from 1000 ml/m^3 or ppm.

In rooms where a large number of people are present and the air quality should meet the hygienic requirements of DIN 1946, Part 2 or DIN EN 13779, the fresh air volume flow is controlled via the *fresh air rate* is determined by the fresh air rate for each person.

As a guide, a *air requirement* (fresh air volume) of 2536 m^3/h per person should be assumed, whereby 20 m^3/h per person is the absolute *minimum air rate* and should not be less than this [55].

The *rate of fresh air supply* indicates how many times per hour the room air is completely exchanged for fresh air/treated recirculated air. Using the data for air demand per person and the corresponding *rate of fresh air supply* (*AS came into force R*) for a specific room, a rough dimensioning of room and ventilation can also be carried out [62]. Table 5.15 gives reference values for the required air exchange rate (fresh air supply rate) according to DIN 1946, Part 2, depending on the use of the room. From this, the required amount of fresh air per hour is determined according to Eq. 5.84

$$\dot{V} = P \cdot \text{ASR}\ [\text{m}^3/\text{h}] \tag{5.84}$$

with

$\dot{V}$ = required air volume in m^3/h,
P = number of persons in a room,
ASR = fresh air supply rate per person to Table 5.15

Example

The bridge of a small KüMos (German: Küstenmotorschiff, engl.: coastal motor vessel) is occupied by three people. How much air must be exchanged every hour?

Solution: The bridge can be regarded as office space. Due to the high level of attention required from bridge personnel, an air exchange rate or fresh air supply rate

Table 5.15 Outdoor air rate (ASR) per hour and person

Species of room	ASR in m^3/h per Person
Office space	40–60
Meeting rooms, conference room	20
Shower	40
Kitchen, galley/pantry	40 (continuous operation) 150–600 (on demand)
Dining rooms and restaurant area	40
Smoking rooms (DIN EN 13779)	90
Theater, cinema	20
Salesrooms	20
WC	10–20 (continuous operation) 30 (on demand)
Living space (crew- and passenger cabins), relaxation rooms	30
Exhibitions and common rooms (canteen)	30

of 60 m³/h per person. According to Eq. 5.84, this results in a required air volume that must be exchanged per hour:

$$\dot{V} = 3 \text{ Pers.} \cdot 60 \text{ m}^3/\text{h pro Person} = 180 \text{ m}^3/\text{h}.$$

◀

In addition to determining fresh air volume via the fresh air supply rate, which is related to the number of people in a room, air volume determination can also be calculated using air exchange rates (AER).[30] Here it is important to first define the type and purpose (use) of a room. To ventilate a passenger cabin, for example, a lower air volume will be required than in a bathroom.

The AER (hourly **air exchange rate** (German: **Luftwechselrate**)—how often the room air is exchanged per hour) is to be taken as a recommendation from Table 5.16 and is inserted in Eq. 5.85:

$$\dot{V} = V_R \cdot \text{AER } [\text{m}^3/\text{h}] \tag{5.85}$$

with

$\dot{V}$ = required air volume per hour in m³/h,
V_R = room volume in m³,
AER = recommended air exchange rate according to Table 5.16

Example of Fresh Air Volume Determination using AER

The bridge in the example above has a volume of $V_R = 45\,\text{m}^3$. How large is the required air volume per hour?

Solution: The bridge can be considered an office workplace. However, since there are increased demands on the concentration of the bridge personnel, an AER of 7/h should be used. From Eq. 5.85 follows:

$$\dot{V} = 45 \text{ m}^3 \cdot 7/\text{h} = 315 \text{ m}^3/\text{h}.$$

This results in a significantly higher air volume per person compared to the air volume determination using ASR. However, it can also be concluded from this that an 45 m³ large (bridge) room can also be occupied by considerably more than three people before unacceptable working conditions, especially due to CO_2 exposure. ◀

Example

How many soldiers can serve in the operations center (OPC) of a warship so that they can enjoy tolerable working conditions? The ventilation system is state of the art, the OPC has a room volume of 150 m³.

[30]To the following [61].

Table 5.16 Recommended air exchange rates. [61]

Type of room	AER/h
Bathroom	5–7
Battery rooms (hazardous-area)	5–10
Meeting rooms	5–8
Office space	4–8
Showers	15–25
Kitchen, galley, pantry	15–30
Cabins, living space	3–6
Smoking rooms	upto 20
Restaurant areas, exhibitions	6–8
Swimming pool	3–4
Gym, gym	4–8
Theater, cinema	5–8
Toilet (in cabin)	5–8
Public toilets	5–15
Workshop with less air deterioration	4–6
Workshop with strong air deterioration	10–20

Solution: The OPC can be regarded as a conference room according to Table 5.15 and 5.16 where high attention is required from the soldiers. Therefore: $AER/h = 8$, ALR = 20 m^3/h per person.

Since both Eqs. 5.84 and 5.85 determine the required amount of fresh air, both can be equated and switched according to the number of persons:

$$P = (V_R \cdot AER)/\text{ALR} = (150\ \text{m}^3 \cdot 8/\text{h})/20\ \text{m}^3/\text{h and person} = 60\ \text{persons.}$$

As a result, a maximum of 60 people could work in the operating theatre under the ventilation conditions mentioned (which is not the case in an operating theatre of this size!). ◄

The required electrical fan power P_{el} is calculated according to the following equation:

$$P_{\text{el}} = \frac{\Delta p \cdot \dot{V}_L}{\eta_L} \tag{5.86}$$

with

Δp = differential pressure (N/m^2),
$\dot{V}_L$ = air volume flow (m^3/s),
η_L = fan efficiency

Since the electrical energy is provided by diesel generator sets or shaft generators (see Sect. 5.8), it is necessary to work toward reducing the electrical requirements for the fans for reasons of environmental protection (reduction of the pollutant emissions of the diesel generator sets) and for reasons of minimizing fuel consumption costs. For this purpose, the fan power must be reduced. From Eq. 5.86 it can be seen that this can be achieved by reducing the pressure loss in the system of ventilation ducts and by increasing the fan efficiency. The air volume flow can only be reduced to a limited extent, as it is predetermined (see above).

On Efficiency

As the fans are mainly continuously operating, good efficiency of the fan and drive motor should be ensured. For conventional engine room ventilation systems, for example, of 60 kW rated output, an efficiency improvement of 5% results in an annual saving of electrical energy of approximately 25,000 kWh.

For Differential Pressure

The geometry of the fan ducts must be optimized for low losses. For this purpose, the cross section should be as large as possible, and the ducts should be as straight as possible. However, compromises must be made here due to shipbuilding conditions.

5.6.2.3 Air Conditioning Systems

Air conditioning systems are used to create the abovementioned required room air conditions (temperature, humidity, purity and CO2 content) on the ship. They have the task of bringing the air in a room to a certain state and keeping it there ("conditioning"). Since extreme air conditions and aggressive sea air are to be expected in ship air conditioning, these systems must meet the highest quality standards. Here, great importance must be attached to material suitability with regard to corrosion behavior. Metallic parts are therefore often powder-coated [26] or even made of stainless steel. In the field of ocean shipping, it is not unusual for air conditioning systems to pass through all the climatic zones of the world—the air conditioning systems must therefore always reliably guarantee the abovementioned room air values for well-being over a wide climatic range.

DIN EN ISO 7547:2009-09 "Ships and marine technology—Air conditioning and ventilation of accommodation spaces on ships—Principles of design and layout" contains specifications for the design and layout of ship air conditioning systems.

The functions of an air conditioning system are therefore:

1. Change the air temperature (heating or cooling),
2. Change the humidity (humidify or dry),
3. Remove air components (filter or replace),
4. Change the local air speed.

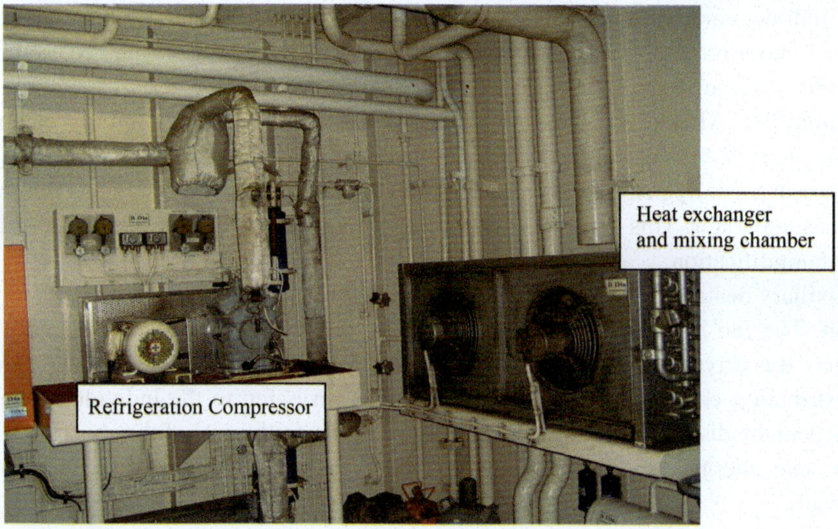

Fig. 5.41 Partial view air conditioning system of a tanker

Smaller air conditioning systems (e.g., decentralized systems installed in individual ship cabins) often do not control all of the above functions.

Design of Ship Air Conditioning Systems[31]

The core component of any air conditioning system is the central unit (Figs. 5.41 and 5.42), where the treatment and adjustment of the supply air parameters takes place. On new ships today, almost only modular units are used. They consist of

- Mixing chamber, here fresh air and return air is mixed, for the crew 100% fresh air and no recirculated air is used,
- Filter (partly also activated carbon filter) for air purification of dirt, dust, bacteria and odors,
- Air preheater, heating media can supply hot water (temperature 80–90 °C), thermal oil, and steam or electric heating elements,
- Air cooler/air dehumidifier: The filtered air enters the air cooler, cooling down to temperatures of approximately 12–14 °C. Dehumidification takes place. The condensate is collected under the cooler and fed to the ship's drainage system. With the air cooler, a distinction is made between the direct and indirect cooling process:
  - Direct cooling process: The refrigerant evaporates directly in the air cooler (mainly used on cargo ships),

[31]For the following see also: [5]; [1, pp. 49–114]; [21].

  - Indirect cooling process: Cold water is fed to the cooler with an inlet temperature of about 6–7 °C, cooling of the cold water in the evaporator (mainly used on ferries and cruise ships).
- Humidifier: After leaving the preheater, the air has a relative humidity of approximately 20–30%; enrichment to 50–60% by means of steam or spray humidification is necessary. For hygienic reasons, steam humidification is preferable.
- Water separator: This separates water droplets entrained from the air cooler or caused by humidification.
- Auxiliary heating,
- Fan: The fan is the only moving part of the air conditioning system (mainly radial fans). It is driven by an electric motor (mostly frequency controlled or pole-changing).
- Distribution chamber: The airflow leaving the fan is fed to the individual supply air ducts in the distribution chamber (number depends on the size of the ship).
- To save energy, heat recovery systems are sometimes installed (regenerator, heat wheel).

Figure 5.42 shows the principle diagram of a ship air conditioning system.

In principle, a distinction is made between single- and dual-duct systems.

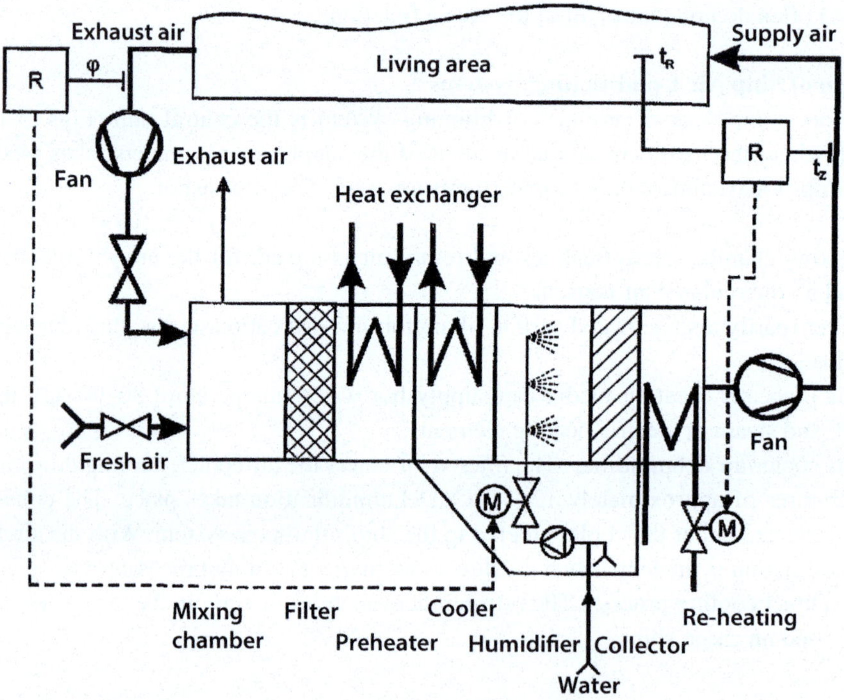

Fig. 5.42 Schematic representation of a typical air conditioning system on merchant ships. *Drawing* Hochhaus, K.-H.

Single-duct System with Electrical Reheating and Heat Recovery (see also Fig. 5.42) Here, an air duct is led from the central air conditioning unit to the air outlet openings in the rooms. This means that all the necessary air treatment (filtering, air cooling/heating, and humidification) takes place in the central unit. The volume flow is controlled in the outlet units of the cabins. Furthermore, electrical reheating can also be carried out there. This means that each room has a variable volume flow rate, and within narrow limits, a readjustment of the temperature.

The exhaust air is sucked out of the cabins via the corridors (and staircases) and remixed as recirculated air with the newly supplied fresh air in the central air conditioning unit. The exhaust air of the sanitary and kitchen areas is discharged separately to the outside.

Dual-duct System with Circulating Air Component Two separate air lines (one each for cold and warm air) supply the rooms. The basic conditioning takes place in the central unit. The hot air flow is raised in temperature by a reheater, while the cold air flow is led directly to the outlet devices. In the outlet unit, hot and cold air flows are then mixed according to the room temperature set on the room thermostat.

Air Conditioning Systems on Passenger Ships and Ferries

Passenger ships and ferries require a higher degree of comfort compared to merchant ships [47]. This requires a greater degree of technical effort, especially in terms of system regulation. Generally, the systems used on merchant ships are also available. As a result of the disadvantages of single and double duct systems, a third variant, the fan coil system, has become established in recent years in the construction of large cruise ships.

This type of system is based on the principle of decentralization of air conditioning. In the air-conditioned room, pre-conditioning is carried out by considerably smaller central units. The preconditioned air enters the rooms via a single-channel system. In each room there is a fan-coil unit instead of the usual outlet units, which can heat and cool the air locally. This consists of a filter for secondary air, an air cooler with control valve, an electric reheater and a fan (at least three-stage).[32] The fan coil units are either mounted vertically in or on the wet cell or horizontally in or on the cabin ceiling and have a condensate and cold water connection. The primary air coming from the ducts is mixed in either on the pressure or suction side.

One of the most important arguments for air conditioning based on decentralized units is the small space requirement of fan coils.

Air Conditioners with Direct Cooling

For air conditioning systems for **cargo ships** is usually used with direct cooling, that is, the evaporator is located directly in the central air conditioning unit.

[32]See also [18].

Air Conditioning Systems with Indirect Cooling
For air conditioning systems above approximately 400 kW and several branched air conditioning units or when using absorption refrigeration systems (see below), the use of central **liquid cooling units** (Figs. 5.42 and 5.43). Water with a flow temperature of approximately +6° C is selected.

Cold Water Network
For the construction of the cold water network, a series connection of the heat exchangers (Fig. 5.44) with a common pump, a parallel connection of the heat exchangers (Fig. 5.43) with a common pump or a parallel connection of the heat exchangers with single pumps are possible.

In addition to refrigeration by means of a compressor refrigeration plant, hybrid refrigeration is also used as an innovative technology (Fig. 5.45).

Hybrid air conditioning systems consist of two refrigeration systems (compressor and absorption system). In compression refrigeration systems, energy is supplied in the form

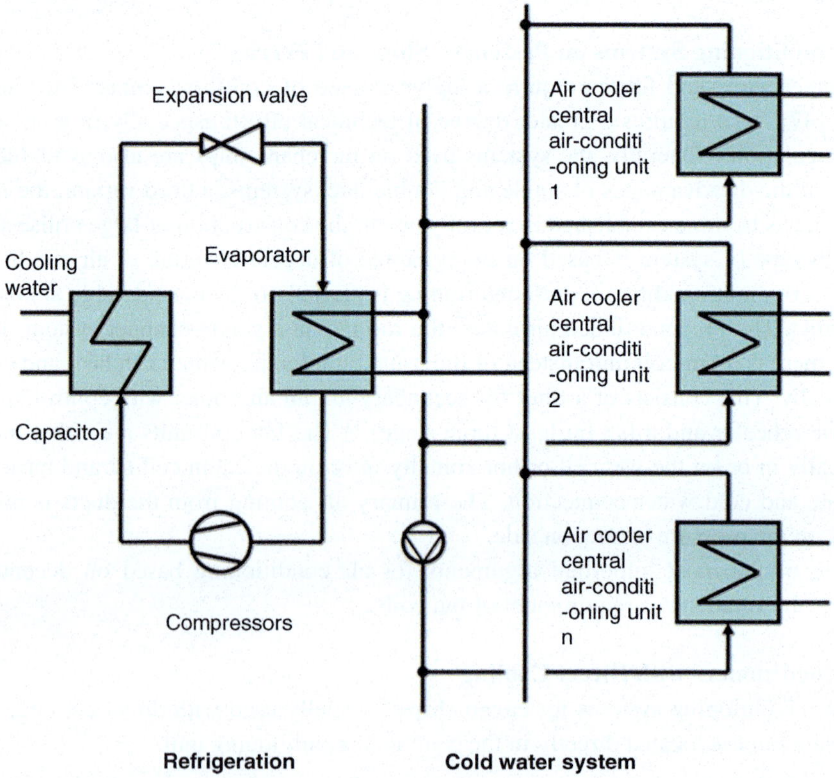

Fig. 5.43 Indirect air conditioning system for passenger ships and ferries. *Drawing* Hochhaus, K.-H

Fig. 5.44 Heat exchanger set of a cooling unit

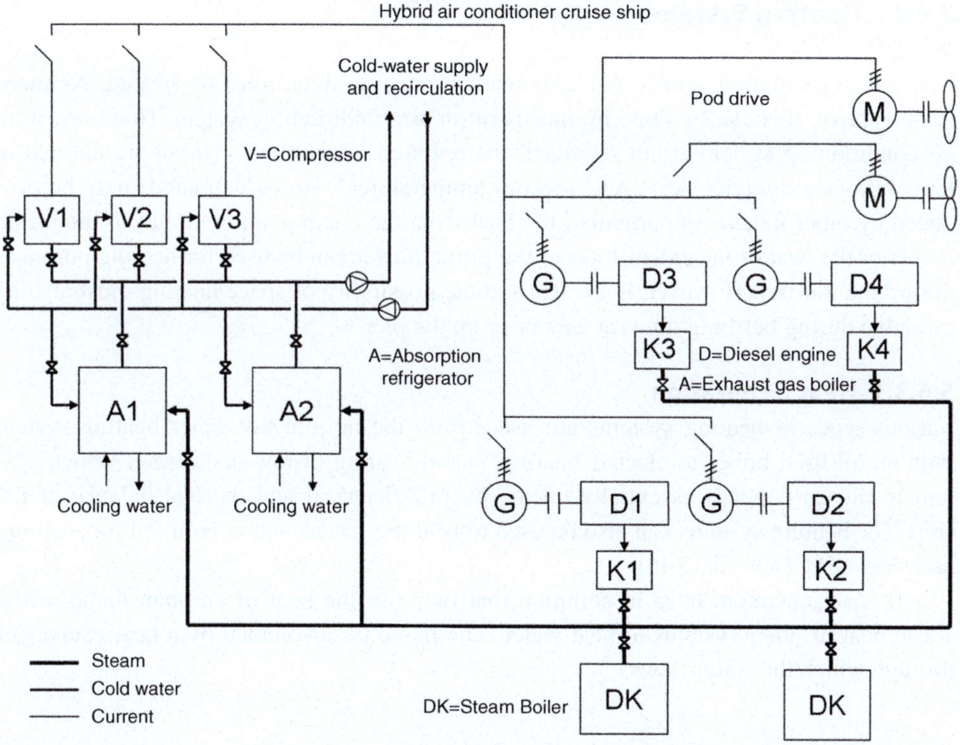

Fig. 5.45 Hybrid air conditioning system consisting of two refrigeration units (compressor and absorption unit) of a cruise ship. *Drawing* Hochhaus, K.-H

of mechanical energy in the refrigerant compressor (see also Sect. 5.6.1 on compressor refrigeration technology). In the absorption chiller, on the other hand, energy is supplied by thermal energy (steam from the exhaust gas boiler, hot cooling water) in the cooker.

In addition to the refrigerant, an absorbent is required that has the ability to absorb the refrigerant vapor (absorb, working fluid pairs). This is done in an additional component, the absorber.

Absorption chillers have a higher energy consumption and thus a lower efficiency compared to compressor chillers. This is not a disadvantage if the waste heat is available free of charge. If the waste heat can be used with short distances, as on a ship, primary energy can be saved and thus CO_2 emissions.

The working material pair lithium bromide/water is used for air conditioning (the working material pair water/ammonia is used for low temperatures). The advantages are the silent operation and the long service life, as the units contain no mechanically moving parts and thus practically no wear parts apart from centrifugal pumps. A prerequisite for trouble-free operation, however, is very careful processing during welding, rinsing and filling of the units to prevent sludge formation and blockage in the pipeline system.

5.6.3 Heating Systems

Living and recreation rooms, but also work rooms on ships must be heated. As mentioned above, this can be done by means of an air conditioning system. However, if an air conditioning system is not provided, the required temperatures (these are defined in the workplace directive ASR A3.5—room temperature[33]—to be extracted) may be produced by other means. In particular, the heat from the cooling water circuit of the prime mover or the heat of the exhaust gas of the prime mover can be used for heating purposes during the journey. However, hot water and the possibility of space heating must also be available during berthing times at anchor or on the pier.

5.6.3.1 Heat Generation

Various types of heating systems are used, from the normal hot water heating system with an oil-fired boiler to electric heating, steam heating or hot air heating. Which system is chosen depends essentially on the overall electrical and thermal balance of the ship. The heating systems can also be used to heat the service water required for washing and showering (see Fig. 5.46).

All heat generators have in common that they use the heat of an open flame or the waste heat of other systems to heat water. The flame is surrounded by a heat exchanger through which the water flows.

[33]The workplace guidelines can be found on the website of the Federal Institute for Occupational Safety and Health [65].

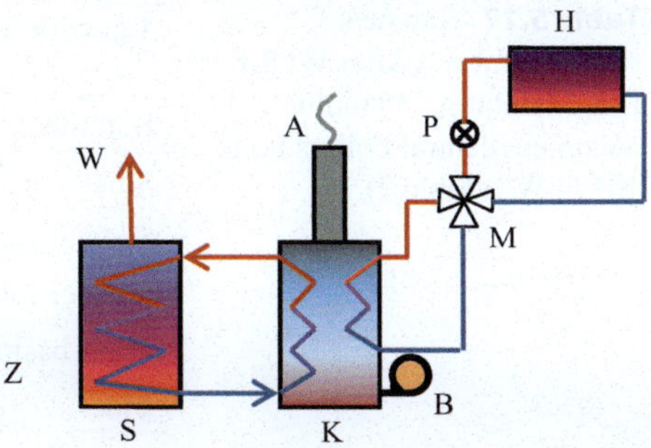

Fig. 5.46 Principle of an oil-fired heating system with domestic hot water heating. *A* Exhaust stack, *B* Brenner, *H* Radiators for space heating, *K* Boiler, *P* Heating circuit pump, *S* Storage tank for service water, *W* Domestic hot water warm, *Z* Cold service water inlet

The heat exchangers in the boiler should be as large as possible in relation to the burner output (this is approximately the heat output of the flame). This ensures that the combustion air of the flame is very strongly cooled. Low exhaust gas temperatures ensure that the water components in the exhaust gas condense. The desired effect is called condensing technology.

For the design of a heating system, the heat requirement that must be provided by the system is the decisive factor. First of all, the power requirement for the *service water heating* is considered [33]:

The heat output requirement for domestic hot water preparation and the design of an installed hot water storage tank depends on the use of the ship (passenger ship, ferry, freighter, naval ship, etc. a.) and the associated requirements for hot water comfort, the total consumption of hot water and the system technology used (with and without water storage in the tank). To calculate the output for hot water preparation, a cold water temperature of 10 °C accepted. Hot water temperatures fluctuate depending on the type of extraction source. Indicative values are 45 °C for wash basins, showers, bathtubs and 60 °C for kitchen purposes and sanitary areas.

The amount of hot water consumed in a day can be determined using various methods, but these are applicable to residential buildings. However, they can be used as a guide for shipping. The DIN 4708 method is commonly used, in which the output for the hot water preparation of a building is determined on the basis of a demand indicator *N*. This is related to an apartment with standard equipment. With an occupancy of 3.5 persons in four rooms, this has a heat requirement of $w = 5.82$ kWh/d and a demand indicator $N = 1$. The demand indicator of a residential building is determined according to Eq. 5.87:

$$N = \frac{\sum (n \cdot P \cdot v \cdot w)}{3.5 \cdot 5.82} \tag{5.87}$$

with

$n =$ number of similar residential units,
$P =$ person occupancy the residential units,

Table 5.17 Tap heat requirement *w*. (After WOLF Ltd, Mainburg, "Planning document central boilers up to 1017 kW", Table 2)

Tap with withdrawal per use (L)		Heat demand *w* (kWh)
Bathtub	140	5.82
Small capacity bathtubs	120	4.89
Shower cabin	40	1.63
Shower cabin	100	4.07
Wash basin	17	0.70
Bidet	20	0.81
Sink	33	1.16

v = number of relevant taps,
w = heat requirement of the taps according to Table 5.17

Example

On a container ship there are three single chambers and five double chambers, each with a corresponding wet room with washbasin and shower. In the galley there is a tap for hot water. How high is the demand indicator N?

From Eq. 5.87 and Table 5.17 follows:

$$N = \frac{(3 \cdot 1 \cdot 1 \cdot 1.63) + (5 \cdot 2 \cdot 1 \cdot 1.63) + 1.16}{3.5 \cdot 5.82} = 21.25.$$

Only the tap with the highest power requirement is taken into account (relevant) per residential unit. Consequently, in Eq. 5.87 only showers are considered, but not washstands. ◀

For residential buildings such as hotels (which may include cruise ships and possibly ferries), increased simultaneous use of the standpipes must be expected. The demand indicator N calculated for residential buildings must in this case be corrected according to practical experience. Reference values for N_{corr} according to Table 5.18 can be used.

According to the calculated demand indicator N, the coefficient of performance N_L of the hot water tank is selected according to the manufacturer's specifications, whereby $N_L \geq N$. If the hot water tank is selected, the heat output in kW required for this tank results from the manufacturer's specifications.[34]

Example

For the container ship mentioned in the above example, a demand indicator of $N = 21.25$ has been determined. An appropriate memory should be installed. Please refer to the data sheet for the memory (extract):

[34]See for example [51].

Table 5.18 Reference values for N_{corr}

N_{corr} to DIN 4708	10	20	30	40	50	100	200
N_{corr} for mainly double cabins with tubs	47	80	111	140	165	280	480
N_{corr} for mainly single cabins with tubs or showers	39	68	90	116	140	240	400
N_{corr} for mainly single cabins with showers	30	54	72	88	105	180	320

Key performance indicator N_L according to DIN 4708:

Storage capacity in liters	300	500	750
N_L at heating water flow temperature			
90 °C	9.7	21.0	40.0
80 °C	9.3	19.0	34.0

According to this data sheet, a storage tank with the key performance indicator N_L 34.0 and a heating water flow temperature of 80 °C must be selected. From the relevant data sheets of the manufacturers, the corresponding storage volumes and their capacity in kW can also be read.

Furthermore, the *heat output for heating purposes* must also be determined when dimensioning the heating system. In the rooms to be heated, the heat is provided by radiators. The heat output of the required radiators depends on the transmission of heat through the room boundary surfaces (walls, ceiling, floors, doors, windows) of the room to be heated (heat transmission). It is calculated according to the equation

$$\dot{Q} = k \cdot A \cdot \Delta T \tag{5.88}$$

with

$\dot{Q}$ = performance in W,
k = heat transmission coefficient in W/m²K,
A = area in m²,
$\Delta T = T_i - T_a$ (T_i or rather T_a = indoor or rather outdoor temperature) in K

The heat transfer must be determined separately for each partial area. For the room under consideration, the individual values must then be added together to form a total heat transfer rate. Furthermore, the dimensioning of heating systems must take into account the ventilation heat losses (free ventilation or air conditioning system), a possible night-time reduction of the heating system, reheating processes etc.

For the heat transfer coefficient k (also called "U-value" in the building industry) through a wall which can consist of several layers (outer skin, insulation, inner lining):

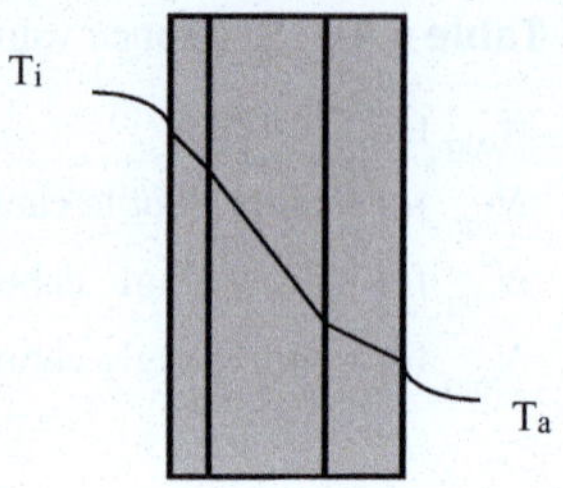

Fig. 5.47 Temperature progression through a multilayer wall

$$\frac{1}{k} = \frac{1}{\alpha_1} + \sum_{i=1}^{i=n} \left(\frac{s}{\lambda}\right)_i + \frac{1}{\alpha_2} \tag{5.89}$$

with

α = heat transfer coefficient inside or rather outside in W/m^2K,
λ = thermal conductivity of respective wall material in $W/m \cdot K$,
s = respective wall thickness in m

The temperature curve through a multilayer wall is shown in Fig. 5.47

Table 5.19 shows λ values of various materials.

Table 5.20 gives reference values for α for different states of the heat transfer between air and wall.

The values in Table 5.21 can be assumed as approximate *k-values* in building industry [25].

The procedure for determining the respective heating capacity of a room by means of the heat transfer is very complex.[35] The location and nature of the room must be taken into account: Inside or outside cabin (in the case of inside cabins, heat passage through walls, ceilings and floors is negligible if the surrounding rooms have the same temperatures), heat loss through ventilation systems, heat supply by persons and/or electrical devices (e.g., in an operations center of a warship) and other things.

In practice, therefore, approximate sizes are often used. An approximate heat output per square meter of space is assumed here, whereby the type of space, in particular its insulation, is decisive. For rough calculations one can probably assume 170 W/m^2 for poorly insulated vessels and 70,100 W/m^2 for well-insulated ships (modern cruise ships). For sanitary or shower rooms a surplus of 10% should be assumed.[36]

---

[35]Under this link [60] there is a "*U*-value calculator" with which the *k*- or *U*-value for various material compositions and air-wall processes can be determined.

[36]In the building industry, a rough estimate is made of 100 W/m^2 for new buildings and 150 W/m^2 for poorly insulated old buildings with single glazing (see [43]).

Table 5.19 λ-Values of selected materials in shipbuilding

Material	Thermal conductivity λ
Aluminum	200
Glass	0.8
Wood	0.2
Fiberboard soft/hard	0.06/0.17
Air, dormant (e.g., between two glass panes)	0.02
Mineral insulation board	0.045
Polyurethane rigid foam (PURE)	0.027
Steel	60
Rock wool	0.04
Vacuum insulation panel	0.006

Table 5.20 Values for the heat transfer coefficient α

		Heat transfer coefficient α (W/m²K)
Air perpendicular to metal wall	At rest	3.5…35
Air perpendicular to metal wall	Moderately moved	23…70
Air parallel to flat metal wall (ship in motion with speed v in m/s)	$v<5$	$5.8 + 4 \cdot v$
Air parallel to flat metal wall (ship in motion with speed v in m/s)	$v>5$	$7.14 \cdot v^{0.78}$

Table 5.21 k-values in construction

$k=1.2$	Good insulation
$k=2.2$	Building from 1975
$k=3$	Old buildings
$k=4$	None or almost no insulation

The guide values refer to a heating system according to the EN 442 standard with a flow temperature of 75 °C and a return temperature of 65 °C. For other supply and return temperatures, the following surpluses must be applied:

- at 70 °C/55 °C factor 1.24,
- at 55 °C/45 °C factor 1.94.

However, approximate values can also be assumed for the spatial volume, for example, for yachts 120…200 W/m³ [40].

Furthermore, for the sake of simplicity, often only the *k*-value of the wall, determined by its thermal conductivity and wall thickness, without heat transfer components inside and outside (α_1 and α_2) is taken into account for the determination of the heat transfer. This simplification is made because the convective components in a room are negligible, while on the outside of a ship—even when it is in motion—they are comparatively low. Surely the heat transfer determined in this way is lower than if it was based on an exact calculation according to Eqs. 5.88 and 5.89. However, this error is taken into account when choosing a radiator by choosing a radiator with the next higher heat output according to the manufacturer's specifications. Any resulting oversizing of the radiator can be accepted, as the desired room temperature can be achieved by a lower heating water flow rate.

If ships still use heavy fuel oil as fuel, a *tank heating* is required to make the viscous heavy oil pumpable (approx. 60…70 °C). For this purpose, heating coils are installed in the heavy oil tanks at the bottom. Furthermore, radiators arranged around the suction port serve to heat the heavy fuel oil. The heating coils at the tank bottom are usually made of steel pipes, in old ships also cast iron pipes. The pipes are fitted with fins to increase the heating surface; cast iron pipes must be loosely supported (2 mm Air) to prevent the pipes from breaking as a result of hull movements. Steel heating coils are welded together in great lengths in order to obtain the smallest possible number of pipe fittings. The individual pipe sections are joined together by means of flange or socket joints and are held together approximately every 2 m. With regard to tank cleaning, they are placed at a height of about 150 mm above the tank bottom. The size of the heating surface depends on the location of the individual tanks (outer tanks require more heating surface, inner tanks less), the type of oil to be transported and the route on which the ship is to be used. The average value for the size of the heating surface can be assumed:

- 0.03 m^2/m^3 of cargo space for central tanks,
- 0.04 m^2/m^3 of cargo space for side tanks.

Tank heaters are usually operated with steam, which is usually taken from a so-called steam converter. The condensate produced in the process flows back to it via control devices and oil separators [14].

Once all heat quantities have been determined, they are added together and thus determine the required output of the heating system. ◄

5.6.3.2 Heat Distribution

In the ship, the energy from the heat generator is fed to the radiators via a pipe system through which the heated water flows [44]. The distribution is controlled by valves. To adjust the temperature of the room to be heated, thermostatic valves, which regulate the water flow on the respective radiator, are used. The heating valves can also be controlled by a thermostat installed in the room (see Fig. 5.48).

Fig. 5.48 Room thermostat for floor heating

With thermostatic valves on radiators (Fig. 5.49), a medium in the cap of the valve is influenced by the room temperature. This makes use of the property that materials stretch depending on temperature. The force of this expansion is transferred to a small pin which influences the pipeline cross section. Because the force for expansion is obtained from the energy of the room air, the thermostatic valve is also referred to as a valve without auxiliary energy, as no additional electrical connection is required.

Also common valves can be found directly at the hot water distribution. These valves usually control less the distribution than the temperature level at which the water flows through the pipes. These control elements are very important for the efficiency of a heating system as a whole, as high efficiencies can be achieved with very low temperatures. The main influencing variable here is the outside temperature; we then speak of a “weather-controlled flow temperature”.

During a cold setback, the outdoor temperature-controlled flow control reacts with higher flow temperatures. At the same time, the burner running times per switch-on are higher.

5.6.3.3 Pipelines

When heated water is distributed by pipelines, steel, copper or stainless steel pipes are used, which are welded or soldered at their joints [44]. Recently, the so-called pressed joint has become generally accepted for stainless steel and copper pipes due to a faster assembly technique. The metals at the connection points are formed in such a way that the seal is ensured by means of an O-ring insert and deformation (see Fig. 5.50). Plastic pipes are also used to some extent, but these are used more in yacht building. Plastics have the advantage that they are more flexible. Screwed and glued connections are common. Plastic pipes are often used for surface heating (underfloor heating)—for example,

Fig. 5.49 Towel radiator with thermostatic valve

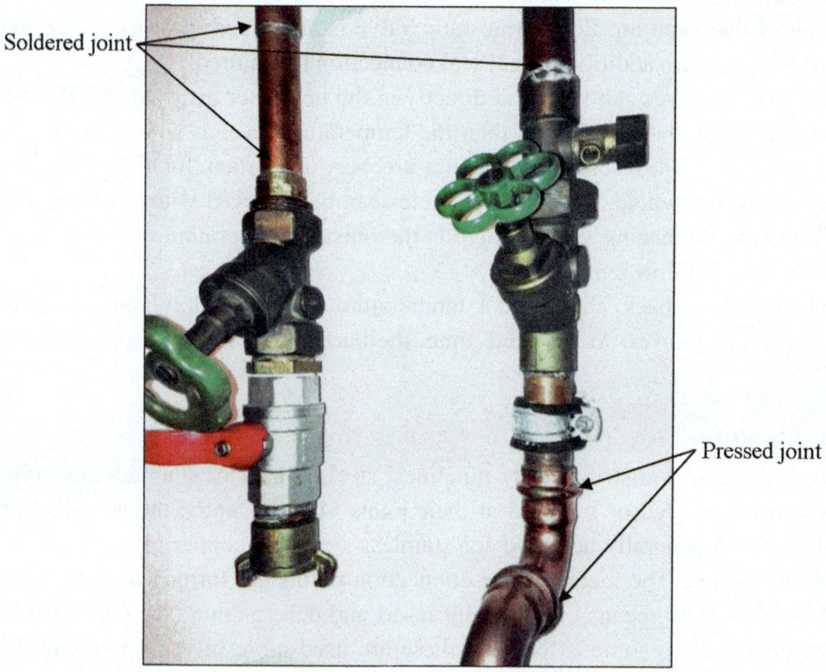

Fig. 5.50 Soldered and pressed joints on water pipe

in the bridge area. Here, the pipeline, in addition to the function of heat distribution, also has the function of a classic heat exchanger or heat consumer.

Excursion "Soldering of copper pipes"
Water pipes are joined together by so-called soft soldering. Here the soldering temperature is max. 450 °C. If soldering is carried out at higher temperatures, this is called brazing. Depending on the different soldering temperatures, different soldering agents—soft or hard solders—are used.

After the tubes have been cut to length, the ends of the burr must be deburred using a file or, for example, a triangular scraper. Then clean the pipe ends at the soldering point with steel wool or sandpaper. It is important that the copper pipe at the soldering point is absolutely clean and free of grease. Then the soldering point is coated with soft solder paste or another soft solder flux. This flux and antioxidant causes the solder to be drawn into the soldering gap between the copper tube and fitting due to the capillary effect and the copper tubes do not oxidize at the soldering points due to the heat effect, which would impair the flow of solder. Then the pipe and fitting are joined together by turning, so that the flux is distributed even better.

Pipe and fitting are heated with the gas flame. Care must be taken to ensure that both workpieces are heated well and evenly. If you hold the burner only against the pipe, the pipe will get hot quickly, but the fitting will not yet have the required temperature. The correct temperature of the workpieces is reached when the solder melts off by itself when it approaches the soldering gap and is drawn into it (if this is not the case, the temperature is still too low; in this case, do not hold the burner flame on the solder, but reheat the soldered joint). The soldering point is sufficiently filled with solder if a clearly visible soldering tin edge forms at the soldering point which is not drawn further into the soldering gap (see Fig. 5.48). With horizontally laid pipes, a small drop of solder can then also form on the underside of the soldering point. Then clean the soldering point with a cloth.

5.6.3.4 Radiators

Among the heat consumers, one finds the classic radiator, the floor heating or the wall heating [44]. But heat exchangers on ventilation systems are also included.

The radiators (heat consumers) transfer the heat to the air in a room. The transfer takes place in two ways: Firstly by convection. This is the direct heating of the air. Secondly, by means of radiation: Here, energy is emitted from the surface of the radiator in the form of radiant heat. Depending on the temperature difference, radiation increases or decreases more than heat transfer by convection.

Therefore, radiators that give off their heat both convectively and through radiation work very effectively. As a rule, approximately 75% of the heat is transferred to the room by convection and approximately 25% by radiation. The optical variations for such radiators are extremely diverse. It is interesting to know that radiators with the same surfaces are more efficient if they are mounted on edge instead of lengthways. This is due to the fact that the room air to be heated is increasingly buoyant along the radiator, thus improving the heat transfer between radiator and air.

Other radiators are multifunctional, such as a towel radiator (Fig. 5.49) or a wardrobe heater, which also functions as a design element. There are no limits to creativity: For example, shapes such as flat, round, symmetrical or asymmetrical are available on the market for designer radiators.

"Design-neutral" are panel heating systems. Usually in the form of underfloor heating systems, they provide pleasant warmth at relatively low temperatures. Underfloor heating systems in particular provide pleasant foot warmth in living areas and allow for a free design of the rooms without having to consider the space and appearance of radiators. Underfloor heating systems work efficiently when the floor covering insulates as little as possible. For this reason, heat-insulating cork or wooden floors are not as suitable as stone or tile flooring, which is otherwise perceived as cold. What all panel heating systems have in common is that all fixing work has to be carefully considered in relation to the drilling depth. Damage to the system due to an incorrectly placed drill hole in the floor results in an enormous repair effort. This type of heating has a higher efficiency than conventional wall radiators. This has to do with the large heat exchanger surface, which allows a considerably lower flow temperature for the same performance.

5.7 Fresh and Drinking Water Production

5.7.1 Introduction

The fresh or drinking water supply consists of water production, water treatment, water storage in special water tanks (bunker systems), and the (drinking) water supply system.[37]

Fresh water (fresh water) is divided into drinking water and process water, whereby a separate system with storage tanks, piping, pressure tanks, and pumps is required for each type of water.

Drinking water serves for food preparation, drinking, washing, and dishwashing. The flushing water of the WC (water closet) facilities is *process water*. The process water used is not only purified drinking water after use, but also partly treated seawater.

In addition, fresh water is required as boiler feed water, for refilling the fresh cooling water and in the onboard laundry.

On new ships, however, only one common fresh water system for drinking and process water is usually installed today. This saves additional piping, tanks, and equipment. For an approximate determination of the fresh water required for the design of the overall system, the demand for technical purposes and the drinking water demand

[37]In-depth [11].

shall be determined: For drinking water, about 200–500 L are assumed per person and day, depending on the type of ship (merchant ship, ferry, cruise ship). For a more precise determination of the water volume, especially for passenger ships, experience values of the shipping companies and shipyards are used.

In ports, fresh water is supplied on board by bunkering the water from land, whereas on the high seas it is supplied by seawater evaporation in fresh water generators specially developed for use on ships [16, p. 63].

In addition, the process of microfiltration (*reverse osmosis*) can be used for fresh water production.

5.7.2 Fresh Water Production by Evaporation

Seawater is evaporated in evaporators (see Fig. 5.51). The steam generated in these plants is then condensed, treated, and supplied to the consumers. Salt dissolved in seawater remains in the so-called brine during the evaporation process.

The seawater sucked in from outboard is mixed with non-evaporated water from the fresh water generator. This recovers the thermal energy of the brine. This mixture then passes through the heat exchanger installed in the fresh water generator. These are used to dissipate the heat of condensation in order to condense the water vapor formed in the fresh water generator. In the heat exchangers, the seawater absorbs the condensation heat, causing it to heat up further. After preheating, the water is transferred to another heat exchanger, which is operated with the cooling water of the main engines. After leaving this heat exchanger, the sea water has a temperature of about 80 °C. The seawater

Fig. 5.51 Seawater evaporator. *Photo* AIDA

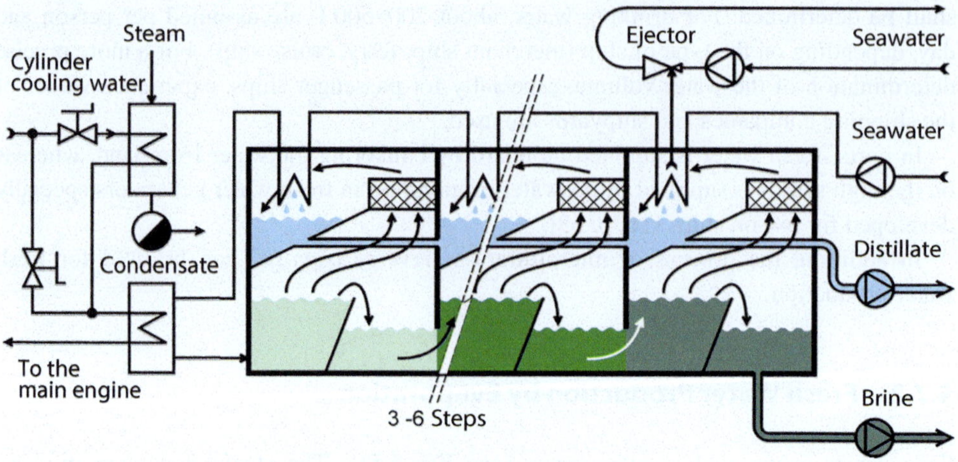

Fig. 5.52 Principle of a flash evaporator. *Graphic* Hochhaus, K.-H.

then passes through several flash evaporators connected in series, whereby the pressure of each stage is further reduced compared to the previous stage. From stage to stage, the sea water begins to boil at lower temperatures. In the last stage the pressure is so low that the water is already $T < 40^{\circ}$ C boils [16, p. 63].

Steam as a heating medium for the evaporators is rarely used on cargo ships today, but is the rule on cruise ships.

The condensate or distillate obtained is treated according to its intended use (drinking or process water). For the treatment to drinking water, disinfection systems (e.g., UV system, chlorination system) and filters for hardening are installed.

In these so-called flash evaporators the heating takes place outside the evaporator in a separate heat exchanger. After heating, the seawater is expanded and evaporated in an expansion valve to the lower pressure prevailing in the evaporator. For better waste heat utilization, several flash evaporators are connected in series as described above (see Fig. 5.52).

Besides the flash evaporator, immersion tube and plate evaporators as well as spray film evaporators are also used. The basic principle, however, is always the same: seawater is evaporated, the condensate is removed and further treated, and the enriched brine can be given outboard.

At *immersion tube evaporator* (Fig. 5.53), the seawater is heated by a heating coil in the evaporator pot. The heating coil is flown through by engine cooling water or superheated steam. When the evaporation temperature is reached, the seawater evaporates. The steam is liquefied in the condenser. The distillate is collected in the distillate collector, from where it is passed on for processing.

In the *plate evaporator* the seawater is evaporated in a plate heat exchanger.

Fig. 5.53 Fresh water treatment. *Photo* AIDA

In the *spray film evaporator,* the seawater is sprayed into the evaporator by means of a nozzle system. Due to the finest water droplets, the seawater to be evaporated now has a very large surface. Due to the large surface, it evaporates in a very short time at the heat exchanger surfaces.

One measure of the effectiveness of seawater evaporation for fresh water production is the efficiency of the plants:

$$\eta = \frac{\dot{m}_R}{\dot{m}_R + \dot{m}_K} \tag{5.90}$$

with $\dot{m}_R$ the pure water mass flow (the distillate) and $\dot{m}_K$ the concentrate, that is, the brine.

5.7.3 Fresh Water Production by Reverse Osmosis

Since the consumption of drinking water on passenger ships is very high, powerful *reverse osmosis systems* are often used (see Fig. 5.54).

The process of *osmosis* [17] occurs when in a system a pure solution (low concentration) is separated from a solution of high concentration by a semi-permeable membrane.

Fig. 5.54 Reverse osmosis system

In such a system, the solution of low concentration will diffuse through the membrane into the concentrated solution and thus dilute it. The diffusion is maintained until the concentration is balanced if the system is left unaffected for some time. The resulting difference between the different liquid columns is a measure of the natural osmotic pressure. This is directly related to the concentration of the enriched solution (see Fig. 5.55).

If the pressure, which must be higher than the osmotic pressure, is exerted against the direction of the natural osmotic pressure (in seawater desalination about 50…100 bar; [54]), the process of reverse osmosis (RO) takes place. Pressure is built up on the high concentration side, pure solution is forced through the membrane, further increasing the concentration of the solution on the high concentration side. This concentrate (salt water brine) is continuously drawn off, the so-called permeate—the fresh water—is fed for further treatment.

The heart of a reverse osmosis plant is the reverse osmosis membrane. The most commonly used membrane is a thin-film polyamide membrane, which is made up of two layers. The "active" ion-depositing component of the membrane is a very thin layer applied to a supporting fabric with an open structure. The membrane is usually used as a spiral wound module, which can be built compactly due to the large membrane surface (Fig. 5.56). The pore size of the membrane depends on the substances or ions to be retained and is between 0.002–0.16 μm for reverse osmosis.

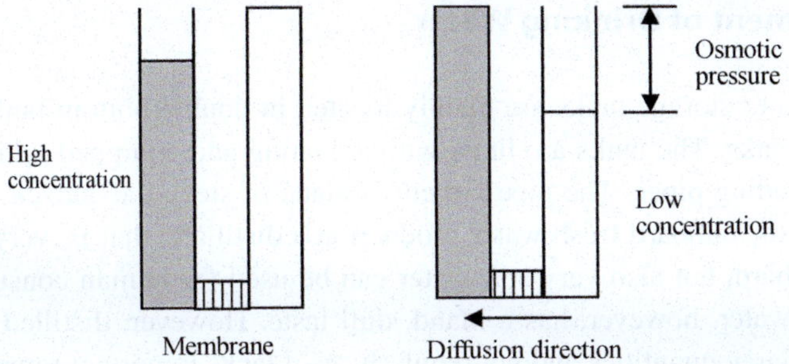

Fig. 5.55 Osmosis process

Fig. 5.56 Reverse osmosis—membrane as a wound module. *Photo* Shankbone, D., CC BY 3.0

The efficiency of a reverse osmosis module is calculated as follows:

a) Ion retention:

$$\%\ \text{Ion-retention} = \frac{\text{Conductivity of fresh water} - \text{Conductivity of permeate}}{\text{Conductivity of fresh water}} \cdot 100\% \tag{5.91}$$

b) Yield:

$$\%\ \text{Yield} = \frac{\text{Flow rate of permeate}}{\text{Flow rate of fresh water}} \cdot 100\% \tag{5.92}$$

5.7.4 Treatment of Drinking Water

The drinking water storage tanks are mainly located in double bottom tanks, but also in special bunker tanks. The tanks are lined with a coating and equipped with suction, filling, air and sounding pipes. The pipe system is made of steel, galvanized steel or stainless steel. With the onboard fresh water production a distillate, that is, very soft water is available. Very hard, but also very soft water can be used for human consumption without harm. Soft water, however, has a bland, dull taste. However, distilled water should not be drunk in large quantities and not regularly, as it lacks important minerals. Calcium bicarbonate is therefore often added to the drinking water through hardening filters to achieve a pleasant refreshing taste.

Activated carbon filters and disinfection equipment are also provided for treatment as drinking water. The following chemical and/or physical processes are used on ships for disinfection:

- Chlorination,
- Ozonation,
- Silver ion treatment,
- UV radiation,
- Filtration,
- Heating.

5.7.5 Drinking and Hot Water System

On cargo ships, drinking water is pumped from the fresh water storage tank to the drinking water pressure tank by the drinking water pump. An air cushion is used to maintain an excess pressure in the pressure tank, which is 4–7 bar, depending on the size of the ship and the extent of the drinking water system. The cold water taps are supplied from this cold water pressure tank. If the pressure drops below a set pressure on the pressure switch, the drinking water pump is switched on by the pressure switch and switched off again when an optimum pressure is reached. With this simple two-point control, a reliable water supply is possible as long as the storage tank contains enough drinking water.

Large passenger ships, such as cruise ships and ferries with high superstructures, have extensive drinking water systems, which usually no longer contain pressure tanks. Here the drinking water pressure pump is constantly in operation. In the simplest case, the consumption volume flows, which usually depend on the time of day, are returned to the storage tank via a pressure-controlled bypass line. However, parallel connections of pumps with the same or different flow rates are often chosen. Another technical possibility for adapting the volume flows to the consumption are pole-changing or speed-adjustable electric drive motors of the pumps. Circulation pumps are also used on passenger ships to avoid stagnant drinking water in the pipes (to prevent the formation of legionella!).

The hot water system is usually heated during the journey with a steam-fired, rarely thermal oil-heated heat exchanger. The desired temperature (approx. 60 °C) set on a thermostat which influences the heating. The pressure is maintained by the cold water system. With this onboard hot water heating system, the waste heat of the exhaust gases is used. In addition, an electric heating system or a heating system with oil firing (see Sect. 5.6.3.1) is installed to ensure the hot water supply in the harbor with the main engine stopped or in the shipyard.

In order to ensure that hot water flows immediately from the hot water taps, the hot water consumers are connected to a ring pipe in which there is constant circulation. A circulation pump is required for this.

5.7.6 Bunkering of Drinking Water

In addition to the production of fresh water, the exclusive bunkering of fresh water in drinking water tanks (made of tasteless plastic, plastic-coated steel tanks or stainless steel tanks) is also an option, especially for ships with a short standing time at sea.

A prerequisite for pure fresh or drinking water on board is a good quality of drinking water. The water must be free of germs and pathogens. In principle, it can be assumed that the drinking water in Northern Europe is of good quality. The German "Drinking Water Regulation" (German: "Trinkwasserverordnung") provides information on the permissible values of various ingredients in our drinking water.[38] In this respect, it must also be applied to the ships' drinking water system.

Some basic precautions should be taken to prevent the contamination of drinking water on board[39]:

- Hoses used for refueling with drinking water must be stored protected against contamination. The hoses must be completely emptied before stowing. The hose couplings must be closed so that no animals, insects or dirt can enter between refueling.
- At the start of refueling, the contents of one hose length should not be fed into the drinking water tank. This is especially true if the filling hose is taken over from the land side and it is not ensured that it was completely empty; in residual amounts of water in the hose, rapid germ growth can occur, especially in warm weather.
- A disinfection plan for regular disinfection of the hose and water tank should be established. Hoses with couplings and drinking water tanks should be disinfected with appropriate (chlorine-free) agents.
- Air inlet filters on the ventilation pipes of the tanks must be changed regularly.
- Drinking water tanks can be equipped with a continuous dosing device for disinfection.

---

[38]Drinking Water Regulation in the version published on March 10, 2016 (BGBl. I p. 459).

[39]More on this in the following [27].

- A drinking water filter (fabric and/or activated carbon) must always be installed on the pressure side of the withdrawal line of the drinking water tank.

For cleaning and disinfecting the fresh water pipe system including the filters, pressure vessels, etc., entire system, if necessary in sections, must be filled with a suitable disinfectant and cleaning agent. The treatment time depends on the data sheets of the manufacturers or distributors (usually twelve hours). Afterward the pipe system must be thoroughly rinsed with clean water.

With regard to the replacement of the water filters, the following should be observed: Clean and disinfect the filter housing with suitable cleaning and disinfectant agent, then rinse with clean water. Only handle new filter cartridges with clean rubber gloves.

Water samples should be taken regularly to check the water quality for compliance with the Drinking Water Ordinance. Various rapid tests are available on the market for the relevant substances, including for the detection of coliform bacteria.

However, these samples cannot replace the prescribed water samples that an accredited laboratory must take. According to the International Health Regulations (IHR), the health authorities must inspect ships every six months to ensure that they comply with ship hygiene. If no defects are found, the authority or a state-authorized body issues a ship sanitation certificate (german: Schiffshygienebescheinigung). In Germany, the port medical services of the federal states (Hafenärztliche Dienste der Länder) are responsible for carrying out ship hygiene inspections [36].

5.8 Transshipment Facilities

Depending on the type of ship, its loading and unloading (charging and discharging) is carried out by different cargo handling equipment. These can be land-based or ship-based. Examples of land-based handling facilities are container gantry cranes for container handling or, for example, also pneumatic conveying equipment and belt conveyors for loading and unloading grain and similar bulk goods. In this section, however, important onboard transshipment facilities will be examined in more detail.

5.8.1 Board Cranes

General cargo, bulk goods or even containers can be lifted on and off board by onboard cranes.[40] Onboard slewing cranes (also known as deck cranes) with rope luffing gear are preferred (Fig. 5.57); slewing cranes with lifting capacity > 150 t are used as *heavy lift cranes* which can carry up to 800 t [23, p. 23].

[40]For further details see also [23].

Fig. 5.57 Onboard cranes of a general cargo freighter

The luffing gear is used to vary the inclination of the crane boom. The statics are designed so that the inclination can be changed under load. With the slewing gear the crane is slewed around the crane axis. If there are several cranes on board, they are arranged in such a way that their conveying areas overlap so that they can serve the entire deck area, as can be clearly seen in Fig. 5.57.

Figure 5.58 shows once again the rope luffing gear, the slewing ring and the load handling attachment—the crane hook—in more detail.

In addition to the rope luffing gear, hydraulic luffing gears are also used (Fig. 5.59).

In order to avoid additional lifting work caused by a vertical load movement when the outreach of the crane arm changes, the load path is kept as horizontal as possible during luffing. This is achieved by a clever use of suspension and luffing ropes.[41]

Crane structures are executed in solid wall or truss construction. The former include both structures consisting of standard sections or web plate girders and structures manufactured in cellular, box or shell construction. Trusses are built up from triangles with a statically determined structure and are considered as plane trusses.

In order to lift the load from or on board, it must be possible to turn the crane around the vertical crane axis. This is done by the slewing gear of the crane. The bearing of the rotating tower as well as the drive of the slewing gear is located in this.

[41]For more details see Vierling in [6, Volume II, pp. 599 f.].

Fig. 5.58 Board crane with rope luffing gear

Fig. 5.59 Onboard crane of an inland vessel with hydraulic luffing gear

Calculation of the Slewing Gear Motor

The highest motor torque $M_{M,\max}$ occurs when the upper part of the crane with boom is rotated under load and headwind.

The *friction force* of *the tower bearing* (rotational resistance) causes a rotational resistance torque M_D:

$$M_D = \mu \cdot (V + H) \cdot \frac{D}{2} \tag{5.93}$$

with

V = vertical force at the tower pivot bearing,
H = horizontal force at the tower pivot bearing,
D = mean rolling circle diameter of the bearing arrangement,
μ = friction coefficient (usual values: 0.005–0.01)

However, this is generally of secondary importance in relation to the more important moments for accelerating the turning masses M_R and for overcoming the moment M_W acting on the *boom due to wind pressure*.

For the calculation of M_R (in relation to the axis of rotation of the crane) only those of the load Q, of the boom Q_A and the sum of other larger crane masses lying together G_m (e.g., counterweight—if available) need be used as gyrating masses. Rotating masses of gear and motor are negligible. For the calculation of $M_{R,}$ the upper part of the crane rotating at the angular speed ω is divided into individual masses m_i determines their center of gravity distances r_i to the axis of rotation of the crane and their mass moments of inertia J_{0i} around its own focus. So the following applies:

$$M_R = \sum \left(m_i r_i^2 + J_{0i}\right) \cdot \omega / t_a \tag{5.94}$$

with $\omega = 2 \cdot \pi \cdot n$ and n the crane speed (0.02–0.05 s^{-1}); usual values for the acceleration time t_a are at 5–10 s.

J_{0i} Can Be Neglected if the Masses m_i Expand Only Slightly in the Horizontal Direction (E.G., Load on Load Hook Q or Counterweight, if Present).In This Respect, the Mass Moment of Inertia Usually Only Remains to Be Considered for the Crane Boom:

$$M_R = \left(Q \cdot r_L^2 + Q_A \cdot r_A^2 + J_{0A} + \sum G_m m_i^2\right) \cdot \frac{\omega}{t_a} \tag{5.95}$$

with

$\omega = 2 \cdot \pi \cdot n$,
r_i = center of gravity distances loads—rotational axis,
m_i = distances Key aspects other loads—rotational axis

The wind moment M_W is determined from a wind pressure w with usually 150 N/m² (unless otherwise specified) to the crane surfaces to the left and right of the axis of rotation A_1 and A_2 ($A_1 > A_2$; center of gravity distances of the surfaces: s_1 or rather s_2):

$$M_W = A_1 \cdot s_1 - A_2 \cdot s_2. \tag{5.96}$$

The maximum motor torque is then:

$$M_{M,\max} = (M_D + |M_W| + M_R)/(i_1 \cdot (1 - i_2) \cdot \eta). \tag{5.97}$$

They are i_1 the ratio of the slewing gearbox and i_2 the standing transmission ratio between the spider and the pinion. The overall efficiency of the spider/pinion and gearbox can be roughly estimated with $\eta = 0.85$. The required motor torque is roughly estimated $M_{M,\text{erf.}} \leq M_{M,\max}/1.7$.

This results in the rated motor power $P_{\text{Motor, nominal}}$ (with the speed n) to:

$$P_{\text{Motor, nominal}} = 2 \cdot \pi \cdot n \cdot M_{M,\text{erf.}}/\eta. \tag{5.98}$$

With the rated motor power $P_{\text{Motor, nominal}}$ and the duty cycle ED = 25% for light grab operation, ED = 40% for floating general cargo operation and ED = 60% for heavy grab operation, the required motor can be selected from the relevant motor catalogs. It must be $P_{\text{Motor ED.}} \geq P_{\text{Motor, nominal}}$.[42]

Calculation of the Hoist Motor

If a load G (kN) is to be lifted with the speed v (m/min), the power to be provided by the engine in steady state (full load steady state power) shall be

$$P = \frac{G \cdot v}{60 \cdot \eta_H} \text{ [kW]}, \tag{5.99}$$

whereby the overall efficiency η_H of the winch unit during lifting is first estimated and later determined as the product of the individual efficiencies of the hoist components for checking purposes; usually an efficiency of $\eta_H \approx 0.9$ can be assumed. The efficiency of lowering η_S is $\approx \eta_H$ if the individual efficiencies of the hoist components do not fall below 0.9.

The permissible temperature rise is decisive for the selection of the motor size. Since the full-load steady state power can only be applied for a part of the time for a working cycle during intermittent operation of the winch units, the motor is selected according to its *relative duty cycle DR*:

$$\text{DC} = \frac{\sum \text{Switch-on time}}{\sum \text{Switch-on times} + \sum \text{currentless pauses}} \cdot 100\%. \tag{5.100}$$

The rated motor power ratings for 20, 40 and 60% DC given in the lists of crane motor manufacturers correspond to the permissible full load holding power.[43]

With regard to *further requirements for design and calculation bases* for cranes and hoists, reference is made, among other things, to the Construction Regulation of DNV GL "Classification and construction regulations: VI Supplementary regulations and guidelines 2—hoists (German: Klassifikations- und Bauvorschriften: VI Ergänzende Vorschriften und Richtlinien 2—Hebezeuge)".

---

[42]To the foregoing: Vierling in [6, Volume II, pp. 589 ff].

[43]To the foregoing: Vierling in [6, Volume II, pp. 556 f.].

Fig. 5.60 Crane on board the Passat

Example of a Crane Task

The crane in Fig. 5.60 is to move a load from 1000 kg by means of the attached pulley block. The pulley block has three rollers each in the upper and lower block. The rope force on the pulling strand must be determined; the frictional forces in the roller bearings can be neglected. Furthermore, the bar forces of the crane as well as the horizontal and vertical forces at the combined support and neck bearing (so-called trivalent bearing) must be determined.

Solution: The number of load-bearing ropes is decisive for the tensile force on which the burden F_L distributed. The weight of the mass to be moved is therefore distributed evenly over all the connections between the lower and upper pulleys, the load-bearing ropes. The tensile force is added:

$$F_Z = F_L/n \tag{5.101}$$

and therefore by insertion:

$$F_Z = F_L/n = 1000\ \text{kg} \cdot 9.81\ \text{m/s}^2/6 = 1635\ \text{N}.$$

The bar forces are determined graphically by the *force corner method*. For this purpose, the rod system is first drawn to scale. Then a force triangle is drawn, whereby the force curves follow the directions of the rod system and the applied load—the

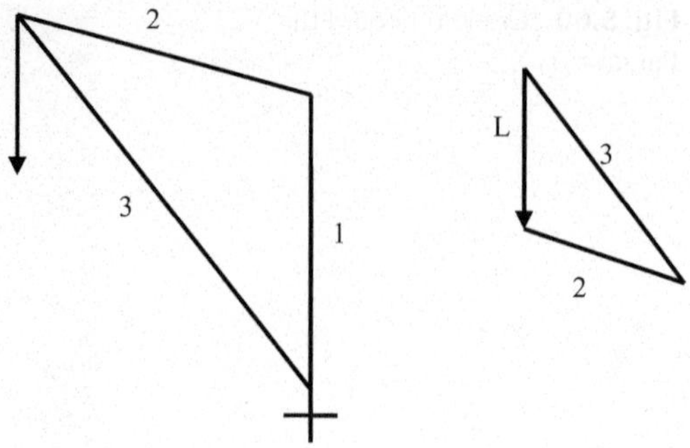

Fig. 5.61 Static system and force corner

forces can then be read directly from the force triangle thus created (Note: The sketch in Fig. 5.61 is not to scale!).

From the polygon of forces can be derived for support 1: Sum of the vertical forces must be zero. It follows that the force in the support must be equal to the load, that is, $F_1 = F_L = 9810$ N. For tension rod 2 a force of $F_2 = 11{,}000$ N and for the pressure rod $F_3 = 15{,}500$ N.

The vertical force at the support and neck bearing is 9810 N, as it must counteract the load. The horizontal force at the neck support is calculated from the cosine of the rod force F_3 which is placed at an angle of 45° acts on support 1:

$$\cos 45^\circ = F_H / F_3.$$

This results in the following $F_H = 10{,}960$ N. ◀

5.8.1.1 Load Hooks, Hook Blocks

In addition to grabs, lifting magnets, etc., the following are used as load handling devices in onboard cranes *load hook* and *hook blocks* in different versions (depending on the load) for application.

If the hook is attached to a bottom block and not directly attached to a rope or chain, it is pivoted. A deep groove ball thrust bearing is fitted between the secured hook nut and the crossbar of the harness or bottom block to allow easy rotation of the hook. In the simple bearing design, the ball raceway is machined into the nut and crosshead and hardened.

To prevent accidents caused by unintentional unhooking of slings (chains, ropes, webbing) from the hook, the hook should be secured with a hook safety device (also called a latch) (Appendix 1, Para. 2.5 of the German Industrial Safety Regulation).For loads > 15 t double hooks predominate (Fig. 5.62).

In order to avoid accidents, the following defects must lead to the hook being taken out of service immediately [23, p. 29]:

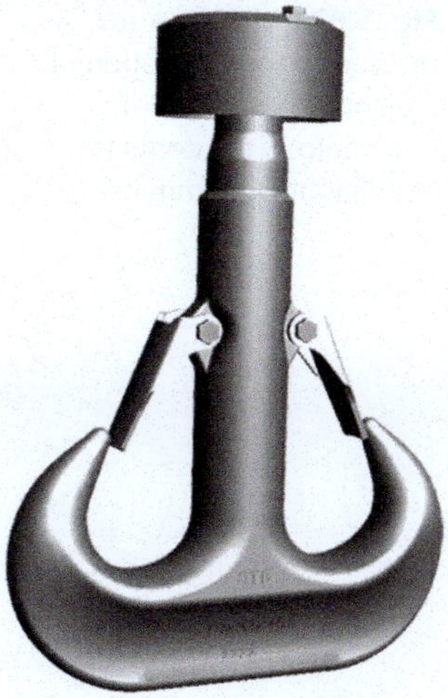

Fig. 5.62 Double hook with hook safety device. *Photo* Stahlhammer Bommern GmbH

- Cracks,
- Signs of wear and tear $>5\%$,
- Coarse deformations and expansion of the hook mouth $\varepsilon > 10\%$.

To determine the expansion of the hook mouth, two markings are attached to the hook, for example, by means of a center punch (Fig. 5.63) and their distance is noted in the crane or hook test book. The elongation ε shall be determined as follows:

$$\varepsilon = \frac{e_{\text{act.}} - e_0}{e_0} \cdot 100\% \tag{5.102}$$

with

e_0 = distance between the marks when the hook is new,
e_{act} = currently measured distance of the markings

Calculation basis: The hook is calculated for tension in the journal cross section and for bending and tension in the strongly curved parts. The hook cross section is designed as a trapezoid with rounded corners.

Example calculation load hook

A load of 1.5 t is attached to the hook made of structural steel St 37 shown in Fig. 5.63. It is for the vulnerable cross section

a) the pure tensile stress distribution,
b) the pure bending stress distribution and
c) to calculate the composite stress distribution.

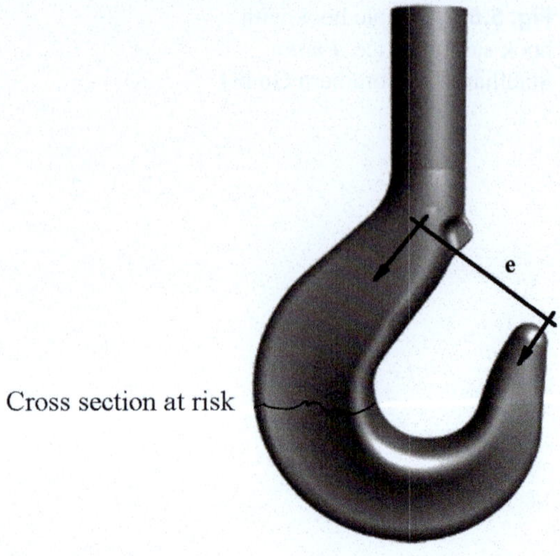

Fig. 5.63 Markings for measuring the elongation of the hook mouth. (Addition of a photo of the company Stahlhammer Bommern GmbH)

The vulnerable cross section can be approximately assumed to be a trapezoid (see Figs. 5.63 and 5.64) with $a = 20$ mm, $b = 45$ mm, and $h = 50$ mm.

Solution:

a) Determination of the pure tensile stress distribution:

The tensile stress is evenly distributed over the cross section

$$\sigma_Z = \frac{F}{A_{\text{Trapez}}} \tag{5.103}$$

with

$$A_{\text{Trapez}} = \frac{a+b}{2} \cdot h = \frac{20\text{ mm} + 45\text{ mm}}{2} \cdot 50\text{ mm} = 1625\text{ mm}^2.$$

Thus, by inserting in Eq. 5.103: $\sigma_Z = 9\text{ N/mm}^2$.

b) Determination of the pure bending stress distribution:

The stress distribution over the considered trapezoidal cross section corresponds to a hyperbola.[44]

Determination of the edge fiber stresses (edge fiber stresses = max. bending stress)

$$\sigma_{b1} = \frac{M_b}{-W_1}, \tag{5.104}$$

$$\sigma_{b2} = \frac{M_b}{W_2} \tag{5.105}$$

[44]See also Vierling in [6, Volume II, p. 546].

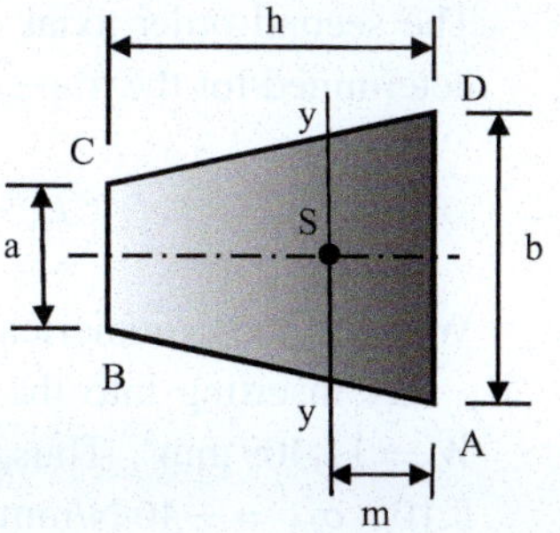

Fig. 5.64 Vulnerable cross section

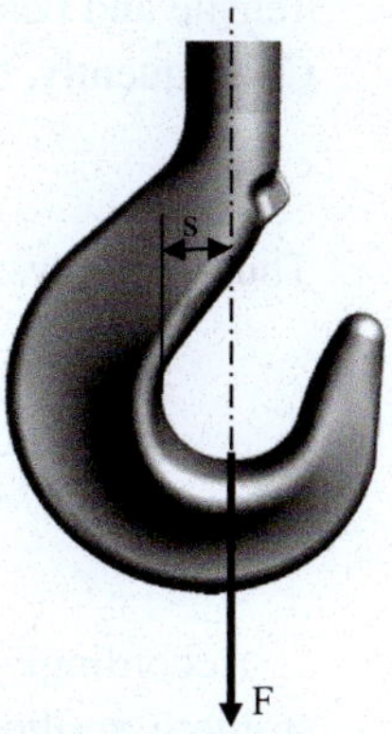

Fig. 5.65 Load applied to the load hook (here $s = 23$ mm). (Addition of a photo of the company Stahlhammer Bommern GmbH)

with the bending moment (see Fig. 5.65)

$$M_b = F \cdot (s + m). \tag{5.106}$$

whereby the center of gravity position m of a trapezoid is calculated as follows:

$$m = \frac{h}{3} \cdot \frac{b + 2a}{b + a}. \tag{5.107}$$

Inserting the above values results in: $m = 21.79$ mm. Thus the bending moment is calculated by inserting in Eq. 5.106: $M_b = 671.85$ Nm.

Determination of the moments of resistance W_1 and W_2:

$$W_1 = \frac{I_{yy}}{m}, \tag{5.108}$$

$$W_2 = \frac{I_{yy}}{(h - m)}. \tag{5.109}$$

The second order axial surface torque in relation to the axis of gravity y–y (I_{yy}) is determined for the trapezoid shown in Fig. 5.63 according to Eq. 5.110[45]:

$$I_{yy} = \frac{6a^2 + 6a \cdot (b-a) + (b-a)^2}{36 \cdot (2a + (b-a))} \cdot h^3. \tag{5.110}$$

With the above numerical values, this results in: $I_{yy} = 321{,}848\,\text{mm}^4$.

By inserting into the Eqs. 5.108 and 5.109 and delivers $W_1 = 14{,}771\,\text{mm}^3$ yields $W_2 = 11{,}409\,\text{mm}^3$. Thus, for the edge fiber stresses, insertion into Eqs. 5.104 and 5.105: $\sigma_{b1} = -46\ \text{N/mm}^2$ and $\sigma_{b2} = 59\ \text{N/mm}^2$.

c) Composite stress distribution

Tensile and bending stresses are normal stresses and add up when they occur together. Consequently, the composite normal stress is given by

$$\sigma_i = \sigma_z + \sigma_{bi}. \tag{5.111}$$

Thus it follows:

$$\sigma_1 = 9\ \text{N/mm}^2 + (-46\ \text{N/mm}^2) = -83\ \text{N/mm}^2$$

and

$$\sigma_2 = 9\ \text{N/mm}^2 + 59\ \text{N/mm}^2 = 68\ \text{N/mm}^2$$

Accordingly, the maximum composite normal stress is $\sigma_{max} = 83\ \text{N/mm}^2$. For strength verification for the load hook must apply: $\sigma_{max} \leq \sigma_{perm} = 160\ \text{N/mm}^2$.[46] In this respect, the load hook under consideration here has about twice the safety factor. ◀

5.8.1.2 Slings

Slings are devices, not belonging to the hoist, which provide a connection between *load bearing equipment* (e.g., the load hook or also grabs, cross beams, tongs permanently connected to the crane) and load bearing equipment and *load-suspension equipment* manufacture. The making of this connection is called "slinging" and is done by the so called slinger. A safe cargo handling depends on his skills and abilities![47]

Load-suspension equipment are also devices which are not part of the hoist and which can be connected to the hoist's load-bearing equipment to pick up the load.

The load handling devices include container harnesses, hangers, grippers, clamps, buckets, lifting magnets, pallet harnesses, traverses, vacuum lifters or also tongs.

---

[45] For more information on the second degree moment of area, see also [3, p. D 25 ff.].

[46] Sigwart in [6, Volume I, p. 534 f.].

[47] For more details on the safe attachment of loads, see BG Information "Anschläger (Slingers)" of the Association of Metalworkers' Liability Insurance [28].

Lifting devices can also be connected to the hoist by means of couplings designed for frequent release.

Slings include chains, ropes (natural fiber, synthetic fiber and wire ropes) as well as lifting slings and slings made of synthetic fiber fabric with and without wire core. Round slings (a band or rope connected to form an endless sling) are also colloquially known as "slips", "strops" or "grummets". All slings must indicate the permissible load capacity (in kilograms or tons) and are internationally recognized as *working load limit* (WLL). It results from the minimum breaking force MBF of the sling, multiplied by a load attachment factor LF:

$$\mathrm{WLL} = \mathrm{MBF} \cdot \mathrm{LF}. \tag{5.112}$$

Lifting Belts and Round Slings

In accordance with the Machinery Directive 2006/42/EC, slings must be specially marked. The manufacturer must therefore provide the following information, which is usually found on a sewn-on label in the case of straps and lifting slings:

- Business name and full address of the manufacturer,
- Designation of the sling gear,
- CE marking,
- Series or type designation,
- Serial number, if applicable,
- Year of manufacture,
- Load capacity.

The load capacity of lifting slings and webbing slings is indicated by a color code (DIN 1492-1 and 1492-2; Table 5.22).

Furthermore the *color of the label* provides information about the material of the lifting sling or the slip:

- **Green** label = PA, polyamide (mainly alkali-resistant),
- **Blue** label = PES, polyester (mainly resistant to mineral acids),
- **Brown** label = PP, polypropylene (good resistance to acids and alkalis, but not to solvents)

Steel Chains, Wire Ropes

Steel chains for attaching loads should comply with the standards DIN 695 or EN 818. Steel chains are versatile. The length of chain strands can be quickly adjusted with shortening claws to align a load horizontally, for example, chains are to be rejected for

- Breakage or deformation of a chain link,
- Cracks or severe corrosion scars,
- Elongation of more than 5%,
- Reduction of the chain link thickness by 10%.

Table 5.22 Color code load capacity of hoisting belts/ round slings

Load capacity [kg]	Color code
500	Pink
1000	Purple
1500	Dark green
2000	Light green
3000	Yellow
4000	Gray
5000	Red
6000	Brown
8000	Blue
> 10.000	Orange

Wire ropes are wear-resistant and relatively flexible. The wear characteristics are easy to see and feel. Wire ropes must be discarded in the case of

- Buckling and pawls in the rope,
- Broken strand,
- Squeezing of the rope in the free length,
- Crushing at the rope eye or at the crimping sleeve at the eye end [23, p. 32].

In the BGI 622 "Loading tables" for slings made of round steel chains, steel wire ropes, round slings, man-made fiber hoisting belts, man-made fiber ropes, and natural fiber ropes, the German Employer's Liability Insurance Association for Wood and Metal (German: Berufsgenossenschaft Holz und Metall) has presented the permissible loads for the above slings in table form depending on the type of sling (see example in Fig. 5.66 for a sling with round steel chain).

- In addition, the manufacturers of slings also provide their own specifications for the load capacity of the various types of sling; these are taken into account by the load factor LF (see above Eq. 5.112 and Fig. 5.67, which shows the application of one or two round slings; the LF values can be used as a guide for slings with ropes and chains). For a slinging angle $\beta \leq 45^{\circ}$, the higher LF can be selected, for angles up to max. 60° the smaller LF must be selected. Slinging angles $> 60^{\circ}$ are inadmissible.

5.8.1.3 Securing the Load by Lashing

To prevent general cargo (boxes, containers, etc.) from slipping in rough seas and thus leading to critical stability situations for the ship, it must be secured accordingly. One way of securing the load is to secure it by lashing it down using tension belts (Fig. 5.68). These are placed over the load to be secured and lashed down with tensioning devices (e.g., ratchet). Care must be taken to ensure sufficient lashing force, which depends both

Round steel chains of grade 4

Nominal chain thickness	Load capacity in kg at tightening and for hoisting chains				
	Single strand	Double strand with angles of inclination of		Hoisting chain	
		0° to 45°	45° to 60°	Single strand	Double strand
mm	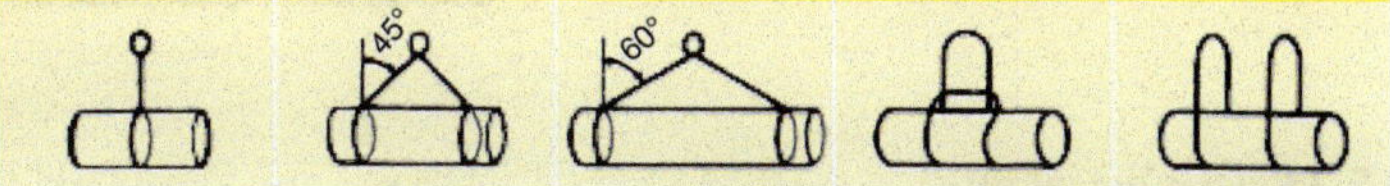				
8	800	1,120	800	1,600	4,000
10	1,250	1,750	1,250	2,500	6,400
13	2,100	3,000	2,100	4,250	10,600
16	3,150	4,400	3,150	6,300	16,000
18	4,000	5,600	4,000	8,000	20,000
20	5,000	7,000	5,000	10,000	25,000
23	6,400	9,000	6,400	13,200	32,000
26	8,500	12,000	8,500	17,000	42,000
28	10,000	14,000	10,000	20,000	50,000
32	12,500	17,500	12,500	25,000	64,000
36	16,000	22,400	16,000	31,500	80,000
40	20,000	28,000	20,000	40,000	100,000
45	25,000	35,000	25,000	50,000	125,000

Fig. 5.66 Load table stop with round steel chain

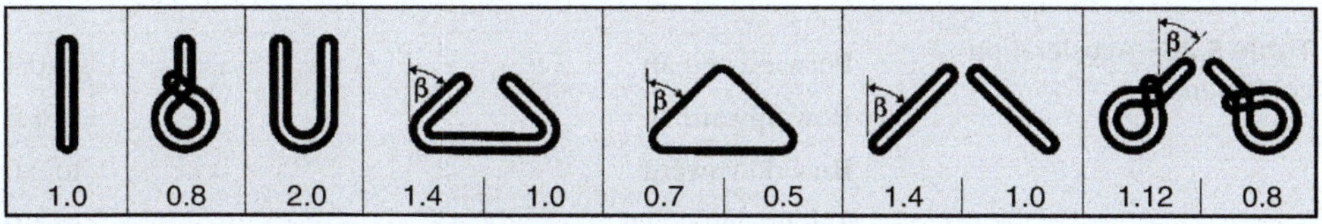

Fig. 5.67 Load stop factors LF for round slings

on the acceleration values of the ship during rolling, pitching and superimposed movements and on the material pairings of the standing surface (e.g., wooden crate on steel floor or crate on steel floor with anti-slip mat in between).

Tables 5.23, 5.24, and 5.25 and Eqs. 5.113–5.115 give corresponding information.

Required preload force for lashing angle α Approximately 90° (see Fig. 5.67):

$$F_V = \frac{C - \mu}{\mu \cdot \sin\alpha} \cdot \frac{F_G}{k} \tag{5.113}$$

with

Fig. 5.68 Load securing by tie-down lashing

Table 5.23 Coefficient of sliding friction μ according to VDI 2700

Material pairing	Dry	Wet	Bold
Wood/wood	0.20–0.50	0.20–0,25	0.05–0.15
Metal/wood	0.20–0.50	0.20–0,25	0.02–0.10
Metal/metal	0.10–0.25	0.10–0,20	0.01–0.10
Concrete/wood	0.30–0.60	0.30–0,50	0.10–0.20

Table 5.24 Further coefficients of sliding friction

Anti-slip mat	>0.50
Big bags (IBC) at wood panel	0.30
Cardboard box at cardboard box or cardboard box at wooden pallet	0.35
Wooden Euro pallet on silk-screen printed floor	0.25
Metal mesh box at screen printed floor	0.25

Table 5.25 Acceleration coefficients

Forward and aft	to 0.4
Bow upward	to 1.0
Bow downward	to 2.0
Tail up and down	to 1.0
Midship up- and downward	to 0.5
Lateral acceleration midships	to 0.8

F_V = preload force at the Ratchet site in daN,[48]
C = acceleration coefficient (see Fig. 5.66 and Table 5.25),
μ = coefficient of sliding friction to Tables 5.23 or 5.24,
F_G = weight force of to secure transport goods in daN (corresponds to the loading weight in kg),

[48] 1 daN = 10 N

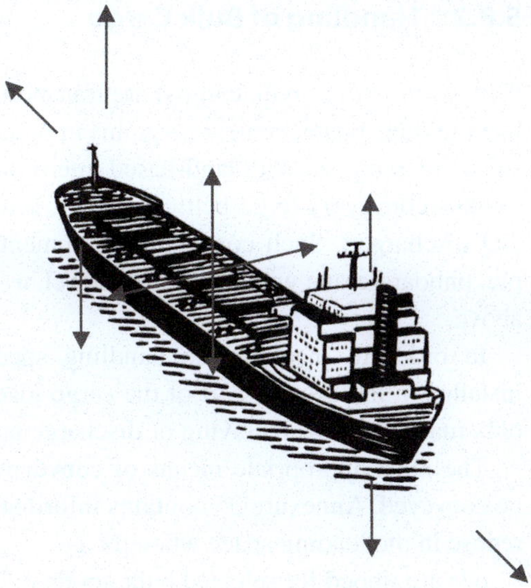

Fig. 5.69 Acceleration coefficients (*C*-values)

k = transmission coefficient (without corner protector 1.5; this factor taken into account, that the preload force at the Ratchet site larger is greater than the other side of the lashing),
$\sin\alpha$ = Sine of lashing angle α (see Fig. 5.68) between lashing means and loading area

The required number n of the tension belts results from the permissible tension force F_V for of a seat belt:

$$n = \frac{F_V}{F_{V\text{perm}}}. \tag{5.114}$$

Simplified calculation (here **both sides** of the lashing are **individually** calculated):

$$F_{V\text{ one side}} = \frac{C - \mu}{\mu \cdot \sin \alpha} \cdot G \tag{5.115}$$

with G = loading weight in kg or in daN.

Figure 5.69 and Table 5.25 give an indication[49] for use of the *C*-values (THS[50]-load ship).

[49] Section 5.3 of the CTU Code [35] contains more precise *C-values* as a function of wave height and sea area.

[50] THS = **T**ransport **H**andling **S**torage.

5.8.2 Handling of Bulk Cargo

Bulk goods (loose bulk cargos) are transported unpackaged as bulk cargo (e.g., ore, coal, bauxite, phosphate, cement or grain) in bulk carriers (also bulk freighters). As a rule, the ships are equipped with land-based intermittent conveyors (e.g., grab excavators) or continuous conveyors (e.g., belt conveyors, screw conveyors, and flow conveyors[51]) loaded and discharged. Such conveying equipment can also be available on board, for example, onboard crane with excavator shovel, as well as the continuous conveyors mentioned above.

In order to increase the handling speed, continuous conveyors are permanently installed in the deepest part of the cargo loading space, which continuously transport the bulk material to the receiving or discharge points of the vertical conveyors.

The most appropriate means of conveying depends not least on the bulk material to be conveyed. Annexure 25 contains information on density, bulk density and the angle of repose in the loading space when the go

ods are tipped for selected bulk goods.

5.8.2.1 Calculated Values

a) Conveying Work (Lifting Work W)

$$W = G \cdot h \tag{5.116}$$

with

G = payload,
h = delivery height

b) Conveying Speed

The instantaneous speed v is the derivation of the path s according to time t:

$$v = \mathrm{d}s/\mathrm{d}t. \tag{5.117}$$

The average speed v is the quotient of distance and time:

$$v = s/t. \tag{5.118}$$

c) Capacity

The conveying capacity P is the product of conveying work W in units of time t:

$$P = W \cdot t. \tag{5.119}$$

[51]Pneumatic delivery or with water as transport medium.

d) Flow and Flow Rate

It is the mass G or quantity V of the material to be moved or conveyed. The volume throughput $\dot{M}$ is the flow rate per time unit within a certain range:

$$\dot{M} = G/t \quad or \quad V/t. \tag{5.120}$$

e) Filling Ratio F for Grabs or Bucket Conveyors

It is defined as the quotient of the actual gripper or cup filling and the maximum gripper or cup volume.

5.8.2.2 Continuous Conveyors

In DIN 15201 "continuous conveyors", the types of conveyed material (bulk material and piece goods) are assigned to continuous conveyors, for which they are suitable alone or together. This results in the classification of continuous conveyors, which can be used,

- Only for bulk material,
- For bulk and piece goods,
- For general cargo only.

According to this provision, the following continuous conveyors are particularly used in ship operation for conveying *bulk material* especially:

- Bucket elevator (vertical and inclined bucket elevators),
- Screw conveyor,
- Pneumatic conveyors.

Belt conveyors are suitable for conveying *bulk and piece goods.*

Roller conveyors or drag chain conveyors are often used as continuous conveyors for loading and unloading unit loads.

Screw Conveyor

As already mentioned in Sect. 5.8.2, continuous conveyors are often used to increase the handling speed in the lowest part of the loading space, which continuously transport the bulk material to the receiving or discharge points of the vertical conveying equipment. Therefore, this type of conveyor will be discussed in more detail here.

The conveying principle is based on the advancing of the material to be conveyed in a semicircular trough or in a tube (similar to a "meat grinder") by means of a partially pressurized spiral surface made of sheet metal, the actual screw, which can be of single or multi-start design. Figure 5.70 shows possible designs of screw conveyors.

The screw is usually driven by an electric motor flanged to the front via a gear unit; the screw shaft is subjected to tensile stress during the conveying process.

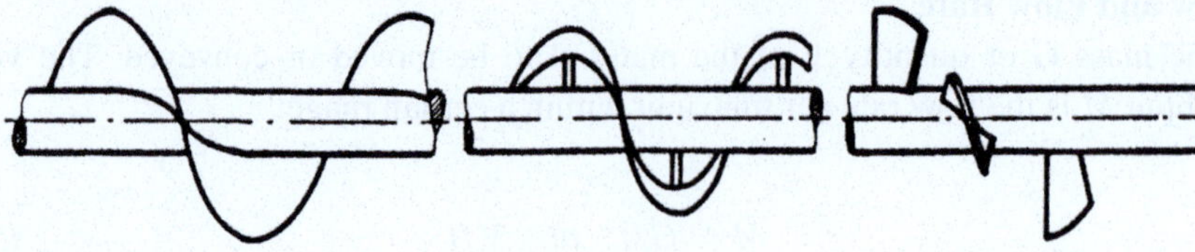

Fig. 5.70 Screw forms. **a** Solid screw, **b** band screw, **c** stirring screw

Fig. 5.71 Filling ratio screw conveyor

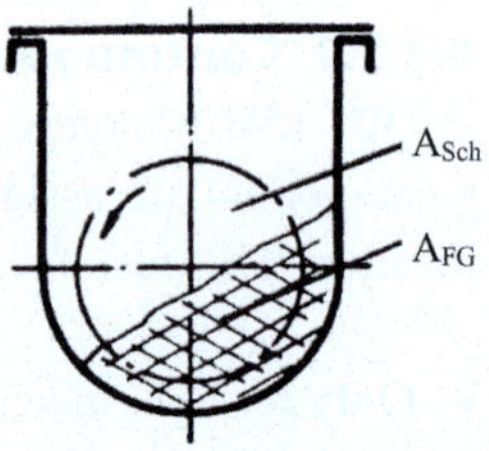

Flow Rate

The flow rate (quantity delivered per time unit) is calculated according to Eq. 5.121:

$$\dot{Q} = \frac{\pi \cdot D^2}{4} \cdot s \cdot \phi \cdot n \tag{5.121}$$

with

$s \cdot n = v$ of the conveying speed (0.2–0.4 m/s),
$D =$ screw diameter,
$s =$ slope of worm thread,
$\varphi =$ filling ratio (0.15–0.45%) $= A_{FG}/A_{Sch}$ (see Fig. 5.71),
$A_{FG} =$ projected area of the material to be conveyed,
$A_{Sch} =$ projected screw surface

On the one hand, the degree of filling is dependent on the *material class* (Table 5.26).

Furthermore, the filling ratio depends on the type of *feed material:* continuous or intermittent.

Drive Power

The drive power at the worm shaft is calculated according to Eq. 5.122:

$$P_a = \dot{Q} \cdot L \cdot k \tag{5.122}$$

Table 5.26 Filling ratio of screw conveyor

A (≈ 45%)	B (≈ 30%)	C (≈ 15%)
Easy, free flowing e.g.,	Fine-grained to small pieces, not free flowing, e.g.,	Tough, fibrous, big pieces, flowing, e.g.,
Cereals	Hard coal	Coke
Flour	Rough Salt	Gravel
Coal dust	Cement	Sand

Table 5.27 Displacement resistance of selected materials

Conveyed material	*K*	Mat. Class
Graphite powder	2.0	A
Barley	2.3	A
Lignite, dry	3.0	B
Soil, dry	4.0	B
Fly ash	5.0	C

with

L = conveying length,
k = displacement resistance (dimensionless number: quotient of resistance force against displacement and weight force of conveyed material; see Table 5.27)

In case of increasing production, a performance bonus is to be provided: For an incline of up to 15° about 25%, with a gradient up to 30° about 30%. For inclinations > 30° conveying can no longer take place due to purely translational displacement; in this case, a minimum screw speed must then be provided in order to hold the bulk material by the centripetal force on the conveying pipe so that it does not slide down.[52]

5.8.3 Vehicle Ramps on Ferries

For loading and unloading, RoRo and RoPax ferries[53] have bow hatches, side hatches or reargates (see also Fig. 5.72), through which the vehicles can drive on board via ramps. The bow ramp is located behind the closed bow flap (also called bow visor; opened in Fig. 5.73). The opening and closing of the bow flap is done hydraulically by means of two hydraulic cylinders.

[52]For more details on the design and dimensioning of steeply inclined screw conveyors, see [41].
[53]RoRo = "roll on/roll off"; RoPax = abbreviation for RoRo ferry with passenger transport.

Fig. 5.72 Closed tailgate

Fig. 5.73 Opened front flap of a ferry

The ramps can be opened and closed by means of hydraulic cylinders or a cable pull system (pulley block principle).

The advantages of the RoRo method are the short loading and unloading times and the gentle cargo handling, as the cargo can remain on the means of transport—truck or train. The disadvantage is the higher construction costs—RoRo ships are special ships and often built to the special wishes of the ship-owner (no ship "off the peg"). In addition, the use of loading space cannot be optimally solved.

Ferries are not unproblematic for safety reasons. Their openings for the loading ramps represent large openings in the outer shell. This can pose a risk in that, in the event of malfunctions or incorrect operation, the ramps are not sealed tightly so that large quantities of water could enter through these openings. In conjunction with the large, difficult to divide holds, this may lead to a critical stability situation and dangerous heeling. Large heels can cause outer skin openings that are not watertight (e.g., ventilation shafts and ducts, supply hatches) to be submerged. This results in secondary flooding, which causes even more water to flow in and can eventually cause the ship to capsize and sink. For example, the sinking of the ferry Estonia on September 28, 1994 in the Baltic Sea should be remembered here. For this reason SOLAS[54] prescribes among other things, multiple redundant backup systems.

5.8.4 Tankers

Tankers are special ships for the transport of liquid or gaseous substances. As examples we can mention: Crude oil, fuels, liquid chemicals, liquid gas, but also water or juices.

Following a decision of the IMO (International Maritime Organization) in 1992 in the International Convention for the Prevention of Pollution from Ships (MARPOL), all tankers built from 1996 onward and carrying 5000 t transport weight must be fitted with a double hull. Furthermore, the IMO has decided that from 2015 only oil tankers with double-walled outer hulls will be allowed to sail the world's oceans.

Tankers are filled and discharged both via onboard and land-based pump and pipeline systems; discharging is usually carried out with onboard submersible pumps, although in the case of so-called supertankers $> 10,000$ t/h the flow rates of the individual cargo pumps are same.

Please refer to the Sect. 5.4 on pumps and piping for details and calculations.

The connection ship-shore is made by means of hose lines. These are supported by the midship mounted *manifold crane* hoisted on board from land (see Fig. 5.74) in order to connect them to the ship's piping system.

The hose connection points are called *manifolds* (see Fig. 5.75).

The loading and unloading activities are monitored by means of loading computers. This allows a prediction of the distribution of forces (buoyancy and weight of the cargo) acting on the ship. It is also used to monitor the level in the tanks. Figure 5.76 shows a view into a tank of an inland vessel for the transport of diesel and heating oil with mechanical level monitoring.

As described above, tankers also carry flammable cargoes which tend to form explosive atmospheres. In these cases, your pump and piping systems must therefore be installed in a so-called explosion-proof design.

[54]International Convention for the Safety of Life at Sea, 1974 (SOLAS; German: Internationales Übereinkommen von 1974 zum Schutz des menschlichen Lebens auf See).

Fig. 5.74 Tanker with midship mounted manifold crane. *Photo* Eweht

Fig. 5.75 Manifold of an inland tanker

In addition, the empty spaces of the tanks, that is, above the cargo, are filled with inert gas. This displaces the previous oxygen-containing tank atmosphere. This prevents the cargo gases from igniting. The inert gas may be a specially treated combustion gas (exhaust gas) produced on the ship. However, any other inert gas which does not form a reactive mixture with the respective charge can also be used (e.g., nitrogen).

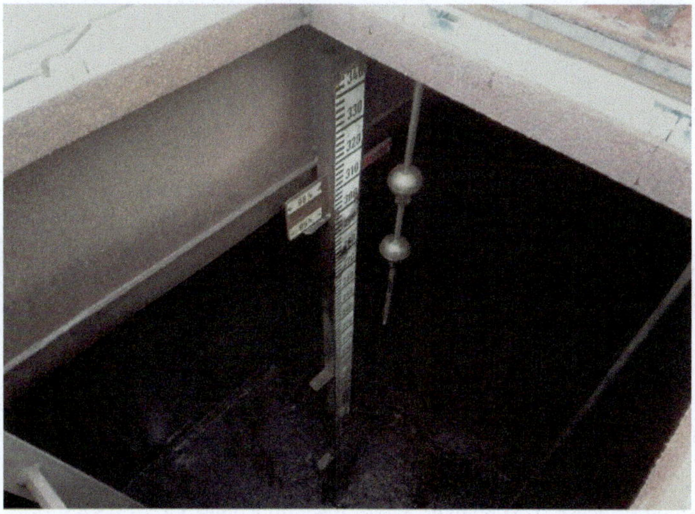

Fig. 5.76 Tank compartment with level indicator of an inland vessel

Fill Level

Furthermore, a certain degree of filling must not be exceeded when transporting dangerous goods. According to Chap. 3, Table C of the European Agreement concerning the International Carriage of Dangerous Goods by *Inland waterways* (ADN) is the maximum degree of filling in inland navigation depending on the dangerous goods transported. This is 97% for ethanol and petrol (UN number 3475), for example. For maritime transport, the classification societies specify the respective maximum degree of filling.

A pre-alarm must be triggered when 94% of the maximum permissible filling level (high load) is reached, and a filling level alarm when 98% of the maximum filling level is reached (overflow; see also Fig. 5.76; [4, p. 843]).

Loading Rate

The loading rate (LR) of the cargo tank must not exceed the following value:

$$\mathrm{LR} = 3600 \cdot U/t \quad [\mathrm{m}^3/\mathrm{h}], \tag{5.123}$$

where

U = the free volume (m^3) at the Fill level at which the overfill protection is triggered,

t = the time (s) required from the triggering of the overfill prevention device until the complete termination of the flow of cargo into the cargo tank; the time is the sum of the individual times required for the measures taken in succession, such as reaction time of the operating personnel, switch-off time for the pumps and closing time of the shut-off valves

Fig. 5.77 "Arctic Princess" LNG tanker. *Photo* Joachim Kohler, CC-BY-SA 4.0

Gas tankers are a special case (Fig. 5.77). They are used to transport liquefied gases in permanently installed cargo tanks. In addition to technical gases, transported gases are mainly liquefied natural gas (LNG *Liquefied Natural Gas*) and liquefied petroleum gas (LPG *Liquefied Petroleum Gas*).By liquefying the gases, a considerable reduction in volume and thus an increase in the mass of the gas to be transported can be achieved (for LPG—liquefaction by compression—about 1/260, for LNG—liquefaction by temperature reduction—1/600).

Example

The natural gas cooled down on land from 20 °C at – 162 °C, is to be transported by the currently largest natural gas tanker Mozah. Its tanks have a total volume of 266,000 m^3 [24].

How many cubic meters of natural gas can be liquefied on land and filled into the ship's tanks?

Solution: Natural gas is a gas mixture which consists mainly of methane. The density ρ_1 of liquid natural gas (boiling point at about – 162 °C) is approximately 450 kg/m^3 [39], in gaseous state at 20 °C for $\rho_2 = 0.66$ kg/m^3. Thus a density ratio of 450: 0.66 = 682 is available. In the case of landside compaction (at approximately 20 °C) a density ratio between gas and liquid phase of approximately 1: 680 can therefore be assumed. Since the volume $V = m \cdot \rho$ and the mass of the gas does not change during liquefaction, about 600 times the amount of liquefied natural gas can thus be fed to the tanker on land:

$$V_1/\rho_1 = V_2/\rho_2. \tag{5.124}$$

Conversion of Eq. 5.124 to VI results in:

$$V_1 = V_2 \cdot \rho_1/\rho_2 = 266{,}000\ \text{m}^3 \cdot 680 = 180.9\ \text{Mio. m}^3.$$

◀

Internationally binding standards on the design and equipment of liquefied gas tankers have been established by the IMO in the *IMO Gas Code* (*IGC*):

- Type 1G tankers: Chlorine, ethylene oxide, methyl bromide,
- Type 2G tanker: Ethane, ethylene, methane (LNG, e.g., LNG tanker)
- Type 2G/2PG tankers: Acetaldehyde, ammonia, butadiene, butylene, dimethylamine (e.g., LPG tankers), ethylamine, ethyl chloride, methyl chloride, propylene, vinyl chloride and butane, propane (LPG),
- Type 3G tanker: Nitrogen, various safety refrigerants.

Each tank must have its own transfer facility. This consists of the charge pump with the connected liquefied gas line, a residual bailing device with the emptying line (this system leads to the lowest point of the tank) and the vapor line for discharging the gaseous phase resulting from the vapor pressure of the gas [4, p. 847].

5.8.5 RAS Device

A special type of cargo or goods handling is the RAS maneuver (Replenishment at sea), also called Underway Replenishment (UNREP). This maneuver, practiced by the world's navies, is used to supply the emergency services at sea *during the trip*. This is about supplying the warships with ammunition, supplies and fuels, water, food, etc. from supply vessels (also called tenders). On the supply ship there is the so-called RAS boom (see Fig. 5.78), to which a suspension rope is attached, which is transferred to the ship to be supplied. Via this suspension rope, for example, tank hoses are led to the ship to be supplied. For the sequence of the maneuvers see below.

The EGV Bonn shown in Fig. 5.78 can simultaneously supply two ships via its RAS boom—both on its starboard and port side (so-called double RAS maneuvers).

During this maneuver, the supplier and up to two other ships are on a parallel course at a speed of about 12–16 kn (22–30 km/h; see Fig. 5.79).

The distance between the two ships is between 30–60 m and depends both on the speed of the journey and on their size (the higher the speed and the larger the vehicles, the greater the distance). A certain minimum distance between parallel ships must not be undercut:

When two ships are travelling in parallel during the RAS maneuver, it is possible that the *Bernoulli effect* interactions occur between the two vessels, which can result in undesired yawing behavior or even the "sucking in" of both vessels toward each other. The extent of the interaction effects depends on several factors, which fall into the categories

Fig. 5.78 RAS boom of the EGV (Einsatzgruppenversorger) "Bonn" of the German Navy

Fig. 5.79 Refueling of two frigates by the French tanker Mame A630. *Photo* © 2016 Bundeswehr/Torsten Kraatz

water depth, *ship* (ship dimensions, draft, displacement, ship shape and ship speed—absolute) and *passage criteria* (difference in size in terms of ship lengths, draft difference between the ships and between draft and water depth).[55]

[55]See also the parallel navigation of ships [57].

Fig. 5.80 Venturi nozzle with pressure curves

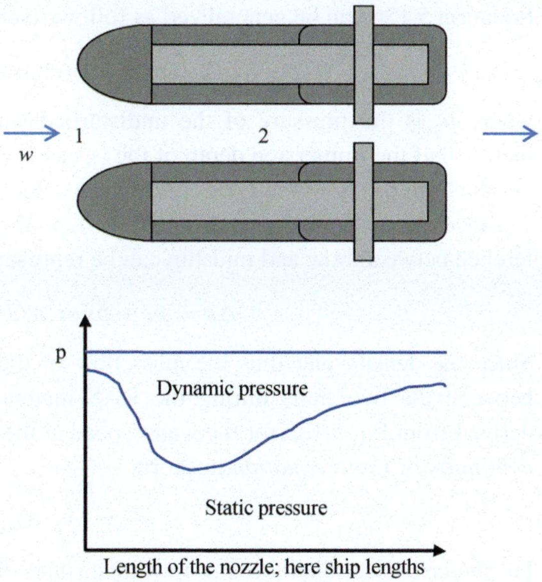

Due to the hull shape of the parallel ships, the shape of a Venturi nozzle is formed between them (Fig. 5.80): At the bow of the respective ships, the flow enters the opening of the "Venturi nozzle", the narrowest point of the nozzle is located approximately in the middle of the ship, only to widen again toward the stern of the ships.

If a fluid (here water) flows through the venturi nozzle, the dynamic pressure (dynamic pressure) is maximum and the static pressure is minimum at the narrowest point of the nozzle (Fig. 5.80). Due to the continuity equation, the velocity of the flowing fluid increases in proportion to the cross sections as it flows through the constricted part:

$$\dot{m} = \rho \cdot A_2 \cdot w_2 = \rho \cdot A_1 \cdot w_1. \tag{5.125}$$

The pressure difference between the areas of undisturbed flow to the nozzle (A_1 and w_1) and in the cross-sectional constriction (A_2 and w_2) is described by the Bernoulli equation, according to which *Bernoulli's law* applies

$$p_1 + (\rho/2) \cdot w_1^2 = p_2 + (\rho/2) \cdot w_2^2 \tag{5.126}$$

with

w_1 = flow rate of water in the area of the bow (simplifies the speed of the ships),
ρ = density of water,
p_1 = static pressure at point 1, that is, in the area of the nozzle inlet (bow area of both ships),
p_2 = static pressure at point 2, that is in the area of the narrowest point of both ships,
w_2 = flow rate of water in the area of the narrowest point between the ships, approximately midships

Equation 5.126 can be generalized as follows (see Fig. 5.80):

$$p + (\rho/2 \cdot w^2) = p_0 \text{ (constant total pressure)}, \qquad (5.127)$$

where p_0 is the pressure of the undisturbed surrounding water, measured at approximately half the immersion depth of the vessels.

The term $(\rho/2 \cdot w^2)$ in Eq. 5.127 indicates the dynamic pressure.

Since there is an incompressible flow here, the density ρ is constant. The pressure difference between bow and midship can be represented by rewriting Eq. 5.126 as follows

$$\Delta p = p_1 - p_2 = \rho/2 \cdot (w_2^2 - w_2^1). \qquad (5.128)$$

Since the density and thus the mass flow of the water remains constant when flowing between the two hulls during the RAS maneuver, the following relationship can be derived from Eq. 5.125 for the water speed at the narrowest point between the hulls—the *continuity* or *Konti equation* in short:

$$w_2 = (A_1/A_2) \cdot w_1. \qquad (5.129)$$

For the cross-sectional areas A_i for the intrinsically round cross-sectional shape of a venturi nozzle, the rectangular areas between the hulls can be used here in a simplified manner, which result from the draft of the ships and the respective distances at the bow and at the narrowest point. Thus, if the speed of both parallel moving ships is known, the flow rate of the water w_2 at the narrowest point can be determined from the distances between the bow of the ships and the narrowest point between the hulls and with the draft of the vehicles.

From Eq. 5.126 it can be seen: If both ships are too close together, the flow rate of the water flowing between the ships increases, as a result of which the water pressure at this point decreases due to *the venturi effect*. This can lead to the two hulls being sucked together despite counteraction. Therefore, a minimum distance between the two ships must be maintained during the RAS maneuver!

Procedure of the RAS Maneuver

The ship to be supplied approaches the supplier from behind; if the ships are at a height with the prescribed distance, the vehicles maintain course and speed (Fig. 5.81; [13, p. 474]).

To establish the ship-to-ship connection, a line is first transferred from one ship to another. This can be done manually with a throwing line. This care line can also be shot over to the other ship with a rifle. This line is used to overhaul a wire rope which is stretched between the highest point of the ship to be supplied and the so-called RAS boom of the supplier. This wire rope serves as a suspension rope, along which the tank hose runs, for example (Fig. 5.82). In addition, a so-called distance line is usually stretched from bow to bow between the two ships, with flags as distance marks. This line with the flags serves the helmsmen during the maneuver as a guide to keep the prescribed distance.

Fig. 5.81 Frigate Hamburg (*below*), EGV Berlin and frigate Hessen (*above*) during the preparation for the RAS maneuver. *Photo* © 2008 Bundeswehr/Ricarda Schönbrodt

Fig. 5.82 Korvette Braunschweig is supplied with fuel by EGV Bonn (clearly visible the wire rope). *Photo* © 2015 Bundeswehr/Matthias Letzin

The following reference values can be used to estimate the time to be planned for a RAS maneuver:

- Approach of the ship to be supplied: 5–11 min,
- Establishment of ship-to-ship connections: 9–12 min for fuel, 12 min for cargo.

Fig. 5.83 Gangway/stelling

The duration of the actual supply activity itself depends on the type and quantity of cargo to be transferred. With regard to the transfer of liquids such as water or fuel, the respective pump flow rate is the decisive factor here. For US ships can be taken as an indication: A US aircraft carrier takes over about 11,241 L aircraft fuel per hour per tank hose. From US supply ships, up to three tank hoses can be transferred in some cases, and one hose is stretched to cruisers and destroyers. For the acceptance of freight by cableway ("Manila highline method"), a service from 35 t/h can be assumed.

People can also be exchanged between the ships using the Manila highline method.

Danger: During the time of fuel and ammunition transfer, smoking is prohibited on the entire ship.

5.8.6 Passenger Ships: Gangway/stelling

Via the gangway, also called "stelling" or "shore leave", passengers and crew get on and off ships (Fig. 5.83). It can be moved manually, for example, with a hand crane, or also hydraulically. As a separate component, similar to a bridge, it can also be placed between the pier and the ship. These entrances must be at least 1 m wide and equipped with railings.[56] A safety net between the ship's side and the quay below the gangway complies with further safety regulations.

[56]§ Sect. 23 of DGUV Regulation 60—Vessels with a type approval on inland waterways (formerly: BGV D 19).

References

Print Media

1. Behrens, V., Hochhaus, K.-H., Wild, Y.: Schiffsbelüftung und Klimaanlagen. In: Handbuch der Werften, Vol. 25, Schifffahrtsverlag Hansa C. Schroedter & Co, Hamburg (2000)
2. Bich, E., Kaselow, D.: Kolbenpumpen—Kreiselpumpen. VEB Deutscher Verlag für Grundstoffindustrie, Leipzig (1978)
3. Böge, A., Böge, W.: Handbuch Maschinenbau. Springer, Wiesbaden (2015)
4. Umschlageinrichtungen. In: Diederichs, H.-O., Meier-Peter, H. (Hrsg.) Handbuch Schiffsbetriebstechnik. Seehafen Verlag, Hamburg (2008)
5. Drews, G.: Schiffslüftung und Warmluftheizung. In: Handbuch der Werften, Vol. 9, Schifffahrtsverlag Hansa C. Schroedter & Co, Hamburg (1967)
6. DUBBEL Taschenbuch für den Maschinenbau Vol. I, II. Springer, Berlin (1974)
7. Fa. Gestra AG: GESTRA-Wegweiser. Fa. Gestra AG, Bremen (2005)
8. Gieck, K.: Technische Formelsammlung. Selbstverlag, Heilbronn (1981)
9. Grundfos GmbH: Pumpenhandbuch. Grundfos, Erkrath (2004)
10. Herner, H.: Entwurf und Einrichtung von Handelsschiffen. Unikum, Barsinghausen (2011)
11. Hesse, T., Hochhaus, K.-H., Mehrkens, J.D.: Trinkwassersysteme auf Schiffen. In: Handbuch der Werften Vol. 26. Hansa, Hamburg (2002)
12. Hochhaus, K.-H.: Vorlesung Schiffshilfsmaschinen, TU Hamburg-Harburg, Stand 10/90
13. Hofbauer, B.-G.: Moderne Seemacht Teil 2. TRUPPENDIENST-Handbuch. Republik Österreich/Bundesminister für Landesverteidigung und Sport, Wien (2016)
14. Illies, K.: Handbuch der Schiffsbetriebstechnik. Vieweg & Sohn, Braunschweig (1970)
15. Kurzhals, H.-A.: Kühlen und Gefrieren von Lebensmitteln, vol. 1. Behr's Verlag, Hamburg (2007)
16. Meyer Werft: Nachhaltigkeitsbericht (2010)
17. Millipore GmbH: Technik-Notizen—Umkehrosmose. Millipore, Eschborn (1991)
18. Sökeland, R.: Fancoil-Regelung in der Praxis. In: DE-Jahrbuch Gebäudetechnik (2007)
19. Teml, A.: Friedrich Tabellenbuch für Metallgewerbe. Ferd. Dümmlers, Bonn (1975)
20. Verband der Chemischen Industrie: VCI-Leitfaden zur Montage von Flanschverbindungen in verfahrenstechnischen Anlagen. Verband der Chemischen Industrie, Frankfurt a. M. (2014)
21. Wille, R.: Lufttechnische Anlagen für Schiffe. STG Jahrbuch. Springer, Hamburg (1953)

Internet

22. Krueger, St: Manövrieren und Manövrierorgane (2001). http://www.ssi.tu-harburg.de/doc/webseiten_dokumente/ssi/vorlesungsunterlagen/manoe.pdf, Accessed: 2. Aug. 2016
23. Sievers, T.: "Entwicklung eines Handbuches zur Aus- und Weiterbildung von Besatzungsmitgliedern bei der Bedienung von Bordkränen", Diplomarbeit, veröffentlicht 2009, unter: https://www.hs-bremen.de/internet/hsb/projekte/maritime/studium/nautik-seeverkehr/diplombachelor/handbuch_20bordkraene.pdf (December 5, 2016)
24. Uken, M.: Gigantische Erdgas-Mengen gehen auf Weltreise (2010). http://www.spiegel.de/wissenschaft/natur/fluessig-verschifft-gigantische-erdgas-mengen-gehen-auf-weltreise-a-668826.html (December 13, 2016)

25. http://www.albrecht-services.de/heizung_berechnung.html (October 30, 2016)
26. http://www.al-ko.de/catalog/de-04-02/files/AL-KO_Prospekt_Marine_dt.pdf (July 21, 2017)
27. http://www.aquarein.de/marineservice/reinigung-von-frischwassersystemen-an-bord-schiffen (November 3, 2016)
28. https://www.arbeitssicherheit.de/media/pdfs/bgi_556.pdf (December 12, 2016)
29. https://ausbildung.stfv-koeln.de/wiki/Mehrzweckstrahlrohr (September 16, 2016)
30. https://www.baua.de/DE/Angebote/Rechtstexte-und-Technische-Regeln/Regelwerk/ASR/ASR-A3-5.html (May 3, 2017)
31. https://www.delta-q.de/export/sites/default/de/downloads/druckverluste_rohrstrecken.pdf (September13, 2016)
32. https://www.delta-q.de/export/sites/default/de/downloads/waermedurchgang_rohre.pdf (July 21, 2017)
33. https://www.delta-q.de/export/sites/default/de/downloads/0.2_Waermeleistung_Trinkwarmwasser.pdf (October 30, 2016)
34. http://www.deutsche-flagge.de/de/download/bau-und-ausruestung/neu-und-umbau/zusaetzliche-Informationen/kennzeichnung-von-rohrleitungen-an-bord-d-6 (September 9, 2016)
35. http://www.deutsche-flagge.de/de/download/sicherheit/ladung/container/ctu-code-2015 (December 18, 2016)
36. http://www.deutsche-flagge.de/de/medizin/hygiene/hygiene-an-bord-und-infektionsschutz#Ueberpruefungen (November 4, 2016)
37. https://www.eberspaecher-marine.de/haendler-und-service/faq.html (October 30, 2016)
38. http://www.ebu-uenf.org/fileupload/2014-09-22%20-%20Farbliche%20Kennzeichnung.pdf (September 9, 2016)
39. https://www.energie-lexikon.info/fluessigerdgas.html (July 21, 2017)
40. http://www.fagross.de/Bootsheizung.htm (October 30, 2016)
41. Günthner, W.-A., Rakitsch, S.: „Ermittlung von Dimensionierungs- und Auslegungsvorschriften für stark geneigte Schneckenförderer", 15. Fachtagung Schüttgutfördertechnik. http://www.fml.mw.tum.de/fml/images/Publikationen/G%C3%BCnthner_Rakitsch%20-%20DimensionierungAuslegungSchneckenf%C3%B6rderer.pdf (2010). Accessed: 27. Dec. 2016
42. http://www.fvlr.de/downloads/FVLR-Richtlinien/FVLR_Rili_10_08.14.pdf (October 19, 2016)
43. http://www.heizkoerper-wissen.de/anleitungen/heizkoerper-berechnung/ (October 30, 2016)
44. http://www.heizsparer.de/heizung/heiztechnik/heizungssystem (Retrieval date October 30, 2016)
45. https://www.hermetic-pumpen.com/system/assets/324/INFO_NPSH_D.pdf (August 29, 2016)
46. http://www.htk-doebeln.de/images/pdf/003.pdf (October 10, 2016)
47. http://www.ikz.de/1996-2005/2004/08/0408078.php (October 19, 2016)
48. http://kaelte-eckert.de/wp-content/uploads/2015/05/Kaeltetechnische_Planung.pdf (October 10, 2016)
49. http://www.leifiphysik.de/mechanik/gravitationsgesetz-und-feld/ausblick/ortsabhaengigkeit-der-erdbeschleunigung (March 12, 2017)
50. https://www.lernhelfer.de/schuelerlexikon/physik-abitur/artikel/bernoullisches-gesetz (March12, 2017)
51. http://www.linear.de/onlinebrowser/VIESSMANN/Pdf/DEU/5368876.pdf (October 30, 2016)
52. http://www.math-tech.at/Beispiele/upload/gra_Druckverlust_in_Rohrleitungen.PDF (September 16, 2016)
53. http://www.mb.uni-siegen.de/tts/personen/juk/wue/kapitel_k_wue.pdf (October 1, 2016)
54. http://www.pca-gmbh.com/tutorial/itms/filter.htm (November 9, 2016)

55. http://www.raumluft.org/fileadmin/dokumente/raumluft.org_-_Info_Luftmenge.pdf (October 20, 2016)
56. http://www.roth-brandschutztechnik.de/img/media/0cacdda6e164326bdfe105ad82644ff6.pdf (September 16, 2016)
57. http://schiw.sf.hs-wismar.de/siw/paper/heft1/beitrag10 (July 21, 2017)
58. http://www.schweizer-fn.de/waerme/waermetauscher/waermetauscher.php (October 1, 2016)
59. http://www.umweltbundesamt.de/themen/eu-verordnung-beschraenkt-klimaschaedliche (July 7, 2017)
60. http://www.u-wert.net/berechnung/u-wert-rechner/ (October 30, 2016)
61. http://www.ventilator.de/luftmengenermittlung-mittels-luftwechselraten (May 3, 2017)
62. http://www.ventilator.de/luftmengenermittlung-nach-der-personenbezogenen-aussenluftrate (October 20, 2016)
63. http://www.witzenmann-speck.de/de/media/Handbuch_der_Kompensatoren_1501de_5_12_12_20_download.pdf (September 20, 2016)
64. http://www.zeno.org/nid/20006115926 (August 8, 2016)
65. www.baua.de (May 3, 2017)
66. www.firedos.de (May 3, 2017)

6 Onboard Power Supply and Electrical Circuit Examples

The electrical equipment is an important but also complex component of the ship, as electrical and electronic systems on board are used in practically all areas. In principle, the regulations of the Verband Deutscher Elektrotechniker (VDE, Association of German Electrical Engineers (AGEE)) 0100 series of standards do not apply to the installation of electrical systems on board; the regulations of the classification societies are more relevant here, for example, the GL classification and construction regulation “I Ship Technology—2 Inland Vessels, 3 Machinery, Systems and Electrical Installations” [8].

However, if these classification and construction regulations do not contain any requirements for specific cases, the VDE standard sheets should be referred to.

A ship has a permanently high power requirement for its systems required for operation, such as lighting, air-conditioning, pumps, fans, bow and stern thrusters, navigation facilities, etc.

An e-balance sheet is used to estimate the expected electrical power requirement for various operating conditions (sea, district, port, possibly summer and winter operation, with and without refrigerated containers). The simultaneity factor also takes into account whether the power consumers are permanent or short term. The simultaneity factors are usually between $g = 0.1$ and $g = 1.0$. A distinction is also made between important, unimportant, and emergency consumers. On this basis, the electricity generators are selected.

6.1 Introduction

The electrical equipment is an important but also complex component of the ship, as electrical and electronic systems on board are used in practically all areas. In principle, the regulations of the Verband Deutscher Elektrotechniker (VDE, Association of

M. Pfaff, *Ship Operation Technology*, https://doi.org/10.1007/978-3-658-32729-3_6

Fig. 6.1 Shaft generator. *Photo* Karl-Heinz Hochhaus, CC BY 3.0

German Electrical Engineers) 0100 series of standards do not apply to the installation of electrical systems on board; the regulations of the classification societies are more relevant here, for example, the GL classification and construction regulation "I Ship Technology—2 Inland Vessels, 3 Machinery, Systems and Electrical Installations" [8].

However, if these classification and construction regulations do not contain any requirements for specific cases, the VDE standard sheets should be referred to.

A ship has a permanently high power requirement for its systems required for operation, such as lighting, air-conditioning, pumps, fans, bow and stern thrusters, navigation facilities, etc.

An e-balance sheet is used to estimate the expected electrical power requirement for various operating conditions (sea, district, port, possibly summer and winter operation, with and without refrigerated containers). The simultaneity factor also takes into account whether the power consumers are permanent or short term. The simultaneity factors are usually between $g = 0.1$ and $g = 1.0$. A distinction is also made between important, unimportant, and emergency consumers. On this basis, the electricity generators are selected.

There are several possibilities for generating the required electricity, some of which can be used in combination:

- EDiEng systems (electric diesel engine system; diesel engine-driven generator),
- Shaft generator (current generator is driven directly or indirectly via the propulsion shaft, Fig. 6.1),
- Exhaust gas-driven turbogenerators,
- Shore power connection (possibility of power supply when the ship is at the pier).

6.2 Onboard Electricity Generation

There are several possibilities for electrical onboard electricity generation depending on the main propulsion system:

- If the main propulsion system consists of a steam turbine with gearbox and fixed propeller, a *shaft generator* and/or also a steam turbine is selected to drive the *three-phase generators*. Ships with nuclear propulsion are also equipped with *turbogenerator sets*.
- If the main propulsion is a diesel engine, the *shaft generator* or one or more *electric diesel aggregates* (*EDiEng systems* or rather called *auxiliary diesel*) are also used here (Fig. 6.2).
- Gas turbine ships are equipped with *EDiEng systems* [4, p. 93].

Usually, three to four *EDiEng systems* are installed. Here, mainly medium-speed (720–900 min^{-1}) and rarely high-speed four-stroke diesel engines (1200–1800 min^{-1}) are used. The speed results from the selected frequency and number of pole pairs. The three-phase alternating current (AC; "three-phase current") generated with these generator sets is required to drive motors; lighting, navigation systems, etc. are operated with 230 V single-phase current or as so-called low-voltage systems (12–24 V).

These generator sets were formerly known as "Jockel." Even today, this still results in the occasional "Jockeln" when talking about electricity generation with a diesel generator set [2, p. 122].

Fig. 6.2 EDiEng system. *Photo* AIDA

Shaft generators are driven by the propulsion shaft via a gearbox. One variant is to integrate the generator directly into the shaft line. The advantage of a shaft generator system is the low specific fuel consumption of the main engine.

Gas turbogenerators are the exception on merchant ships, but have been installed, for example, on the *Millennium class* cruise ships of the shipping company Celebrity Cruises.

In **exhaust gas-driven turbogenerators**, the energy of the exhaust gas is used in a turbo set to generate electrical power. For this purpose, an exhaust gas or waste heat boiler is installed downstream of the ship propulsion diesel engines. In this, superheated steam is generated by means of the exhaust gas heat, which generates electricity in a steam turbogenerator.

6.2.1 Generators

A generator is a machine for converting mechanical energy—which is supplied to the generator in the form of rotation of the rotor—into electrical energy by induction. The generator works on the opposite principle to the electric motor—see Sect. 4.3.4 "Electric motor." Similar to the electric motor, the generator can also be designed as an external pole machine (induction coil on the rotor, excitation winding, or permanent magnet on the stator) or as an internal pole machine (induction winding on the stator, excitation winding, or permanent magnet on the rotor). The excitation winding can be supplied by an auxiliary generator (external excitation) or by self-generated current (self-excitation) [5, p. 385 f.].

It is advisable to design generators in such a way that the electrical energy requirement in sea operation corresponds to about 60–80% of the power of a generator. This allows the auxiliary diesel to operate at sea in a working range with good efficiency [4, p. 94].

In **AC and three-phase generators,** the current is supplied via slip rings (external pole generator) or directly at the stator winding (internal pole generator). The frequency of the generated AC or three-phase current depends on the speed and the number of excitation windings.

The most common types of AC generator are the synchronous generator, which corresponds to the synchronous motor in its design, and the asynchronous generator.

In **direct current (DC) generator**, which is always designed as an external pole machine, a commutator (current reverser) is used instead of slip rings, which constantly reverses the polarity of the generated AC according to the frequency and thus rectifies it. Multipole DC generators produce a relatively smoothed technical DC.

Likewise, as with the electric motor, there are different types of circuitry for the winding of the generators:

- The shunt circuit—here the generated voltage is relatively load-independent,
- The series circuit—here the voltage increases with increasing load,
- The double circuit—it is a combination of the shunt and series circuit, which can be designed so that the output voltage of the generator is completely independent of the load.

The electrical power generated P_{el} is equal to the mechanical power P_{mech} minus the losses incurred P_i due to mechanical friction, copper and iron losses. From this follows the power equation of an electric generator:

$$P_{\mathrm{el}} = P_{\mathrm{mech}} - P_i. \tag{6.1}$$

6.3 Shore-Side Power Supply

The shore-side power supply [3, p. 105 ff.][1] of ships, also known as cold ironing, sometimes also referred to as alternative maritime power (AMP), shore power, high-voltage shore connection (HVSC), or onshore power supply (OPS), is intended to reduce air pollution caused by emissions from the onboard power generators while in port. However, the problem with shore-side power supply at present is that there are currently no uniform interfaces for the cable connections between land and ship. A standard is being intended to remedy this obstacle: IEC/PAS 60092-510:2009 Electrical Installations in Ships—Special Features—High Voltage Shore Connection Systems (HVSC-Systems) describes specifications for shore connection systems for ships

6.3.1 Background

In ports, Directive 2005/33/EC of the European Parliament and of the Council of July 6, 2005 amending Directive 1999/32/EC as regards the sulfur content of marine fuels, which has been in force since January 1, 2010, together with MARPOL Annex VI, sets the objective of using marine fuels with a maximum sulfur content of 0.1% or using a shore-side power supply system available at the port. In Californian ports, the use of shore-side electricity has been mandatory since 2014. By 2020, most ships are expected to have switched to shore-side electricity [1, p. 22].

[1] Supplementary to the following, also [10].

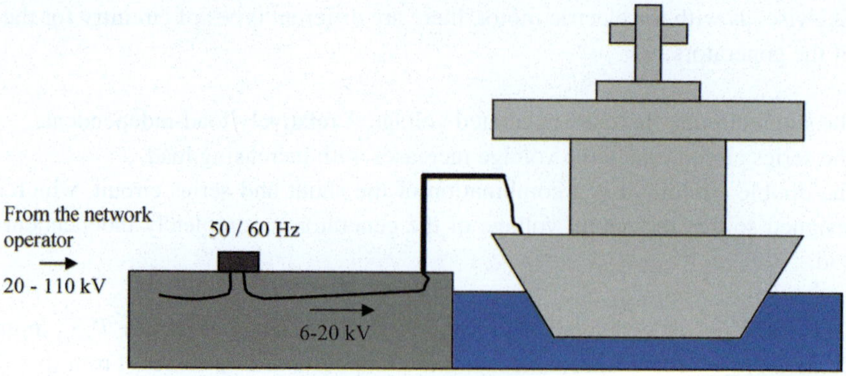

Fig. 6.3 Principle of shore-side power supply

6.3.2 Technology of Shore-Side Power Supply

The electricity supplied from the public grid is converted in the converter on the pier to the voltage and frequency required on board the ship (see Fig. 6.3). On land, appropriate transformer stations must be set up, from which the appropriate shore power cables are brought on board and connected to one or more feed-in points.

The connection capacities for cruise ships vary depending on the size of the ships and are currently around 10 to 12 MVA (megavolt ampere) for ships with 3000–4000 passengers, up to 6.5 MVA for refrigerated and large container ships with 6000–12,000 TEU (twenty-foot equivalent unit), and 2–4 MVA for large roll-on/roll-off (Ro-Ro) ships and ferries. In the case of ferries, the demand for electrical energy depends on the number of refrigerated containers and trailers with refrigerated cargo, as these are often connected to the onboard power supply.

In the port of Hamburg, the liquefied natural gas (LNG) hybrid barge Hummel has been in operation since 2015, on which special Caterpillar engines powered by LNG can generate electricity with a capacity of 7.5 MW (50/60 Hz) via five generators from Zeppelin Power Systems. This barge can be brought close to the berth of the ships to be supplied (initially the AIDA fleet). Via a cable connection barge ship, the power is led to the ship to be supplied [6].

6.4 The Electrical System

The onboard power supply system is mainly designed as a three-phase system with a frequency of 60 Hz. As a rule, the onboard network of a seagoing vessel is designed in such a way that power consumers are supplied directly or via subdistribution (SD) boards from the main switchboard. The grid voltage in the distribution networks is 400 V for

Fig. 6.4 Main switchboard with an older type of bus bar behind it

a grid frequency of 50 Hz or 440 V for a grid frequency of 60 Hz. Ships with onboard power supplies of >8 MW have an additional medium voltage level of 3.3, 6.6, or 10 kV. These are used on passenger ships and large container ships with many refrigerated containers. If other voltages are required, they must be downconverted by transformers. There are also many electrical consumers on board that require DC. For this purpose, the AC voltage must be rectified via rectifiers.

The electrical power is distributed to the power consumers by means of a main switchboard (Fig. 6.4) via a bus bar located in the electrical system.

For reasons of ship safety, there is at least one main and one emergency network. This is supplied by an emergency power generator and an emergency or auxiliary switchboard outside the engine room. All electrical consumers relevant to ship safety are connected via the emergency network. These include, for example, emergency lighting and navigation devices. However, during normal ship operation, the safety-relevant consumers are also fed via the main power supply system. If the main power supply fails, the emergency power supply starts automatically. Modern control panels are used to monitor and switch the individual circuits (see Fig. 6.5).

Figure 6.6 shows a schematic diagram of a simple wiring system.

With the **DC electrical system** for ships, ABB AG claims to have developed the most flexible power supply and propulsion system for ships to date. The system combines the various DC connections on board the ship and distributes the electrical energy through a single 1000 V DC circuit. This means that the usual alternating current switchgear (AC switchgear), decentralized rectifiers, and power converter transformers can be dispensed with.

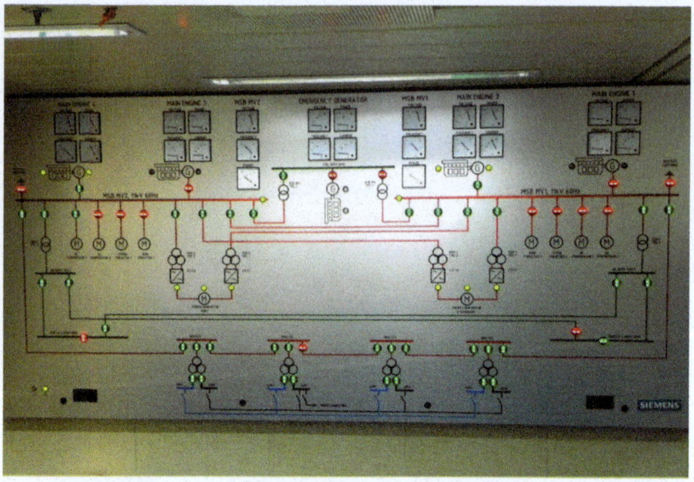

Fig. 6.5 Display panel of an onboard electrical system. *Photo* AIDA

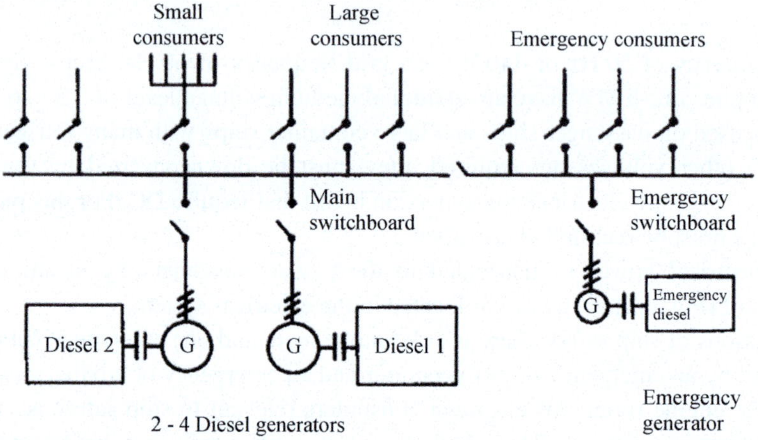

Fig. 6.6 Schematic representation of a simple wiring system on board. *Source* Karl-Heinz Hochhaus, CC BY 3.0

The DC onboard power supply system combines the advantages of AC and DC components and systems, complies with the relevant provisions and regulations for selectivity and equipment protection, can be used for electrical power up to 20 MW, and operates with a rated voltage of 1000 V DC [7].

Example

The SEATEC chart plotter NAV-6/NAV-6i is to be operated with 24 V DC. On the bridge, you will find 230 V AC. What technical measures are required?

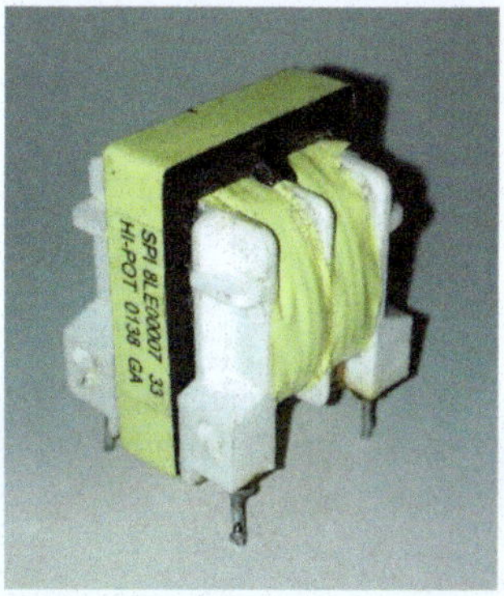

Fig. 6.7 Miniature transformer

Fig. 6.8 Flat bridge rectifiers

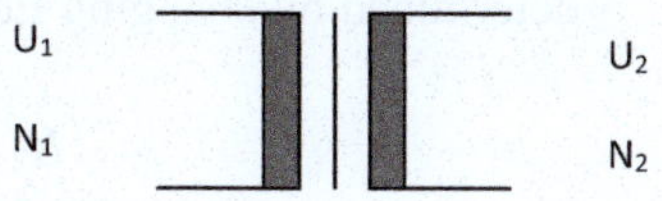

Fig. 6.9 Transformer circuit diagram

The voltage must first be transformed from 230 V down to 24 V. The AC voltage must then be rectified. This is usually done with a power supply unit in which the transformer (Fig. 6.7) and rectifier (Fig. 6.8) are integrated. ◀

Example: Design of transformer and rectifier for the above chart plotter

The number of windings and the voltage on the primary side (N_1, U_1) of the transformer are related to the number of windings and voltage on the secondary side (N_2, U_2) (Fig. 6.9) as follows:

$$N_1/U_1 = N_2/U_2. \tag{6.2}$$

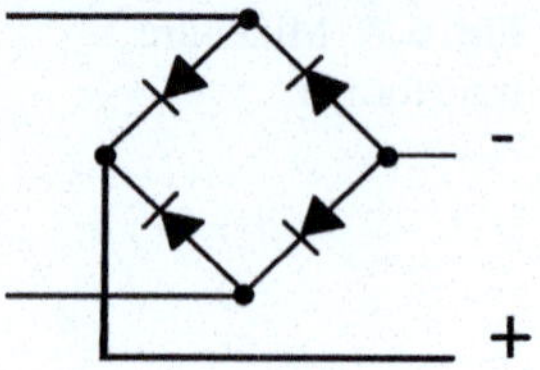

Fig. 6.10 Bridge rectifiers

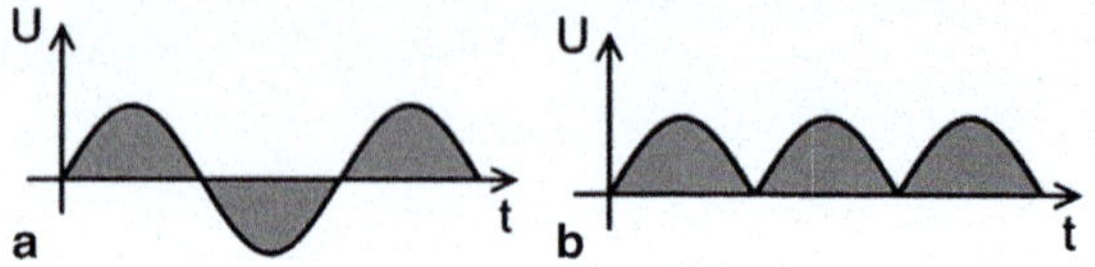

Fig. 6.11 Legend placeholder. **a** AC voltage, **b** pulsating DC voltage

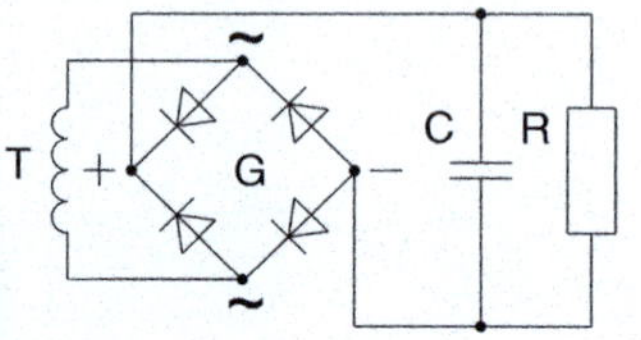

Fig. 6.12 Bridge rectifier with smoothing capacitor. *T* transformer secondary winding, *G* bridge rectifier, *C* smoothing capacitor, *R* load resistance. Picture: Saure, CC BY-SA 3.0

If, for example, a transformer with 958 windings on the primary side is selected, the required number of windings on the secondary side for the above chart plotter is determined by changing Eq. 6.2 to N2 and inserting the given values:

$$N_2 = \frac{U_2 \cdot N_1}{U_1} = \frac{24\mathrm{V} \cdot 958}{230\mathrm{V}} = 100.$$

The voltage from 24 V, which is applied to the secondary side of the transformer, is an AC voltage. It must therefore still be rectified. This is done by a rectifier. The standard rectifier for single-phase AC is the bridge rectifier, also called Graetz circuit or two-pulse bridge circuit. The circuit is formed by four diodes (Fig. 6.10).

Figure 6.8 shows the structural design of a bridge rectifier in the form of a so-called flat bridge rectifier. The component shown here with the dimensions 2.3 cm × 1.8 cm (without connecting wires) is designed for 1000 V and 6 A on the AC side.

The AC voltage (Fig. 6.11a) applied on the left in Fig. 6.10 is converted into **pulsating DC voltage** (Fig. 6.11b).

As this is a two-way rectification, the negative half oscillation of the AC voltage in the DC circuit appears positive. The lower half-wave is quasi "folded up" by the rectification. The ripple has twice the frequency of the input voltage, which reduces the subsequent filter effort. This is done by an electrolytic capacitor (Fig. 6.12).

Bridge rectifiers for AC and three-phase current are often offered as diodes already connected together in a common housing. For higher currents, they contain a cooling surface and a hole for mounting on a heat sink [11].

Which diodes are used depends on the applied voltage U and power P of the connected device. The electrical power is the product of voltage and current I:

$$P = U \cdot I. \tag{6.3}$$

Assuming that the plotter has a power consumption of 200 mA, the device requires an electrical power of $P = U \cdot I = 24\ \text{V} \cdot 0.2\ \text{A} = 4.8\ \text{W}$.

So here, four Zener diodes **ZD-5W 24 V** could be considered [12].

In a bridge rectifier (also B2U rectifier), the smoothing capacitor is always charged after half a mains period, that is, at 50 Hz every 10 ms. If the voltage during this time Δt only drops to ΔU, the capacity C of the system must be calculated according to Eq. 6.4 (with "I" the electric current):

$$C = I \cdot \frac{\Delta t}{\Delta U}. \tag{6.4}$$

Assuming that ΔU is to be 0.1 V, the capacity for the smoothing capacitor in the present circuit is as shown in Eq. 6.4

$$C = I \cdot \frac{\Delta t}{\Delta U} = 0.2 \cdot \frac{0.01}{0.1} = 0.02\text{F} \tag{6.5}$$

◀

6.5 Electrical Circuit Examples

In the following, common electrical circuits for power consumers found on board, for example lighting, are presented.

6.5.1 Switch-off

This circuit is used wherever a power consumer is to be switched on and off. For onboard installation, toggle switches with two switch positions (on/off) are usually used, for example, to switch a lamp.

The circuit diagram and the connection of the circuit breaker in the AC network are described below (see also Fig. 6.13).

Starting from the junction box 3, in which phase (L), neutral (N), and protective earth (PE) conductors are connected, the wiring is carried out. The circuit breaker 2 has two connections. The phase is connected to the terminal marked there with the letter L, and the switched cable to the light is connected to the contact marked ↑. As can be seen from

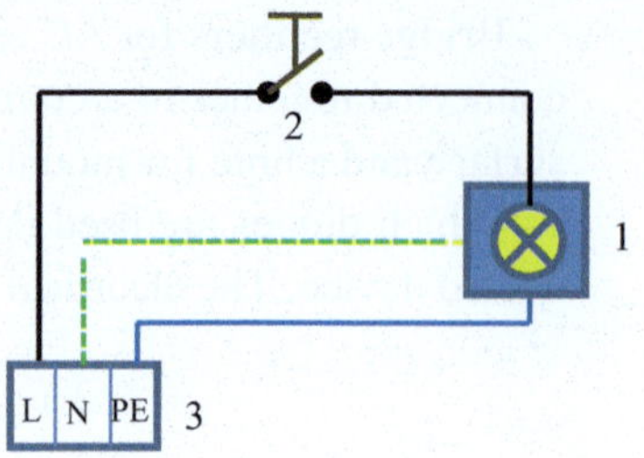

Fig. 6.13 Switch-off

the circuit diagram shown, switch 2 interrupts phase L. The neutral conductor N and the protective conductor PE are connected to lamp 1. Only phase L should be switched at all times, otherwise the lamp will be live even if the switch is turned off. On the lamp, the switched L-conductor should be connected to the foot contact, while the neutral conductor N is connected to the lamp socket, so that the lamp socket is not live when the lamp is changed. The protective contact PE is connected to the metal housing of the lamp. The cables are normally color coded. Normally, the phase L is black, the neutral conductor N blue, and the protective conductor PE yellow/green.

6.5.2 Alternating Circuit

If a power consumer, for example a lamp, is to be able to be switched on and off from two different places (e.g., at the cabin door and at the bed), this can be realized by means of the alternating circuit.

Wiring instructions

A changeover switch has three terminals. A contact (L) is used for the live cable (called external conductor or phase). With the second changeover switch, this contact is used for the wire going to the luminaire; this is why it is often referred to as the "lamp wire." The other two contacts on the changeover switch are marked with an arrow or with the letter "K." These two contacts are connected to the two contacts of the second changeover switch with two wires (corresponding).

A supply line, for example cable NYM-J 3 × 1.5^2, is pulled to the first junction box. From there, a second cable, for example NYM-J 5 × 1.5^2, is pulled to the second junction box. From there, a cable, for example NYM-J 3 × 1.5^2, is led to the lamp, cable NYM-J 5 × 1.5^2 is led from the junction boxes to the changeover switches, the brown wire of the supply lines is connected to terminal L of the changeover switches, the black and the grey wire are used as "corresponding" by the NYM-J 5 × 1.5^2, and the terminals marked "arrow" or "K" are used on both changeover switches.

In the junction boxes, the wires are connected as shown in Fig. 6.14 and the lamp is connected accordingly.

Fig. 6.14 Alternating circuit

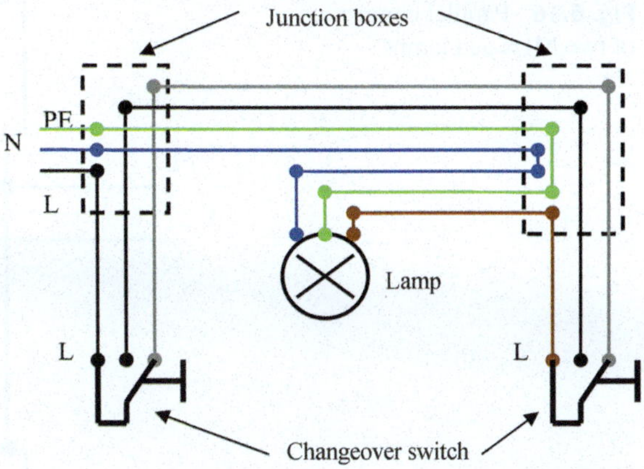

Fig. 6.15 Switching a lamp via motion detector (MD)

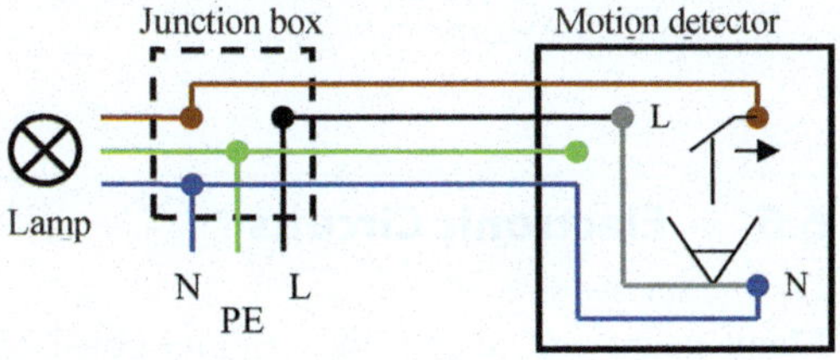

6.5.3 Motion Detector

A **motion detector** (MD) is an electronic sensor that detects movements in its immediate surroundings and can therefore function as an electrical switch. It is mainly used to automatically switch on lighting (e.g., when entering rooms or corridors) or to trigger an alarm (e.g., in case of unauthorized access). The circuit diagram in Fig. 6.15 shows the switching of a lamp by means of a MD.

Among other sensors, the passive infrared (PIR) sensor is the most commonly used type of MD.

MDs are usually adjustable in sensitivity and coupled with a twilight switch that can also be adjusted. Attachment panels can be used to cover part of the visible sectors and the sensor module can also be swiveled if necessary.

MDs can be coupled together via cables (directly or via a bus system) or by means of a radio module. If one of the networked MDs responds, all connected detectors switch on their power consumers (lighting, alarm) [9].

Especially when entering longer corridors and hallways, the detection range of one motion sensor may not be sufficient, so that a second one is required. This must then be connected in parallel with the first one. The circuit diagram in Fig. 6.16 shows the parallel connection of two MDs—starting from the SD—and two lamps connected in parallel.

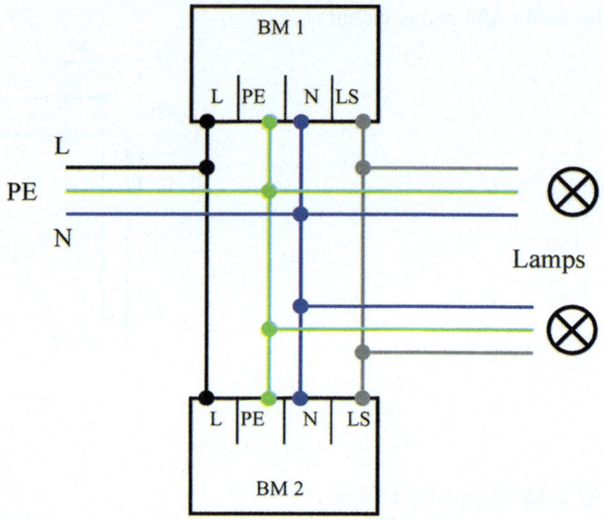

Fig. 6.16 Parallel connection of two MDs and lamps

6.6 Electronic Circuits

Many years ago, electronic circuits consisted of a multitude of electronic components such as transistors, resistors, diodes, and capacitors. With these components, circuits for communication (transmitting/receiving systems, power units, circuits for control and regulation purposes, etc.) were built.

Today, in many cases, specially developed *integrated circuits* (ICs) are used, which contain several of the aforementioned components combined in one component. Figure 6.17 shows the circuit board of an electronic circuit with IC and other components such as resistors, diodes, capacitors, and transistors.

A purely electronic circuit for temperature monitoring is described and shown below (Fig. 6.18).

With this circuit, temperature changes, for example in cooling loads, can be safely monitored. The temperature sensor, a negative temperature coefficient (NTC) resistor (the resistance of this component is temperature-dependent), is mounted by means of a cable connection at the point where the temperature monitoring is to be carried out. The light-emitting diode (LED) lights up when the temperature falls below the set temperature. This means that the LED lights up as long as the set temperature (set point) is maintained or is below this value. If the temperature rises above the set point in the cold store, the LED first starts flashing and then goes out completely. The use of a green LED is suitable for this purpose.

On the other hand, this circuit can also be used to detect impermissible cooling: The LED does not light up as long as the target temperature is maintained at a certain level; if it drops, the LED first starts to flash and then lights up completely. In this case, the use of a red LED is recommended.

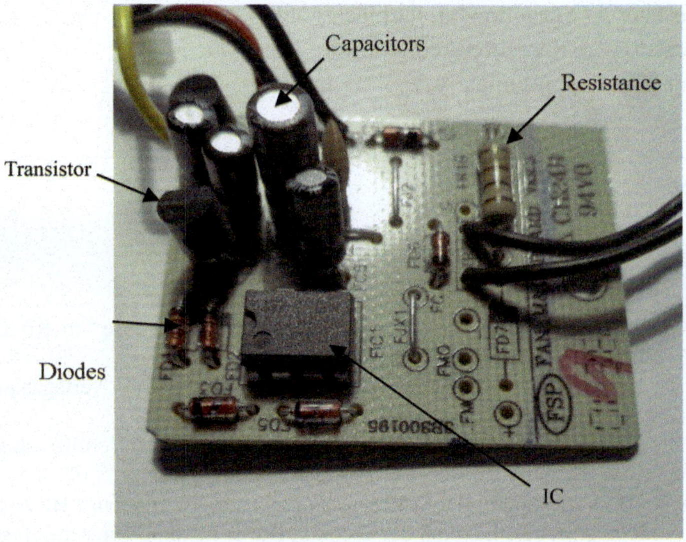

Fig. 6.17 Board with integrated circuit (IC) and other electronic components

Fig. 6.18 Temperature monitor

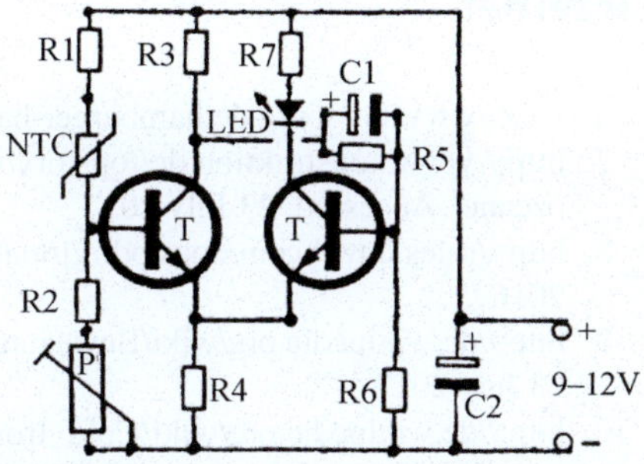

With this circuit it must be taken into account that the NTC resistor needs approximately 30 s until it has adjusted to the current temperature.

Parts list:

- T1, T2: two transistors BC 237B,
- LED: one light-emitting diode,
- P: one trim potentiometer 50 K,
- C1: one electrolytic capacitor 100 μF, 25 V,
- C2: one electrolytic capacitor 47 μF, 16 V,
- NTC: one heat-sensitive resistor (approximately 47 K at 20 °C),
- R1, R2, R5: three resistors 2.87 K (red/grey/violet/brown/red),
- R3, R7: two resistors 150 Ω (brown/green/brown/gold),

- R4: one resistor 47 Ω (yellow/violet/black/gold),
- R6: one resistor 2.2 K (red/red/red/gold),
- One circuit board approximately $27 \times 55\ mm^2$.

References

Print media

1. Astoria Kreuzfahrten-Zentrale (Hrsg.): „Energie aus der Steckdose" in Kreuzfahrt-Zeitung, Osnabrück (2016)
2. Gebauer, J., Krenz, E.: Marine-Enzyklopädie. Brandenburgisches Verlagshaus. Tosa, Wien (2003)
3. Hildebrandt, M., Reinicke, H.: Reduction of emissions from cruise ships in ports – Perspectives from Hamburg. Z. Hansa **6**, 105 (2011)
4. Hochhaus, K.-H.: Vorlesungsskript Schiffshilfsmaschinen, TU Hamburg-Harburg, Stand 10/90
5. Institut, Lexikographisches: Lexikon der Technik, Bd. 1. Lexikographisches Institut, München (1986)

Internet

6. http://www.hamburg.de/hamburger-hafen/4385644/lnghybridbarge/. Accessed: 28 Dec. 2016
7. http://www.konstruktion.de/topstory/containerschiff-emma-maersk-mit-110-000-ps-uber-die-ozeane/. Accessed: 23 July 2017
8. http://rules.dnvgl.com/docs/pdf/gl/maritimerules2016Jan/gl_i-2-3_d.pdf. Accessed: 6 Jan. 2016
9. http://de.wikipedia.org/wiki/Bewegungsmelder, mit weitergehenden Ausführungen. Accessed: 23 July 2017
10. http://de.wikipedia.org/wiki/Cold_Ironing. m. w. N. Accessed: 23 July 2017
11. http://de.wikipedia.org/wiki/Gleichrichter. Accessed: 23 July 2017
12. www.reichelt.de. Accessed: 23 July 2017

7 Occupational Safety and Ship Safety, Fire Protection

Accidents have been happening at sea ever since the advent of shipping. Spectacular shipping accidents such as those of the Titanic, Pamir, Exxon Valdez, Estonia, and Prestige or Pallas have caused a great deal of suffering for people and the environment, but without them the great progress in safety regulations would not have been possible.

Whereas, in the past, safety regulations often differed from flag state to flag state, today ship safety is essentially determined by global rules laid down by the International Maritime Organization (IMO). The *International Convention for the Safety of Life at Sea* (SOLAS Convention) has already been in existence since 1913.[1] This convention has since been developing further and further. Furthermore, the relevant international regulations include, for example, the *International Safety Management Code* (ISM), which ensures safety aspects of operation of ships, and the *International Ship and Port Facility Security Code* (ISPS), which since 2002 has specified compliance with global security requirements against external threats (e.g., piracy, terrorism) [25].

In addition to the abovementioned international legal regulations, however, an almost incalculable number of national legal and technical standards also serve to protect crews, passengers, and the ship itself. A detailed explanation would go beyond the scope here. In this chapter, the essential aspects of occupational health, safety on board, and ship safety, which includes fire protection on board, are presented.

[1]International Convention for the Safety of Life at Sea, 1974 (SOLAS; German: *Internationales Übereinkommen von 1974 zum Schutz des menschlichen Lebens auf See*).

M. Pfaff, *Ship Operation Technology*, https://doi.org/10.1007/978-3-658-32729-3_7

7.1 Occupational Health and Safety, Safety at Work, and Ship Safety

With regard to the question of sources of knowledge concerning aspects of occupational health and safety and ship safety, the principle of "lex specialis before lex generalis" applies. This means that where special technical and legal regulations on occupational health and safety and ship safety exist, these are to be applied; the other, particularly national regulations on occupational health and safety are only applicable in a subsidiary manner.

For example, the Ship Safety Act (SSA; "Schiffssicherheitsgesetz, SchSg"), concretized by the Ship Safety Ordinance (SSO; Schiffssicherheitsverordnung, SchSV), stipulates that the international and EU regulations on ship safety are generally applicable.

In addition, occupational safety is also regulated by the Occupational Safety and Health Act (OHSA, German: "Arbeitsschutzgesetz"). According to Section 1 para. 2 OHSA, this does not apply to the occupational health and safety of employees on seagoing ships. However, this is only the case if and insofar as corresponding legal provisions exist in other regulations, for example, in the Ship Safety Act (German: Schiffssicherheitsgesetz). This means that the OHSA, only, does not apply if another regulation lex specialis for maritime shipping contains corresponding regulations on occupational safety.

In this respect, the Ship Safety Act lex specialis seems to be applied. However, on the question of whether and to what extent the OHSA or the Ship Safety Act is to be applied, a closer look at the regulatory content of both provisions is required.

Thus in Section 1 para. 1 *OHSA* The aim and application of this provision is formulated as follows:

> The purpose of this law is to ensure and improve the safety and health protection of employees at work through occupational safety measures. It applies in all areas of activity…

This makes the following clear: The *OHSA* aims to improve the safety and health protection **exclusively for employees**.

Through the *Ship Safety Act* (and by the provisions adopted pursuant to this Act "SchSV" [German: Schiffssicherheitsverordnung]) the **safe operation of the ship** should be aimed at. This includes in particular that the ship and its accessories are kept in a safe operating condition and are guided safely; the necessary precautions are taken to protect third parties against hazards arising from the operation; to protect the marine environment; and the air against hazards or unlawful interference arising from the operation. This also includes the effective selection, instruction, monitoring, and support of persons who are assigned to this task in the shipping company and on the ship (see § 3 Ship Safety Act).

The **destination** is therefore not only aimed at occupational health and safety, but also essentially **far-reaching**: Through safe ship operation, which includes not only technical but also organizational requirements (in particular selection and monitoring of management personnel), the Act also aims to protect the marine environment and the air. It also aims to ensure safe ship operation, which includes not only personnel requirements but also requirements for navigation systems. In addition, *Third parties* are to be protected against hazards arising from the ship operation. *Third parties* are not only the *employees* on board, but also *other persons* (passengers, other road users). As already mentioned above, this is to be achieved by **mandatory application of international standards**—see the wordings of § 1 para. 1 Ship Safety Act:

> This law determines the measures to be taken in implementing the international regulations on ship safety and environmental protection at sea in force from time to time…

In § 1 para. 2 Ship Safety Act defines which international regulations are covered with this:

> These are the provisions of nationally applicable international law listed in Sections A to C of the Annex to the Ship Safety Act and the Acts of the European Communities or of the European Union listed in Section D of the Annex, as amended. For the purposes of this Act, International Ship Safety Standards are the regulations listed in Section E of the Annex and published in Germany as applicable recognized rules of technology or maritime practice, in the version specified in each case. The ship security regulations within the meaning of the first sentence should also include the international regulations governing the prevention of external dangers, insofar as reference is made to these regulations in accordance with the following provisions in the Annex.

In this respect, against the background of the speciality principle, this means that the OHSA is then to be applied on a mandatory basis, insofar, as the Ship Safety Act does not contain any regulations there.

Essential provisions of the OHSA that are not contained in the Ship Safety Act are the regulations for the preparation of a risk assessment (§ 5 OHSA) and its documentation (§ 6 OHSA).

The risk assessment according to § 5 OHSA[2]

The legislator gives the employer a wide scope for implementing the Occupational Health and Safety Act. For example, the law does not stipulate on how the risk assessment is to be carried out; only principles are named. This means that a risk assessment has to consider the respective workplaces and activities individually.

General rules

The scope of risk assessment is based on the operational requirements and circumstances. All foreseeable work processes on the ship must be taken into account. This also

[2]For further information and assistance in preparing a risk assessment, see [23].

includes events and tasks that take place outside the "normal" work processes, such as operational disruptions.

The risk assessment should be structured in such a way that all the identifiable hazards and risks are investigated. The OHSA refers to the following sources of danger, for example: work procedures, work processes, working hours, insufficient qualification, and instruction of employees. Once a hazard is identified, the risk assessment must describe how the hazard can be eliminated or reduced.

A risk assessment is required for each activity or workplace on board. In the case of similar activities, the same work procedures and the same workplaces, the assessment of one workplace, procedure or activity is sufficient.[3]

Furthermore, the obligation to instruct according to §§ 9, 12 OHSA and § 11 "Occupational Medical Precaution" are part of the applicable regulations. Obligations and rights of employees (§§ 15, 17) are also part of the applicable regulations of the OHSA.

§ 7 of the OHSA regarding the transfer of tasks to employees is essentially also regulated by § 7 of the Ship Safety Act, so that lex specialis § 7 of the OHSA is not applicable to seagoing ships.

Here it should be noted that the OHSA is concretized by several regulations. The following are of particular importance for shipping:

- Occupational Health and Safety Ordinance on Artificial Optical Radiation(German: "Arbeitsschutzverordnung zu künstlicher optischer Strahlung"),
- Workplace Ordinance (German: "Arbeitsstättenverordnung"),
- Industrial Safety Regulation (German: "Betriebssicherheitsverordnung"),
- Ordinance on Work with Visual Display Units (German: "Bildschirmarbeitsverordnung"),
- Ordinance on Biological Substances (German: "Biostoffverordnung"),
- Ordinance on Hazardous Substances (German: "Gefahrstoffverordnung"),
- Noise and Vibration Work Protection Ordinance (German: "Lärm- und Vibrations-Arbeitsschutzverordnung"),
- Load Handling Ordinance (German: "Lastenhandhabungsverordnung"),
- PSA Use Ordinance (German; "PSA-Benutzungsverordnung"),
- Ordinance on occupational health precautions (German: "Verordnung zur arbeitsmedizinischen Vorsorge").

In view of the above, it is clear that the provisions in question contain only partially identical provisions. Only then does lex specialis have priority over the Ship Safety Act. In the vast majority of cases, however, both laws have different regulatory contents. Thus the OHSA is applicable in any case, as far as it concerns an assessment and documentation of the hazard at the workplace and a resulting obligation to instruct.

[3]In depth [27].

Furthermore, the regulations mentioned above on the basis of the OHSA are applicable, unless—exceptionally—regulations with the same content are contained in international regulations, which would then be binding via the Ship Safety Act. The Ship Safety Act contains structural/technical and organizational regulations for safe ship operation, which also protect employees; this law is then relevant in this respect. The OHSA may then only be referred to again in the event of gaps in the regulations in this respect.

In addition to the abovementioned laws and regulations, however, a large number of accident prevention regulations, leaflets and guidelines of the German Accident Insurance (Deutsche Gesetzliche Unfallversicherung, formerly: Hauptverband der gewerblichen Berufsgenossenschaften, German Federation of Institutions for Statutory Accident Insurance and Prevention) are also relevant. Here in particular the **UVV Sea** (accident prevention regulations for maritime companies) are of importance. A further example of accident prevention regulations to be taken into account in shipping is the *DGUV 52 (BGV D6) Cranes* which contains requirements for the operation and testing of cranes. An overview of all regulations of the German Accident Insurance can be found under www.dguv.de.

7.1.1 SOLAS

As explained in section, international conventions on ship safety are declared binding in Germany by the Ship Safety Act. In this context, the SOLAS Convention is particularly important. It contains minimum standards for safety on merchant ships. The last amendment entered into force in 2016 [24].

SOLAS describes in 14 chapters technical and organizational measures for occupational safety, ship safety, and maritime environmental protection:

1. General provisions: types of ships, issue of documents,
2. Construction: Subdivision of the hull, stability, machinery and electrical installations, fire protection, fire detection, and firefighting,
3. Rescue equipment and facilities,
4. Radio equipment: Implementation of the *Global Maritime Distress and Safety System* (*GMDSS*) with ultrashort wave radio (USW radio) Digital Selective Call (DSC controller), satellite Emergency Position-Indicating Radio Beacons (EPIRBs) and Search and Rescue Radar Transponders (SART),
5. Navigation safety: Crew,
6. Carriage of cargo (excluding liquids and gases),
7. Transport of dangerous goods: compliance with the *International Maritime Dangerous Goods Code (IMDG Code)*,
8. Nuclear ships (ships with nuclear propulsion): compliance with the *Code of Safety for Nuclear Merchant Ships*,

9. Management of safe navigation: implementation of the *International Safety Management Code (ISM Code)*,
10. Safety measures for high-speed craft: obligation to comply with the *International Code of Safety for High-Speed Craft (HSC code)*,
11. Special measures to improve maritime security (including implementation of the International Ship and Port Facilities Security Code (ISPS Code),
12. Additional security measures for bulk cargo: structural requirements for cargo ships longer than 150 m in length,
13. Verification of compliance: Since January 1, 2016, IMO member states have committed themselves to participate in an audit scheme.
14. Security measures for ships in polar waters: Mandatory compliance with the Code for ships operating in polar waters (*International Code for Ships Operating in Polar Waters—The Polar Code*).

In the following, some aspects of the SOLAS Convention are examined in more detail.

7.1.1.1 Lifeboats, Life Rafts

Lifeboats and life rafts are *collective rescue facilities* (group rescue). Lifeboats are launched using a special launching device (boat davit, Fig. 7.1; free-fall device, Fig. 7.2).[4] The launching appliance must be designed so that the lifeboat's cruising speed complies with the following formula:

$$v = 0.4 + 0.02 \cdot H, \tag{7.1}$$

where v the numerical value of the ferry speed measured in m/s, and H is the numerical value of the height difference between the boat deck and the ballast line measured in m. For v a deviation of +/−10% is permissible.

With regard to the required number and arrangement of these collective rescue facilities, the following applies in principle[5]:

Lifeboats on cargo ships must provide space for all persons on board on each side or at the stern. On passenger ships there should be room for people in boats and life-rafts (Fig. 7.3) on both sides. On cargo ships under 1600 GRT[6] there must be at least one lifeboat on board. Life-rafts with room for everyone on board should be on either side of the ship if the lifeboats are arranged at the stern. On cargo ships <1600 GRT life-rafts with room for people on board must be provided on each side; life-rafts must be equipped with an approved water pressure release device (Fig. 7.4).

[4]Deepening to lifeboats and their launching appliances [9].

[5]Details can be found in the SOLAS Convention.

[6]The gross register ton is an obsolete unit of measurement for seagoing ships. It corresponds exactly to 100 English cubic feet and about 2.83 cubic meters.

Fig. 7.1 Lifeboat to boat davit (*right*), rescue boat (*left*)

Fig. 7.2 Freefall lifeboat

For cargo ships of 500–1600 GRT, with the exception of tankers, the following equipment with life-saving appliances should be considered as equivalent substitutes for life-saving appliances:

Fig. 7.3 Life-rafts

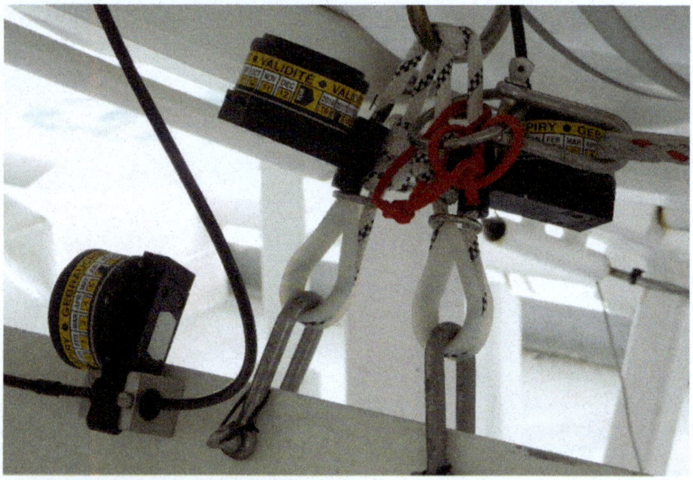

Fig. 7.4 Water pressure release of the life rafts

- A motor rescue boat for people, which meets the requirements for rescue boats (Fig. 7.1; rescue boats are mainly used for the rescue and recovery overboard of fallen persons), on one side,
- On the other hand, one or more fissile life-rafts for all and,
- Free floating life-rafts for all [16].

In addition, life rafts (Fig. 7.3) must be available on board for another 25% of passengers and crew members [26].

The fact that life-rafts are carried along with the lifeboat as an extra means of rescue is due to the fact that there are situations where not all lifeboats can be deployed, such as the accident of *Costa Concordia* showed.

For example, the ship *My Ship 1* (1924 passengers, 780 crew members) has following capacities [21]:

- 12 lifeboats for 150 persons each,
- 2 lifeboats for 61 persons each,
- 4 tender boats for 150 persons each,
- 63 life-rafts for 25 persons each.

Further requirements are placed on the construction of lifeboats on tankers because of the potentially high-fire temperatures and toxic gases on board. Free-fall lifeboats are generally found on freighters in general, but especially for dangerous goods carriers, whose design allows the hazard area to be left in a protected lifeboat via a *free-fall rescue boat* (Fig. 7.2). Such boats are located at the stern of the deckhouse, often one lifeboat to starboard and one to port.

After boarding and locking the free-fall lifeboat, the crew fastens their seat belts. Upon releasing the boat's holding device, the boat slides on an inclined plane in free fall with its bow forward into the sea, thus reducing the impact on the water surface [33].

Fig. 7.5 Life ring with lamp

Table. 7.1 Minimum number of lifebuoys

Ship length (m) Cargo ship	Minimum number Lifebelts	Ship length (m) Passenger ship	Minimum number Lifebelts
<50	4		
≥50	6	<60	8
<100	8	60<120	12
100<150	10	120<180	18
150<200	12	180<240	24
≥200	14	≥240	30

7.1.1.2 Life Belts, Life Jackets

Half of all lifebuoys (but on passenger ships at least six) should be fitted with self-lighting lights, which cannot be extinguished by water and should remain lit for at least 45 min (Fig. 7.5).

The number of lifebuoys required depends on the type of ship and its length (see Table 7.1).

Life jackets must be provided for every person on board; additional suitable jackets must be kept available for children. Passenger ships must also carry life jackets for 5% of all persons on board, which must be stored in a clear place on deck [16].

Life jackets approved for SOLAS use on European ships must be marked with the *steering wheel symbol*. It confirms that the requirements of the regulations are met, such as floats designed in double-chamber systems, the provision of life jacket lights or the required number of reflective strips [32].

7.2 Fire Protection

7.2.1 Introduction

Fire on board is always a very serious case of damage [8]. If a small incipient fire can perhaps still be extinguished in time, scenarios that expand to become a major fire can lead to catastrophic consequences. Even firefighting by means of water is not unproblematic in the event of a major fire on board, as the ship is flooded with more or less large quantities of water which—if the bilge pumping system is not able to pump this water out again—can quickly bring the ship into a critical stability condition. Fire protection on board is therefore of great importance. A distinction is made between structural and defensive fire protection (Fig. 7.6).

Measures of the *structural fire protection* (use of nonflammable and hardly inflammable building materials, division of the ship into fire compartments, etc.) are primarily intended to prevent the occurrence of a fire. A fire that has already broken out is

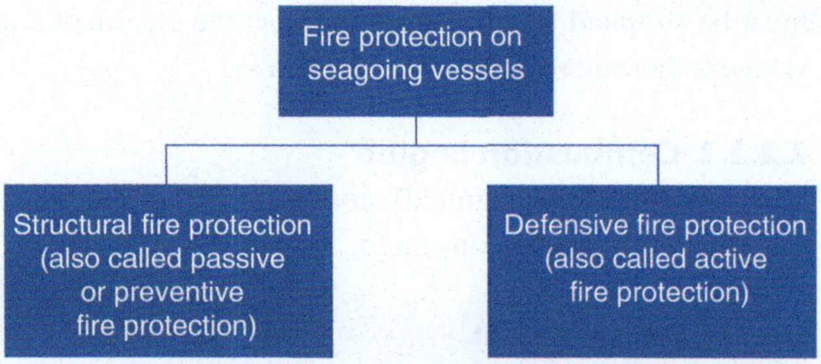

Fig. 7.6 Fire protection measures

prevented from spreading any further by structural measures. Furthermore, in the event of a fire, these measures are intended to enable safe evacuation of crew and passengers from the affected areas.

The *defensive fire protection* does not only include the actual firefighting by firefighting teams, but also includes the installation of fire detection technology and fire extinguishing systems (sprinklers, HI-FOG) and equipment (hand-held fire extinguishers, wall hydrants, fire pumps, etc.).

The fire protection measures are regulated in extensive technical and legal standards. However, the main points to note are as follows: At *International Convention for the Safety of Life at Sea, 1974* (International Convention for the Safety of Life at Sea—SOLAS) are described in Chap. 2 Rules and regulations.[7] In addition, the relevant construction regulations of the classification societies also contain supplementary regulations.[8] Special requirements for naval ships may be contained in construction regulations of the German Federal Armed Forces (Bundeswehr).

According to the Occupational Safety and Health Act, the requirements resulting from special regulations for shipping are relevant. However, if areas of occupational health and safety or occupational safety—including fire protection—are not regulated there, other standards and regulations for fire protection must be used.

7.2.2 Introduction to Fire Training

7.2.2.1 Exothermic Reaction

Combustion is a chemical reaction in which the burning material combines with oxygen to form the combustion products.[9] Like any chemical reaction, combustion is also associated with an energy conversion; it will be between *exothermic* and *endothermic* reactions. With an *exothermic* reaction energy is released, in an *endothermic* reaction, energy

[7]For this purpose also [28].

[8]Such as [3].

[9]For fire training see [18].

must be supplied to the system to allow the chemical reaction to take place. Combustion is an exothermic reaction: heat is released.

7.2.2.2 Combustion Engine

After being initiated (ignited), combustion continues independently until the combustible material or oxygen is consumed.

Here is an Experiment

If a glass is placed over a burning tea light, the flame will extinguish after a short time because the combustion oxygen under the glass has been used up. The following chemical reaction took place (without consideration of stoichiometric conditions):

$$AC + O_2 \rightarrow A' + CO_2 + H_2O. \tag{7.2}$$

In this reaction equation, AC is any compound of one or more substances (A) with carbon (C); the combustion product is A′, carbon dioxide (CO_2), and water (H_2O). Any complete combustion under stoichiometric conditions produces the combustion product, CO_2, and H_2O.

The driving force of such an independent reaction is two fundamental laws of nature:

- Nature strives to adopt the most energy-poor state.
- Nature strives to assume the state of greatest possible disorder.

In order to describe the driving force of a reaction more precisely, the two facts must be linked together.

The energy state of a system is described by its enthalpy (*H*), the state of disorder by entropy (*S*). Both act against each other, as can be seen in the example of the evaporation of a liquid at the boiling point (e.g., water at 100 °C). The liquid evaporates because *S* increases strongly during the transition from the liquid to the gas phase. At the same time, however, energy must be continuously supplied from outside, which increases the enthalpy content of the system. If the enthalpy content of the system is reduced (energy extraction by switching off the heat supply), part of the vapor condenses, which simultaneously reduces the entropy content. A state variable, that describes this interaction of *H* and *S*, is called free enthalpy, also Gibbs' enthalpy *G*. In a chemical reaction, the enthalpies and entropies of the end products are different from those of the starting materials. Whether a reaction takes place independently depends on how large the change, that is, the difference between these values, is. The change of the free enthalpy ΔG is described by the Gibbs–Helmholtz equation

$$\Delta G = \Delta H - (T \cdot \Delta S) \tag{7.3}$$

with *T* the temperature of the system (in case of large fires, the temperature *T* between 800 and 1000 °C) [13, p. 17].

A reaction runs independently as long as ΔG is negative, that is, free enthalpy can be withdrawn from the system (negative sign of ΔG). If the difference of the free enthalpies is zero, there is no more drive for the reaction, that is, the reaction comes to a standstill—there is equilibrium.

7.2.2.3 Rate of Combustion

As reaction speed v_{RG} is the change in concentration of the substances involved in the reaction as a function of time. The following applies to the concentration dependence of the combustion rate

$$v_{RS} \sim c_n \tag{7.4}$$

with

c = Concentration of the substances involved in the combustion,
n = Exponent, that describes the reaction category.

In principle, a distinction is made between the following cases with regard to the rate of combustion [13, p. 26 f.]:

- Combustion: exothermic reaction involving oxygen,
- Deflagration = deflagration: to a certain extent a weak explosion, which proceeds only with low-pressure development; ignition speed in the order of cm/s (slower than the speed of sound),
- Explosion: Extremely fast oxidation reaction under sudden, strong heat and pressure, ignition speed in the order of m/s,
- Detonation: Combustion speed in the order of km/s (faster than the speed of sound).

7.2.2.4 Conditions of Combustion

As a prerequisite for combustion to take place the following must be present (see also Fig. 7.7).

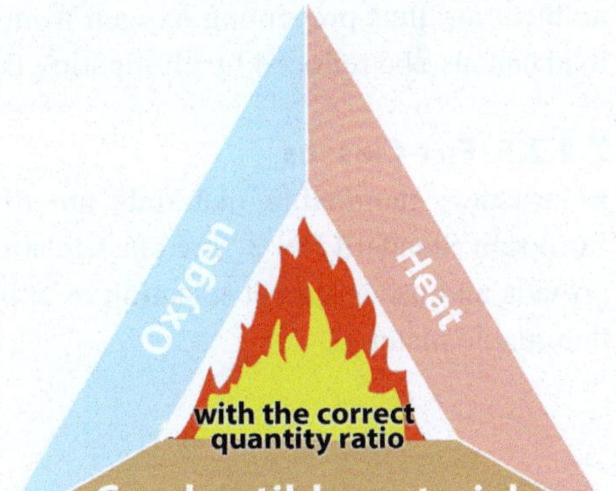

Fig. 7.7 Burn triangle. *Photo* Stefan-Xp, CC BY-SA 3.0

Table 7.2 Examples of ignition temperatures in air

	Ignition temperature in °C
Gases	
Acetylene	305
Butane	630
Carbon oxide	605
Fumes	
Diesel fuel	220…350
Fuel oil	approx. 250
Toluene	535
Solid materials	
Spruce wood	280
Hard coal	350
Newsprint	185

- A flammable substance,
- Oxygen,
- Ignition temperature and minimum combustion temperature.

The combustion is initiated by the process of ignition. Ignition occurs when a flammable substance is heated to a certain temperature, the ignition temperature (for examples see Table 7.2; [13, p. 29 f.]). It is the lowest temperature of a heated wall or surface at which a combustible substance in contact with atmospheric oxygen is just about to burn after a maximum of 5 min.

In addition, the combustible material and the combustion oxygen must be present in the correct proportion to each other (stoichiometry).

In this respect, the goal of any firefighting operation is to work toward eliminating one of the three conditions! The use of water, for example, is essentially aimed at extracting heat from the fire, that is, cooling it down to an extent that it falls below the ignition temperature. Foam and extinguishing gases (N_2, CO_2) cause a displacement of ambient air, thus preventing oxygen from reaching the source of the fire. Finally, the fire load can also be reduced by eliminating flammable substances.

7.2.2.5 Fire Classes

In practice, flammable materials are divided into five fire classes according to the European Standard EN 2 "fire classification" (see Table 7.3). This classification serves to provide and use a correct selection of appropriate extinguishing agents depending on the flammable material.

Table 7.3 Fire classes according to EN 2 (Pictograms: KØLUMBUS)

Fire class	Description	Example	Extinguishing agents	Pictograms
A	Fires of solid materials, mainly organic, which normally burn with embers	Wood, paper, coal, some plastics (especially duroplastics), textiles	Water, aqueous solutions, foam, ABC powder, extinguishing gel, various small extinguishing devices, for example, fire blanket or fire pad	
B	Fires of liquids or substances becoming liquids	Petrol, ethanol, tar, wax, many plastics (especially thermoplastics), ether, paints, resin	Foam, ABC powder, BC powder, CO_2	
C	Fires from gases	Acetylene, hydrogen, natural gas, methane, propane, and butane	ABC powder, BC powder, CO_2 only in exceptional cases (here there are specifically constructed special extinguishing systems with gas jet nozzle), interrupt gas supply by pushing off	
D	Fires from metals	Aluminum, magnesium, sodium, potassium, lithium, and their alloys	Metal fire powder (D-powder), dryer sand; **at no case water or foam!**	
F	Fires of edible oils/fats (vegetable or animal oils and fats) in frying and fat frying appliances and other kitchen equipment and appliances	Edible oils and fats	Fat fire extinguisher with special extinguishing agents (extinguishing agent tending to saponification), appropriate extinguishing spray, fire blanket; **in case no water!**	

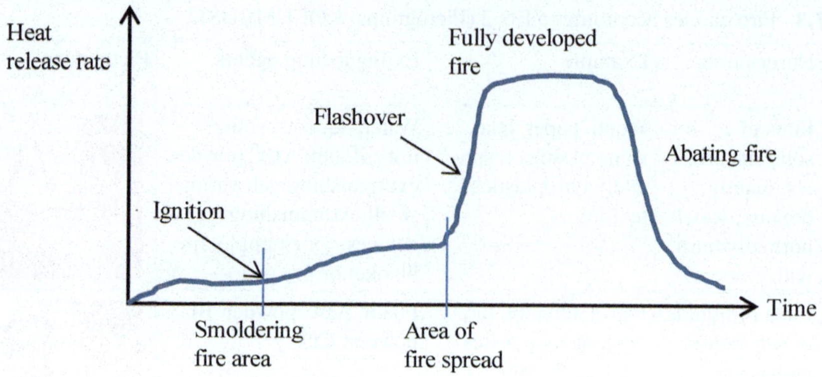

Fig. 7.8 Phases of a fire

7.2.2.6 Stages of Combustion

Fires can be divided into several fire phases (Fig. 7.8).

The following aspects can be used to characterize a fire:

- Releasing energy,
- Fire propagation speed and fire spread,
- Amount of toxic components in the fire smoke,
- Amount of smoke,
- Temperature (maximum temperature, temperature curve),
- Certain time parameters such as smoldering fire phase, incipient fire phase, time to flashover,
- Heat release rate.

The heat release rate Q is often used as the most important parameter and is described in the Eq. 7.5 [15]:

$$Q = \alpha \cdot t^2 \tag{7.5}$$

with

Q = Heat release rate (kW),
α = Parameters of the increase mechanism of heat release (kW/s^2),
t = Fire duration without smoldering fire phase (s)

The parameter α describes the rate of increase of the heat release rate and is basically divided into four levels: slow, medium, fast, and very fast. The cases "very slow" and "extremely fast" mentioned in the Table 7.4 were extrapolated based on the four standard levels [6].

Table 7.4 Parameters of the mechanism of increase of heat release α

Case	α (kW/s²)	$t_{character}$ (s)	Examples
Very slow	0.0007	1200	–
Slow	0.0028	600	Densely packed wood products
Medium	0.0111	300	Polyester mattress, solid wood furniture
Quick	0.0444	150	Stacks of wooden pallets, filled mailbags
Very quick	0.1778	75	Light curtains, some upholstered furniture
Extreme quick	0.7111	37.5	–

7.2.3 Structural Fire Protection, Requirements for Components and Materials

SOLAS formulates requirements for components and building materials with regard to their fire resistance [15, p. 20 f.].

SOLAS Chapter II-2, Section A, Regulation 3 classifies separating components (walls, doors, bulkheads) into three different classes. Requirements for their resistance are listed in the Table 7.5. Depending on how long the criteria *maximum temperature rise* and *maximum temperature* must be met, the type designation for the component is supplemented by a time in minutes: A-0, A-15, A-30, B-0, B-15. Due to missing requirements regarding these criteria, an extension is not necessary for type C.

Connecting doors and other openings in class A and B walls must have the same fire resistance as the partitions themselves and be self-closing.

The building regulations of the federal states for example, which also include fire protection requirements for *buildings* use the following German building authority designations for classification and gradation:

- *Fire-retardant* corresponds to a functional integrity of at least 30 min (F 30),
- *Highly fire-retardant* corresponds to a functional integrity of at least 60 min (F 60),
- *Fire-retardant* corresponds to a functional integrity of at least 90 min (F 90).

Paints, varnishes, and other indoor surface treatment products should not release excessive amounts of smoke and other toxic substances. Only floor coverings that do not increase the risk of fire, explosion or release of toxic substances may be used in recreation rooms and service areas (SOLAS II-2, Section B, Regulation 6).

In addition, SOLAS requires ships to be divided into Main Vertical Zones (MVZ) in order to contain and prevent the spread of fires: The length between these sections should not exceed 40 m. In order to ensure sufficient fire resistance, subdivisions should be made using components of type A-60 (see above) (SOLAS II-2, para. C, Regulation 9). On ships with several decks, the length of the main fire compartments may be extended to a maximum of 48 m. This will ensure that they are in line with the watertight

Table 7.5 Overview of fire resistance classes according to SOLAS II-2, Section A, Regulation 3

Resistance class	Building material	Suitable Stiffening required	Heat transfer			Prevention of passage from		SOLAS Chapter II 2, Part A, Regulation 3
			Isolation	Max. temperature rise (average)	Maximum temperature (selectively)	Smoke	Fire	
A	Steel or equivalent. material	X	With allowed noncombustible building materials	Δ140°C	Δ180°C	X (60 min)	X (60 min)	0.2
B	Allowed noncombustible building materials	–	–	Δ140°C	Δ225°C	–	X (30 min)	0.4
C	Allowed noncombustible building materials	–	–	–	–	–	–	0.10

Fig. 7.9 Smoke detector on cabin ceiling

transverse bulkheads. This extension is also for the creation of large public spaces (e.g., restaurants or even theatres on cruise ships) up to a maximum of 1600 m^2 permissible (SOLAS II-2, Section C, Regulation 9).

7.2.4 Fire Detection and Alarm

In order to ensure effective *fire detection* in accordance with SOLAS II-2, Section C, Regulation 7, fire detectors and alarm devices should be appropriate for the particular area of application, its size and damage potential. In addition to automatic fire detectors, manually operated alarm devices should be installed in an easily accessible location. The alarm should be visual and audible (in the ship's technical control room and on the bridge) and should be clearly distinguishable from other signals.

SOLAS distinguishes between ships carrying more and less than 36 passengers. In the following, the regulations that apply to ships with more than 36 passengers are discussed.

Fire detection and alarm systems should be installed in recreation rooms, corridors, stairwells, and escape routes (see Fig. 7.9).

Furthermore, requirements for areas with no or low-fire risk are facilitated by the fact that no fire detectors need to be present. An indication of the number and positioning of the smoke detectors is given in the FSS Code[10] and displayed in Table7.6.

Within stairwells, smoke detectors must be installed at least at the highest point and on every second floor.

[10]International Code for Fire Safety Systems.

Table 7.6 Number of required smoke detectors according to FSS code

Detector type	Max. area per detector (m^2)	Max. distance between the centers (m)	Max. distance from scots (m)
Heat/Fire	37	9	4.5
Smoke	74	11	5.5

7.2.5 Firefighting Equipment and Installations

As the target of fixed and portable firefighting equipment, SOLAS II-2, Section C, Regulation 10 states *Suppression and extinguishing the fire in the room of origin*. For this purpose, it is planned to install sprinkler systems that are appropriate to the fire-propagation potential in the respective room. If the ship has more than 36 passengers, an automatic sprinkler system should be installed in all lounges, service rooms, corridors, and stairways. Requirements for areas with no or low-fire risk[11] are reduced to the extent that no automatic fire extinguishing system is required [15, p. 30].

In addition to *fixed firefighting systems,* easily accessible *hand fire extinguisher* and *wall hydrants* should be provided. Hydrant connections should be easily accessible and easily connected with hoses. Protection against frost must be ensured.

Suitable portable extinguishing equipment should be available in sufficient numbers. If the gross tonnage (GT) is more than 1000, a minimum of five extinguishers is required. These must be easy to operate, clearly visible, and accessible at all times. Any restriction in use due to weather, vibrations or other factors should be excluded. One extinguisher should be located at the entrance for each room provided. The stocking of certain quantities of spare fillings is due to the mostly remote location of ships.

The construction regulation "Rules for Classification - Ships - Part 4: Systems and Components, Chapter 11: Fire Safety, Edition October 2015" of the classification society DNV·GL differentiates the fire protection requirements according to the size and type of the ships: Cargo ships <500 RT or cargo ships >500 RT and passenger ships; here are a few remarks from this construction regulation[12]:

> Section 2 - Fire protection measures for cargo ships of less than 500 RT
>
> 2.1 Fire pumps
> Ships above 150 GT should be provided with at least one independent power driven fire pump. Ships of 150 GT and below shall be provided with at least one power driven fire pump, which may be driven by the main engine.
>
> The capacities of the main fire pump (see Fig. 7.10) shall be as required for ships of 500 GT and above (ref. SOLAS Chap. II-2/10.2.2), but should not be less than 25 m3/h and each such pump shall in any event be capable of delivering at least the two required jets of water with the required flow rates and pressures.

---

[11]Empty rooms, toilets, CO_2 rooms, etc.

[12]Taken from the English original and translated into German.

Fig. 7.10 Fire pump. *Photo* AIDA

The pressure head of the main fire pump shall be so chosen that the requirements of SOLAS Chap. II-2/10.2.1.6 are met. Alternatively, on ships of less than 300 GT every nozzle must be capable of delivering a water jet of at least 12m length horizontally.

On ships above 150 GT an additional power driven fire pump should be provided in a position outside the space containing the main fire pump. On ships less than 150 GT, this pump can be then hand operated.

The additional pump should have sufficient capacity and pressure to provide a 6-m jet throw with nozzles not less than 9-mm diameter.

The jet throw should be capable of being directed on to any part of the ship.

On ships of less than 100 GT, no additional fire pump is required.

2.2 Fire mains and hydrants

The fire main shall have a diameter of sufficient size to maintain a steady distribution and pressure.

A sufficient number of fire hydrants should be provided, and so located that at least one jet of water can reach any normally accessible part of the ship. At least one hydrant must be provided in the machinery space.

Not less than three (3) fire hoses of at least 15 m in length, complete with couplings and nozzles, should be provided.

The nozzles should be of dual purpose (spray/jet) type with 12 mm jet and integral shut-off. The jet may be reduced to 10 mm and shutoff omitted for ships with hand-operated fire pumps.

2.3 Fire extinguishers

Each powder or carbon dioxide extinguisher should have a capacity of at least 5 kg and each foam extinguisher should have a capacity of at least 9 L.

Not less than three (3) portable fire extinguishers should be provided in the accommodation and service spaces.

Not less than two (2) portable fire extinguishers, suitable for extinguishing oil fires,[13] should be provided in each boiler room, cargo pump room, and spaces containing any part of any oil–fuel installation.

In machinery spaces containing internal combustion machinery, one (1) portable fire extinguisher should be provided for every 375 kW of engine power.

2.4 Fixed firefighting systems
For ships with >150 GT, a fixed fire extinguishing system shall be installed in the engine room and cargo pump rooms. For ships with ≥500 GT, such a system is required (see SOLAS Chapter II-2/10.4).

Naval combat ships often have an outside spray system. On the one hand, this is used to fight fires caused by weapons in combat. In this way, the crew assigned to damage prevention does not have to leave the protecting ship's hold during firefighting operations. On the other hand, the spraying device is also used for the decontamination of NBC warfare agents that have been brought to the surface as a result of combat operations.

Water sprinkler systems are used as fixed fire-extinguishing systems, as they are also used in buildings on land. However, the use of water also causes water damage, so that especially in sensitive areas (electrical engineering, control room, engine room) CO_2-extinguishing systems can be installed. The gas is carried on board in compressed gas cylinders. Depending on the spray head design, the gas is discharged directly as gas to flood the room in question, and thus displace the atmospheric oxygen (extinguishing effect by suffocation), or as snow (extinguishing effect by cooling and suffocation).[14] Since CO_2 is not unproblematic for humans due to its suffocating effect (possibility of leaving the area to be flooded in time), the so-called HI-FOG system has recently become established, especially on cruise ships. This is a *water mist firefighting system*.[15]

Water misting systems represent a relatively new technology in the development of automatic extinguishing systems. In contrast to conventional sprinkler systems, the extinguishing effect of very small water drops (water mist, or "fog") is used.

The water mist consists of the smallest drops of water. The diameter of a fog drop is about 50 μm. Under high pressure (>35 bar, hence the designation "HI" from "high") extinguishing water is pressed through stainless steel pipes to the extinguishing nozzles. The special design of the nozzles (Fig. 7.11) generates the water mist.

The necessary pressure is generated by high-pressure pumps or alternatively by compressed gas cylinders (nitrogen) (Fig. 7.12).

---

[13]These are fire extinguishers of fire class B.

[14]For detailed information on dimensioning of CO_2 extinguishing systems see [19].

[15]For use in shipping see also [31].

Fig. 7.11 Spray nozzle of a HI-FOG system

Fig. 7.12 Cylinder battery HI-FOG system. *Photo* AIDA

The combination of optimum droplet size and distribution of the droplets under high pressure ensures deep penetration of the water mist into the fire source and rapid cooling of the smoke gases. In addition, smoke gases are washed out of the air that allows an early entry to firefighters from the fire brigade to enter the fire room.

The extinguishing effect is based on the following effects:

- Cooling,
- Absorption of radiant heat (heat shielding),
- Inertization (local oxygen displacement).

The advantage of fogging systems compared to sprinkler systems is the low-water consumption; that is, compared to a sprinkler system, a smaller water supply is necessary. In the case of activation, therefore, considerably less water damage is caused [22].

There is also no additional water ingress into the ship, as the required firefighting water is already carried on board in tanks.

In principle, the HI-FOG fire protection system [30] can be divided into four system types:

- Wet-pipe system,
- Spray-flood system,
- Dry-sprinkler system,
- Pilot operated fire-suppression system.

Typically *wet-pipe systems* are used in accommodation and similar areas where the flammable media are made of solid materials. As soon as the ambient temperature reaches the specified limit, the glass barrel in the sprinkler breaks and the water mist is triggered by this special sprinkler.

A *spray-flood system* usually has open spray heads; the water passage is controlled by valves in the water distribution network. When a valve is opened, the water mist is released from each spray head in the section controlled by the valve. Spray flooding systems are normally used in areas where fuel fires are more likely to occur.

Dry-sprinkler systems work similar to wet-pipe systems, but in this case the water is held by a closed valve on the pump side.

Instead, the pipes are filled with compressed air and monitored: a loss of compressed air indicates a defective sprinkler glass barrel. The water is then released from the activated sprinkler(s).

Dry-sprinkler systems are typically used in areas where there is a risk of frost.

The *pilot operated fire-suppression system* is essentially similar to a dry-sprinkler system, except that it is connected to a fire-alarm system. A triggering therefore requires both, the bursting of the glass barrel and the alarm triggered by the independent fire-alarm system.

Pilot-operated systems are used for the protection of premises where the risk of false triggering or leakage must be kept to an absolute minimum.

7.2.6 Firefighting by Firefighting Teams

Where fires cannot be extinguished in the initial phase by hand fire extinguishers or automatic firefighting systems, firefighting should be carried out by firefighting teams. These are crew members specially trained for firefighting. They are particularly suitable for wearing heavy respiratory protection device. This requires regular training under heavy respiratory protection device and passing the basic occupational health examination G 26 for wearing respiratory protection device [20]. The equipment of the fire protection troops is based on that of the fire brigades on land; for an internal attack, the equipment is usually required:

- Bucket syringe/fire extinguisher,
- Hose carrier basket,
- C-jet pipe,
- Portable spotlight,
- Radiotelephone for radio communication on board,
- Fire brigade rope (among other things, to secure the way back),
- Dead man's detector,
- Helmet light,
- Crowbar,
- Fireman's axe,
- Ram to open doors,
- Escape hood for persons to be rescued,
- Compressed air breathing apparatus.

7.2.6.1 Instructions for Firefighting

The vast majority of fires occur inside the ship; fires on the upper deck are the exception. Therefore, the firefighting attack inside the building is of particular importance: the so-called indoor attack in closed rooms is a complex and difficult task. The specially trained crew members learn the necessary tactics and techniques in courses for firefighting. The following is only a brief description of general aspects of intrinsic safety when opening a door or bulkhead behind which a fire is suspected.[16]

If closed doors or bulkheads have to be opened, the following procedure for intrinsic safety should be followed:

- Slowly approach the back of your hand to the door/bulkhead and carefully check for heat from bottom to top.
- Also check the latch/lever with the back of your hand.

[16]For firefighting in internal attack [29].

- If the door opens toward the opening side, it should be opened in a squatting position by leaning against it from the hinge side.
- If the door opens away from the person opening it, it should be opened in a squatting position, protected by the adjacent wall. If available and possible, a pointer should be placed around the handle so that the door can be closed again without having to reach into the fire room.
- As a rule, the ceiling of the room should be cooled with spray jet to avoid a flashover.

A "foam attack" is usually used to fight liquid fires. Unlike water, foam is lighter than all flammable liquids and is, therefore, suitable for covering and sealing the surface of liquids in a floating manner. The main extinguishing effect is based on suffocation; the dense foam cover prevents the further development of fuel vapors. The fire is therefore extinguished due to lack of combustible material. Due to its water content, however, the foam is also capable of cooling down to a certain extent, that is, it is also able to extinguish fires of solid, glowing substances such as wood, paper, etc.

Extinguishing foam basically consists of the components *water*, *foaming agent* (carried on board in canisters) and *filling gas* which inflates the bubbles. In the simplest case, air serves as the filling gas. First of all, a certain percentage (3–5%) of a foaming agent is added to the water to be foamed by means of a "proportioner". This mixture is sprayed at a pressure of at least 5 bar in the foam jet pipe and sucks in the required air from outside by injector action. The foam is created by swirling the water/foam concentrate mixture with the air sucked in inside the foam jet pipe (Fig. 7.13; [13, p. 57 f.]).

In the following, the handling of portable fire extinguishers is briefly described:

- Portable fire extinguishers should be tested in general once a year and at least every 2 years. A note of the test must be affixed to the extinguisher (test mark).
- The functional life depends on the filling quantity; Table 7.7 shows the minimum spraying times that must be achieved according to DIN EN 3-4: 1995 Annex A.

Warnings for firefighting with hand fire extinguisher:

- *Caution with electrical systems*: Water and foam extinguishers only up to 1000 V, minimum distance 3 m; ABC and D powder extinguishers only up to 1000 V, minimum distance 1 m; BC powder extinguishers up to 1000 V, minimum distance 1 m (pay attention: > 1000 V VDE 0132!),
- *CO_2 extinguisher*: Caution for use in narrow, poorly ventilated rooms; minimum distance 1 m for electrical installations up to 1000 V.

Rules for firefighting:

- Always attack fire in the direction of the wind, never against the wind direction, as you will then have to fight the flames and cannot reach the source of the fire (embers).

Fig. 7.13 Modern middle foam pipe. *Photo* Magnus Mertens, CC BY-SA 2.0 de

Table 7.7 Functional life depending on the filling quantity

Filling quantity (kg or L)	Minimum functional life (s)
Up to 3	6
About 3–6	9
About 6–10	12
About 10	15

- Extinguish surface fires starting from the front, do not spray into the flames, but into the source of the fire from below. You are thus protected from flames.
- Extinguish dripping and flowing fires from top to bottom! Extinguish from the exit point to the ground.
- Use enough extinguishers at once! Never use fire extinguishers one after the other, otherwise the extinguishing success is not guaranteed due to the high spreading speed of the fire.
- Beware of reignition! Reignition can cause the fire to flare up again.

For the practice of internal attack under heavy respiratory protection device (Fig. 7.14) it is of interest to know how long the air supply in the cylinders of the troop man's respiratory protection device is sufficient.[17] For this purpose the **Boyle–Mariotte's Law** provides the answer. Based on the ideal gas law

$$p \cdot V = m \cdot R \cdot T \tag{7.6}$$

[17]The following with the collaboration of Pfaff, N.

Fig. 7.14 Compressed air bottles of the respiratory protection device wearers

with

R Gas constant,
p Pressure,
V Gas volume,
T Gas temperature,
m Mass of the gases

Equation 7.6 can be paraphrased as follows:

$$p \cdot \frac{V}{T} = m \cdot R. \tag{7.7}$$

If a gas is contained in a closed system (e.g., air in a bottle of a respiratory protection device), it can be seen when looking at Eq. 7.7 that the volume cannot change when the system heats up, as long as the gas cannot expand (as in the closed breathing air bottle). The mass remains constant, as does R. Therefore, when the temperature rises, the pressure in the system must increase (the pressure in the breathing air cylinder increases!). So one can say:

$$p \cdot \frac{V}{T} = \text{const.} \tag{7.8}$$

In other words:

$$p_1 \cdot \frac{V_1}{T_1} = p_2 \cdot \frac{V_2}{T_2}. \quad (7.9)$$

This shows that for a temperature change at constant pressure ($p_1 = p_2$) again the law of *Gay–Lussac* applies:

$$\frac{V_1}{T_1} = \frac{V_2}{T_2}. \quad (7.10)$$

The same applies to a pressure change under constant temperature ($T_1 = T_2$):

$$\frac{p_1}{p_2} = \frac{V_2}{V_1}. \quad (7.11)$$

Example for Calculating the Operating Time Under Heavy Respiratory Protection Device

Often respiratory protection devices with two cylinders of 4 L each with a respective filling pressure of 200 bar are used. The air consumption (breathing minute volume) can be assumed to be 50 L/min in use.

It has to be determined how many liters of air are available to the firefighter at an ambient pressure of about 1 bar. According to Boyle–Mariotte, Eq. 7.11:

$$\frac{p_1}{p_2} = \frac{V_2}{V_1}.$$

The cylinder pressure p_1 is 200 bar, the cylinder volume V_1 is 2×4 L, which is 8 L. The ambient pressure is $p_2 = 1$ bar, the amount of air available for breathing is V_2. Thus, by changing the Eq.: 7.11

$$V_2 = \left(\frac{p_1}{p_2}\right) \cdot V_1.$$

The insertion of the above numerical values result in $V_2 = 1600$ L.

If the firefighter breathes 50 L/min while on duty, he can therefore only use a maximum of

$$1600\text{L}/50\text{L/min} = 32\text{min}$$

Since an alarm signal for withdrawal already sounds at a residual pressure from 50 bar, its operating time is reduced accordingly.

7.3 Safety Marking at the Workplace, Ship Safety Guidance System

7.3.1 Safety Marking at the Workplace

The purpose of any safety marking is to draw attention quickly and unambiguously to objects and situations that may cause danger. The system of safety marking is regulated in several regulations on occupational health and safety. Basically, the marking is carried out in accordance with BGV A 8 "Safety and Health Protection Marking at the Workplace" (BGV = Berufsgenossenschaftliche Vorschrift (Employers' Liability Insurance Association Regulation, ELIAR)), issued by German Federation of Institutions for Statutory Accident Insurance and Prevention). Furthermore, DIN 81220 "Ships and marine technology - signs and symbols for fire protection, life-saving equipment, and escape routes" issued by the DIN Standards Committee Shipbuilding and Marine Technology must also be taken into account. For the area of German naval ships, the special features in the VG standards sheets of series 81226 of the DIN Standards Committee Shipbuilding and Marine Technology are relevant.

According to the accident prevention regulation BGV A 8[18] of the National Association of Institutions for Statutory Accident Insurance and Prevention, colors are chosen for safety marking that offer optimum contrast and have significant psychological properties. The safety color conveys danger, commands, prohibitions or instructions. The signal effect and informative value are concretized by the form of the safety signs and the symbols used on them. With regard to marking, a distinction should be made between

- Safety sign,
- Prohibition sign,
- Warning sign,
- Commandment sigs,
- Rescue sign,
- Fire protection sign,
- Hint sign,
- Auxiliary sign,
- Combination sign.

It is only in the combination of geometric shape and safety color together with the pictogram that the safety sign, which is unmistakable in its message and therefore unambiguous, is created (see also Fig. 7.15). Supplementary information is also provided in the Employers' Liability Insurance Association Rules 216 "Optical Safety Guidance Systems" (including safety lighting)—of German Federation of Institutions for Statutory Accident Insurance and Prevention.

[18]Safety and health signs at work.

Fig. 7.15 Behind this door are: hydrant with water hose and powder fire extinguisher 6 kg

Table 7.8 clearly shows the system of safety marking at the workplace. Although this accident prevention regulation is not applicable to public inland and maritime transport, it is nevertheless applicable to purely commercial inland and maritime transport.

According to Annex 1.3 of the Workplace Ordinance, safety and health protection markings must be used if hazards to the safety and health of employees cannot be avoided or sufficiently limited by technical or organizational measures; this requirement is implemented by the Workplace Directive ASR A1.3. This directive follows the abovementioned marking system and contains in its Annex 1, a compilation of the safety markings used in the commercial sector in accordance with DIN EN ISO 7010 and DIN 4844-2.

The latter DIN standards as well as ISO 7001, ISO 24409 are also relevant for public shipping (transport of passengers), but essentially follow the abovementioned system. In particular, there are additions to the symbols used to describe the special features of shipping (see Fig. 7.16).[19]

A selection of identification plates can be found in Annex 25.

[19]An overview of the currently valid safety markings can be found, for example, at the company A-SPE [37].

Table 7.8 System of safety marking at the workplace

Safety color	Form	Meaning	Application example
Red		Prohibition sign	Access denied 
		Fire protection sign	Fire extinguisher 
Yellow		Warning sign	Danger warnings (here: explosion) 
Green	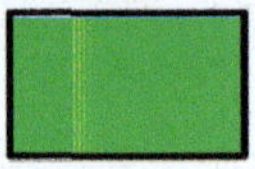	Rescue sign	Identification from escape routes 
			Collection point 
Blue		Command sign	Wearing hearing protection 

7.3.2 Safety Guidance System

The **safety guidance system (SGS)** makes the escape route to the secured area and the medical, safety and fire protection equipment visible in the event of a failure of the main and replacement lighting, and in the event of smoke and fumes (see Figs. 7.17 and 7.18). For markings close to the ground (Fig. 7.19), the standard sheets of the DIN 81,230 series "Safety guidance system close to the ground" apply, and also, the ISO 15,370 standard "Ships and marine technology - Safety guidance system close to the ground on passenger ships – Arrangement" of the DIN Standards Committee Shipbuilding and Marine Technology apply. In addition, the standard VG 81226-10 "Signs and

Fig. 7.16 Marking of assembly point "U" on deck 4, the lifeboat deck

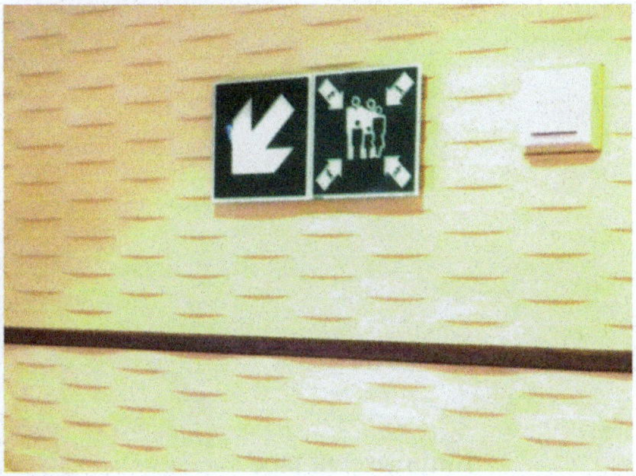

Fig. 7.17 Escape route direction to the assembly point

identification plates - Part 10: Safety guidance system near the ground SGS" of the DIN Standards Committee Shipbuilding and Marine Technology regarding defense technology requirements must be taken into account.

The SGS consists of sufficiently long photoluminescent guide markings, escape signs, signs (Figs. 7.17 and 7.18) and signs for ship's safety and for medical facilities. The guide markings show the shortest way to a secured area with continuous lines on the

Fig. 7.18 Escape route marking under the ceiling

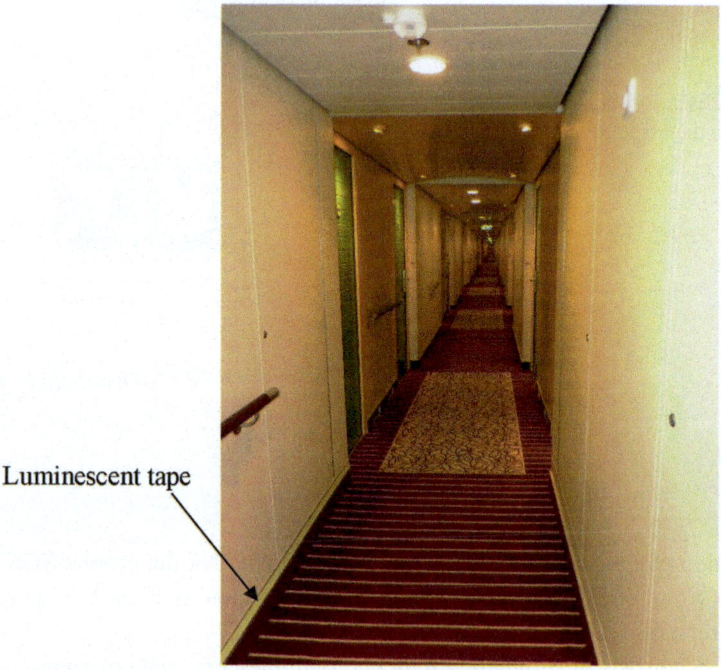

Fig. 7.19 Ground level safety guidance system

walls at approx. 30–40 cm above the corridor or on the floor and with individual symbols on the floor. The integrated arrows in the lines and individual arrow symbols on the floor indicate the only escape route direction; in lines without arrows the escape route can be taken in both directions.

Doors, door frames, stairs, obstacles etc. must be marked along the escape route. The door opener of the central locking system and the sash fasteners should be made recognizable by a flat background; the opening direction must be indicated.

In the case of safety and fire protection equipment, the operating elements must be designed in a photoluminescent version or made visible by means of a flat background.

The storage locations of the ship safety and medical equipment are marked with photoluminescent signs and symbols.

The luminance must comply with ISO 15370.

The safety guidance system also includes an escape and rescue plan, which is posted at exposed locations (on cruise ships, e.g., also in each cabin), on which the respective location and escape route are marked. Information on rescue and firefighting equipment completes the plan (see Fig. 7.20).

Note: Escape and rescue routes must not be narrowed by parked objects and must not be locked. Escape and rescue doors must open in the direction of escape. Fire protection doors must be self-closing and must not be held open by wedging.

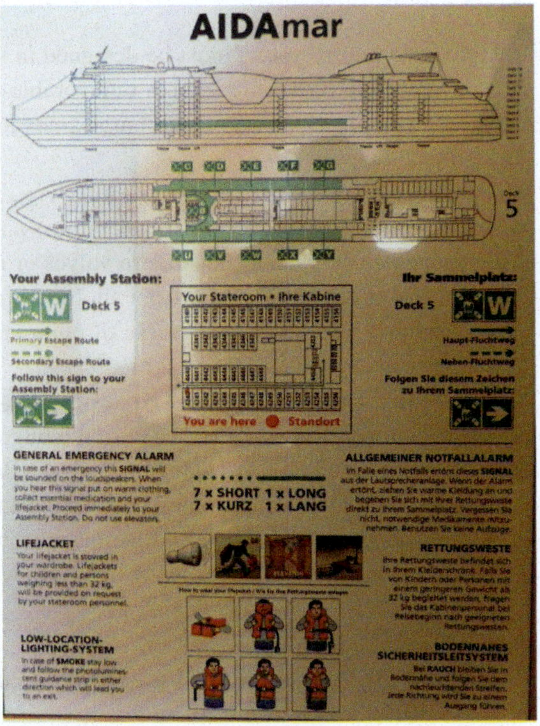

Fig. 7.20 Escape and rescue plan

7.4 Bilge Pumping Systems

7.4.1 Introduction

Although, from a purely systematic point of view, bilge pumping and drainage systems are to be counted among the ship's operating and auxiliary systems and could therefore have been dealt with in Chap. 5, they will be dealt with in this section, however, as they are of essential importance for ship safety [11, p. 79]: Water ingress due to leakage or by taking over water in heavy seas as well as in the context of firefighting always carries the risk that the ship will reach a critical stability situation, which may ultimately lead to sinking. Therefore, a sufficiently dimensioned, redundant bilge pumping system must be installed for each watertight compartment of the ship, via which water that has penetrated into the ship's interior is pumped outboard.

The system is used to drain the so-called bilge wells and bilges in engine rooms and cargo holds. For environmental reasons, the bilge water must be passed through a so-called bilge water separator to separate oil. The bilge system extends over the entire length of the ship. The water that collects in the bilge wells originates from leaks and condensation processes on the outer skin.

7.4.2 Essential Requirements, Design Notes

The pipe and pump system must be designed in such a way that the water can be drained from any section of the ship even in unfavorable trim. Suction baskets in the suction cups and sludge boxes in the main discharge lines are designed to retain dirt particles, and thus protect the pumps.

The bilge pumps must be self-priming (e.g., submersible pumps, piston pumps, vacuum system, self-priming centrifugal pump) and the shutoff devices of the individual drains must be designed as non-return valves or non-return flaps to prevent one compartment from running full of water from another through the bilge system.

Each power bilge pump is capable of pumping water through the required main bilge pipe system at a speed of at least 2 m/s.[20]

Bilge pumping systems should be designed as automatically operating systems: When an upper point is reached (water level in the bilge well) the shutoff valves are opened and the pumps are switched on (display on bridge and/or control station). When the water level reaches the lower point, the pumps are switched off and the shutoff devices are closed again. It is important that the ballast pumps are integrated into the bilge pumping system in order to be able to handle larger water ingresses with the support of these pumps.

[20]See Chap. 1, Part C, point 3.2.4 of Directive 2010/36/EU, Official Journal of the EU L 162/1 of June 29, 2010.

The dimensioning of pipelines and pumps[21] is carried out in accordance with the relevant legal provisions and the rules of the classification societies. The basic rule is that the bilge pumping system must be designed larger than the fire extinguishing system, otherwise there is a risk of self-draining in the event of firefighting.

In Chap. 8 of Annex II to the Annex to the Inland Waterways Inspection Ordinance[22] requirements for bilge pumping systems for *inland ships* defined:

1. Each watertight compartment must be capable of being drained separately. This does not apply to watertight compartments, which are usually operated in an airtight manner.
2. On ships for which a crew is required, there has to be two independent bilge pumps, which must not be placed in the same space and at least one of which should be motor driven. However, if these ships have a propulsion power of less than 225 kW or a deadweight capacity of less than 350 t or, in the case of ships not intended for the carriage of goods, a displacement of less than 250 m^3, one hand or motor bilge pump is sufficient.
   Each of the prescribed pumps must be suitable for each watertight compartment.
3. The minimum flow rate Q_1 of the first bilge pump must be calculated using the following formula:
   $$Q_1 = 0.1 \cdot d_1^2 \quad \text{[L/min]}. \tag{7.12}$$
   d_1 shall be calculated according to the following formula:
   $$d_1 = 1.5 \cdot \sqrt{L \cdot (B + H)} + 25 \quad \text{[mm]}. \tag{7.13}$$
4. The minimum flow rate Q_2 of the second bilge pump is calculated in the same way as Eq. 7.12, *d2* will be determined according to the following formula:
   $$d_2 = 2 \cdot \sqrt{L \cdot (B + H)} + 25 \quad \text{[mm]}. \tag{7.14}$$
   However, the dimension d_2 need not be greater than the dimension d_1.
   When dimensioning Q_2, L refers to the longest watertight compartment. In these formulas are:
   L = Length the relevant watertight compartment in m,
   d_1 = Calculated inner diameter of the main bilge pipe in mm,
   d_2 = Calculated inner diameter of the twin bilge pipe in mm.
5. If the bilge pumps are connected to a bilge system, the inner bilge pipe diameters should be at least d_1 in mm and the inside diameters of the twin-tube steering columns are to be at least d_2 in mm.

[21] See also Sect. 5.4 for the calculation.

[22] In the version dated December 6, 2008, BGBl. I. 2450.

For ships with L of less than 25 m the dimensions d_1 and d_2 may be reduced to 35 mm.

6. Only self-priming bilge pumps are permitted.
7. In each flat-bottomed, steerable compartment with a width of over 5 m there should be at least one suction cup each on the starboard and port side.
8. It can be possible to drain the aft peak to the main engine room via an easily accessible self-closing valve.
9. Twin bilge pipes of individual compartments shall be connected to the main bilge pipe by a non-return valve which can be shut off.
   Compartments or other rooms designed as ballast cells need only be connected to the bilge system via a simple shutoff device. This does not apply to cargo spaces which are equipped to receive ballast. The filling of such cargo spaces with ballast water should be effected by means of a permanently installed ballast line separate from the bilge line or by means of branch lines, which may be connected to the main bilge line as flexible lines or by means of movable intermediate pieces. Bottom valves are not permitted for this purpose.
10. Cargo hold bilges shall be fitted with level measuring devices.
11. Where a bilge pumping system with permanently installed pipes is provided, the bilge pipes for bilges intended for the collection of oily water must be fitted with shutoff devices and also be sealed in the closed state by an inspection body. The number and position of these shutoff devices must be entered in the ship's certificate.

With regard to the calculation of the bilge pipe diameter on *seagoing ships,* for example, reference is also made to Sect. 11 of GL's construction regulation "I Ship Technology, 1 seagoing vessel, 2 machinery".

7.5 Navigation Equipment, Light Guidance, Radio

To ensure safe navigation, safety, and ease of navigation, watercrafts must be equipped with navigation and radio equipment as well as appropriate position lights, depending on type, size, and sailing area. The following sections provide an overview.

7.5.1 Navigation Facilities

Navigation systems are required to determine the course and location of a ship [14, p. 78 ff.].

The **compass** (Fig. 7.21) is used to determine the direction of the ship (the course). In addition to a gyrocompass, the installation of a magnetic compass is still mandatory today, as it still indicates the course safely in case of a power failure.

Fig. 7.21 Liquid bearing magnetic compass. *Photo* PaterMc Fly, CC BY 3.0

Automatic pilot systems are now standard. Computer technology, which processes global positioning system (GPS) signals in conjunction with an electronic sea chart, allows waypoints to be entered, which the ship then steers automatically. The system automatically compensates for misalignment due to wind and current.

The **echo sounder** is a device for measuring water depth indirectly. The mode of operation of the echo sounder is based on the measurement of the transit time between the emission of an ultrasonic pulse and the reception of the echo signal reflected by the sea floor (Fig. 7.22). If the propagation speed of the sound in the water is known, the depth of water under the keel can be determined. The depth is recorded by means of an echograph; the echosounder can also be coupled with optical and acoustic indicators [7].

Similarly, a sonar (sound navigation and ranging) system is used to locate objects in space and underwater. This system is used by the navies for submarine hunting (Fig. 7.23). This is not a vertical, but a horizontal echo sounder. It has a distance and direction display. It thus offers the possibility of detecting objects in the vicinity of the

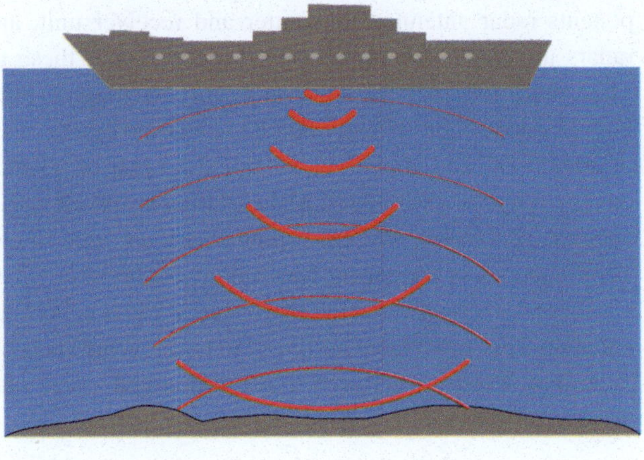

Fig. 7.22 Sound propagation echo sounder (schematic). *Graphic* Sgbeer, CC BY-SA 3.0

Fig. 7.23 Sonar dome of the USS Cowpens. *Photo* Alan Warner, US Navy

U-hunting ships. The sonar is located under the hull of the ship, usually placed near the bow in the so-called *Sonar Dome*.

Submarines also have sonar to detect underwater obstacles as well as enemy surface and underwater vehicles.

For the determination of further navigation parameters also **wind gauge**, **Rudder angle indicator**, **inclinometer** and **log** is used for speed measurement.

The position is determined today via **GPS**, which works with geostationary satellites. The determination of the northern or southern latitude by means of sextant is in principle only possible if the GPS system fails.

Radar equipments are used for navigation near land and for collision avoidance. Radar stands for ***Ra****dio* ***D****etection* ***a****nd* ***R****anging*. A radar installation consists of the components radar antenna, transmitter and receiver unit, and screen. Short- and long-range radars are used. Common to all devices is that these are omnidirectional pulse radars. This means that they transmit via a rotating antenna (Fig. 7.24; speed 25–30 min^{-1}) emits high frequency and strongly bundled electromagnetic waves in very short pulses. On ships of 10,000 GT and above, the so-called S-band or 10 cm band, which operates in the frequency range of 2920–3100 MHz and has a wavelength of about 10 cm, is used. X-band devices, also known as 3 cm band devices, are used in pleasure craft and on smaller ships. The frequency of the carrier waves here lies between 9320 and 9500 MHz [5, p. 53].

These waves are reflected by objects ("radar echo") and received again by the rotating antenna. This modulates and amplifies the echo and delivers the electrical signal to the radar unit. The time between transmission and reception of the pulses is a measure of the distance of the detected object:

Fig. 7.24 Radar antennas

$$s = c \cdot t/2 \tag{7.15}$$

with

s = Distance to the object or obstacle (m),
c = Speed of light (299,792,458 m/s),
t = Running time of the pulse (s).

At the same speed as the antenna rotates, the writing beam, which represents the emitted pulse, rotates on the screen and displays all incoming echoes (pips = echo points) on the screen: Land as wide strips, barrels as small pips, ships as small or larger pips depending on the size. In the middle of the screen is your own ship. Concentric measuring rings (two in Fig. 7.25) represent distance steps. The measuring range can be changed; with it, the distances between the measuring rings ("range") also vary. At the outer edge of the screen there is a 360° angle scale, which allows fast, rough bearing. Precise bearings can be taken with the aid of a rotating bearing ruler [5].

A problem with the display on the screen is the "noise" caused by sea states: high waves can lead to echoes that are displayed on the radar screen; smaller objects such as small watercraft can be lost in the noise. These interferences can be partly suppressed by "noise suppression".

The basis of navigation is the **sea chart** in paper form. In addition to the paper sea chart, the data can also be stored in digital form as **electronic sea chart**[23] are available and displayed (*Electronic Chart Display and Information System—ECDIS*). The radar image can be superimposed on the screen that displays the electronic nautical chart.

[23]See also [1].

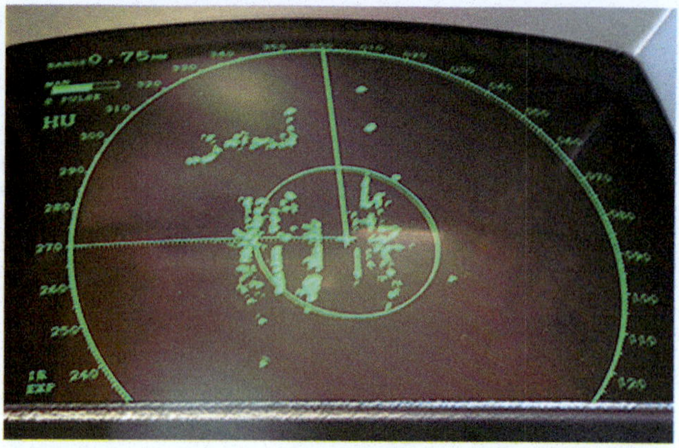

Fig. 7.25 Panoramic radar image of the Elbe river west of Hamburg. *Photo* Barbara Eckholdt

Technical development has led to an ever-increasing convergence of all the above-mentioned nautical devices into an "integrated navigation system".

7.5.2 Light Guidance

The collision avoidance rules (CAR, Rules 20–31 and Annex I1.–14.), the Maritime Shipping Ordinance—(Subsection 8 to 10 SeeSchStrO; Seeschifffahrtsstraßen-Ordnung) and the Ems Estuary Shipping Regulations (EmsSchO) require all vehicles to use **lights** at night and in reduced visibility. For the area of inland navigation there are regulations for light guidance amongst others in the following regulations: Inland Waterways Ordinance (IWO; Binnenschifffahrtsstraßen-Ordnung, BinSchStrO), Rhine Shipping Police Ordinance (RSPO; Rheinschifffahrtspolizeiverordnung, RheinSchPV), Danube Shipping Police Ordinance (DSPO; Donauschiffahrtpolizeiverordnung, DonauSchPV), and Mosel Shipping Police Ordinance (MSPO; Moselschiffahrtpolizeiverordnung, MoselSchPV). This should make it possible to identify the type of ship approaching and its course toward the observer, which helps to prevent collisions. Ships that are entitled to fly the German flag may only carry lights (position lights) that have been type tested by the German Federal Maritime and Hydrographic Agency (FMHA; Bundesamt für Seeschifffahrt und Hydrographie, BSH) or a comparable European body. For this purpose the lanterns are provided with a test number in case of a national approval (e.g., BSH/00/01/00) or with a steering wheel in case of an EU-wide approval (Fig. 7.26). Instruments approved by the German Hydrographic Institute (GHI; Deutsches Hydrographisches Institut, DHI) remain valid (Fig. 7.27). A CE mark does not replace the approval according to the CAR! Lanterns approved by other Member States of the EU are treated as equivalent if they achieve the same level of protection on an equally permanent basis [2].

Fig. 7.26 Marking of a starboard lantern

Fig. 7.27 Registration marking of the DHI

The legally required approvals for navigation lights, in relation to the sailing area (sea or inland waterways) and the application (pleasure craft or commercial shipping), can be found in the Table 7.9 [2].

The most important position lamps/navigation lights are [5, p. 96 f.] (see also Fig. 7.28):

- Two sidelights red and green, each shining over a sector of 112.5°,
- A white tail light, shining over a sector of 135°,

Table 7.9 Overview of the required approvals for navigation lights

	Maritime shipping routes	Inland waterways
Legal basis	SeeSchStrO	Inland Waterways Ordinance (BinSchStrO), Rhine Shipping Police Ordinance (RheinSchPV), Danube Shipping Police Ordinance (DonauSchPV), Mosel Shipping Police Ordinance (MoselSchPV) and so on
Professional vehicles	EU steering wheel approval	EU steering wheel approval with protection of existing buildings for already built-in lights with "Anchor license" (see Fig. 7.26)
Sport boats	EU steering wheel approval or national EU approval (e.g., national BSH- or DHI approval)	Like professional vehicles

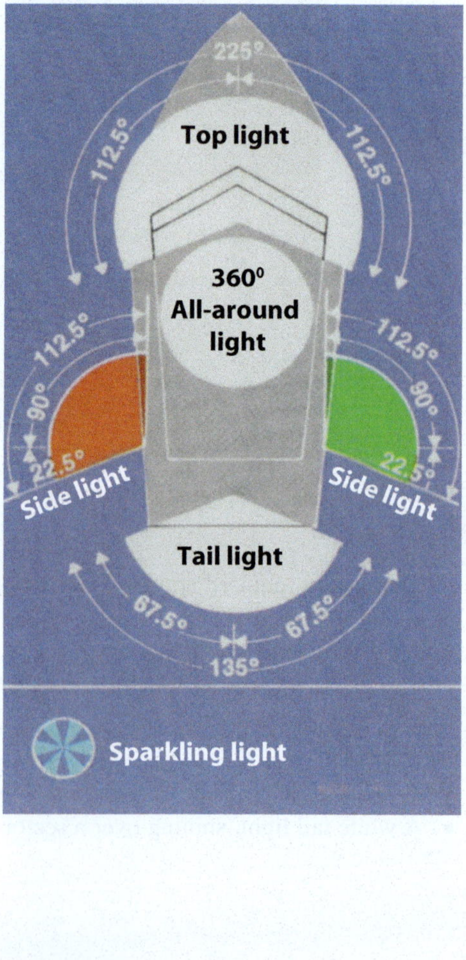

Fig. 7.28 Position lamps, machine vehicle <50 m [17]

- One or two white top lights—depending on the length of the ship (also called "steamship lights", as ships under sail do not carry these lights)—shining over a sector of 225°,
- White, red, green or blue omnidirectional lights that shine or sparkle firmly over the entire horizon (360°) (blue sparkle light: Federal Police, Water Police, Customs, Water and Shipping Administration or boats of the rescue service in action).

Only electrically operated lights may be used; LEDs have become the preferred light source today due to their long service life and low-energy consumption.

According to Sect. 8 SeeSchStrO, the so-called minimum range (headlight range) of all position lights must be at least 2 nm (nautical mile). According to rule 22 of the CAR, the following ranges of lights are prescribed (Table 7.10).

In order to ensure these minimum carrying distances, minimum light intensities are required (Table 7.11).[24]

An overview of the light guidance required in inland and maritime navigation can be found in Annex 27.

7.5.3 Radio Equipment

The equipping of ships with radio equipment depends on the type, size, and sailing area (A1 to A4) of the ship.[25] Chapter IV of the SOLAS Convention governs the implementation of the *Global Maritime Distress and Safety System (GMDSS)* with radio equipment Digital Selectiv Call (DSC-Controler), satellite Emergency Position-Indicating Radio Beacons (EPIRBs) as well as Search and Rescue Radar Transponders (SART) and satellite communication equipment. For example, ships in area A1 (which covers an area defined by the competent administration within the radiotelephony range of at least one USW[26]-coastal radio station, which is continuously available for DSC alerts—in Europe these are coastal areas up to approximately 30 nm), carry the following radio equipment:

- USW DSC maritime radio installation,
- USW DSC monitor receiver,
- NAVTEX-receiver (Navigational Text Messages), serves worldwide for the distribution of safety and weather information (Maritime Safety Information) on 518 kHz. This service works only in radioteletype mode.

[24] Annex I No. 8 of the CAR.

[25] Also deepening [10].

[26] Ultrashort wave radio.

Table 7.10 Range of the position lanterns

Vehicle	Light	Scope (sm)
≥50 m Length	Top light	6
	Sidelight	3
	Taillight	3
	Towing light	3
	White, red, green or yellow rotating beacon	3
≥12 m<50 m Length	Top light	5, at Vehicles <20 m Length: 3
	Sidelight	2
	Taillight	2
	Towing light	2
	White, red, green or yellow rotating beacon	2
<12 m Length	Top light	2
	Sidelight	1
	Taillight	2
	Towing light	2
	White, red, green or yellow rotating beacon	2

Table 7.11 Minimum luminous intensity of the lamps from position lamps

Scope (nm)	Light intensity (Cd)
1	0.9
2	4.3
3	12
4	27
5	52
6	94

- Emergency satellite beacon (EPIRB[27] -INMARSAT; EPIRB-COSPAS-SARSAT) (Fig. 7.29),
- Radar transponder,
- USW handheld radio.

[27]EPIRB = Emergency Position Indicating Radio Beacon; these beacons (INMARSAT: transmitter 1.5 GHz, direction-finding transmitter 121.5 MHz; COSPAS/SARSAT 406 and 121.5 MHz) transmit distress signals. The distress call is triggered mechanically or by dipping the beacon into the water [4, p. 140 f.].

Fig. 7.29 Various marine distress beacons. *Photo* US National Oceanographic and Atmospheric Association—NOAA

Nationally, the equipping of the seagoing ships under the German flag with USW marine radio systems is regulated in the German SchSV. According to this regulation, all passenger ships and cargo ships of 300 GT and above are required to be fitted with radio equipment.

An overview of the mandatory radio equipment in accordance with SOLAS depending on the shipping areas can be found in Annex 26.

7.5.3.1 Technical Characteristics of Marine USW Radio Equipment

A USW radiotelephone system consists of the transceiver unit, the control unit (Fig. 7.30), and the antenna system with the associated power supply (12 V or 24 V). The control unit can also be provided as a remote hand-held unit. Some radio manufacturers also offer the option of connecting several control units. This allows the radio system to be operated from several places on board (bridge and radio room).

Fig. 7.30 DSC maritime and inland radio installation of the company ICOM

One or more backup power sources must be available for the power supply. In the event of failure of the main and emergency power sources, the emergency battery should be capable of powering the following equipment (if any):

- USW maritime radio installation (frequency range: 156–174 MHz),
- MW/SW maritime radio installation[28] (Frequency range MW: 1606–3800 kHz, minimum range during the day 150 nm, at night even further; frequency range SW maritime radio: 4–30 MHz range, quasi-worldwide radio traffic is possible due to reflection of radio waves at the ionosphere; [4, p. 135]),
- INMARSAT marine earth station,[29]
- Illumination of the operating elements of the marine radio systems,
- Gyrocompass.

The *maximum output power* in the USW range must not exceed 25 W for a marine radio station. In principle, the system should be operated with an output power of 0.5–1 W, only in case of communication difficulties should the system be switched to 25 W.

For *mutual monitoring of two channels* the Dual Watch (DW) function must be switched on. Only one working channel (e.g., Channel 70) in connection with channel 16 (the distress and call channel) can be monitored at any one time. As soon as radio traffic is operated on one or both channels, the device automatically switches to this channel.

Through the *digital selective call procedure (digital selective calling)* it is possible to inform individual or specific groups of maritime radio stations that radio traffic is available for them even if the maritime radio station is not occupied. A radio station can thus be "dialed" directly via its phone number, similar to a telephone.

Compared to other radio waves, ultrashort waves have a completely different propagation characteristic, comparable to light—they propagate in a completely straight line. There are no reflections or reflections caused by air layers of the atmosphere, as is the case with shortwave radio (SW radio). Since USWs propagate in a straight line, USW radio traffic is virtually only possible within the range of vision between two radio stations (20–30 nm—"quasi-optical range"). In this respect the antenna system requires special attention:

Wavelength and length of the antenna are approximately related as follows:

$$L = \lambda/2. \tag{7.16}$$

This means that the length L of the antenna is approximately half the wavelength λ of the corresponding frequency. For the maritime radio-USW range—at a frequency of

[28]MW = marginal wave, SW = shortwave.

[29]INMARSAT is a British company that operates a mobile communications service via satellites that covers most of the earth's surface.

Fig. 7.31 USW radio antenna

about 156 to about 174 MHz [4, p. 134]—the wavelength is about 2 m, so the length of the antenna is about 1 m (Fig. 7.31).

With regards to the *wavelength*: electromagnetic waves propagate at the speed of light (approximately 300,000 km/s). Wavelength and frequency of the oscillation are related in the following way:

$$\lambda = 300{,}000/f \qquad (7.17)$$

with

λ Wavelength in m,
f Frequency of the vibration in kHz

Due to the straight-line propagation of the USW, the antenna should be placed as high as possible on the ship to achieve better radiation. It must also be mounted vertically. Metallic objects in the area of the radiation field considerably impair the transmission power.

7.5.3.2 Technical Background to USW Radio Equipment

Low-frequency useful signals, such as speech, cannot be transmitted directly over a radio channel.[30] For transmission, the wanted signal must be shifted to another frequency range (high frequency).

[30]Section with the collaboration of Pfaff, N.

In high-frequency transmission technology (radio technology), information—roughly differentiated—can be *analog* or *digital*. However, a high-frequency carrier is always required (in the frequency range from 30 to 300 GHz and in a wavelength range from $\lambda = 10$ km to 1 mm; [35]), onto which the information is modulated.

The common *analog modulation types* are amplitude modulation (AM) and frequency modulation (FM). In digital information transmission, too, the digital data must be prepared and modulated onto a carrier.

Difference Between Analog and Digital Information/Data or Signals

Data or information or even signals can be available in *analog* and *digital* form [36]. "Analog" means that any intermediate value between maximum and minimum is possible. "Digital" only provides very concrete values, no arbitrary intermediate values. A distinction is made between "quantized" and "binary" in this respect. A good example of this is the clock. With an analog display of the time (clock face with clock hand), there are virtually no unoccupied spaces in the analog data information "time". This is not the case with a digital clock, which displays the time to the minute, for example. Only the full minute can be read here, but not whether the space between 2 min has taken up one, two or 59s.

With signals, one speaks of an *analog signal,* if there is a value to be assigned exactly at any given time.

Digital signals are made up of concrete values. These can be stepped (quantized = several different concrete values), whereby the amplitude is transmitted by a "numerical value" (e.g., −2 V, −1 V, 0 V, +1 V, +2 V), or *binary,* where there are only two values ("bi" = two) (0 and 1, or low and high, or on and off).

Converting Analog to Digital Signals

In order to be able to transmit analog information, such as Speech or music digitally via radio (Fig. 7.32), this information must be digitized in the transmitter. For this purpose, the analog signal is first divided (quantized) into individual stages by the analog-to-digital converter (AD converter) in the transmitter. The individual step values are then converted into a binary code. This binary signal is modulated onto a carrier frequency and transmitted.

On the receiver side, the binary signal is received via the antenna and converted back to the original analog signal by a digital-to-analog converter (DA converter) and subsequent filters.

In terms of hardware, the converters are designed as integrated circuits (IC) (Fig. 7.33).

Parallel and Serial Data Transmission

When the analog signal has been quantized with the AD converter and then binary coded, it is available as a digital value with the corresponding voltage states. If you want

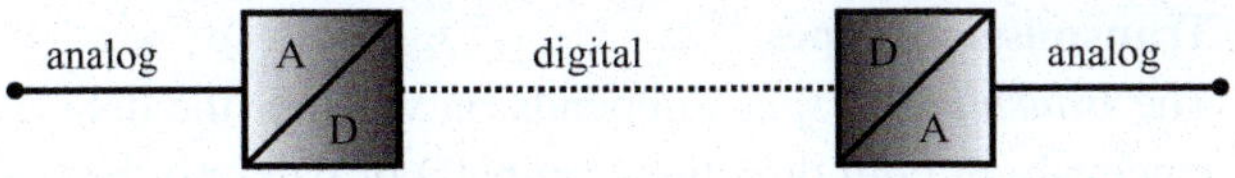

Fig. 7.32 Principle of digital transmission

Fig. 7.33 IC component of a DA converter

to transmit this value, for example, 8 bits at once, you need 8 frequency channels (parallel data transmission).

To get by with one channel (one frequency), the individual bits are transmitted one after the other. This is called *serial communication.*

The disadvantage of serial data transmission compared to parallel data transmission is immediately apparent: It takes at least 8 times as long with an 8-bit system and 32 times as long with a 32-bit system. However, you can get by with one channel.

In addition, care must be taken to ensure that the two rotating "switches" of the transmitter and receiver start at the same time (start signal) and then rotate at the same speed (synchronous). The same step speed (*baud rate* and *transmission rate*) between the transmitter and receiver, must be set.

Modulation/Modulation Method

Signal transmission by radio is all about transmitting as much information as possible without loss. When transmitting different signals on the same transmission path, signal transmission is hardly possible without prior signal processing. Therefore, different modulation methods are used to convert information and data into electrical signals so that they are suitable for transmission.

What is modulation?

Modulation is:

- Frequency adjustment,
- Multiple use of the transmission medium,
- Increase of the interference immunity.

How does modulation work?

Every electrical signal has three characteristics (signal parameters): *amplitude*, *frequency,* and *phase (polarity)*. During modulation, one or more of these signal parameters (*of the carrier signal*) is changed or modulated by the information signal (speech). The **information signal** also called modulation signal, is "impressed" on the carrier signal.

Transmission Types

The *transmission speed* depends on whether the data (signals) can be transmitted simultaneously in both directions (*duplex*) or can only be transmitted and received alternately in one direction (*simplex*) (see also Fig. 7.34). In the field of radio technology, therefore, the *duplex* spoken when transmitter and receiver use different frequencies; often in this context it is also spoken of transmitting and receiving in the lower and upper band.

To be able to perform duplex communication, the radios must be equipped with **duplex switch** in order to connect the device antenna to the transmitter and receiver simultaneously. The prerequisite is the use of a duplex channel. This is a frequency pair where the transmit and receive frequencies are offset by a fixed duplex spacing. These two frequencies are then called upper and lower band [34].

7.5.3.3 Marginal Wave and Shortwave Radio Installations

Due to the limited range of USW radio, SOLAS requires the use of marginal wave or shortwave radio equipment in some areas of operation (see Annex 26). However, shortwave radio is only of minor importance due to satellite communication (INMARSAT, COSPAS-SARSAT, GPS).

7.5.3.4 Automatic Identification System (AIS)

The *automatic identification system* (AIS) characterizes an automatic ship identification and monitoring system for global shipping (Fig. 7.35). With AIS, ships identify themselves and clearly communicate their position, course, speed and other data to others. AIS serves

- the prevention of collision at sea,
- the automatic exchange of information between ships and with land stations,
- the traffic centers on the coast as a complementary means of maritime traffic safety,
- via special USW transmitters and receivers, for example, driving data is automatically exchanged at short intervals. Automatic identification system also allows a view over obstacles and thus supplements the radar image display or the display in combination with the electronic nautical chart ECDIS. Depending on the antenna height an AIS station has a range of 20–30 nm.

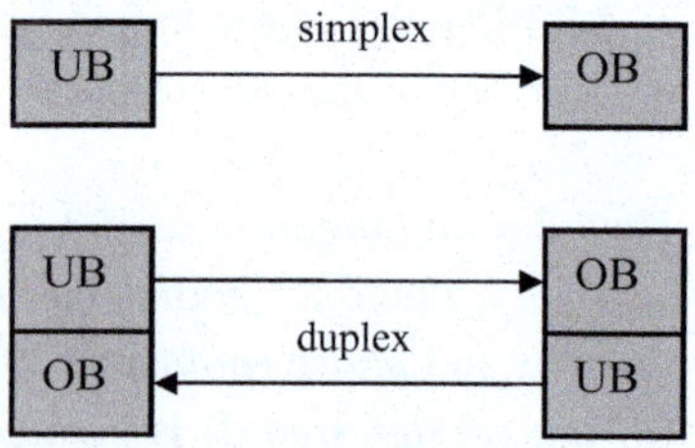

Fig. 7.34 Simplex and duplex method

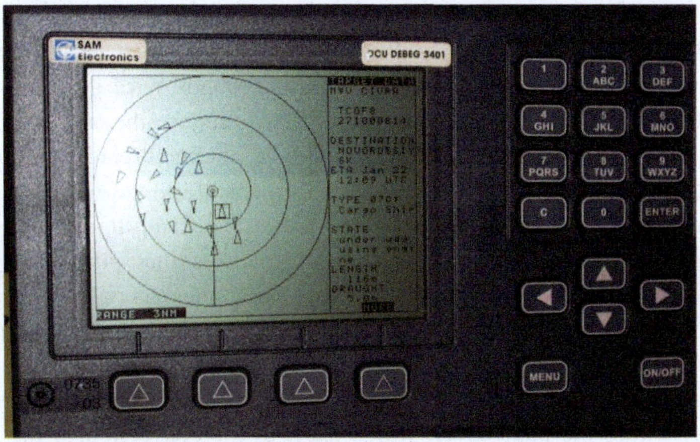

Fig. 7.35 AIS control panel. *Photo* Clipper, CC BY 2.5

7.6 Survivability of Warships

With regard to the aspects of occupational safety, ship safety, and fire protection described in the previous sections, these are broadly applicable to warships [12]. However, as these are not comparable with civilian ships due to their peculiarities (weapons, explosive ordnance on board, special technical equipment) and especially in the event of an armed conflict 1: 1 (although today, especially with regard to the construction of the hull, the merchant ship standard is already partly used), there are some special features here.

For warships, the danger of fires and explosions as a result of hostile weapon action or even accidents without hostile action is permanently present. Today the danger for warships is much higher than in former times, because the ships have a high density of essential equipment, which cannot be easily repaired by the crew after a hit. Weapon systems today have a very high impact; even a single hit can result in the total failure of a warship.

The most important factors to describe the survivability of a warship are:

- Sensitivity,
- Vulnerability,
- Repairability.

The *sensitivity* of a warship is determined by how, for example, attacks can be repelled in principle or how the concept of a target can be prevented at all. This can be achieved through a variety of technical measures. Early detection of enemies can be achieved with the appropriate sensor technology (near and far radar, optical reconnaissance, infrared detection, sonar). Shipbuilding measures are used in order to avoid being recognized as a target at all or at least only very late. First of all, this includes the so-called stealth construction (see Fig. 7.36). This serves to reduce the probability of detection by radar. To avoid reflection of radar beams to the transmitter or to reduce the energy of the reflected beams, the surfaces of the hull and superstructure are inclined relative to the radar beam and angle reflectors and curves are avoided. The shape alone can reduce the radar cross section by a factor of 10–100. An even higher factor is difficult to achieve, since the so-called Huygens principle states that even an extremely inclined plate will reflect part of the radar energy back to the transmitter. Radar-absorbing materials are used to further reduce this, but these are usually less effective against radar in the low-frequency range. For these coating materials, materials under the name "iron ball paint" are used. This paint contains tiny balls coated with carbon or iron oxide. Radar waves generate molecular vibrations within the paint layer, which leads to the conversion of the radar energy into heat. The thermal energy is then dissipated into the structure. Another variant of corresponding coatings is the use of neoprene polymer layers with iron oxide or carbon particles. Coatings with carbon fiber reinforced plastics are also used.

The signature detection by infrared detection of the propulsion systems can be reduced by placing the exhaust system close to the water surface and close to the stern of the ship. Good insulation of the chimney system is just as conceivable as cooling the exhaust gases with sea water or blowing the exhaust gases out under water.

In order to reduce noise radiation, special ship's propellers with the lowest possible tendency to cavitation are used. Furthermore, care must be taken to ensure good

Fig. 7.36 German frigate of the class F124 in stealth construction. *Photo* Brian Burnell, CC BY-SA 3.0

decoupling of noisy machinery toward the hull in order to minimize structure-borne noise, which can be amplified by the hull as a resonating body.

Measures to reduce electromagnetic radiation also serve to reduce early detection.

With regard to the sensitivity of a warship, the nature and extent of possible countermeasures (interference with enemy detection capabilities) as well as the maneuverability of the ship itself must also be understood.

In summary, the *sensitivity* factor follows the approach: "avoid being discovered, and if you do, avoid being hit!".

The factor *vulnerability* includes the ability to withstand hits and minimize their effects. Here, the aspect of ballistic protection is important: Particularly sensitive areas are especially protected. For example, a 19 mm strong Kevlar lacquer was applied to ships of the Oliver-Hazard-Perry class.

Depending on the threat situation, temporary ballistic protection measures can also be taken.

Also, warships are usually divided into sections with considerably more bulkheads than merchant ships. On the one hand, this serves the purpose of limiting the water ingress to a small area in the event of damage below the waterline, for example, by torpedo or mine impact. But even in the event of ballistic hits leading to explosion or fire, small areas to be sealed off can locally limit the expansion of fire and warfare agents.

To protect the crew and sensitive, important systems, for example, appropriate chairs on the bridge and operations center as well as system components are shock-absorbingly mounted (spring or rubber elements) to reduce the consequences of the explosion waves. The effectiveness of these measures is tested by so-called blast tests.

The *repairability* includes the ability to restore operational capability with on-board resources after damage.

Furthermore, for combat ships the *NBC Defence* has a special significance. This does not only mean protection against NBC weapons; due to the new deployment scenarios, the ships and crews on humanitarian missions are also exposed to the threat of epidemics such as swine fever or smallpox. The danger should not be underestimated whenever the situation in ports called at for the purposes is not fully clarified.

Asymmetric attacks can be carried out with comparatively simple means such as chlorine gas at the ship's berth. Dirty bombs or improvised explosive devices pose a similar threat.

Essential for the detection and defense of NBC warfare agents are analysis systems with which the air can be analyzed and alarms given in good time. Air filter systems (NBC filters) should be capable of filtering the corresponding hazardous substances and preventing them from spreading inside the ship.

A slight overpressure should be maintained in the ship by means of ventilation systems, at least in areas to be specially defined, so that no pollutants can enter the interior. These areas must be entered and exited via air locks.

Soldiers who have to leave the interior of the ship in the event of an NBC alarm must be equipped with adequate protective equipment (overgarment or full protective suit); a decontamination facility must be provided after the external deployment.

Spraying systems for firefighting, which are also used for decontamination during or after exposure to NBC weapons, have already been described in Sect. 7.2.5.

References

Print media

1. Bundesamt für Seeschifffahrt und Hydrographie: Elektronische Seekarten. Bundesamt für Seeschifffahrt und Hydrographie, Hamburg (2017)
2. Bundesamt für Seeschifffahrt und Hydrographie: Lichterführung und Schallsignalanlagen. Bundesamt für Seeschifffahrt und Hydrographie, Hamburg (2017)
3. DNV·GL, "RULES FOR CLASSIFICATION – Ships – Part 4: Systems and components, Chapter 11: Fire safety", Edition October 2015
4. Donat, H.: Yacht Bordbuch. Delius Klasing, Bielefeld (2004)
5. Dreyer, R., Huth, W.: Sportküstenschifferschein und Sportbootführerschein See. Delius Klasing, Bielefeld (2001)
6. Forell, B.: Niveau der Personensicherheit von Versammlungsstätten – Nachweis nach vfdb-Leitfaden. TU-Braunschweig, Braunschweig (2016)
7. Gebauer, J., Krenz, E.: Marineenzyklopädie. Tosa, Wien (2003)
8. Hahne, J. (Hrsg.): Handbuch Schiffssicherheit, p. 56. Seehafen, Hamburg (2012)
9. Hank, O.: Unfallverhütung bei Aussetzvorrichtungen und Rettungsbooten. Diplomarbeit, Hochschule Bremen (2010)
10. Hannemann, U.: Beschränkt gültiges Funkbetriebszeugnis (SRC), UKW-Sprechfunkzeugnis für den Binnenschifffahrtsfunk (UBI). Delius Klasing, Bielefeld (2003)
11. Hochhaus, K.-H.: Vorlesungsskript Schiffshilfsmaschinen, TU Hamburg-Harburg, Stand 10/90
12. Hofbauer, B.-G.: Moderne Seemacht Teil II. Republik Österreich, Bundesminister für Landesverteidigung und Sport, Wien (2015)
13. Kaufhold, F.: Verbrennen und Löschen. Kohlhammer, Stuttgart (1993)
14. Mayer, C.-F., Marquardt, R.-S.: Schiffstechnik und Schiffbautechnologie. Seehafen, Hamburg (2006)
15. Milek, S.: „Sicherheitrelevante Maßnahmen im Land- und Schiffbau – Brandschutztechnische Betrachtung einer schwimmenden Versammlungsstätte", Bachelorarbeit an der Hochschule f. Angewandte Wissenschaften Hamburg, Fakultät Life Sciences (2014)
16. Müller, J., Krauß, J.: Handbuch für die Schiffsführung. Springer, Berlin (1980)
17. Overschmidt, H., Gliewe, R.: Sportbootführerschein Binnen – Segel Motor. Delius Klasing, Bielefeld (2017)
18. Rempe, A., Rodewald, G.: Brandlehre. Kohlhammer, Stuttgart (1993)
19. Verband der Sachversicherer e. V., „Feuerlöschanlagen mit Kohlenstoffdioxid, Planung und Einbau", Köln (2017)

Internet

20. http://publikationen.dguv.de/dguv/pdf/10002/i-504-26.pdf. Accessed 2 July 2017
21. https://tuicruises.com/faq/sicherheit/. Accessed 23 Jan 2017
22. http://www.accuro.at/de/technologie/hochdruckwassernebel-loeschanlagen.html. Accessed 26. Jan 2017
23. https://www.baua.de/de/Themen-von-A-Z/Gefaehrdungsbeurteilung/Gefaehrdungsbeurteilung.html. Accessed 11. Feb 2017
24. http://bw-ladungssicherung.de/glossar/solas/. Accessed 28 Jan 2017
25. http://www.deutsche-flagge.de/de/sicherheit. Accessed 7. Jan 2016
26. http://www.e-hoi.de/sicherheit-auf-kreuzfahrtschiffen/. Accessed 23. Jan 2017
27. http://www.gefaehrdungsbeurteilung.de/de/einstieg/definition. Accessed 11 Feb 2017
28. https://www.hs-bremen.de/internet/hsb/projekte/maritime/studium/nautikseeverkehr/diplom-bachelor/diplarb_riecke.pdf. Accessed 5 Feb 2017
29. https://www.lfs-sh.de/Content/Ausbildung/Documents/LeitfadeInnenangriff_102011.pdf. Accessed 29 Jan 2017
30. http://www.marioff.com/de/fire-protection/hi-fogr-brandschutz-systemtypen/nassrohrsystem. Accessed 26. Jan 2017
31. http://www.marioff.com/de/water-mist/wassernebel-brandbekaempfungssysteme. Accessed 26 Jan 2017
32. http://www.mt-boehlen.de/downloads/International%20Convention%20for%20the%20Safety%20of%20Life%20at%20Sea.pdf. Accessed 23 Jan 2017
33. http://www.schiffslexikon.com/rettungsboot-65.html. Accessed 23 Jan 2017
34. http://de.wikipedia.org/wiki/Duplexweiche. Accessed 24 July 2017
35. http://de.wikipedia.org/wiki/Frequenzband. Accessed 24 July 2017
36. www.amateurfunkpruefung.de. Accessed 28 Dec 2017
37. www.a-spe.com. Accessed 28 Dec 2017

8 Environmental Protection in Maritime Transport

The topic of "marine environmental protection" has been of great international importance for many years. Environmentally hazardous chemicals in ship painting, the introduction of nonnative organisms with the ballast water, the discharge of wastewater and waste into the sea, and the pollutants from exhaust gases or oil pollution can have a lasting impact on the state of the marine environment.

8.1 Maritime Environmental Legislation

Due to their global nature, maritime issues are primarily regulated by the International Maritime Organization (IMO). Aspects of environmental protection are dealt by the Marine Environmental Protection Committee (MEPC) and are mainly laid down in the *International Convention for the Prevention of Pollution from Ships* (**MARPOL**). This convention was implemented in Germany by the *Marpol law*, which has been transposed into national law.

The international agreement MARPOL consists of the original convention, two additional protocols and six annexes. The systems (also with *Annex* I to VI) of the convention regulate the various types of potential environmental damage associated with the operation of ships:

- Annex I: Prevention of pollution by oil,
- Annex II: Prevention of pollution by noxious liquid substances,
- Annex III: Prevention of pollution by pollutants transported in packaged form
- Annex IV: Prevention of pollution by ship-generated wastewater,
- Annex V: Prevention of pollution by ship-generated waste,
- Annex VI: Regulations for the prevention of air pollution from seagoing ships.

M. Pfaff, *Ship Operation Technology*, https://doi.org/10.1007/978-3-658-32729-3_8

Other important IMO conventions relating to environmental protection in maritime transport are:

- The International Convention on the Prevention of Marine Pollution by Dumping of Wastes and Other Matter, 1972 (London Convention) and its 1996 Protocol (updating the Convention) (London Protocol),
- The International Convention on the control of harmful antifouling systems on ships (AFS-Convention, 2001),
- The International Convention for the Control and Management of Ships' Ballast Water and Sediments (Ballast Water Convention, 2004),
- The International Convention on the Safe and Environmentally Sound Recycling of Ships (Hong Kong Convention, adopted in 2009)

At the *European level* environmental protection in maritime transport is also addressed in the convention for the Protection of the Marine Environment of the North-East Atlantic (OSPAR Convention, 1992) and the Baltic Sea(Helsinki Convention, 1992). The Green Paper (2006) and the Blue Paper (2007) on a future maritime policy for the EU identify the existing problems in the maritime transport sector and set out strategies to address them. Among other things, the Blue Book contains an action plan for the EU Commission, which describes urgently needed measures for the shipping sector (including the reduction for air pollution). The EU Marine Strategy Framework Directive (MSFD), which entered into force in 2008, is derived from the Blue Book. It sets out an integrated approach to promote sustainable use of European seas and to protect and conserve marine ecosystems. The objective is to achieve or maintain good marine status by 2020.

Furthermore, at European level the *Directive on port reception facilities for ship-generated waste and cargo residues* (*Directive 2000/59/EC*) should also be mentioned. Since overboard dumping of waste is generally not permitted, the waste produced on board must be stored temporarily and sent ashore for disposal in the next port. This Directive aims to improve and establish the availability and use of port reception facilities for ship-generated waste and cargo residues. The Directive applies in principle to all ships calling at a port of an EU country, irrespective of their flag. Only official vehicles and warships are excluded from it. EU countries must ensure that their ports are equipped with appropriate reception facilities. The Directive also stipulates that masters of ships (other than fishing vessels or recreational craft authorized to carry up to 12 passengers) wishing to call at an EU port must give advance notification of:

- The last port and time of the last discharge of ship-generated waste,
- The type and quantity of ship-generated waste and cargo residues to be discharged and/or retained on board, and the percentage of maximum storage capacity.

All ship-generated waste must be delivered to a port reception facility before leaving an EU port, unless the master can demonstrate that there is sufficient storage capacity on the ship to reach the port of delivery. Nevertheless, the EU country may require that the waste be unloaded before leaving the port if:

- No appropriate facilities are available in the port of call,
- Or this port is not known,
- Or there is a risk that the waste will be thrown overboard.

This Directive was adopted by the Federal States of the Federal Republic of Germany, which have ports called at by the corresponding ships, by means of Land Port Waste Management Acts for their respective areas.

In addition, the *convention on the collection, discharge, and acceptance of waste* by *Rhine and inland waterway transport from September 9, 1996* (*CDNI*), which came into force on November 1, 2009.[1] As the disposal of ship-generated waste is carried out from or to land, this convention defines rules that are binding on the parties involved. These are aimed at:

- The promotion of waste prevention,
- The organization of waste disposal facilities on the entire waterway network,
- Ensuring adequate funding in line with the "polluter pays" principle, and
- To facilitate compliance with the prohibitions on inland waterway transport with regard to the discharge of waste by dumping overboard into waters.

This convention was transposed into national law by the Federal Republic of Germany by Act of December 13, 2003, Federal Law Gazette (FLG; Bundesgesetzblatt, BGBl.) II p. 1799.

In addition to the international regulations, more extensive German regulations on environmental protection, such as the Federal Immission Control Act (BImSchG—Bundes-Immissionsschutzgesetz), the Recycling Management Act (KrWG—Kreislaufwirtschaftsgesetz), or the Water Resources Act (WHG—Wasserhaushaltsgesetz), may apply to ships under German flag and also to ships under foreign flag in German territorial waters.

8.2 Possible Environmental Damage

In the following sections, some scenarios through which shipping can affect the environment, in particular the marine environment and also the air, are presented and what basic measures can be taken against them.

[1]Further details [79].

8.2.1 Contamination by Oil

Oil disasters, such as those caused by the accidents of the tankers Erika (1999 off the French coast of Brittany) and Prestige (2002 off the northwest coast of Spain), must be reduced in number and scale. Oil pollution and spills can cause lasting damage to marine ecosystems. With the exception of a few heavy oils, oil initially floats at the water surface. Low-viscosity oils (lighter crude oils and most oil products) spread very quickly and form a thin layer. Tough oils, like heavy oil, spread more slowly and form thicker oil slicks. With sufficient wind, the swell splits thin oil into drops and tough oil into larger lumps. While the larger ones remain directly at or below the surface and merge again in the thin case, droplets below a certain size remain stable in the water and are slowly distributed into deeper water layers or washed up on the coasts.

Oil slicks dampen the swell, so that seabirds often consider the smooth surfaces as a resting place and land there. As a result, the viscous oil sticks to their plumage, destroying their thermal insulation, in some cases, their ability to swim, and is absorbed when the birds try to clean themselves. This usually leads to mass mortality of seabirds through suffocation, hypothermia, drowning, and poisoning.

If the oil reaches the coast, it first settles there. In the course of time, it is washed away again, especially on coasts with strong surf, and so with changing weather and tidal conditions it may be relocated several times. Oil that is still relatively liquid and deposited on beaches poisons or suffocates soil organisms such as worms or snails.

It is not only the nature that is threatened; fishing, tourism, and coastal protection are also affected. Especially in sensitive marine areas, such as the Wadden Sea or polar waters, an oil spill would be an ecological disaster with long-lasting consequences.

For example, following the accident of the oil tanker "Exxon Valdez" off Alaska in 1989, the IMO's Environment Committee tightened the age limits for *single-hull oil tankers* and the ban on transporting heavy fuel oil in single-hull tankers with a deadweight tonnage of over 5000 t. Since then, tankers with a deadweight tonnage of more than 5000 t must be fitted with a double hull. This second outer hull is intended to prevent oil leaking after a collision [79].

8.2.2 Pollution from Ship Sewage

For the discharge of ship sewage into marine waters, the regulations contained in MARPOL Annex IV, to which other conventions and national legislation refer, are primarily relevant. In general, the discharge regulations according to MARPOL Annex IV apply to ships on international voyages which are either of a gross tonnage (GT) $\geq$ 400 or < 400 for more than 15 persons. Persons are considered to be both crew members and passengers. Recreational craft are excluded from the regulations of MARPOL Annex IV.

The unregulated discharge of wastewater is generally prohibited. Under certain conditions, however, wastewater may be discharged into the sea (see Table 8.1). For example,

Table 8.1 Discharge conditions of ship sewage according to MARPOL Annex IV

Discharge of ship sewage		
Untreated	Mechanically treated and disinfected (Regulation 11 Para. 1 No. 1)	From wastewater treatment plants (Regulation 11 Para. 1 No. 2)
– $\geq$ 12 nm from nearest country – Discharge rate authorized by the competent administration – Ship maintains course at initiate and – A speed $\geq$ 4 kn	– $\geq$ 3 nm from nearest country	– Approved wastewater treatment plant meeting the requirements of Regulation 9.2.1, and – Does not make solids and discoloration visible in the surrounding water

mechanically treated and disinfected wastewater may be discharged from a distance of 3 nautical miles (nm) from the coast (baseline). In special areas, such as the Baltic Sea, stricter discharge rules apply to passenger ships. A passenger ship is any ship carrying more than 12 passengers. Discharge, there, is permitted only if the passenger ships have an IMO-certified *wastewater treatment plant,* where the wastewater is treated in such a way that the discharge does not cause visible floating solids or discoloration.

The North Sea and the North-East Atlantic are specially protected under the OSPAR Convention, the regulations under MARPOL are adopted with regard to the discharge of wastewater. A special feature is that in 2011 the IMO will be the first and so far only *special area*[2] in accordance with MARPOL Annex IV, which lays down stricter rules for the discharge of wastewater from passenger ships in port. However, these regulations will only come into force when all riparian states have reported "sufficient" port reception facilities for wastewater. In the Baltic Sea area the MARPOL regulations are also applicable to smaller ships and recreational craft through HELCOM (Annex IV), provided they have a toilet on board. Then, for example, the equipment of wastewater retention systems on board will be required, so that the wastewater must be discharged in the port.

Nationally, the discharge of ship sewage is derived from the abovementioned regulations and is specified in the Maritime Environmental Behavior Ordinance (Sect. 9 Para. 1 SeeUmwVerhV—See-Umweltverhaltensverordnung). The ban on discharge of wastewater concerns "in accordance with the MARPOL regulations, all ships of a certain size and above on international voyages and all ships and recreational craft on voyages between German ports, in the estuaries of navigable rivers, in territorial waters, and all ships and recreational craft flying the German flag on voyages between German ports." In the Baltic Sea in particular, it is added that "recreational craft which, contrary to Sect. 6b (1) of the Ship Safety Ordinance (SSO; German: Schiffssicherheitsverordnung, SchSV), are not equipped with a tank and are not subject to any exemption, may not navigate the territorial sea" [79].

[2]For a list of currently existing special areas, see [54].

In addition, it applies to the discharge of wastewater from passenger ships within a special area (such as the Baltic Sea):

The discharge of waste water into the sea is prohibited as from January 1, 2016 for new passenger ships and as from January 1, 2018 for existing passenger ships, unless the following conditions are met:

- An approved on-board sewage treatment plant complying with the requirements of Regulation 9.2.1 of Annex IV of MARPOL, and
- No solids and no discoloration are visible in the surrounding water.

8.2.3 Pollution by Ship-Generated Waste

Planned measures under the EU MSFD, such as consistent waste documentation and separation on board or stricter controls at sea and in ports, are intended to achieve a significant reduction in waste discharged into the sea by 2020 in order to help restore or maintain the "good environmental status" of Europe's marine regions. This will be supported by appropriate measures within the regional action plans under the North-East Atlantic (OSPAR) and Baltic Sea (HELCOM) marine conventions. This includes, for example, guidelines for a coordinated approach to the partial or complete integration of waste fees into the regular port fees to prevent illegal shipments.

The discharge of waste into the sea is regulated internationally by MARPOL Annex V. With defined exceptions (food waste, harmless cargo residues, cleaning agents and additives, and animal carcasses), no waste may be discharged from ships into the sea. Since January 2013, these requirements have been in force worldwide. Stricter regulations apply when staying in particularly sensitive areas, which have been designated as special areas (e.g., the Baltic and North Seas) due to their specific oceanic and ecological conditions. No animal carcasses produced during transport are to be disposed of in these areas. The introduction of food waste that has not been pulverized, and cargo residues that are not found in the washing water are also not permitted [73].

An overview of the restrictions on the discharge or discharge of waste into the sea in accordance with MARPOL Annex V, Regulations 4–6, is given in Annex 28.

8.2.4 Air Pollution by Ship Exhaust Gases

Exhaust gases on board ships are primarily produced by the operation of the propulsion engines. Furthermore, especially during berthing times in port, the emissions of the Electric diesel plants[3] should not be neglected. The exhaust gases from ship waste incineration plants are also of further relevance.

MARPOL Annex VI is dedicated to this complex of issues.

[3]Electric diesel engine aggregates.

8.2.5 Carryover of Organisms by Ballast Water

When ballast water is taken in, organisms are regularly ingested, which could be a small fish, benthic and planktonic organisms or even pathogenic germs [52]. These are released when the ballast water is pumped out of the ballast water tanks elsewhere (see also Fig. 8.1). Due to the increasing speed of shipping traffic, the probability that organisms will survive the passage in the ballast water tanks is growing. In the meantime, numerous alien species have settled along this path, for example in the North Sea and the Baltic Sea, where, the shipworm causes damage, especially to the coastal protection structures such as groynes, by eating the wood of these structures.

The *Ballast Water Convention* that was adopted by the IMO in February 2004, is an attempt to counteract the carryover of non-native organisms through ballast water. This convention calls for a ballast water management system that largely dispenses with the previously customary uncontrolled water exchange when ballast water is taken in and discharged. Instead, the ballast water on board of every ship must be treated by appropriate ballast water treatment systems prior to discharge into the marine environment in such a way that a standard prescribed by the convention (D2 standard) is achieved. This is achieved by the appropriate ballast water treatment plants. For a transitional period, the convention allows ballast water exchange according to the less demanding D1 standard under certain conditions.

Since the Ballast Water Convention is an international agreement, it requires ratification and implementation in national law. Germany joined the convention on February 13, 2013 with the Ballast Water Act.

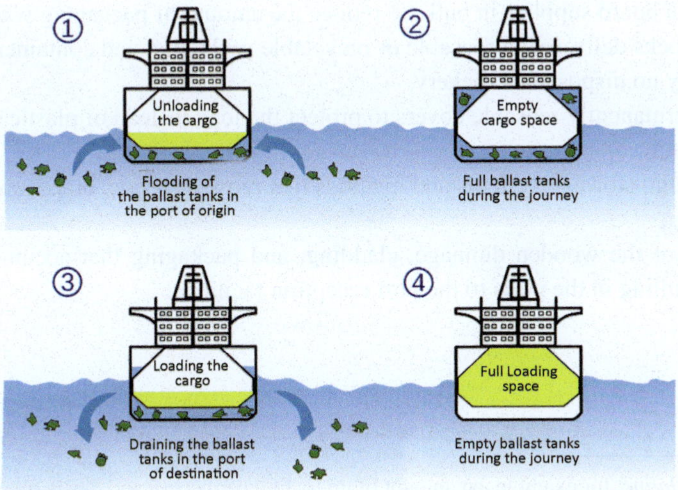

Fig. 8.1 Filling and draining the ballast water tanks. *Graphic:* Maxxl2, CC BY-SA 3.0

8.3 Technical Measures for Marine Environmental Protection

In the following sections, possible technical solutions are presented to meet the legal and technical requirements for marine environmental protection.

8.3.1 Waste Management on Board

A ship can be considered like a private household, a small business or even a large village, depending on the type of ship and its number of passengers and crew. It is, therefore, not surprising that a ship produces similar amounts of waste as on land. Numerical values of about 1.5 kg per person and day [23] up to 5.3 kg per person and day (2.5 kg food waste, 1.8 kg packaging waste, 1 kg glass and can waste; [58]) can be taken into account.[4] As explained above, overboarding of the waste produced is not permitted in principle—with a few exceptions, see Annex 28. The waste must be temporarily stored on the ship until it is handed over to shore or until it is disposed of by shipboard waste incineration plants. This requires appropriate waste management on board.

8.3.1.1 Sorting and Storage on Board

The *2012 Directive for the implementation of MARPOL Annex V* (VkBl 20/2012 No. 181 p. 795 ff.) contains information and concrete details on waste management. With regard to waste treatment on board, the Directive initially aims at *reduction of waste* or rather *waste prevention*. Shipowners and operators should reduce the amount of material they take on board that could be generated from waste, thereby avoiding the generation of waste. In this context it is advised:

- To take on board supplies in bulk (to reduce the amount of packaging waste),
- Use of stocks delivered in reusable or recyclable packaging and containers,
- Preferably no disposable crockery,
- Use of permanently reusable covers to protect the load instead of plastic tarpaulins for single use,
- Use of cargo stowage systems and methods that reuse dunnage, bracing, cladding, and packaging,
- Delivery of the wooden dunnage, cladding, and packaging that accumulates during cargo handling in the ports to the port reception facilities.

[4]In the *Time Online*, for example, an amount of up to 3 kg per person and day is mentioned [78].

Fig. 8.2 Waste shredder

Waste that cannot be avoided must be disposed of properly and without causing damage. This may be done by incineration in on-board incinerators or by temporary storage on board and transfer to land.

For example, food waste, especially from the kitchen area of cruise ships, is sucked into a waste area using negative pressure, where it is shredded and dried in various steps. This produces a peaty mass, which burns well and can be fed to any existing on-board incinerator. The same applies to shredded waste from wood, paper/cardboard, plastics (shredder; see Fig. 8.2), and sewage sludge from the wastewater treatment plant and oil sludge from the engine room.

If a ship waste incineration plant is not operated on board, however, the question arises as to how the waste, in particular with regard to the degree of separation, must be temporarily stored. For the MARPOL Annex V, the Directive on port reception facilities for ship-generated waste and cargo residues (Directive 2000/59/EC) and the convention on the collection, delivery, and reception of waste by Rhine and inland waterway of September 9, 1996 (CDNI) require the provision of port reception facilities for delivery of ship-generated waste. There the waste is transferred to land. Further disposal is then carried out according to the respective regulations of the port states. For example, it cannot be ruled out that in an Asian port, a separation into individual types of waste that

Fig. 8.3 Waste compactor

is typical in Germany may differ. Even in German ports, theoretically different sorting depths may be required.

This is primarily due to the different disposal concepts of the port municipalities, to which the waste is delivered directly or to a third party commissioned by the municipality. Consequently, different requirements for the separation and sorting of waste fractions are sometimes necessary both on ships and in ports.

Regardless of the fractions into which the waste is to be separated and temporarily stored, storage should be preceded by pretreatment. These are the essential treatment measures aimed at reducing the amount of waste in order to save space. In addition to shredders (Fig. 8.2), for example, for shredding glass, there are also compactors (Fig. 8.3) for compressing paper, beverage cans or similar loose waste.

The answer to the question into which fractions the waste on board should be separated is based on the KrWG. According to Subsection 11 and 14 KrWG, the Battery Act and the Electrical and Electronic Equipment Act, biowaste, paper, metal, plastic and glass waste, used batteries, and old electrical equipment must be collected separately for the purpose of promoting recycling and other material recovery, insofar as this is technically possible and economically reasonable. All other waste must then be collected as residual waste.

Against the background of the possible introduction of a recycling bin throughout the country, as already practiced by several municipalities, waste separation could be simplified even further; what used to be collected via the yellow sack would be placed in it: Plastics and composites such as yoghurt pots, care product bags, plastic bags, foams, shampoo bottles, dishwashing and washing agent bottles, polystyrene trays for food, toothpaste tubes. In addition, the following are also placed in the recycling bin: Plastic and metal articles of daily use such as shower curtains, buckets, watering cans, children's toys, transparent covers, laundry baskets, aluminum foil, aluminum lids, aluminum bowls, tin cans, crown caps, deposit-free drinks cans, wrapping foil, fittings, baking tins, cutlery, wire scraps, dustpans, clothes hangers, nails, pans, pots, tools, and similar items [50].

However, these sorting requirements are primarily aimed at waste disposal on land. This means that in Germany—and also in the other EU countries—waste from port reception facilities is disposed of according to these sorting depths. In fact, however, the local authorities and other disposal companies responsible for waste disposal may make different arrangements, so that theoretically different disposal structures can be found in each port. Nevertheless, it is advisable to keep waste containers on board for the above-mentioned waste fractions. They can always be tipped together, but deeper sorting in waste treatment plants on land is also possible. The sorting depth described above, however, achieves a legal minimum level of waste separation, with which the waste can generally be transferred to the port reception facilities without any problems.

With regard to the implementation of practicable waste management systems, the crews of ships must also be considered. These are inevitably confronted with the issue of waste. To do this, they must be efficiently informed, motivated, and actively involved in the organizational processes. Now imagine what it must be like for the crew member to witness how the waste he has meticulously separated is tipped back into a large container in the port, while in the other port the waste separation carried out is not yet sufficient.

It goes without saying that no influence can be exerted on waste disposal in ports of other countries. However, urgent efforts must be made to ensure that waste separation systems as described above are available at least for ships flying the German flag and also in Germany's port reception facilities. A uniform waste disposal concept should be established and operated throughout Germany for **all** the ports. The advantages are obvious:

- Cost savings are achieved through standardization. If there is no link to municipal waste disposal, there is no cost commitment by the respective municipality. The port operator can achieve calculable price negotiations if he commissions suitable contractors on his own.
- The shipping companies can create "tailor-made" and thus practicable concepts for their ships.
- The motivation of the crew to accept and "live" the concept that applies to them is increased.

Aspects of waste storage on board
According to the BGR 111 "Working in Kitchens," No. 3.1.5.23, the following must be observed with regard to the storage of foodwaste and catering waste:

- Collecting containers for food waste that is stored temporarily until disposal must be set up outdoors or in suitable rooms so that no health hazards can arise from them.
- Rapid decomposition of food waste is counteracted if the collection containers are placed in suitable rooms or boxes with a temperature of max. 10 °C.
- Outdoors, collection containers for food waste and leftovers should be placed in the shaded area if possible, but not in the immediate vicinity of openings to rooms where food is processed.
- In order to avoid health hazards from microorganisms, collection containers for food waste and food scraps must be cleaned and, if necessary, disinfected after emptying.
- In accordance with Sect. 24 Para. 1 of the Accident Prevention Regulation "General Regulations" (BGV A1), waste collection containers may only be set up in such a way that traffic routes and workplaces are not restricted.

Hazardous waste, that is, waste that is subject to the Hazardous Substances Ordinance (such as solvent and paint residues) must be stored in specially ventilated rooms. The technical rules for hazardous substances TRGS 520 provide more detailed information on this.

Waste from the medical and hospital sector also requires special storage; aspects of infection protection are of primary importance here. Communication No. 18 of the State Working Group on Waste (LAGA—Länderarbeitsgemeinschaft Abfall) formulates corresponding requirements.

Basically, the waste on board should be stored in a separate load. In order to avoid annoyance or even danger for crew and passengers, the rooms should be ventilated in such a way that no odor nuisance is caused on board by the waste. The formation of odors and germs through decomposition and rotting processes can be significantly reduced by cooling the waste storage room to around 10 °C.

8.3.1.2 Waste Incineration on Board

(a) Introduction to the topic and problem
As described above, outboard disposal of ship's waste is generally not permitted according to the international law—and also according to national regulations.[5] However, there is often not enough space on watercraft for intermediate storage of **all** waste generated on board, proper disposal is available only at the next port. From the point of view of the shipowners, the storage space required for waste storage can certainly be used for the

[5]See, e.g., the High Seas Injection Act.

original purposes for which the ship is designed, if it is not necessary to keep it available—namely as cargo space or as berth space for accommodating passengers. Passenger shipping is particularly affected here. With a passenger capacity of often more than 2,000 people, plus about 700 crew members, the daily volume of waste is quite comparable to that of a small village. For example, the amount of waste (combustible materials) on cruise ships is quantitatively approx. 2.5 kg per capita and day and consists qualitatively of various biomass fractions, packaging, paper, glass, metals, wood, sludge from fuel, and lubricating oil processing, whereby the **flammable share** is about 1.2–1.5 kg per person and day.[6]

Thus, for example, on the AIDAdiva—a modern cruise ship of the AIDA-Cruises from Rostock—with a maximum of 2500 passengers and 607 crew members [77] thus produces about 4.2 t waste per day. Therefore, for example, the combustible waste of all AIDA ships is incinerated on board according to information from AIDA-Cruises [2, p. 19, 21].

The required throughput capacity D in kg/h of a ship waste incinerator (called incinerator) is calculated according to Eq. 8.1:

$$D = P \cdot A/24 \text{ h/d} \tag{8.1}$$

with

A = average amount of waste per person and day,
P = number of persons on board.

Example

With 3000 persons on board with an average combustible waste generation of 1.2 kg per person per day, a ship waste incineration plant would have to have the following throughput capacity:

$$D = 3000 \cdot 1.2 \text{ kg/d}/24 \text{ h/d} = 150 \text{ kg/h}.$$

◀

And it is not only a space problem; also from a hygienic and health point of view, waste storage on board with the associated formation of germs (bacteria, viruses, and moulds) is not harmless. Especially on board a ship—as a more or less closed system—the formation and spread of germs is a factor that should not be underestimated: The reproduction of germs occurs rapidly, exponentially within a certain time window. If they spread through the ship's ventilation system—especially in the case of improper waste storage—this can lead to serious health problems for the crew and fellow passengers.

---

[6]Hellwig et al. assume 1.2 kg per capita and day of combustible waste [19]; own research at the German Navy provides 1.5 kg per capita and day as average combustible waste.

Waste incineration is therefore a sensible option, especially for ships with long periods at sea. The advantages can be summarized as follows:

- Reduction of fire load and explosion hazard on board,
- Avoidance of odor nuisance,
- Reduction of the hygienic-health burden on the crew and possibly fellow passengers,
- Greater space availability for the actual target use of the ship by reducing the amount of waste stored,
- Environmentally compatible waste disposal (prerequisite: the installed system corresponds to the state of the art),
- Significant simplification of waste sorting on board by reducing the waste fractions,
- Reduction of the technical effort for waste compactors, waste shredders, and cooling of waste storage rooms, etc.,
- Simplification of the ship's internal operational procedures with regard to waste management.

Table 8.2 gives an overview of the typical composition of ship's waste that is incinerated. However, it is not possible to formulate exact numerical values for the composition of waste; ship waste, just like household waste, is characterized by a high degree of heterogeneity with regard to both its contents and moisture. In this respect, it is almost impossible to precisely characterize the ship's waste suitable for incineration. The data material presented here ultimately only has the character of orientation values. In this respect, Table 8.2 only provides a rough orientation with regard to the percentage composition of waste.

However, the waste gas from the waste incineration plant contains pollutants which must be limited. Therefore, requirements must be imposed on the incineration of ship-generated waste with regard to waste burnout and air pollutants contained in the waste gas of the incineration facilities.[7] The legal requirements of national and international regulations to be observed in this context—and possibly also their competition with each other—are outlined below.[8]

(b) International law arrangements

First *MARPOL* to name. In Annex VI to MARPOL, Regulation 16 formulates requirements for the construction and operation of waste incineration plants installed on ships since 2005. Pursuant to Regulation 16, para. 6.1, these incinerators should comply with the provisions of Annex IV to this Annex VI; exceptions are specified in Regulation 16, para. 6.2.

[7] On the subject of sustainability in particular [33, p. 113]; [55].

[8] On the overall topic, see [32, p. 72 ff.].

Table 8.2 Reference values composition of ship waste

Types of waste for incineration	In weight %
Paper, cardboard, cardboard boxes (PPK)	55–75
Plastics	17–23
Wood	5–6
Textiles, also contaminated with oil	3–5
Galley waste	2–3
Inert components, such as metals	2
Waste from the hospital area	1
	100

In addition, each installation must be approved by the competent authority—in Germany this is the Employer's Liability Insurance Association for Transport and Traffic—in accordance with a standard specification drawn up by the IMO. This standard specification is IMO Resolution MEPC. 76(40).[9]

The MEPC. 76(40) contains relatively detailed requirements for waste incineration plants on ships, some of which correspond to the contents of the Waste Incineration Plant Regulation (17. BImSchV, as of January 27, 2009) applicable to German waste incineration plants. This applies in particular to requirements for the operation of the plants (see table in Annex 29).

(c) European legal acts

First, the European "Industrial Emissions Directive" (Directive 2010/75/EU) should be mentioned here. This directive lays down requirements, among other things:

- The operating conditions of the combustion devices,
- Emission limit values into the air,
- The handling of residues from the plant,
- Control and monitoring equipment,
- Measurement conditions.

This Directive has been implemented in Germany by the "Waste Incineration Plants Regulation," 17. BImSchV, which is based on immission control law.

The "Marine Equipment Directive"[10] provides for a **"type approval"** for plants and equipment covered by this Directive. In Annex A.1, under A.1/2.7, "on-board

[9]International Maritime Organization, Resolution MEPC.76(40), "Standard specification for on-board combustion plants" of February 23, 2007 (VkBl. No. 6 of March 31, 2007, p. 174).

[10]Commission Directive 2008/67/EC of June 30, 2008 amending Council Directive 96/98/EC on marine equipment, OJ EC No. L 171, p. 16.

incinerators" are mentioned, which then meet the requirements of IMO Resolution MEPC. 76(40) to obtain a type approval.

As a **result** it should be noted that ship waste incinerators are subject to international law both under MARPOL Annex VI and the Marine Equipment Directive due to its reference to Resolution MEPC. 76(40). The same material requirements must be imposed on ship waste incinerators with regard to the construction and operation of ship waste incineration plants, and that these **must have a type examination certificate**.

(d) National regulations

At the **national level**, the immission control requirements for ship waste incineration plants are first of all the immission control law.

According to § 38 para. 1 of the –BImSchG, watercraft must be designed in such a way that their emissions caused by participation in traffic do not exceed the limits to be observed for protection against harmful environmental effects when operated as intended.

The basic obligation according to § 38 para. 1 sentence 1 BImSchG refers to the condition of the ships. The condition is *comprehensive* to understand the condition of the vehicle, namely its construction, equipment (e.g., ship waste incinerator), and maintenance condition. § 1 sentence 2 BImSchG generally regulates the immission control requirements for the operation of vehicles. Not only the driving process but also all activities accompanying it are to be attributed to the company [17, § 38, marginal no. 15 ff.].

The operation of a ship's waste incineration plant should be assigned to the accompanying operation of the ship in order to dispose of the waste produced in an orderly and harmless manner. If the incinerator fails, proper waste disposal is not ensured. Consequence: The ship could no longer be operated and could no longer participate in traffic, unless it has other possibilities for proper waste disposal (emergency disposal).

However, it is questionable whether the BImSchG can be applied at all if the incineration takes place outside the German territory, that is, extraterritorially. Operation on the open sea outside the 12-nm zone is not covered by the scope of the BImSchG, as this area is extraterritorial.[11] Consequently, application of German law would not be possible there. However, it cannot be the case that a lawless space exists on a ship which is outside the 12-nm zone. All German legal regulations of occupational health and safety at sea, ship safety law, etc., applicable to the shipping sector would then not apply. This is countered by the so-called flag principle [38, p. 10].[12] The flag principle or flag law states that a ship is granted the right to fly the national flag by registration in the ship

[11]See Articles 2 and 3 of the Convention—United Nations Convention on the Law of the Sea and Agreement on the implementation of Part XI of the Convention; Official Journal of the European Communities, L 179/3 of June 23, 1998, German translation.

[12]See also "Shipping Dictionary" in [56] and Article 3 of the Convention on the Reduction of Statelessness in [74].

register of a state. Through this registration it acquires the nationality of the state, that is, it is subject to its legal system and enjoys its diplomatic protection.[13]

In this respect, German law is also applicable outside the 12-nm zone on a ship under German flag, including § 38 BImSchG with its consequences.

Emission-limiting requirements for ships and also for waste incineration plants installed on them are, thus, based on § 38 BImSchG "Characteristics and operation of vehicles." However, § 38 BImSchG does not conclusively regulate the requirements for the condition and operation of vehicles, in this case ships. It only defines the general requirement that **applicable limit values** to be observed must not be exceeded. This general mandatory allocation needs to be made more specific. More specific immission control requirements should therefore be sought elsewhere [17, before § 38, para. 9, § 38, para. 6].

Shipping law must also be taken into account here.

(e) Immission control requirements under shipping law

In connection with the shipping law, the German Maritime Safety Act (German: Schiffssicherheitsgesetz—SchSG) must also be taken into account. In § 1 para. 1 SchSG it is stipulated that this law determines which measures are to be taken in the implementation of the respectively applicable **international regulations** on ship safety and the *environmental protection* to be carried out at sea. According to § 1 para. 2 SchSG, refers to the internationally valid ship safety standards according with the Annex to the SchSG, which are mentioned in Part A II. This explicitly includes the international MARPOL Convention. With regard to waste incineration at sea, Annex VI to MARPOL is mentioned there. The SchSG, therefore, provides for compliance with the requirements of Annex VI to MARPOL as a necessary measure with regard to the construction and operation of ship waste incineration plants.

Taking into account the above, the requirements of MARPOL Annex VI, Regulation 16, in conjunction with IMO Resolution MEPC. 76(40) for incinerators on ships are relevant. Such plants must comply with the requirements set out in the Resolution MEPC. 76(40). A type approval according to Directive 2008/67/EC in conjunction with Directive 96/98/EC is required as a necessary concession.

Furthermore, § 38 para. 2 of BImSchG states that the Federal Government may, by statutory order, lay down requirements for the condition and operation of vehicles to protect against harmful effects on the environment. In Annex 1 of the SchSV, the Federal Government has stipulated that, among other things, ship waste incineration plants comply with the Ship Equipment Directive and must be tested and operated accordingly. Thus, ship waste incinerators meet the environmental protection requirements when they are subject to the MEPC. 76(40).

[13]"Shipping Dictionary" in [56].

Finally, further stricter emission-limiting requirements with regard to § 38 BImSchG only result from the immission control law itself.

(f) Concretization of standards through immission control law

It should be taken into account that § 38 para. 1 BImSchG comprehensively regulates the condition of vehicles (i.e., also of ships). In terms of content, the obligation from § 38 para. 1 BImSchG refers to the compliance with emission limits during operation of the vehicles. It is sufficient that any substantive legal provision contains corresponding provisions [17, § 38, paragraph 15 ff.].

In this respect, the emissions resulting from the operation of a ship waste incineration plant must be limited according to an existing legal provision. Through the SchSG and the SchSV with reference to MARPOL Annex VI and IMO Resolution MEPC. 76(40), such legal rules exist. In this respect, a ship waste incineration plant could already meet the immission control requirements according to § 38 para. 1 sentence BImSchG, provided that it meets the requirements of international law already described above according to the SchSG.

In addition, however, it must be taken into account that a central concept of the BImSchG *state of the art* (see §§ 3 para. 6, 22 para. 1 BImSchG).

With regard to the material requirements for the description of the state of the art in waste incineration, the 17. BImSchV is relevant in Germany.[14] In order to concretize § 38 para. 1 sentence 2 BImSchG, according to which vehicles must be operated in such a way that avoidable emissions are prevented and unavoidable emissions are kept to a minimum, the question of which air pollutants can be limited and to what extent by technical measures, could bebest answered by "cautious borrowings" from the 17. BImSchV.

The question of the state of the art for emission reduction of ship waste incineration plants is rather derived from the BImSchG itself. This states that the *state of the art* is the state of development of advanced processes, facilities or operating methods which makes the practical suitability of a measure for limiting emissions into air, water, and soil, for ensuring plant safety, environmentally compatible waste management or otherwise for preventing or reducing effects on the environment, appear to be ensured in order to achieve a generally high level of protection for the environment as a whole (§ 3 para. 6 BImSchG). The state of the art is to be defined in each case under consideration of the individual case (different ship types, retrofitting of an existing combustion plant or new ship construction, stability criteria, influence on the ship's technical design, etc.).[15] Among other things, economic criteria must also be taken into account (the so-called principle of proportionality under administrative law).

---

[14]Note: In addition to the BImSchG, the Recycling Management Act also requires that the *state of the art* be taken into account in waste disposal.

[15]More detailed information on the state of the art [22, Sect. 3, No. 93 ff].

(g) Technical and emission-limiting requirements for ship waste incineration plants
Comprehensive waste gas purification processes enable waste gas values to be achieved in domestic waste incineration plants, which are far below the requirements of the 17. BImSchV. However, such complex systems cannot be implemented on board ships. Other possible solutions must be sought here. One approach to a solution is to consider the *specific particularities* of a ship.

The emission limit values must be determined according to the condition of the individual ship type, which must be complied with during normal operation; however, they must not be lower than the minimum requirements of MARPOL Annex VI [32, p. 76 f.].

The requirements to be formulated must also take into account that any impairment of the objects of protection according to § 1 BImSchG can be regarded as extremely low due to the comparatively low combustion performance of the plants compared to land-based plants and operation at changing positions. Furthermore, the space available on ships is limited. Moreover, the implementation of flue gas cleaning systems may affect the stability of the ship. In addition, it must also be taken into account that ships primarily have a different task from that of a "waste incineration ship." It should also be borne in mind that some ship waste incineration plants are operated for only a few hours a day; in this respect, operation is predominantly discontinuous. The emission behavior differs from that of a household waste incineration plant operating continuously on land. Also, in plant designs for ship waste incineration, the ship's own movements due to the sea state will also have to be examined and taken into account in the process engineering. In this respect, requirements are to be defined for ship waste incineration plants which take these aspects into account, but nevertheless comply with the above mentioned relevant national and international regulations as far as possible.

(h) Minimum technical requirements and emission limits
The table in Annex 29 compares the essential requirements of the above mentioned national and international regulations on waste incineration on ships. The less "stringent" emission-limiting requirements of MARPOL/MEPC are striking 76(40) as compared to the 17. BImSchV.

In compliance with the legal explanations given above, the following requirements apply to the ship waste incineration plants *practicable* to formulate requirements for the operation and emission behavior of these facilities, which form a "reasonable" compromise between maritime law and immission control law. The "requirement profile" formulated in Table 8.3 is to be understood as a **proposal** and, in particular, has taken into account the fact that the IMO already has a role in the MEPC. 76(40) in A1.7 a *effort clause* with regard to further emission limitations for incinerators with a thermal input > 1500 kW (see the table in Annex 29): According to this, it is intended that plants with a thermal input > 1500 kW operated on passenger ships and cruise liners "should" be equipped with exhaust gas cleaning systems in order to limit HCl, SO_x, and "other individual components." However, no concrete limit values are proposed. This option is justified by the fact that a high proportion of plastics and other organic substances is to

Table 8.3 Technical and emission-limiting requirements for ship waste incineration plants (guidance values)

Technical requirements for the firing systems	Emission values[a]	Waste not to be incinerated
Minimum temperature of the gases in the post-combustion chamber with a residence time of at least 2 s and $O_{2min} = 6\%$: 850 °C (1,100 °C when using waste containing halogen); auxiliary or support burner; automatic locking of the waste supply if the temperature falls below the abovementioned 850 °C	Quenches of flue gas to minimizedioxin emissions; CO: 100 mg/Nm^3; SO_x indicated as SO_2: 200 mg/Nm^3; Total dust: 30 mg/Nm^3 or Ringelmann 1; HCl: 60 mg/Nm^3; C_{tot}: 20 mg/Nm^3; (each has half-hourly average values)	Wastes listed in MARPOL Annex V containing more than traces of heavy metals; refined petroleum products containing halogens

[a]Emission standards refer to a volume content of oxygen in waste gas in the standard state (1,013 mbar, 273.15 K) of 11% (reference oxygen content), after deduction of the moisture content—Sect. 5 para. 2 of the 17. BImSchV -; in this respect, the concentrations are based on the so-called standard volume (Nm^3)

be expected on these units in particular and that the incineration plants could be operated for longer periods of time even in sensitive coastal areas.

With regard to the technical requirements, an automatic locking of the waste feed if the temperature falls below 850 °C is already a technical standard today. It is also important to fit additional burners in order to reliably prevent the exhaust gas temperature in the combustion chamber from falling below 850 °C, for example, when charging with very moist waste. Furthermore, these auxiliary burners are used to preheat the combustion chamber, as waste may only be disposed of at a combustion chamber temperature of 850 °C or higher.

The formation of dioxins by the de novo synthesis, which takes place during slow cooling of the flue gas in a temperature window between 400 and 200 °C, is avoided by quenching or rapid cooling of the flue gas.

In addition, the flue gas cooling is used for thermal conditioning of the flue gas in order to cool it down to a temperature level of about 150–200 °C. In this temperature window, the adsorption of SO_x and HCl takes place optimally.

In principle, flue gas cooling could be carried out in three ways:

- Directly by injecting fresh air into the flue gas flow,
- Directly by injecting cold water into the flue gas flow,
- Indirectly by means of a heat exchanger.

Although the injection of air or water represents the more effective variant of flue gas cooling (called Quenching), this results in an increase in the flue gas volume flow, which ultimately leads to a significantly larger dimensioning of the subsequent assemblies and apparatus. Cooling by fresh air injection is also prohibited, as this would lead to an impermissible dilution of the flue gas from an immission control point of view, unless this air volume is not taken into account when determining the emission values within the framework of exhaust gas measurement.[16]

CO limitation is necessary for fire and explosion protection reasons alone, as CO is a very reactive gas. Furthermore, the CO value is a measure of the burnout of the combustible material: the lower the CO value, the better the burnout. Compliance with the limit value is determined by optimizing the firing system itself. In this context, particular attention should be paid to ensure sufficient excess of air (excess air figure $\lambda > 1$; [34, p. 40 f.]) and a clean burnout of the waste. For example, the literature describes that carbon monoxide emissions as well as emissions of nitrogen oxides and even hydrocarbon compounds can be significantly influenced, already, in the combustion chamber [34, p. 38 ff.].

According to this, a minimum temperature of 850 °C, a residence time of the flue gas in the combustion chamber of 2 s and a minimum oxygen content in the flue gas of 6% by volume are considered to have an influence on the combustion. In this respect, a reduction of the pollutant components can be achieved by a combustion-optimized furnace design and combustion air supply.

Total dust, HCl, SO_x, and the emission of organically bound carbon (C_{tot}) should be minimized using current exhaust gas cleaning technology. Due to the process engineering effort required for this, largely reduces the other pollutants according to the 17. BImSchV.

The limitation of the abovementioned pollutants to half hourly mean values does justice to the fact that ship waste incineration plants are often not in transit but only in operation on an hourly basis (e.g., Plants on warships of the German Navy).

After the MEPC. 76(40) would also exclude the incineration of PVC waste. As a chlorine carrier, PVC is responsible, among other things, for the formation of HCl and dioxins. In my view, however, a waste incineration plant that complies with the comments made here can dispose of the quantities of PVC normally produced on a ship without the infringing the abovementioned emission values (especially for hydrochloric acid). In addition, significant dioxin formation can be avoided by rapid flue gas cooling and by the flue gas cleaning technology described below. The incineration of PVC waste should not be ruled out for the simple reason that according to general life experience, household waste and household-type commercial waste—including waste from ships—also

[16]See also no. 5.1.2 of the First General Administrative Regulation for the Federal Immission Control Act (Technical Instructions on Air Quality Control—TA Luft) of July 24, 2002 (GMBl. p. 511).

generates PVC, which otherwise has to be sorted out and temporarily stored at great expense (personnel, space, and fire load problem!). In addition, the PVC content of ship waste plays a rather minor role.

(i) Emissions from ship waste incineration plants

The type and quantity of substances contained in the waste gases from waste incineration plants depends to a large extent on the composition of the waste, the incineration system, the flue gas routing, and the incineration conditions.

In order to reduce the emissions of the incinerators to such an extent that the orientation values mentioned in Table 8.3 are reliably maintained, three process principles are suitable:

- Wet process,
- Quasi-dry process,
- Dry process.

At *wet process* the reactant (adsorbent) for the sorption of pollutants to be chemically bound is fed wet into a reactor (e.g., lime milk). The reaction products are also wet. The flue dust is separated before the reactor (adsorber).

For the *quasi-dry process* the adsorbent is also introduced wet into the adsorber, but the reaction product is dry due to subsequent evaporation. The solids separation takes place here after the chemical cleaning stage.

In the *dry process* the adsorbent is present in a dry state. Depending on whether an entrained-flow adsorption or a fixed-bed adsorption is selected, the solids separation takes place after or before the adsorber.[17]

Insofar as waste incineration plants are operated on ships without waste gas purification facilities, it is not technically possible, to achieve the limit values specified in Table 8.3, in particular for SO_x, HCl, and C_{tot} merely by optimizing the incinerators. For this purpose a **flue gas cleaning** is required.

The following raw gas loads can be assumed for ship waste incineration plants (Table 8.4).

This comparison clearly shows that pure combustion furnaces without any exhaust gas purification are not able to comply with the formulated emission limits! Even the emissions limited under MARPOL Annex VI, Regulation 16 (highlighted in bold in Table 8.4) obviously cannot be complied with without emission control technology; this applies in particular to dust emissions.

It is striking, however, that the fluctuation margin of the CO value shows both a shortfall and an excess of the limit value. This can be explained in so far as this value in particular depends strongly on the combustion conditions in the furnace and especially on the excess air number.

[17]In-depth [34, p. 7 ff. and p. 72 ff.].

Table 8.4 Raw gas load of ship waste incineration plants (based on the standard condition and on 11% O_2)

Pollutant	Literature[a] (mg/Nm³)	Limit value[b] (mg/Nm³)
C_{tot}	300–500	20
CO	**20–600**	**100**
SO_2	200–800	200
HCl	400–2,000	60
Dust	**800–15,000**	**30 or Ringelmann 1**

[a][32, p. 9]; [4, p. 22]; [24]; [42, p. 5], all with partly different dates
[b][32, p. 72 ff.]

The following example for the *conception of a flue gas cleaning system* for ship waste incineration plants, is **a** solution to safely comply with the limit values mentioned [30, p. 27]; [31, p. 43 f.].

(j) Flue gas cleaning systems for ship waste incineration plants: Introduction
With regard to the complexity of the apparatus, the flue gas components to be eliminated determine the complexity. In this respect, a look must be taken here at the substances to be separated; these are listed above in Tables 8.3 and 8.4.

Therefore, only the following process components are required: flue gas cooling by a heat exchanger/flue gas cooler; dust separation by centrifugal and/or filtering separation, a limitation of SO_2, HCl, and C_{tot} by adsorption on lime and activated carbon.

Essential boundary conditions for the planning of flue gas cleaning are also the following aspects:

- as low weight as possible (aspect of additional load input into the steel ship structure, position of the metacentric height)
- in case of retrofitting, if possible, no additional or new operating effort for the on-board personnel,
- in the case of retrofitting, no or only minimal interference with the ship's system,
- the lowest possible maintenance effort by the crew,
- no additional or new sources of danger (e.g., fire hazard, hygiene).

When planning ship waste incineration plants, it makes a significant difference whether a new ship is planned or whether an existing incineration plant on a ship is to be retrofitted with flue gas cleaning.

In particular, the aspects of space requirements and stability issues can be sufficiently taken into account when planning a new ship. Here, the common process technology with Quenschen or flue gas cooler, **entrained-flow adsorption** and filtering separation is suitable (Fig. 8.4).

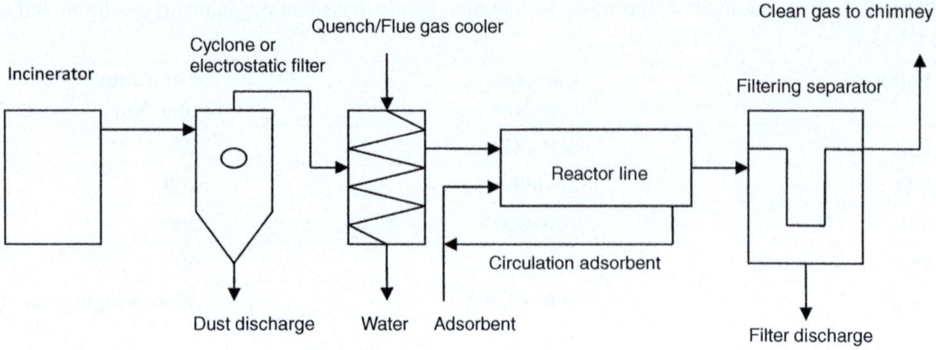

Fig. 8.4 Entrained flow adsorption

However, the equipment required for this cannot always be implemented in **retrofits**, which requires other solutions. In the case of entrained-flow adsorption in particular, storage space must be provided for the fresh adsorbent and for the filter discharge. In addition, apparatus required for conveying the adsorbent into the reactor (e.g., pneumatic conveying) is also necessary.

For retrofitting existing incineration plants, **a possible** variant the exhaust gas cleaning based on is the **fixed-bed adsorption** consisting of the following components: dust separator (cyclone), flue gas cooler (heat exchanger), fixed-bed adsorber, andN_2 addition for inerting the adsorber (Fig. 8.5).

An essential component of this variant is a **fixed-bed adsorbers** filled with an adsorbent mixture of hydrated lime and activated carbon. It is also possible to fill the adsorber vessel with the two adsorbent materials in layers, for example, first activated carbon, then lime. An upstream **cyclone** is only used for dust pre-separation to prevent the adsorber from losing its service life due to the accumulation of dust particles and to prevent the pressure loss in this apparatus from becoming too high.

Due to the higher separation efficiency as compared to the cyclone, dedusting by means of an electric or fabric filter would be preferable [6, p. 19].

However, the choice of dedusting system must be based on detailed engineering. Here, space and stability issues of the ship as well as considerations of pressure loss in the system are of primary importance. In addition, the flue gas temperature is of decisive importance in filtering separation; the upper application range of synthetic filter media (e.g., needle felts made of PTFE) is 250–280 °C. Glass fiber media in the form of fabrics or needle felts can be provided with PTFE and/or graphite coatings, resulting in maximum operating temperatures of 300–320 °C. Above 300 °C, only ceramic or metallic filter media can be used [40, p. 362].

A **flue gas cooling** is required to cool the exhaust gas down to a temperature of about 150 °C. At temperatures around 150 °C, the adsorber develops its greatest range of

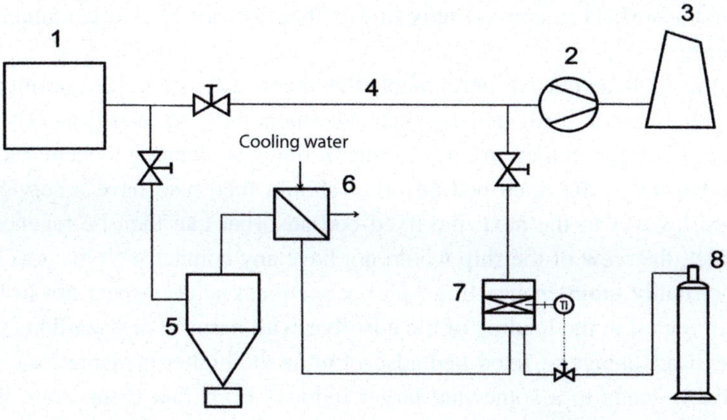

Fig. 8.5 Fixed-bed adsorption for flue gas cleaning: *1* incinerator, *2* suction blower, *3* chimney, *4* bypass line, *5* dust separation (here cyclone; but filtering is also possible), *6* flue gas cooler/quench, *7* fixed-bed adsorber, *8* nitrogen supply for inerting the adsorber, triggering temperature controlled

activity on one hand, while on the other hand de novo synthesis is avoided by the rapid flue gas cooling.[18]

Airborne or fixed-bed adsorption?

As described above, there are basically two conceivable processes: First, the adsorbent can be injected into a reaction section (*entrained-flow adsorption*). The adsorbent is then separated from the flue gas flow again via a fabric filter. Adsorption takes place in the reaction section and, if necessary, in the filter cake of the filter, before the adsorbent is "knocked off" from there.

A second possibility is the adsorption by means of *fixed-bed adsorbers*. The flue gas, which has already been cleaned of dust by a separator, flows through a fixed-adsorbent bed in which the adsorbable substances are separated.

Advantages and disadvantages of both methods

A disadvantage of the *entrained-flow adsorption* on ships, is the larger space and room requirement as compared to the *fixed-bed adsorption*. Storage capacities must be provided for fresh adsorbent and also for the discharged adsorbent. In addition, the equipment required for the adsorbent dosing device, reaction section, and subsequent filtering

[18]Dioxin recovery.

of the loaded adsorbent is considerably higher than for one or two containers for fixed-bed adsorption.

Due to the more extensive equipment, the increased storage requirements and the associated handling of fresh and residual adsorbent in entrained-flow adsorption, the overall plant is more complex to operate than a flue gas cleaning system with fixed-bed adsorption. However, the fixed-bed adsorber should then also have a service life from one shipyard layover to the next; the fixed-bed adsorber can then be regenerated there. In this respect, the crew of the ship would not have any contact with the adsorber, as this plant runs virtually maintenance-free—an occupational safety aspect not to be underestimated with regard to the loading of the adsorber with harmful or hazardous substances.

A major disadvantage of fixed-bed adsorption is the higher pressure loss in the plant, which probably leads to a somewhat larger induced draft fan; to measure the pressure loss, a differential pressure measurement at the adsorber is required.

However, whether this disadvantage is as serious as shown by the detailed engineering, especially since the higher power consumption required for such a blower on ships should not normally be the problem. It must be taken into account that the additional expenditure for equipment and longer pipes, fittings, and elbows also cause a pressure loss during entrained-flow adsorption, which should not be underestimated. In fixed-bed adsorption, the pressure loss is mainly caused by the bedding in the adsorber, which can continuously increase during operation due to the introduction of flue dust from combustion. However, upstream dust separation significantly reduces the load on the fixed bed, so that service lives of more than 1 year can be achieved [72].

The large buffer capacity against pollutant peaks and the low sensitivity to load fluctuations are further arguments in favor of the use of fixed-bed adsorbers [72], since precisely such operating conditions can be expected in a ship waste incineration plant—possibly caused by intermittent operation.

It must also be borne in mind that space is scarce on ships, is an essential point in favor of fixed-bed adsorption. In addition, this method is also easy to use. These mentioned important advantages make the aspect of possibly somewhat higher pressure losses fade into the background.

A disadvantage of fixed-bed adsorption using activated carbon that should not be underestimated is the danger of spontaneous combustion of the carbon [3, pp. 51–54].

This makes it possible to inert the adsorber—usually with N_2—and a permanent temperature monitoring of this apparatus is required.

Design criteria fixed-bed adsorber

The design criteria required for a fixed-bed adsorber are presented below, together with the associated calculation steps, as examples for an **example plant** introduced of a sample plant. This should be a retrofit for an existing incinerator in which the waste produced is incinerated for 6 h a day. The adsorber should have a service life of 200 days, that is, 1,200 operating hours. It is also assumed that about 240 kg of waste are to be incinerated during the 6 h of operation (i.e., 48 t in 200 days or 40 kg/h).[19] For the incineration of 1 kg waste per hour, it can be assumed that on average 6 Nm^3/h flue gas is produced[20] (see further below), insofar as in this example plant 240 Nm^3/h combustion gases are produced.

As a boundary condition for the design of the adsorber, the.

- Service life,
- The daily operating combustion period, and
- To take into account the quantity of waste to be incinerated.

With regard to the quantity of flue gas produced, the following consideration is based on the following: The quantity of flue gas can be determined via combustion calculations.[21] For "simple"fuels such as coke, gas or oil, this also provides sufficiently accurate results, since the individual combustible constituents can be determined with sufficient accuracy and at reasonable cost or are known. Waste, on the other hand, is an extremely heterogeneous mixture of substances, which is present with a large number of varying ingredients. Precise calculations to determine the flue gas quantity are therefore not possible. In this respect, it makes sense to use empirical values when determining the amount of flue gas produced in the course of waste incineration.

Usually, when burning 1 kg of waste with a calorific value H_u from about 10 MJ/kg[22] in relation to the standard condition[23] an O_2 content of 11% and after deduction of the moisture content in the flue gas with a specific flue gas quantity of about 6 Nm^3/kg can be calculated. With this—perhaps somewhat exaggerated—value, however, one is on the safer side within the framework of the basic design of a flue gas cleaning system, since the literature values for this are fluctuating.[24]

Furthermore, the raw gas data are required for the design of an adsorber. In addition to moisture content and raw gas temperature, the loads of the pollutants to be separated are of particular importance. It is difficult to obtain reliable values for this because the waste to be incinerated is usually a very heterogeneous mixture of materials. This leads to certain fluctuation margins in the raw gas composition during incineration; Table 8.4 lists various raw gas data from the incineration of household waste.

---

[19]Based on a quantity of waste to be incinerated of 1.2 kg per person and day, and a crew of 200 persons.

[20]Reference oxygen content 11%, exhaust gas dry, H_u about 9–11 MJ/kg.

[21]See, for example, Riediger "combustion" in [11, p. 509 ff.].

[22]The lower calorific value H_u of municipal waste is about 9–11 MJ/kg.

[23]273.15 K and 1,013 hPa.

[24]For example, Achternbosch and Richers name a value of 5 Nm^3/kg [1, p. 8].

Determination of the required amount of adsorbent

For the determination of the required adsorbent quantity, the flue gas standard condition is assumed in the following. However, the actual exhaust gas volume flow is then taken into account when dimensioning the flue gas cleaning equipment.

Furthermore, it is also important to take into account that there is a mixture of pollutants in the flue gas, which should be separated—at least partially—by the adsorbent. In this context, it is important to note that the acidic components SO_x and HCl are better bound by chemical absorption on lime rather than to activated carbon by adsorption. For hydrocarbons, however, activated carbon is suitable. However, this does not mean that portions of HCl and in particular of sulfur compounds from the flue gas are not also bound to activated carbon. It is therefore advisable to choose a mixture of lime and activated carbon as the adsorbent—be it for entrained-flow adsorption or for fixed-bed adsorption.

This method of using an adsorbent mixture is also chosen for stationary waste incineration plants on land.[25]

However, the fact that the flue gas is not to be cleaned from a single substance but a mixture of substances is to be separated, makes an exact mathematical design difficult.

In terms of process technology, two adsorber variants are possible: (a) one container with the adsorbent mixture or (b) two containers, one each for lime and activated carbon in series. Furthermore, the adsorber vessel can be filled with the adsorbents in layers. With regard to the amount of adsorbents required in each case, however, this is initially of secondary importance.

For the design of the vessel adsorber, but also of the entire flue gas cleaning system, the following options are available [8]:

1. Compute,
2. Consideration of findings from existing plants,
3. Obtaining data through laboratory testing.

To *arithmetical interpretation* of flue gas cleaning, be it the determination of the required adsorbent quantity or the dimensioning of the required apparatus, it should be noted that only rough reference values can be obtained. This follows from the problem that in the field of ship waste incineration, as a rule, nonstationary states can be assumed with regard to both the waste composition and the temperatures in the process and also frequently discontinuous incineration processes. This makes it difficult to calculate with exact numerical values.

However, flue gas cleaning in waste incineration is well established; therefore, it is common practice in the planning of new plants to *experience* of existing installations.

[25]See for example [7, p. 213 ff.]; [24, 26, 83].

The respective project to be designed is then usually based on these two data sources. In the case of previously unknown process projects, verification can then be carried out by trial and error.

Determination of the required adsorbent quantity by calculation

While the "acidic" components are bound to the lime by chemisorption, the total C deposition takes place by physisorption on activated carbon.

In chemisorption, atoms or molecules attach themselves to the surface of a solid body, forming a chemical bond between the adsorbent and the adsorbing particles.

The absorption capacity (adsorption capacity, load) resulting from the size and structure of the *internal surface* of the **activated carbon** during physisorption for **a** specific pollutant is usually determined for the equilibrium state at constant temperature T as a function of the concentration c in the raw gas by the *adsorption isotherm* $x=f(c)T$. There are different models for the quantitative description of adsorption isotherms,[26] which will not be explained further here.

In practice, the graphic representations of the adsorption isotherms thus serve to estimate the loading of the activated carbon filter. This results in the service life of the activated carbon filter.[27] Here, the adsorbed mass of pollutants is related to the mass of the unloaded activated carbon; the loading is given in percentage:

$$\text{Loading} = (\text{Mass}_{\text{Pollutant}}/\text{Mass}_{\text{Activated Carbon,Unloaded}}) \cdot 100\%. \tag{8.2}$$

For separation of **any** substance from the exhaust gas, the above approaches are still relatively easy to apply.

However, as already mentioned, several components have to be adsorbed in the flue gas cleaning of ship waste incineration, which makes the design very complex.

In multicomponent adsorption, the individual substances compete for the adsorption sites in the activated carbon. In the adsorber, gas chromatography effects occur quasi [62]. This phenomenon can be handled in the design of the adsorber with the aid of the "theory of the ideally adsorbed solution," but this requires knowledge of the individual adsorption isotherms of the substances to be separated [64].

For the practice-oriented design of an adsorber for the flue gas cleaning of a ship waste incineration plant, however, this effort is not very pragmatic. In this case, it is advisable to make an approximate dimensioning with simplifying approaches [80].

As described above, the acidifying substances such as HCl and SO_x bound to lime by chemical absorption. The activated carbon filter should therefore only contain a total C mixture and, if necessary, traces of unreacted HCl and SO_x connections.

[26] The different models are explained in [44] and others.

[27] The following calculations for adsorption are based on company GUTH's designs [65].

Design for total C separation

The flue gas is not only *a* organic substance; rather, there is a *complex mix* of organic components. Therefore, a so-called multicomponent adsorption on the activated carbon must be considered. Here, the basic relationships, as for the adsorption of a single substance, remain the same, but become quite complex due to the different binding forces of the individual substances.

The competition between the different substances in multicomponent adsorption for the adsorption sites in the activated carbon means that the load for the individual components is reduced when other substances are adsorbed simultaneously. The better the other substances are adsorbable, the lower the load.

As already mentioned, the calculation of the mutual influence in multicomponent adsorption is carried out with the help of the "theory of the ideally adsorbed solution." It requires knowledge of the individual isotherms of the substances involved and the composition of the mixture. Precise results can, therefore, only be obtained empirically. In practice, a simplistic approach is appropriate [64].

It is assumed that all total C components participate in the adsorption and that the substances lead to an average load of the activated carbon of about 20 wt.%.[28]

The pollutant load of total C can be calculated for ship waste incineration with 400 mg/Nm^3 which is in line with an average assumption regarding the quantities of plastics used by weight and the raw gas loads mentioned in the literature.

The quantity of pollutants introduced into the activated carbon is calculated according to Eq. 8.3:

$$S_m = \dot{V} \cdot c \cdot \frac{24\ \text{h/d}}{1{,}000{,}000\ \text{mg/kg}} \tag{8.3}$$

with

S_m = Pollutant quantity in kg/d,

$\dot{V}$ = Flue gas volume flow in Nm^3/h,

c = Concentration of pollutant in the flue gas in mg/Nm^3 (400 mg/Nm^3 can be assumed for ship waste incineration—see above)

Example

What is the pollutant load entered in the fixed-bed activated carbon adsorber under consideration here S_m (organic components), if a maximum load of 20% by weight can be assumed?

[28]See [64]; further assumption from the evaluation of various literature references (e.g., [69]). as mean value.

Solution: According to Eq. 8.3:

$$S_m = 240\ \text{Nm}^3/\text{h} \cdot 400\ \text{mg}/\text{Nm}^3 \cdot \frac{24\ \text{h/d}}{1{,}000{,}000\ \text{mg/kg}} = S_m = 2.304\ \text{kg/d}.$$

With 6 h of operation per day:

$$2.304\ \text{kg/d} : (24\ \text{h}/6\ \text{h}) = S_m = 0.576\ \text{kg/d}.$$

The **activated carbon consumption** A_{KV} for total C separation is determined with an assumed maximum coal load of 20%:

$$A_{KV} = \frac{S_m}{20\%} \cdot 100\% = 2.99\ \text{kg/d}. \tag{8.4}$$

Over the intended service life of 200 days, this results in a required filling quantity of approximately 580 kg activated carbon. ◀

However, it is questionable whether the approach of 400 mg/Nm^3 at C_{tot} is realistic. It is quite likely that in smaller combustion units, a much better conversion of the organic component to CO_2 and H_2O takes place than in large furnaces, such as land-based waste incineration plants. The above mentioned literature values refer to such furnaces.

For example, the literature also states that, according to the current state of the art, emission values according to the 27. BImSchV[29] for carbon monoxide and *organic substances can generally be undercut by combustion engineering measures alone* [72].

Enhanced flue gas cleaning techniques would therefore be primarily aimed at limiting PCDD/F emissions from such plants. However, these substances are destroyed by the combustion temperatures required by the 17. BImSchV and MARPOL. In order to prevent a regression of dioxins and furans in the flue gas duct due to de novo synthesis, the flue gas is quenched or rapidly cooled down by means of a flue gas cooler. In this respect, an activated carbon filter is only to be regarded as a so-called police filter to buffer the dioxin or furan slip in the exhaust system.

On the basis of these considerations, it seems justified to reduce the calculated value of the amount of activated carbon for a ship waste incineration plant.

In this respect, in the first approach, about half of the 580 kg activated carbon identified in the example above could initially be used.

Design for the HCl and SO_x separation

These acidic components can be eliminated from the flue gas by chemisorption with calcium hydroxide. A stoichiometric approach will be followed initially. For this purpose, the molar masses of the components involved in the reaction must be taken into account in the corresponding reaction equations; Table 8.5 contains the molar masses for the relevant substances.

[29]27th Ordinance to the Federal Immission Control Act—Ordinance on installations for cremation.

Table 8.5 Molar masses

Substance	Molar mass in g/mol
Ca	40.08
O	16.00
H	1.01
Cl	35.45
C	12.01
S	32.07
$Ca(OH)_2$	74.28
HCl	36.46
SO_2	64.07

The *HCl conversion* takes place according to the following reaction equation:

$$Ca(OH)_2 + 2HCl \rightarrow CaCl_2 + 2H_2O,$$
$$74.28\,g/mol + 72.92\,g/mol.$$

Assumption: The flue gas contains 400 mg/Nm^3 HCl (see Table 8.4). This assumption is based on the fact that chlorine input in ship waste incineration plants, for example, through PVC incineration, can be assumed to be rather low. Therefore, the assumption of a raw gas load at the lower limit of the literature seems justified. With a flue gas volume flow of 240 Nm^3/h · 1,200h = 288,000Nm^3 is therefore associated with an HCl accumulation of

$$288{,}000 Nm^3 \cdot 400 mg/Nm^3 = 115{,}200 g\ HCl$$

to go out.

According to the abovementioned reaction Eq. 74.28 g $Ca(OH)_2$ are required to convert 72.92 g HCl; 117 kg $Ca(OH)_2$ are then required to convert 115,200 g HCl.

The limitation of the *sulfur oxide compounds* is considered above for SO_2; in this respect the determination of the required quantity of hydrated lime is also carried out for SO_2 in simplified manner. The SO_2 *implementation* on "FGD gypsum"[30] is simplified by the following reaction equation [61]:

$$Ca(OH)_2 + SO_2 + \tfrac{1}{2}O_2 \rightarrow CaSO_4 + H_2O,$$
$$74.28\,g/mol + 64.07\,g/mol.$$

Assumption: The flue gas contains 500 mg/Nm^3 SO_2 (see Table 8.4). This assumption corresponds to the average literature reference of the raw gas load of SO_2. With a flue gas volume flow of 240 Nm^3/h · 1200 h = 288,000 Nm^3 is in this respect comparable to an SO_2 accumulation of

[30]Flue gas desulfurization plant (REA—**R**auchgas**e**ntschwefelungs**a**nlage).

$$288{,}000\ \mathrm{Nm}^3 \cdot 500\ \mathrm{mg/Nm}^3 = 144{,}000\ \mathrm{g}\ SO_2$$

to calculate.

According to abovementioned reaction Eq. 74.28 g $Ca(OH)_2$ are required for the conversion of 64.07 g SO_2; 177 kg $Ca(OH)_2$ are then required for the conversion of 144,000 g SO_2.

In total, for the incineration plant under consideration for the separation of the "acidic components" a quantity of hydrated lime of

$$117\ \mathrm{kg}\ (\text{for HCl}) + 177\ \mathrm{kg}\ (\text{for } SO_2) = 294\ \mathrm{kg}\ Ca(OH)_2,$$

of about **300 kg calcium hydroxide.**

Calculated an adsorbent quantity, consisting of approximately 300 kg lime and 300 kg activated carbon, is required for the incineration plant under consideration.

Consideration of findings from existing plants

As already described, the purification of exhaust gas by means of adsorption is state of the art for waste incineration on land as well as for the operation of crematoria. The emitted substances are comparable. Therefore, it is obvious to focus on the operating conditions to be investigated here by comparing the waste gas or pollutant flows with the quantities of adsorbents used.

Table 8.6 shows the quantities of adsorbents used for **1 t** of incinerated waste in kg for various incineration plants. For the plant there under No. 4, that is, the ship waste incineration plant under consideration here, the values calculated above—also related to 1 t of waste to be incinerated—are entered.

From this table it can be seen that there is at most an approximation for the use of activated carbon between the plant according to No. 1 and the calculations. It is striking

Table 8.6 Comparison of the amount of adsorbents from different incineration plants

Annex no	Lime quantity ($Ca(OH)_2$)	A-carbon crowd, for example, Hearth furnace coke	Total quantity	Cleaning procedure
1[a]	27.6 kg (92%)	2.4 kg (8%)	30 kg	Entrained-flow adsorption
2[b]	0.7 kg (70%)	0.3 kg (30%)	1 kg	Entrained-flow adsorption
3[c]	23.3 kg (93.2%)	1.7 kg (6.8%)	25 kg	Entrained-flow adsorption
4	6.25 kg (50%)	6.25 kg (50%)	12.5 kg	Fixed-bed adsorption

[a]Waste incineration plant Hameln (Energos plant)
[b]Crematories in Bielefeld and Minden
[c]Waste incineration plant for 5,000 t/a, Company Michaelis Environmental Technology, Schleehofstr. 12, 97,209 Veitshöchheim [82]

that the lowest adsorbent consumption is reported in the crematoria (Annex 2). This is also explainable in so far as according to the fact that 27th BImSchV only CO, dust, C_{tot} and dioxins and furans are limited. The "acidic" components, HF, Hg, etc. are not considered here, but are nevertheless considered in the 17. BImSchV. In this respect, a larger quantity of adsorbent must also be required to reduce a larger "pollutant cocktail." This is also confirmed by the amount of adsorbent obtained from the above calculation: In the ship waste incinerator under consideration here, fewer pollutants are to be eliminated than in conventional waste incineration plants, but more than in crematoria. Consequently, the amount of adsorbent required here must lie between the amount required for the two types of plant according to the 17. and 27. BImSchV—and this applies to the calculated amount of lime!

There is an uncertainty regarding the amount of activated carbon required. In addition to the considerations described below, the required quantity is strongly dependent on the type and quality of the coal.

From the above, it is clear that the theoretically required amount of activated carbon differs between the 6.25 and 0.3 kg calculated here in actual operation. Here it is obvious that the calculated value is considerably higher than that of all other investments. This may be due to the fact that the calculation approach is based on the assumption of complete total carbon separation from the flue gas—see above under "Design for total carbon capture"—and the A-carbon is only loaded to 20%. In real terms, however, the concentration in clean gas only needs to be limited at 20 mg/Nm^3. With an assumed raw gas load of 400 mg/Nm^3, this means a pollutant load of 95% is to be separated. In this respect, the required amount of A-carbon can be reduced to 95% of the calculated value, that is, 6 kg/t of ship's waste to be incinerated. A lower value for the amount of activated carbon compared with the calculation is certainly also justified by the consideration that not inconsiderable quantities of C_{tot} are already bound to the flue dust that is separated in the cyclone and to the support screen for the adsorbent in the adsorber. This justifies a further reduction of the required quantity of A coal to $\ll$6 kg/t of ship-generated waste to be incinerated.

This consideration also applies to the amount of lime required: a stoichiometric approach has been followed here. This means that the quantity added to the lime is completely chemically sorbed. However, complete separation is not required for SO_2 and HCl either. SO_2 is at 200 mg/Nm^3 and HCl at 60 mg/Nm^3 to be limited. With the corresponding raw gas values for 500 mg/Nm^3 SO_2 and 400 mg/Nm^3 HCl (see Table 8.4), this makes up 60% SO_2 quantity and 85% of the HCl load to be separated. In this respect, the amount of hydrated lime for the SO_2 separation by 40% to 106 kg and for HCl separation by 15% to approximately 100 kg, that is, in total to approximately 206 kg. Converted to 1 t of waste to be incinerated, 4.3 kg of hydrated lime would then be required for this ship waste incinerator.

The question is, however, which quantities are realistic. Evaluation of Table 8.6 shows that the amount of activated carbon used is always only a fraction of the amount of hydrated lime. This fact can also be taken into account in the case of the incinerator under consideration here: As the calculated value for the quantity of activated carbon is very high in percentage terms, in this case an averaged AK quantity from Annexes 1–3 is considered.

From the **comparative consideration** of existing plants with the calculated quantities of adsorbents, the following quantities of hydrated lime and activated carbon can therefore be considered for the combustion plant under consideration, taking into account the above considerations:

Lime hydrate:	4.5 kg/t of waste incinerated,
Activated carbon:	1.5 kg/t of waste incinerated

This corresponds to a mixing ratio of 75% lime hydrate and 25% activated carbon. This percentage approach provides an adsorbent composition that lies between the percentage composition for land-based waste incinerators and crematoria. This is plausible in so far as the emission limits to be observed for a ship waste incineration plant also lie between the requirements of the 17. BImSchV and the 27. BImSchV. Therefore, it can be stated for the example plant considered here: If 48 t waste were to be incinerated in 200 days or within 1200 operating hours, this would mean

$$48\ \text{t} \cdot 4.5\ \text{kg/t} = 216\ \text{kg}\ Ca(OH)_2$$

and

$$48\ \text{t} \cdot 1.5\ \text{kg/t} = 72\ \text{kg Activated Carbon}$$

in total as adsorbent mixture 288 kg required. The mixing ratio is here is 75% for $Ca(OH)_2$ and 25% for activated carbon.

However, in regard to the filling of a fixed-bed adsorber with hydrated lime as a reaction agent, it should be noted that this is difficult to obtain on the market in a granular structure (grain diameter about 1–4 mm)—as a rule, it is only available in powder form. This is excellent for entrained-flow adsorption, since lime with a very large surface area is present here. However, powdered lime is not suitable for fixed-bed adsorption, as the pressure loss in the adsorber becomes too high. This means that the exhaust gas purification system can no longer function properly. In this respect, it should be pointed out that adsorption with hydrated lime in the example plant considered here is only possible under the conditions described if $Ca(OH)_2$ is available and used in a granular structure!

However, exhaust gas cleaning is also possible with calcium carbonate ($CaCO_3$), also known as carbonated lime.[31] Calcium carbonate is available in the market in different grain sizes.

The reactions proceed according to the following reaction equations; the required molar masses can be taken from Table 8.7.

[31] See, for example, the product information of the Dammann company [57].

Table 8.7 Molar masses

Substance	Molar mass in g/mol
Ca	40.08
O	16
H	1.1
Cl	35.45
C	12.01
S	32.07
$CaCO_3$	100.1
HCl	36.46
SO_2	64.07

Separation of HCl:

$$\mathrm{CaCO_3 + 2\ HCl \rightarrow CaCl_2 + CO_2 + H_2O},$$
$$100.10\ \mathrm{g/mol} + 72.92\ \mathrm{g/mol}.$$

The exhaust gas contains 1,15,200 g HCl over the period considered (see above).

According to the above reaction equation, 100.1 g $CaCO_3$ are required to convert 72.92 g HCl; 158 kg $CaCO_3$ are then required to convert 1,15,200 g HCl.

At the *deposition of* SO_2 calcium carbonate is converted to FGD gypsum; this reaction proceeds in a simplified manner according to the following reaction equation:

$$\mathrm{CaCO_3 + SO_2 + \tfrac{1}{2}O_2 \rightarrow CaSO_4 + CO_2},$$
$$100.10\ \mathrm{g/mol} + 64.07\ \mathrm{g/mol}$$

According to the above reaction equation, 100.1 g $CaCO_3$ are required for the conversion of 64.07 g SO_2 in the example plant considered here; for the conversion of 1,44,000 g SO_2 in the period considered (see again above), 225 kg $CaCO_3$ are then required.

For the purely stoichiometric conversion, the HCl and SO_2 separation by *Calcium carbonate* in total about 383 kg calcium carbonate is required. Here, the same considerations apply again with regard to the reduction of the adsorbent quantity due to the fact that the 100% separation of the substances is not required: only 60% of the assumed raw gas load of SO_2 and 85% of the assumed crude gas load of HCl have to be separated. In this respect, the amount of calcium carbonate for the SO_2 separation can be reduced by 40% to 135 kg and for HCl separation by 15% to 134 kg, that is, in total to about 270 kg. Converted to *1t* of ship's waste to be incinerated, *5.6 kg calcium carbonate* is then required for this plant. As a result, an adsorbent mixture of approximately 270 kg calcium carbonate and 72 kg activated carbon (adsorbent mixture from 342 kg) has to be selected for the example plant under consideration.

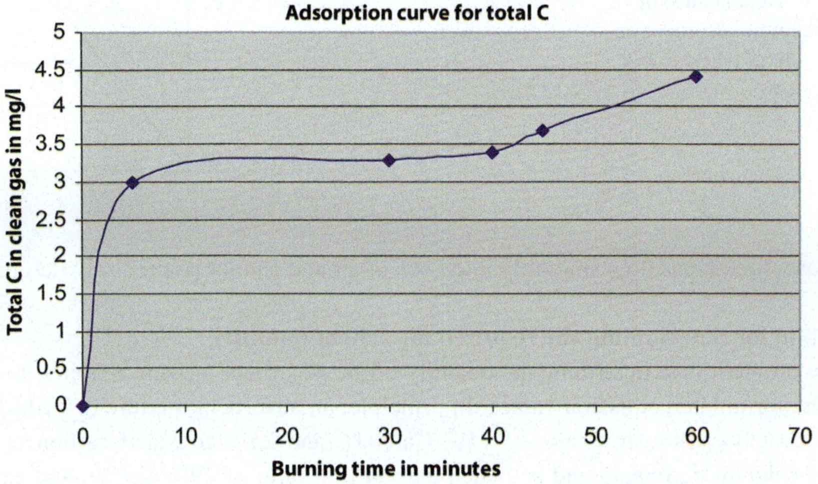

Fig. 8.6 Breakthrough curve of an activated carbon filter

Determination of the required amount of adsorbent by laboratory test

The third possibility for determining the required adsorbent quantity and thus for dimensioning the adsorber and the entire flue gas cleaning system is to carry out laboratory tests. For this purpose, the waste is incinerated on a laboratory scale in the composition in which it is normally produced on the ship (comparable to household waste and household-type commercial waste). The cleaning components of the plant to be designed (in the example considered here a cyclone and a fixed-bed adsorber, which is filled with a defined amount of adsorbent) are integrated into the exhaust gas section (see Fig. 8.5). The laboratory test is used to verify data obtained by calculation or comparison.

Figure 8.6 shows a concentration curve for C_{tot} in the clean gas obtained in the laboratory tests.

This “typical” concentration curve can be described as follows:

- At low concentrations approximately linear (controlled by Henry’s law),
- With increasing concentration the clean gas concentration seems to approach a limit value (explainable by the monomolecular occupancy of the adsorbent surface),
- The further increase of the clean gas concentration is explained by the increase of the concentration of the substance in the adsorbent bed, which approaches the saturation concentration (multilayer coating and capillary condensation), until the adsorber “breaks through.”

Fig. 8.7 Dependence of the amount of adsorbent on the amount of waste to be incinerated

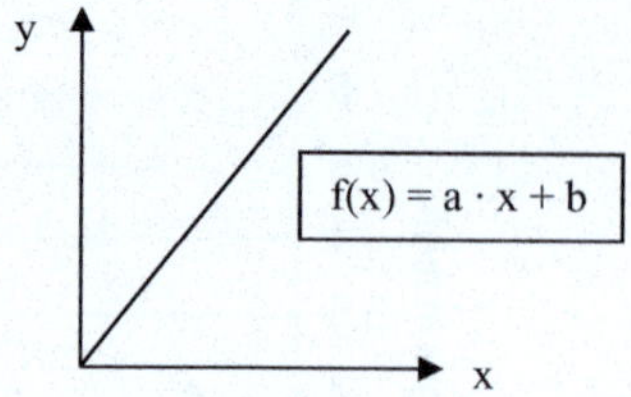

These three areas flow smoothly into each other and cannot be precisely delimited.[32]

Equation for determining the required adsorbent quantity

A core problem is to determine the quantity of the adsorbent mixture required to comply with the formulated emission values. In principle, an adsorbent mixture of hydrated lime and activated carbon in a ratio of 75% $Ca(OH)_2$ and 25% activated carbon or a mixture of calcium carbonate and activated carbon in a ratio of 79% $CaCO_3$ and 21% activated carbon should be chosen for flue gas cleaning for ship waste incineration plants. If hydrated lime is used, 6 kg adsorbent mixture per tonne of waste to be incinerated; if calcium carbonate is used, 7 kg adsorbent mixture per tonne of waste to be incinerated is required.

The required amount of adsorbent quantity M consisting of lime and activated carbon, is linearly dependent on the amount of waste A to be incinerated within a certain period, for example,, from one shipyard layover to the next; this means that the dependence of the required adsorbent quantity on the waste quantity to be incinerated can be represented with a linear function (Fig. 8.7).

Derivation of the straight-line equation for the required adsorbent quantity: In the general straight-line equation $y = a \cdot x + b$ is "y" the required amount of adsorbent M and "x" the variable, the quantity of waste to be incinerated A. The slope of the straight line a is an adsorbent-specific constant. The point of intersection of the straight line with the x-axis (zero point) $f(x_0) = 0$ is determined as follows.

With $y = 0$ follows: $0 = a \cdot x + b$.

If no waste is incinerated, the amount of waste $x = 0$; thus for "b": $b = 0$ and thus for the zero of the equation:

$$\text{Zero } N\ (0|0).$$

This reduces the general straight line equation here to "$y = a \cdot x$." In this respect the following applies: The required adsorbent quantity M is a linear function of the waste quantity A to be incinerated:

$$M = f(A) = k \cdot A \tag{8.5}$$

[32]For this purpose, Röhr and Holzapfel [63].

with "k" = adsorbent-specific constant.

The linearity of this relationship is explained as follows: Chemistry absorption is idealized purely stoichiometrically; the reaction equations are linear in nature.

This also applies to the loading of the activated carbon:

$$\text{Loading} = (\text{Mass}_{\text{pollutant}}/\text{Mass}_{\text{Activated Carbon, Unloaded}}) \cdot 100\%.$$

This shows that the required quantity of *A*-carbon quantity is linearly dependent on the mass of the pollutant to be separated. The load is therefore the proportionality factor for this linear equation:

$$\text{Mass}_{\text{Activated Carbon, Unloaded}} = \frac{\text{Mass}_{\text{Pollutant}} \cdot 100\%}{\text{Loading}},$$

where the load is to be entered in %.

In the above explanations, a ship waste incineration plant with a flue gas cleaning system is considered in which 48t of waste are to be incinerated over the intended lifetime of the fixed-bed adsorber under consideration. For this purpose, an adsorbent quantity of 288 kg is required when using *lime hydrate* (75% $Ca(OH)_2$ and 25% activated carbon) is required; accordingly, *M* determines to

$$M_{\text{Lime hydrate}} = 6\ \text{kg/t} \cdot A. \tag{8.6}$$

Here the adsorbent-specific constant is $k = 6$ kg/t waste to be incinerated.

When using *calcium carbonate* (79% $CaCO_3$ and 21% activated carbon) an adsorbent mixture from 342 kg is required for this waste quantity; accordingly *M* is determined to

$$M_{\text{Calciumcarbonate}} = 7\ \text{kg/t} \cdot A \tag{8.7}$$

with $k = 7$ kg/t waste to be incinerated.

Note: In the above equations, "*M*" should be in kilograms and "*A*" in tons.

If the factor "k" is represented as a dimensionless quantity, the dependency $M = f(A)$ must be represented as follows for the two adsorbent mixtures:

a) Activated Carbon/Lime Hydrate : $M = 0.006 \cdot A$,
b) Activated Carbon/Calicumcarbonate: $M = 0.007 \cdot A$,

Appendix 30 Contains Diagrams for Both Adsorbent Mixtures Which Show the Above Dependencies Graphically. This Enables the Planner of Such Plants to Read off the Required Mass of Adsorbent Mixture Directly. The Diagrams Differentiate According to the Respective Waste Volume to Be Considered Per Person and day (1.2–1.5 kg Per Person and day), Depending on How It is Specified as a Boundary Condition for the Individual Case Within the Framework of Project Planning.

Design information for a fixed-bed adsorber

The particle size of the adsorbents should be > 1 mm < 5 mm.

From the flue gas volume flow $\dot{V}_G$ and the cross-sectional area A of the apparatus, the flue gas velocity w_g through the apparatus is calculated by changing Eq. 8.8. The quotient n indicates the number of individual adsorber vessels, if necessary connected in parallel. With only one adsorber vessel, $n = 1$ must be set.

$$A = \frac{\dot{V}_G}{3600 \cdot w_G \cdot n}. \tag{8.8}$$

From the gas velocity and flue gas density[33] $\rho_G = 0.85$ kg/m³ (at 150 °C) results in the permissible gas load factor F for vessel adsorbers with adsorbent fixed bed results from

$$F = w_G \cdot \sqrt{\rho_G} \quad \left(\text{perm. approximately } 0.2 - 0.4 \frac{\text{m}}{\text{s}} \sqrt{\frac{\text{kg}}{\text{m}^3}}\right). \tag{8.9}$$

The flue gas velocity in the apparatus must not be too high in order to guarantee a reasonable mass transfer within the apparatus. This is taken into account by the gas load factor.

The required dumping height Z in a apparatus is calculated from the weight of the adsorbent mixture G_{Ads} and the bulk density of about $\rho_s = 600$ kg/m³ ([44]; Table 8.4) is calculated with Eq. 8.10:

$$Z = \frac{G_{\text{Ads}}}{\rho_S \cdot A}. \tag{8.10}$$

To ensure that the adsorbent bed is evenly loaded, ensure that there is sufficient diffuser chamber height above and below the bed.

The adsorbent rests on a support grid, which must be adapted to the grain size of approximately 1 mm. Here, for example, metal filters can be used, as they are also commonly used in fume hoods. The fill should be placed in the center of the apparatus. The abovementioned sieve should be placed above the fill. It is intended to prevent adsorbent material from being carried away with the flue gas flow. In order to check whether the metal filters are excessively clogged by flue dust and adsorbent abrasion, a differential pressure measurement at the adsorber must be provided. The test operation must show how high the permissible differential pressure may become. If the permissible differential pressure is exceeded, the dust must be knocked off and sucked out of the adsorber.

A drain for any condensate that may accumulate must be provided at the adsorber bottom.

In addition, an inlet for a nitrogen line must be installed at the bottom of each apparatus. The nitrogen line connects an N_2 gas cylinder with the adsorber. Nitrogen is used for inerting the adsorber in case of gluten formation in the activated carbon.

[33]Flue gas at 150 °C and 1 bar [12, p. 161].

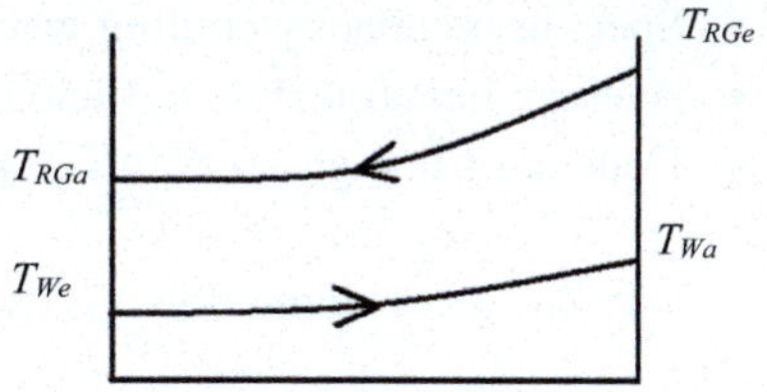

Fig. 8.8 Temperature curves in the counterflow heat exchanger

To detect any formation of gluten, a thermometer must be provided on the apparatus which measures the temperature in the bulk material. At temperatures > 400 °C, inerting must be activated, the flue gas treatment must be sealed off and the bypass line be opened (see Fig. 8.5).

Dimensioning of the heat exchanger for flue gas cooling

Since the adsorption is to take place at a temperature of about 150 °C and the flue gas leaves the incinerator at a temperature of about 850 °C, the heat exchanger must be dimensioned accordingly. However, the following calculations for the design of this apparatus are only approximate. This applies *in particular* for determining the specific heat capacity of the flue gas. An exact determination of the average specific heat capacity is impossible due to the changing and therefore not exactly determinable individual components of the flue gas as well as due to—even if only slightly—fluctuating flue gas temperatures. To want to make an "exact" calculation of the average specific heat capacity for the flue gas, e.g., according to Witte [48, p. 80], would pretend to be inadmissible accuracy. Within the framework of the design of components for such plants, as they are considered here, only—as the only serious approach—assumptions remain to be made. Furthermore, depending on the season and sailing area, the cooling water temperature will not always be exactly at the 20 °C assumed here.

In order to reasonably compensate for the many imponderables regarding the exact design of the apparatus, the plant must be equipped with a corresponding control system (e.g., proportional control). With regard to the required dimensioning of the flue gas cooling system, the *manipulated variable* "cooling water quantity" must therefore be determined for the *controlled variable* "flue gas outlet temperature."

The following boundary conditions can be assumed when designing the flue gas cooler (see also Fig. 8.8):

- Flue gas cooling from 850 to 150 °C $\rightarrow \Delta T_{FG} = 700$ °C,
- Heating of the cooling (sea) water from 20 to 25 °C $\rightarrow \Delta T_W = 5$ °C,[34]
- Spec. mean heat capacity flue gas[35] $c_{pFG} = 1.185$ kJ/kgK,

[34]Also only a rough estimate; it depends on sailing area and season.

[35]Flue gas at 500 °C and 1 bar: 1.185 kJ/kgK, density of 0.46 kg/m3 [12, p. 161].

- Spec. heat capacity cooling water[36] $c_{pW} = 4.18$ kJ/kgK,
- Average heat transfer coefficient gas/water[37] $k \approx 60$ W/m^{2}K (unpolluted),
- Density of flue gas at 850 °C[38] $\rho \approx 0.315$ kg/m^3.

The flue gas volume flow $\dot{V}_{FG}$ is determined by changing the ideal gas equation:

$$\frac{p_N \cdot \dot{V}_N}{T_N} = \frac{p_{FG} \cdot \dot{V}_{FG}}{T_{FG}}, \tag{8.11}$$

where the indices "*N*" characterize the flue gas in the standard state,[39] p_{FG} can also be assumed to be 1,013 mbar and T_{FG} is the temperature at the heat exchanger inlet (850 °C).

First of all, to determine the *cooling load* $\dot{Q}_{re}$ for the design of the heat exchanger:

$$\dot{Q}_{re} = \dot{m}_{FG} \cdot c_{pFG} \cdot \Delta T_{FG} \tag{8.12}$$

with $\dot{m}_{FG} = \dot{V}_{FG} \cdot \rho_{FG}$.
This determines the required cooling water quantity $\dot{m}$

$$\dot{m} = \frac{\dot{Q}_{re}}{c_{pW} \cdot \Delta T_W}. \tag{8.13}$$

The required heat transfer surface A is calculated from the mean logarithmic temperature difference Δt_m (here for a countercurrent heat exchanger):

$$\Delta t_m = \frac{\Delta T_{gr} - \Delta T_{kl}}{\ln \frac{\Delta T_{gr}}{\Delta T_{kl}}}. \tag{8.14}$$

By changing the equation for the heat flow over a heat transfer surface A

$$Q = k \cdot A \cdot \Delta t_m \tag{8.15}$$

is obtained for A

$$A = \frac{\dot{Q}}{k \cdot \Delta t_m}. \tag{8.16}$$

For reasons of space, the choice of an appropriate plate heat exchanger is recommended here, which is also easy to maintain.

The condensate produced during flue gas cooling must be removed from the system by design.

---

[36]At 20 °C [12, p. 154].

[37][12, p. 111].

[38][12, p. 161], averaged from the densities at 900 °C and 800 °C.

[39]p_N=1,013 mbar, T_N=273,15 K.

Cyclone dedusting

The dust load is low. The primary purpose of cyclone dedusting is to protect the heat exchanger and adsorbents from excessive dust loading. By selecting the waste gas purification process shown here, the required dust limit value is reliably met: Dust loads which pass the cyclone and the heat exchanger are retained in the adsorber apparatuses at the latest.

A cyclone dust collector is a centrifugal separator. The particle-laden gas is set into a rotating flow in the apparatus. In this process, the centrifugal force transports the dust particles to the funnel-shaped wall of the cyclone, where they are separated downward due to their greater mass than that of the gas (see Fig. 8.9).

The design of the cyclone and its geometrical configuration is based on VDI 3676 [45].

According to Fig. 13 of VDI 3676, a gas-tight ash box must be provided below the apex opening (outlet of the separated dust). Its volume is determined by the amount of dust G separated during the day of operation. On average, a raw gas load of dust with about 5 g/Nm^3 can be calculated (see Table 8.4). The assumption made here is based on the fact that modern incinerators have a clever combustion and flue gas routing system, whereby the aim is already to achieve as complete a combustion as possible with extremely low-dust content.

$$G_{\text{Ashes}} = \dot{V}_{\text{Gas}} \cdot t \cdot 5\text{g/Nm}^3 \cdot 0.9 \tag{8.17}$$

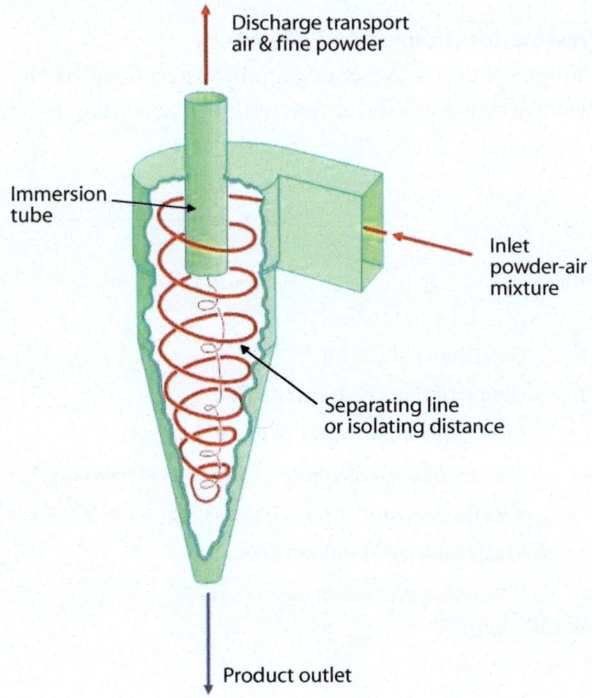

Fig. 8.9 Cyclone dust collector principle. *Picture:* Toto601

with

$\dot{V}_{Gas}$ = Flue gas volume flow at standard state (Nm^3/h),
t = Daily operating time in hours (h)

From the density of the fly ash (which is calculated with about 0.9 kg/dm^3 can be assumed), the volume of the ash box is

$$V = \frac{G_{Ashes}}{\rho}. \tag{8.18}$$

In order to not empty the ash box manually every day of operation, it is recommended to empty the cyclone via a screw conveyor or via a rotary valve. The flue dust can then be conveyed directly into a container, which must be dimensioned accordingly from one maintenance interval to the next.

Pressure loss in the system
Due to the installation of the flue gas cooler, the flue gas slide valves, the cyclone, and the adsorbers as well as in the pipes, pressure losses occur which have to be handled by the induced draft fan. In the following, the course of the pressure loss calculation is shown as an example for the above considered retrofit of an incineration plant. It shows the pressure losses to be expected with such measures **in order of magnitude**.

Pressure loss in fixed-bed adsorbers
The pressure loss Δp is essentially determined by the filling level and the geometry of the adsorbent particles; it is determined according to VDI 3674:

$$\Delta p = \frac{3 \cdot \psi \cdot \rho \cdot w^2}{2} \cdot \frac{H}{d} \cdot \frac{1-\varepsilon}{\varepsilon^3} \tag{8.19}$$

with

ε = Gap degree of the fill,
H = Height of fixed beds (m),
d = Medium grain diameter of adsorbent (m),
ρ = Density flue gas at approx. 150 °C: $\rho = 0.86 kg/m^3$,
ν = Kinetic viscosity of fumes: $\nu = 32.8 \cdot 10^{-6} m^2/s$,[a]
w = Mean velocity of inflow (m/s),
ψ = Resistance coefficient (see Eq. 8.20)
[a]At 150 °C [68].

The above smoke density should be determined according to *GESTRA*-sign [16] as follows:

$$\rho = \frac{270 \cdot p \cdot p_0}{T} \tag{8.20}$$

with

ρ = Density at operating status, here at 150 °C (kg/m³),
ρ_0 = Standard density (1.34 kg/m³; [16, p. 60]),
p = Absolute pressure, here roughly assumed with 1 bar,
T = Operating temperature ($T = 273 + 150^{\circ}$ C; K)

Resistance coefficient according to VDI 3674:

$$\psi = \frac{150}{Re} + 1.75. \tag{8.21}$$

The dimensionless Reynolds number *Re* is determined—taking into account the degree of gap ε of the fill—according to the following equation:

$$Re = \frac{1}{1-\varepsilon} \cdot \frac{w \cdot d}{\nu}. \tag{8.22}$$

Pressure loss in the cyclone
The pressure drop in the cyclone is calculated from the equation ([45]; Table 8.3):

$$\Delta p = \xi \cdot \frac{\rho}{2} \cdot w^2 \tag{8.23}$$

with the mean axial immersion tube speed $w = \frac{\dot{V}}{A_i}$.

Pressure loss in the fittings
According to *GESTRA* [16, p. 14 ff.] the pressure loss in a fitting is calculated according to the following equation:

$$\Delta p = \xi \cdot a \cdot \frac{\rho \cdot w^2}{2}. \tag{8.24}$$

The so-called "body factor" *a* amounts after *GESTRA* [16, p. 12] for valves and fittings "1"; in this respect the product of the resistance coefficient and the body factor

$\zeta \cdot a = \zeta = C$ (C is called proportionality factor),[40] whereby the individual proportionality factors of the fittings in a pipeline under consideration are added together:

$$C = \sum C_i. \tag{8.25}$$

When calculating the pressure loss in the valves and fittings, the installation location of these components must be taken into account (raw gas side and clean gas side), as different temperatures are present here and these have an influence on the flue gas density, which is an essential parameter in determining the pressure loss.

Pressure loss in the pipelines

The pressure loss in pipelines is calculated according to the following equation:

$$\Delta p_{\text{pipe}} = \lambda \cdot \frac{l}{d} \cdot \frac{\rho \cdot w^2}{2} \tag{8.26}$$

with

λ = Pipe friction number,

l/d = Specific pipe length with l = pipe length and d the inner pipe diameter

Equation 8.26 shows the temperature-dependent factors of flue gas density ρ and flow velocity w of the gas. Strictly speaking, in order to determine the exact pressure losses in the pipeline system, it would have to be divided into corresponding sections. However, since the actual pipe runs in the installation room of the plant are not yet known in the framework of a basic design, it is advisable to first of all calculate with approximations. Therefore, it seems appropriate to use a mean flue gas density over the considered temperature range of the flue gas and also a mean flue gas velocity in Eq. 8.26. In this respect, for rough considerations of flue gas cleaning systems for ship waste incinerators for $\rho = 0.59 \text{kg/m}^3$ and for $w = 10 \text{m/s}$ can be selected.

The proportionality factor C is intended for a pipeline according to Eq. 8.27:

$$C = \lambda \cdot l/d. \tag{8.27}$$

Pressure loss in the flue gas cooler or heat exchanger

To determine the pressure loss in the flue gas cooler or heat exchanger, detailed manufacturer's specifications are required. However, this apparatus will certainly make the largest contribution to the pressure loss compared to the other pipelines, fittings, and valves due to its design surface size and shape.

[40]Proportionality factors from [16, Fig. 4, p. 16] for nominal size DN 150.

Table 8.8 Disposal options for residues from flue gas cleaning

Procedure	Application
Solidification with additives	Landfilling
Solidification with fly ash (from cyclone) and water	Backfill material at mining
Blending with grate ash or fly ash	Landfilling
Without treatment	Landfilling in salt caverns, underground or hazardous waste landfills
Production from aluminum cement	Portland cement replacement

Total pressure loss

The total pressure loss in the system is the sum of the individual pressure losses, that is,

$$\Delta p_{\text{tot}} = \sum \Delta p_i. \tag{8.28}$$

Fate of the reaction products from the adsorber

During the shipyard lay days, the dust separated in the cyclone as well as the loaded adsorbent from the fixed-bed adsorber will have to be disposed of. This is hazardous waste as defined in Sect. 3 of the AVV,[41] which must be disposed of in accordance with waste disposal regulations. In the following, the possibilities for this will be shown.

The residues from the dry sorption processes (whether from the entrained flow or the fixed bed process) partly still contain free lime hydrate or free calcium carbonate, which can be used for further treatment, such as solidification. In addition to soluble salts (chlorides), all residues contain in particular heavy metals, possibly traces of dioxins and proportions of flue dust. The possibilities and practical treatment techniques for such residues are shown in Table 8.8.

Alternative waste gas purification systems: dust separation

Alternatively, instead of the cyclone, a filter separator, which is to be dimensioned and designed according to the flue gas temperature, can be used for predusting. It would have to be equipped with, for example, teflon, mineral wool or stainless steel mesh. Thomé-Kozmiensky et al. present in a table very clearly different filter materials with corresponding possible temperature impacts and resistances against flue gas contents [41, p. 62].

As a rule, however, such a deduster will again require more space than a cyclone under comparable operating conditions.[42]

[41]Waste Catalogue Ordinance.

[42]See, utility model protection by the German Patent and Trademark Office dated August 17, 2006, No.: 20 2006 009 138.6.

As a further alternative to prededusting, an electrostatic precipitator is also an option for flue gas cleaning in ship waste incineration plants. The electrical energy required for this should be available as a rule in sufficient size due to the vehicle electrical system.

In entrained-flow adsorption, the adsorbent is regularly precipitated on a filtering separator as the last purification stage (see Fig. 8.4). However, this does not render prededusting upstream of the flue gas cooler obsolete.

Which dedusting system is ultimately chosen must be clarified during project planning on the basis of the existing boundary conditions:

- Retrofitting or new construction,
- Available electrical energy,
- Stability issues,
- Room or space availability.
- What effort can be put into the measurement and control technology?

Alternative waste gas purification systems: Adsorbent composition

As an alternative solution to the project planning described above, especially in the case of retrofitting and flue gas temperatures around 300 °C after cyclone dedusting, flue gas cleaning can only be carried out with a flue gas cooler, cyclone and adsorber, which are exclusively filled with **lime**.

This variant can be implemented for the following reasons and is also justifiable from an immission control point of view:

a) When retrofitting existing installations, compromises must usually always be made. In particular, the administrative law to be taken into account is *principle of proportionality* of importance. In this respect, it must not be possible to achieve the desired objective by an excessive (financial and/or technical) effort alone. The use of lime alone can already achieve a significant improvement in emission values with a minimum of equipment.
b) If exhaust gas temperatures around 250 °C are used, it is likely that desorption processes on the activated carbon will already predominate, which would make the use of activated carbon obsolete.
c) This cleaning process makes use of the special chemical properties of mercury: This forms mercury(II) oxide at temperatures above 300 °C. This compound is retained in the adsorber. Furthermore, the organic carbon content binds very well to the flue dust. As this is largely separated by the cyclone and further particles are retained in the adsorber, it can be assumed that the emission values for C_{tot} despite reduced equipment requirements.

Furthermore, the N_2 gas storage can be omitted. This was intended for the inerting the adsorber in case of smoldering fire in the activated carbon. This is not to be expected

Table 8.9 NO_x emission limit values according to MARPOL Annex VI [67]

Animal	Date	NO_x limit value (g/kWh)		
		$n<130$	$130 \leq n < 2{,}000$	$n \geq 2{,}000$
I	January 1, 2000	17.0	$45 \cdot n^{-0.2}$	9.8
II	January 1, 2011	14.4	$44 \cdot n^{-0.23}$	7.7
III	January 1, 2016	3.4	$9 \cdot n^{-0.2}$	2.0

with "n" the operating speed of the combustion engine

with the existing maximum flue gas temperatures, the design of the adsorber and the absence of activated carbon. Any flying sparks in the flue gas pipe to the adsorber are retained by the metal support screen of the adsorber bed.[43]

8.3.2 Exhaust Emissions of the Propulsion and EDiEng Systems, Liquefied Natural Gas (LNG) Propulsion

During the operation of combustion engines, substances such as sulfur oxides (SO_x), nitrogen oxides (NO_x), carbon monoxide (CO) and carbon dioxide (CO_2), hydrocarbons, soot, and fine dusts are emitted.

The steady rise in fuel prices in recent years has led to container ships sailing more slowly and thus consuming considerably less fuel, for example, efforts are also being made to reduce propulsion power by means of propulsion-improving measures. A low specific energy consumption also means a low CO_2 emissions; the CO_2 emissions are proportional to fuel consumption by a factor of about 3 for drives with internal combustion engines.

8.3.2.1 NO_x Reduction

The NO_x emissions and sulfur oxide emissions depend on the combustion process and the sulfur content of the fuel respectively; they can be directly influenced by the type of fuel and by optimizing the combustion process.

According to MARPOL Annex VI, the NO_x emission limit values specified in Table 8.9 apply. This table also indicates the date from which the requirements must be met.[44]

The IMO emission standards are usually referred to as Tier I…III standards. The Tier I standards were established in the 1997 version of Annex VI, while the Tier II/-III standards were introduced by the amendments to Annex VI adopted in 2008:

---

[43]Utility model protection by the German Patent and Trademark Office from August 17, 2006, No.: 20 2006 009 138.6.

[44]The following from [60].

- The "1997 Protocol" to MARPOL Annex VI entered into force on May 19, 2005. It applies retroactively to new engines, with more than 130 kW, installed on ships built on or after January 1, 2000 or which have undergone major conversions after that date. In anticipation of the ratification of Annex VI, most marine engine manufacturers have since 2000 been designing engines that comply with the standards mentioned above.
- Amendments to Annex VI in October 2008:
  1. New fuel quality requirements from July 2010,
  2. Tier II and III NO_x emission standards for new engines, and
  3. Tier-I-NO_x requirements for existing engines before 2000.

The revised Annex VI entered into force on July 1, 2010.

Annex VI now makes two distinctions in terms of emission and fuel quality requirements: (1) global requirements and (2) stricter requirements for ships in Emission Control AreasECA.[45] The ECAs have correspondingly stricter requirements.

Existing emission control areas are [25, p. 17]; [81]:

- Baltic Sea,
- North Sea, including the English Channel,
- 24 nm off the Californian coast, offshore islands are part of the coastline,
- 200 nm off the North American coasts of Canada and the USA, including the "Great Lakes" and Hawaii,
- The coastal waters around Puerto Rico and the American Virgin Islands (the so-called U. S. Caribbean Emission Area).

The designation of further ECAs is planned:

- The entire Mediterranean region,
- Japanese waters,
- Norwegian waters.

are still under discussion:

- Alaska,
- Australia,
- South Korea,
- Black Sea.

[45]Emission Control Area (ECA).

The emission values for NO_x are determined as a function of the maximum operating speed of the engine n (min^{-1}) is defined. The Tier I and Tier II limits are global, while the Tier III standards apply only in the ECAs.

However, the strict Tier III emission limits are probably not achievable by optimising combustion processes alone; here, more extensive processes for NO_x reduction is necessary.

The NO_x fractions in the exhaust gas of diesel engines are approximately 95% NO and 5% NO_2. The abovementioned NO_x values in the ECA can be calculated using the *selective catalytic reaction with a DeNO$_x$ catalyst* (*SCR catalytic converter*) can be achieved.

Here, the nitrogen oxides react stoichiometrically with ammonia or urea as reducing agent to form nitrogen (N_2) and water (H_2O). These substances are natural components of the atmosphere.

In the SCR process (Fig. 8.10), urea is first injected into the exhaust gas flow of the combustion engine and mixed with the exhaust gas flow. The exhaust gas flow then passes the DeNO$_x$ reactor. This contains extruded ceramic blocks with a honeycomb structure as catalyst. These ceramic blocks are coated with the actual catalyst made of vanadium pentoxide, titanium dioxide, aluminum oxide or others. At temperatures between 280 and 600 °C the reaction takes place [9].

8.3.2.2 SO$_x$ Reduction

Despite optimized combustion processes in combustion engines, the actual emissions of **sulfur oxides (*SO$_x$*)** is still very high, especially when operating heavy oil machines. The frequently used heavy oil has a sulfur content of about 2.5%. The exhaust gases of marine engines running on heavy fuel oil, therefore, contain a high proportion of SO_2.

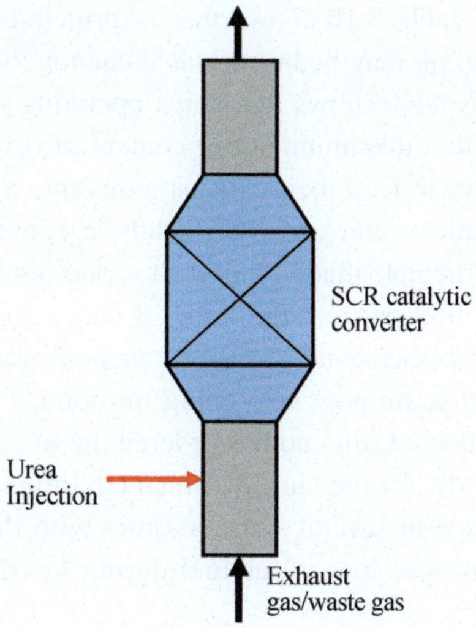

Fig. 8.10 Scheme SCR catalytic converter

Table 8.10 Sulfur content in fuel [9, p. 6]

Execution date	Type of vessel	Operating area	Sulfur content in % (mass sulfur/mass fuel)	Regulation
May 19, 2006	All	Baltic Sea EEA	1.5	MARPOL
August11, 2006	All	Baltic Sea EEA	1.5	EU
August 11, 2006	Passenger ships	All EU territories	1.5	EU
August 11, 2007	All	North Sea and English Channel ECA	1.5	EU
November 22, 2007	All	North Sea and English Channel ECA	1.5	MARPOL
January 1, 2010	All	All EU ports	0.1	MARPOL
January 1, 2010	Inland vessels	All inland waterways	0.1	EU
July 1, 2010	All	ECA	1.0	EU
January 1, 2012	16 especially Greek Ferries	Greek ports	0.1	MARPOL
January 1, 2012	All	Worldwide	3.5	MARPOL
January 1, 2015	All	ECA	0.1	MARPOL
January 1, 2020	All	Worldwide	0.5	MARPOL

In accordance with MARPOL Annex VI and EU Directive 2005/33/EC, sulfur limits are therefore set in the fuel, depending on the area of operation; Table 8.10 shows the required maximum quantities and the corresponding date of execution.

Table 8.10 shows that in principle, from 2020 onward, no more conventional heavy fuel oil may be burned, as it cannot comply with the required sulfur content. Furthermore, this table shows that ships operating in all the ECA's are obliged to use low-sulfur fuel with a maximum sulfur content of 0.1% from January 1, 2015. Alternatively, they would have to feed their exhaust gases into a desulfurization plant. A combination of diesel particulate filter and SCR catalytic converter can achieve the urgently needed reduction of particulate matter, diesel soot, and nitrogen oxides [25, p. 14].

In Asia, too, the issue of ship emissions has moved to the top of the agenda against the background of serious air pollution in large parts of China. For example, strict sulfur limits are now also being introduced in Hong Kong. The Chinese government has also followed suit and has ordered the use of low-sulfur fuel in the Pearl River Delta off Hong Kong, for the Gulf of Bohai (Beijing) and the Shanghai region until 2019. This is to take place in several steps, starting with the regulation that from January 1, 2017 ships may only use low-sulfur fuel during lay days in a total of eleven leading seaports. In early

2018, this regulation will be extended to all ports in the so-called Pearl River Delta, the Yangtze estuary, and Bohai Bay. Stage 3 will make the use of low-sulfur fuel mandatory from 2019 onward, even when approaching Chinese ports [25, p. 15 f.].

The limit value of max. 0.1% sulfur content in fuel can only be met by so-called marine diesel (similar to car diesel). However, exhaust filters and scrubbers, the so-called scrubbers, as an alternative are not yet available in the market in series—especially not for large ships. In addition, their installation costs 1.5–2 million € even for small ships. The consequence is that more and more shipping lines are switching to marine diesel [46, p. 7]. First, this contributes to environmental protection, and second, the extensive facilities on board for processing the heavy fuel oil for combustion can be dispensed with, thus creating additional shipping space.

If installations for sulfur emission control are to be considered, a distinction can be made between *wet* (Fig. 8.11) and *dry* desulfurization.

The **wet** desulfurization plants known as *scrubber* work with *lime milk* as a reactant. The exhaust gas flow of the machine is led in counterflow to the scrubbing liquid, which is injected into the scrubber in a finely distributed manner. The chemical reaction, the mass transfer, takes place at the phase boundary of the small drops. This can be improved by installations in the scrubber. In order to eliminate droplets from the further exhaust gas flow by means of droplets entrained with the exhaust gas at the outlet of the scrubber, a droplet separator is usually located at the head of the scrubber.

The flow velocity of the exhaust gas in the scrubber is between 1 and 4 m/s. The residence time of the gas in the scrubber ranges between 10 and 30 s. For the design of a scrubber without internals, a pressure drop of 100– 200 Pa can be assumed [10, p. 1106].

Special *scrubbers* use seawater or fresh water with caustic soda solution to reduce sulfur oxides in the exhaust gas (Fig. 8.12).

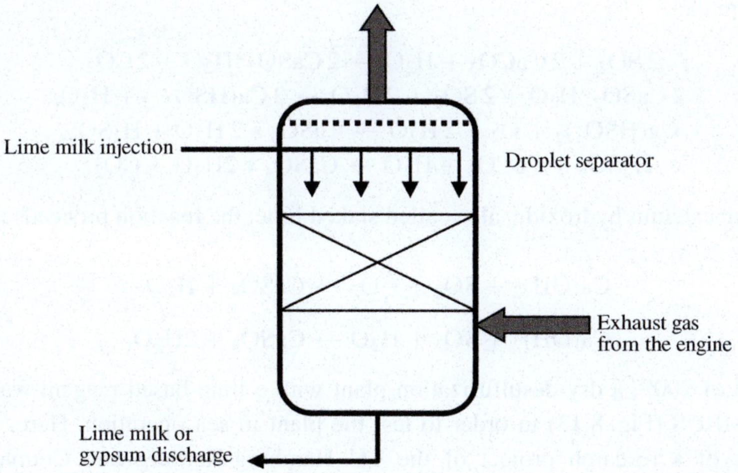

Fig. 8.11 Schematic diagram of a simple exhaust gas scrubber

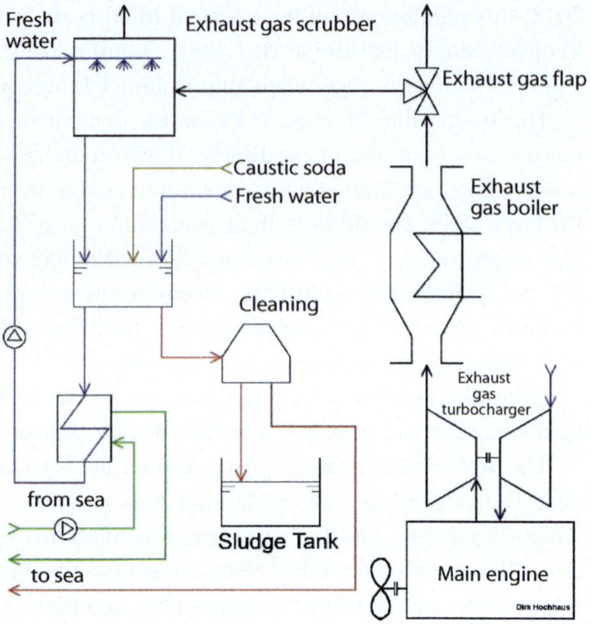

Fig. 8.12 Schematic representation of a hybrid system for sulfur purification. Exhaust gas turbocharger (ATL = Abgasturbolader). The "sludge tank" is the sludge container in which the sludge resulting from the purification of the scrubbing liquid is collected. (Graphic: K.-H. Hochhaus, CC BY 3.0)

The **dry** desulfurization plants bind the sulfur compounds by *chemiesorption* of granulated lime (calcium carbonate ($CaCO_3$) or calcium hydroxide ($Ca(OH)_2$)) with a large inner surface, whereby plaster is produced. The granulate (grain size: 2–8 mm) with a specific surface area of about 18–20 m^2/g [9, p. 12] is located as a fixed bed in a fixed-bed absorber.

When using calcium carbonate, which is finely ground and used as an aqueous suspension in a scrubber (wet process), the chemical reaction proceeds according to the following formula:

$$2\,SO_2 + 2\,CaCO_3 + H_2O \rightarrow 2\,CaSO_3{*}H_2O + 2\,CO_2,$$
$$2\,CaSO_3{*}H_2O + 2\,SO_2 + 2\,H_2O \rightarrow 2\,Ca(HSO_3)_2 + H_2O,$$
$$Ca(HSO_3)_2 + O_2 + 2\,H_2O \rightarrow CaSO_4 * 2\,H_2O + H_2SO_4,$$
$$H_2SO_4 + CaCO_3 + H_2O \rightarrow CaSO_4 * 2H_2O + CO_2,$$

When using calcium hydroxide, also called slaked lime, the reaction proceeds as follows:

$$Ca(OH)_2 + SO_2 + \frac{1}{2}O_2 \rightarrow CaSO_4 + H_2O,$$
$$Ca(OH)_2 + SO_3 + H_2O \rightarrow CaSO_4 + 2H_2O.$$

At the end of 2009, a dry desulfurization plant with a lime-based reagent was installed on the TIMBUS (Fig. 8.13) in order to test the plant in sea operation. Here, within the framework of a research project of the TU Hamburg-Harburg and Couple Systems GmbH, investigations were carried out in real ship operation for flue gas desulfurization.

Fig. 8.13 Installation of the dry desulfurization plant on the Timbus. (Photo: Olaf Knüppel, Couple Systems GmbH—Ursula Horn)

At the beginning of 2010, the manufacturer Alfa Laval Aalborg B. V. installed a desulfurisation plant based on the *wet systems* installed on board the "DFDS Tor Line" Tor Ficaria of the shipping company. This system can be operated with fresh water as well as with sea water, which is why it is also referred to as a hybrid system. The wet scrubber cleans the passing exhaust gases of the main MAN B&W 9L60MC-C engine.

8.3.2.3 LNG Propulsion

Alternatively, propulsion concepts using LNG are already being implemented today. For example, in 2013 the United Arab Shipping Co. (UASC) will have container ships (with 18,000 TEU[46]), which are designed in such a way that they can later be converted to LNG use in a comparatively uncomplicated manner.

As the first German seagoing vessel, the new Helgoland has *dual-fuel propulsion* This means one that can be supplied with both conventional diesel fuel and LNG for operation. This means that the exhaust gas regulations that will apply from 2015 onward can be complied with.

The shipping company AG "EMS" has also converted its ferry East Frisia to operate with the LNG. At the beginning of 2014, the shipping company had already signed the first supply contract in Germany for LNG as marine fuel with Bomin Linde LNG. Since the natural gas is only available at approx. -163° C has to be cooled down (at this

[46]Twenty-foot equivalent unit (20-ft container).

temperature the gas liquefies), appropriate tank facilities must be provided both on board and ashore [46, p. 13 ff.].

Following the ports of Hamburg and Brunsbüttel, Bremen's ports are also following the trend of reducing pollutant emissions in the ports by using LNG as propulsion energy for working ships. In addition, the construction of an LNG port filling station is also being accelerated [46, p. 33].

The world's first LNG hybrid barge was built at the SAM SHIPBUILDING AND MACHINERY shipyard in Komarono, Slovakia, in 2014. The customers, Becker Marine Systems and AIDA Cruises, cooperated in the development of this barge with the aim of supplying cruise ships with environment friendly electricity during their time in port in Hamburg, so that they can dispense with their on-board auxiliary power units for power generation. Becker Marine Systems has also initiated the construction of an LNG-powered Elbe ferry between Wedel and Jork under the project name *Elblinien* [46, p. 38 f.].

In addition, AIDA Cruises has placed an order with Meyer Werft in Papenburg for the construction of two innovative passenger ships for delivery in 2019 and 2020. With their 180,000 GT they will then not only be the largest cruise ships built in Germany, but also the first to be operated with LNG [47, p. 45 ff.].

8.4 Wastewater Management

8.4.1 Introduction

In recent years, the protection of the marine environment and coastal waters has come increasingly into the public eye [39, Sect. 3, para. 18b]; [18, p. 154 ff.]

The Baltic Sea is particularly polluted by ship sewage. Due to their comparatively small connections to the North Sea and thus to the Atlantic Ocean, a complete water exchange takes about 20–40 years [15].

The coastal areas in particular are at risk from ship sewage. This applies in particular to the use of these areas as bathing waters with regard to the risk of coliform germs for bathers.

Marine and coastal protection is increasingly taken into account at the international level through international agreements (e.g., through the Helsinki, MARPOL, and OSPAR Conventions), in the EU area through various directives (e.g., the Water Framework Directive and the Bathing Water Directive), and through national legal acts (in Germany through the Federal Water Act (Wasserhaushaltsgesetz—WHG) with the regulations adopted on it).

Annex IV to the MARPOL Convention **ship sewage** defined as:

- Drainage and other waste from any type of toilet, urinal, and WC scupper,
- Drainage from the medical area (pharmacy, hospital, etc.) through washbasins, washtubs and scuppers located in this area,

Table 8.11 Discharge conditions for black water

Initiate from black water		
Untreated	Mechanically treated and disinfected	From processing plants
– Stored in collection tanks – At least 12 nm from the nearest country – Discharge rate authorized by the administration – Ship follows its course – Minimum speed 4 kn	– At least 3 nm from the nearest country – Investment approved by the management	– Administrative approved wasterwater treatment plant meeting the requirements of Regulation 9.2.1 – No solids and no discoloration visible in the surrounding water

- Drainage from rooms where live animals are present, or
- Other wastewater, if mixed with the effluent defined above.

In terms of content, the wastewater defined here corresponds to the generally known *black water*. According to this definition, wastewater from galleys, laundries, galleys and showers, which is referred to as *gray water*, is then *not* ship wastewater within the meaning of Annex IV of the MARPOL Convention as long as it is not mixed with black water. In principle, MARPOL Annex IV provides *prohibition to discharge black water*, but formulates exceptions.

8.4.2 Discharge Regulations for Ship Sewage According to MARPOL Annex IV

The discharge conditions according to Annex IV to MARPOL apply to ships engaged in international voyages with a GT $\geq$ 400 or to ships with a GT of $<$ 400, which are licensed to carry more than 15 persons (see Regulation 2):

a) Discharge of wastewater (black water) by ships, other than *no passenger ships*, *at all levels* (= special areas and outside special areas) and of *passenger ships outside special areas.*

The discharge of sewage into the sea is prohibited under Regulation 11 para. 1 of Annex IV of MARPOL, unless the conditions in Table 8.11 are met.[47]

b) Discharge of wastewater (black water) from passenger ships and ferries within a special area.

[47]Federal Law Gazette (Bundesgesetzblatt; BGBl.) Volume 2012 Part II No. 32, issued in Bonn on October 23, 2012.

The discharge of wastewater into the sea is:

- New passenger ships from January 1, 2016,
- Existing passenger ships from January 1, 2018

are prohibited unless the following conditions are met:

- Accredited **wastewater treatment plant** on board which meets the requirements of Regulation 9(2.1), and
- An surrounding water no solids and no discoloration are visible.

Through Sect. 9 of the **See-UmwVerhV**, it is stipulated that.

1) The discharge of ship sewage into the sea is prohibited in accordance with the provisions of Annex IV, Regulation 11, paragraph 1, part of the sentence before paragraph 1.1 and paragraph 3, part of the sentence before the second sentence of MARPOL
   1) Outside the water areas for ships sailing from a German port to a German port as described in Sect. 3 paragraph 1 number 1
      a) For ships on sea waterways,
      b) For ships flying the federal flag, also seawards of the boundary of the sea waterways,
   2) In the Baltic Sea
      a) For ships not listed in Regulation 2(1) of Annex IV to MARPOL, including recreational craft, provided that such ships have a toilet equipped with a wastewater retention system, on sea waterways,
      b) For vessels flying the German flag referred to in point (a), also seawards of the boundary of the sea waterways.
2) The master of the ship or the person otherwise responsible for the operation of the ship may not navigate on sea waterways in the Baltic Sea with a ship, including a recreational craft, which has a toilet and is not equipped with a wastewater retention system, contrary to Sect. 6b paragraph 1 of SchSV.

Wastewater treatment plants for maritime shipping must undergo an aptitude test and be approved in Germany by the Berufsgenossenschaft für Transport und Verkehrswirtschaft (Employer's Liability Insurance Association for Transport and Traffic); formerly Seeberufsgenossenschaft (Ship Safety Division). For this purpose, the smallest type of a series is selected and tested according to the specifications of IMO Resolution MEPC.227(64) [49].

The requirements listed in Table 8.12 above are made on the effluent quality.

The **biochemical oxygen demand** is a measure of the content of biodegradable substances in wastewater. It indicates how much dissolved oxygen is required in a certain time for the biological degradation of the organic wastewater constituents. Usually the BOD_5 is used. This value is the amount of oxygen in mg/L that bacteria and all other

Table 8.12 Requirements for the discharge of ship sewage treatment plants

Parameters	Concentration
BSB_5	≤ 25 mg/L
CSB	≤ 125 mg/L
Swim- and suspended solids	≤ 35 mg/L
Chlorine content	≤ 0.5 mg/L
pH value	6.0–8.5
Nitrogen, total TN	20 mg/L[a]
Phosphorus, total (TP)	1 mg/L[a]
Fecal coliforms germs	≤ 100/100 ml

[a]In order to reduce nutrient inputs into the Baltic Sea, binding discharge limits for these substances have been set for the first time. According to these limits or a 80% reduction of TP and 70% reduction of TN. These limit values are based on the discharge values of municipal sewage treatment plants

microorganisms present in the water consume within 5 days at a temperature of 20 °C, from which the amount of organic matter decomposed in the process is deduced.

The **chemical oxygen demand** is a measure of the sum of all chemically oxidizable substances in wastewater. It indicates the amount of oxygen in mg/L that would be required for their oxidation if oxygen were the oxidant.[48]

Filterable substances are substances that are not dissolved in wastewater, that is, solids that can be removed from the wastewater by filtration [75].

With the **pH value** the hydrogen ion concentration in the water is given. It should be in the neutral range (neutral range at pH 7, e.g., distilled water).

The **free chlorine** present in the drain is limited to 0.5 mg/L. Essentially, the chlorine in the effluent is caused by chlorination of the water for disinfection. If other disinfection methods are used, such as UV radiation, this limit will not cause any problems.

The **fecal coliform germs** are bacteria normally found in the intestines of humans and mammals (*Escherichia coli* and other lactose-cleaving enterobacteria).

Regionally, stricter requirements already apply today than those in Table 8.12. For example, cruise ships operating in Alaska have to comply with requirements for ship sewage disposal that exceed the requirements of Annex IV of the MARPOL Convention. In addition to compliance with stricter discharge limits, requirements are also imposed on gray water disposal.

[48]On BOD and COD, see also [43, p. 10].

8.4.3 Wastewater on Board

Knowledge of the volume of wastewater is necessary for dimensioning the storage tanks—and also the wastewater treatment plants (see Sect. 8.4.5). This is particularly dependent on user behavior, the drainage technology installed on board, the region being traveled through and, in particular, the type of ship. The main source of black water is the cabin area. There are mainly **vacuum toilets** to the mission. The information given in the literature regarding the amount of black water varies greatly. They fluctuate between 15 and 74 L per person and day; in the literature, an average value of 31 L per person and day is assumed for cruise ships [49, p. 530 f.].

In the case of gray water, the data on daily occurrence fluctuates between 172 and 243 L per person and day. As an average value for the gray water accumulation can be assumed from 220 L per person and day. With regard to the situation on cruise ships, the gray water accumulation is as follows: 61% of the gray water comes from the cabin area, 25% from kitchens, and 14% from the laundry.

If one now takes the average values mentioned, the occurrence of black water to gray water on cruise ships is in the ratio 1: 7. The total occurrence of black and gray water is thus on average 251 L per person and day. In purely mathematical terms, the volume of wastewater to be stored or treated on a cruise ship with 2500 passengers can vary between 80 m^3/d (excluding black water) and 630 m^3/d (black plus gray water) [449 p. 531].

BG Transport provides values for the dimensioning of collection tanks and the design of wastewater treatment plants, which are shown in Tables 8.13 and 8.14 [5, Appendix 4].

8.4.4 Wastewater Storage

As an alternative to wastewater treatment, passenger ships can dispose of their wastewater in the ports. The special area status of the Baltic Sea under Annex IV of the MARPOL Convention is now fully effective, as all Baltic Sea Member States have also notified the IMO that their ports have sufficient capacity to receive wastewater from passenger ships [59].

Against this background, tanks for storing wastewater are often installed on board ships. Until the next port, the wastewater is collected in them and then disposed of by facilities located in the port (Figs. 8.14 and 8.15).

In accordance with Regulation 10 of Annex IV to MARPOL, both pipes of the reception facilities must be equipped with a standardized discharge connection piece according to Table 8.15 so that the pipes of the reception facilities can be connected to the ship's discharge pipe; ships engaged in national voyages may be equipped with quick-connect couplings instead of the abovementioned sewage shore connection (Fig. 8.15).

Direct branches to sea in the black water system should be provided with blind hole flanges. The wastewater collection tank must have a level indicator or a direction finder [5].

Table 8.13 Design of collection tanks and requirements for wastewater treatment

		<400 BRZ	<400 BRZ and > 15 persons[c]		Passenger ships ≥ 400 BRZ[c]		Passenger ships ≥ 1,000 BRZ	Seagoing vessels except passenger ships ≥ 400 BRZ	
			a	b	a	b		a	b
Approved wastewater treatment plants according to IMO resolution MEPC.159(55) or MEPC.2(VI)[a]		–	–	X	X	–	X	X	–
Total capacity of wastewater collection tanks[b–e]	Black water	–	70 L/(P·d)[d]	–	–	70 L/(P·d)[d]	–	–	70 L/(P·d)[f]
	Gray- and black water	–	180 L/(P·d)[d]		–	P/(P·d)[d]	–		180 L/(P·d)[f]
Internal. sewage land connection[g]		–	X	X	X	X	X		

Legend:
X = present
[a]From January 1, 2010, only wastewater treatment plants which are type approved according to Resolution MEPC.159(55) and from January 1, 2016 according to Resolution MEPC.227(64) may be installed on ships
[b]Discharges of untreated wastewater at a distance of more than 12 nm from the nearest country, taking into account resolution MEPC.157(55)
[c]The equipment should be designed alternatively according to columns a or b
[d]If vacuum systems are used for wastewater collection tank systems, the total capacity of the tanks should be designed in accordance with ISO 15,749
[e]Design of collecting tanks for passenger ships operating close to the coast
It should be taken as a basis:
$VS = VPT \cdot Fg \cdot x$
VS = volume of the collection tank,
VPT = 20 L per person and day for black and grey water,
Fg = number of passengers according to safety certificate,
$x = =$ number after x days ($x = 1, 2, 3 \ \ldots$)
for bathing boats and sport fishing vessels
Bathing boats and recreational fishing vessels should be equipped with a wastewater collection tank of 1 m^3 volume, unless the 1 m^3 tank cannot be provided for reasons of space or it can be demonstrated that the wastewater production is lower. According to the sailing permit, these ships may not sail for more than 10 h and there must be a confirmation from the shipowner must confirm that the wastewater is discharged regularly
[f]A holding period of 3 days should be taken as a basis
[g]For the vessels mentioned in No. 5 of the legend, a quick coupling can also be used

Table 8.14 Wastewater quantities per person and day in litres for the design of wastewater treatment plants

	Without vacuum system		With vacuum system	
	Black water	Gray and black water	Black water	Gray and black water
Passenger ship	70	230	25	185
Seagoing vessels except passenger ships	70	180	25	135

Fig. 8.14 Wastewater levy in the port …

For vessels with a side height of 5 m and less, the inside diameter of the drain connection may be 38 mm. On ships engaged in certain commercial operations, such as passenger ferries, the ship's discharge hose may instead be fitted with a discharge connection, such as quick-connect couplings (Fig. 8.15).

The storage of wastewater on board therefore requires a lot of space and, given the appropriate size, represents an additional weight for the ship that should not be neglected. Therefore, in addition to storage, wastewater treatment has now become established on new cruise ship constructions.

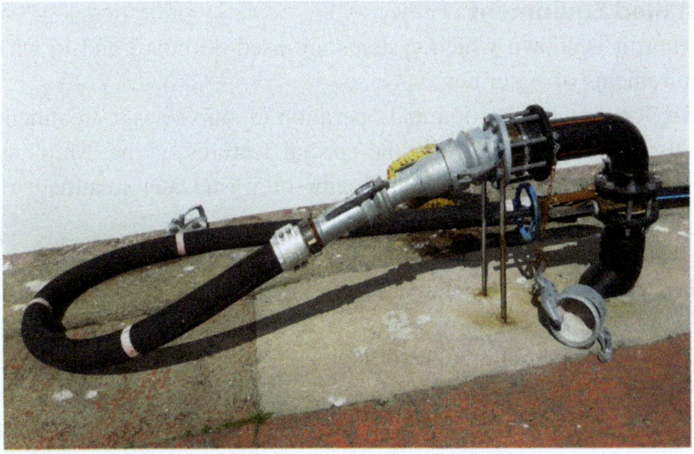

Fig. 8.15 … and transfer to underground collection tank on land

Table 8.15 Standard dimensions of flanges for discharge connections according to MARPOL Annex IV, Regulation 10

Description	Dimension
Outside diameter	210 mm
Inner diameter	According to the outside diameter of the pipe
Bolt circle diameter	170 mm
Slots at flange	4 holes Ø 18 mm in same distance from each other at a bolt circle with the mentioned diameter arranged and to edge of flanges open, with a slot width from 18 mm
Flange thickness	16 mm
Screws and nuts, quantity, and diameter	4, each with Ø 16 mm and appropriate length

The flange is designed to be suitable for pipes up to a $Ø_i$ 100 mm, it must be made of steel or other equivalent material with a smooth surface. The flange together with the gasket should be suitable for a working pressure of 600 kPa

8.4.5 Wastewater Treatment Plants

In the light of the above, it is important to consider the possibility of treating the wastewater produced on board in such a way that it cannot be stored but can be discharged into sea or port waters. Furthermore, the above explanations also made it clear that such discharges may only be made from ship sewage treatment plants that are appropriately approved or certified (responsible institution in Germany: BG Transport).

8.4.5.1 Installed Equipment

In the following, it is shown which systems are used on board and to what extent they meet the requirements of water law.

Problematic for the construction and operation of ship sewage treatment plants compared to plants operated on land are the special features of the "ship" system, which means that known processes and technologies of wastewater treatment of land plants cannot simply be transferred 1: 1.

Different designs have to be chosen for a ship, for example, for settling and aeration tanks than those usual in the municipal sector—an open tank design is out of the question. Also, due to the ship's own movements caused by the swell, a sufficient sedimentation rate of sewage sludge will not be easy to achieve. Added to this is the space problem on board, which forces compact designs. A further problem, especially for the formation of a functional biology, are possibly fluctuating inflow rates, which must be compensated for by "clever pump and aerator control." Furthermore, the plants should be easy to operate, since—especially in the area of naval vessels—frequently changing crews must be expected, who may not have the time to deal sufficiently with a difficult plant operation.[49]

In order to ensure that the wastewater treatment plant (AWA—Abwasseranlage) works safely with a healthy biology, it must be maintained regularly and carefully. Since the content of the AWA as a biological substance within the meaning of the BioStoffV[50] the specific requirements of this regulation should be taken into account. This includes, among other things, a written risk assessment, extensive instruction of the maintenance personnel, and the provision of appropriate protective clothing.

For all activities with biological substances, at least the general hygiene measures must be observed. In particular, the employer must ensure that workplaces and work equipment are kept in a clean condition appropriate to the work process and are cleaned regularly, floors and surfaces of work equipment and work surfaces should be easy to clean, to ensure that washing facilities are available, and that the changing facilities should be available away from the workplace; work clothing must be changed and cleaned regularly and as necessary.

The sensitivity of the systems and the particular hazards posed by the contents of an AWA require particularly well-informed on-board personnel for maintenance work. In this respect, these installations must only be opened and maintained by trained personnel and under supervision, and appropriate safety measures.

The *basic proceedings* of wastewater treatment commonly used today in the maritime sector:

---

[49]See also Altmann [51].

[50]Bio-substances Ordinance of July 15, 2013 (BGBl. I p. 2514, last amended by Article 146 G v. March 29, 2017 I 626).

- Conventional aerobic activated sludge plants for black water treatment (activated sludge/clarification) with disinfection,
- Conventional aerobic activated sludge plants for the black **and** gray water treatment (activation/clarification) with disinfection,
- Membrane bioreactors (micro- and ultrafiltration) for black or black and gray water treatment as well as wastewater from the medical and hospital sector,
- Reverse osmosis plants for gray water treatment (cruise ships).

In addition to membrane filtration, **disinfection (sterilization)** of the wastewater is mainly achieved by chlorination, UV light treatment (Fig. 8.16), and ozonization [23, p. 9/6].

In some cases, mechanical pre-treatment (sieve, screen) of the wastewater is carried out. However, this is essential in membrane filter technology processes in order to prevent the membrane from being unnecessarily burdened by coarse materials.

(a) Mechanical treatment with disinfection[51]

The wastewater (black water, gray water, and galley wastewater) produced on the ship is passed through a round collection tank and then through a fine screen. Here the undissolved ingredients are retained and stored in the collection tank, also called sludge or coarse material tank. To avoid turbulences in the collection tank and to use the effect of centrifugal separation, the feed into the collection tank is tangential. Via an ascending pipe with circumferential filter openings, in which a rotating brush system is located, the wastewater passes through the screen element into the drain pipe and into the homogenization tank. Here, disinfection can be carried out by means of chlorination—for example, chlorine bleaching lye—or UV irradiation (Fig. 8.17).

The screen element located in the riser pipe retains the solids that have not already settled in the coarse material tank. The rotating brush in the form of a screw permanently cleans the filter surface and guides the separated particles back into the coarse material tank. The screen hole size grading can be different, for example, in steps of 3, 1, and 0.5 mm.

(b) Aerobiological processes

After mechanical preliminary clarification and, if necessary, precipitation of dissolved substances by adding precipitants (which is the exception in the area of ship sewage treatment plants),[52] the biological clarification takes place in the so-called activated sludge stage (see Fig. 8.18). Here, with the aid of their enzyme systems, microorganisms ensure that organic impurities are converted into CO_2 and H_2O. The bacteria are particularly active at relatively low temperatures between 5 and 33 °C, where they are

[51]See also data sheet "Mechanical wastewater pretreatment," company Martin, http://www.siclaro.ch/de/produkte/downloads/Maritim/ProspektRechen2004.pdf (call date April 3, 2017).

[52]Information from the company Ocean Clean GmbH, Rostock.

Fig. 8.16 Wastewater treatment plant with UV disinfection

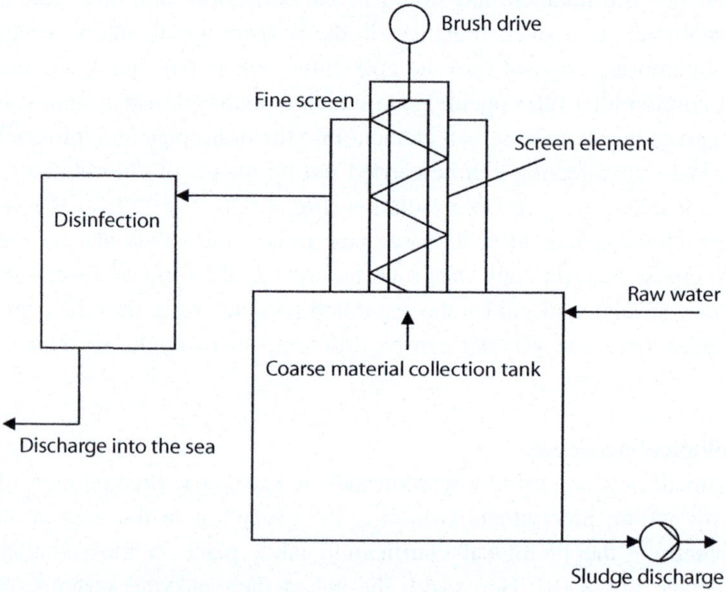

Fig. 8.17 Functional diagram "mechanical treatment with disinfection"

able to break down chemically different classes of compounds. In the activation stage, the organisms are forced to adapt to the fluctuating environment in terms of temperature, organic load, and possibly intermittent inflow rates through metabolic adaptation; that is, to adapt to fluctuating operating conditions. The bacteria cultures make demands on

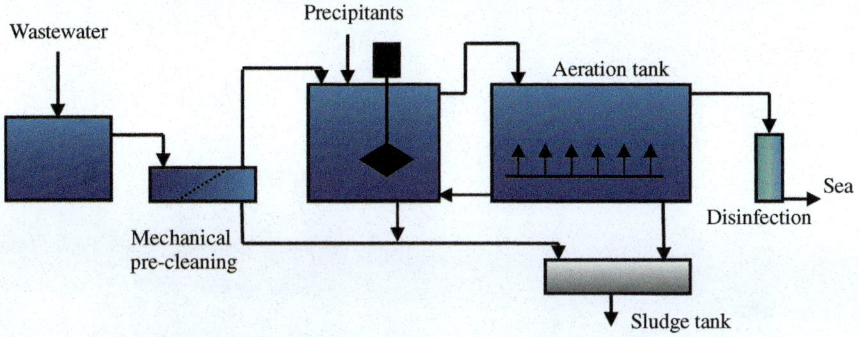

Fig. 8.18 Principle of a mechanical–biological wastewater treatment with disinfection stage

their living conditions: The optimum pH range is between pH 6 and8. There must also be sufficient nitrogen and phosphate, but this is normally supplied in sufficient quantities by the raw water [13, p. 21].

In principle, the biological purification process is divided between the *activated sludge process* and the *fixed-bed processes* differentiated.

In the *activated sludge process* the biomass is suspended in the wastewater to be treated and is supplied with oxygen via aerators. The wastewater passes through the clarifier continuously (continuous flow biology) or discontinuously. Due to the special design of the activation (aeration), microbubbles with a diameter of 30–50 μm are achieved. The advantage over the fixed-bed process is the high concentration of microorganisms with a likewise very large contact surface. The "flakes" of the activated sludge are surrounded by the wastewater, while at the same time the necessary oxygen and wastewater components are continuously mixed. Disadvantage: without activated sludge the AWA does not function. If the plant is completely emptied, for example, for repair or if the biomass is destroyed/pumped out in case of an accident, new activated sludge has to be procured to "inoculate" the plant again.

In the *fixed-bed processes* the biomass is fixed as a biofilm on growth carriers. The bacteria take the oxygen they need from the water flowing around them. In trickling filter systems, the growth medium consists of an open grid structure, the biofilm is simultaneously washed around by wastewater and atmospheric air.

Figure 8.19 shows an activated sludge plant. The plant in compact design consists of a buffer tank and an activated sludge stage. The fine screen is connected upstream as an external component of the plant (not shown in the picture).

After the biological treatment stage, coliform germs can be killed off as far as possible by chlorine or UV disinfection (whenever black water is treated in addition to gray water; Fig. 8.18).

Fig. 8.19 Aerobiologically working ship sewage treatment plant

Instead of chlorine or UV disinfection, the system shown in Fig. 8.19 operates with a downstream membrane filter stage (Fig. 8.20), which eliminates coliform bacteria from the water due to its small pore size.

(c) Biomembrane technology

The process ([28]; see Fig. 8.21) is similar to the two technologies described above. The three-step procedure is structured as follows:

The plant essentially consists of three tank systems. In the first stage coarse material is separated by a self-cleaning fine screening by means of a brush system and electrically driven self-cleaning screening cylinder. The substances separated here are collected in the system-integrated tank.

In the second stage, biological activation takes palce by atmospheric oxygen (activated sludge tank) where microorganisms in the activated sludge metabolize the organic components into CO_2 and water. For this purpose, air is blown into the activation stage by means of a blower with a capacity adapted to the size of the AWA. In addition to the enrichment of the wastewater with atmospheric oxygen, a continuous mixing in the aeration tank is thus achieved.

The discharge of excess activated sludge as well as the separated coarse materials is carried out by the circulation pump (see Fig. 8.22). This pump delivers continuously from the activated sludge into the membrane chamber. There, clean water is pumped off via the membrane (filtrate pump). Because the liquid in the membrane chamber would thicken when water is separated, the circulating pump must be designed stronger than

Fig. 8.20 Container of a membrane filter unit

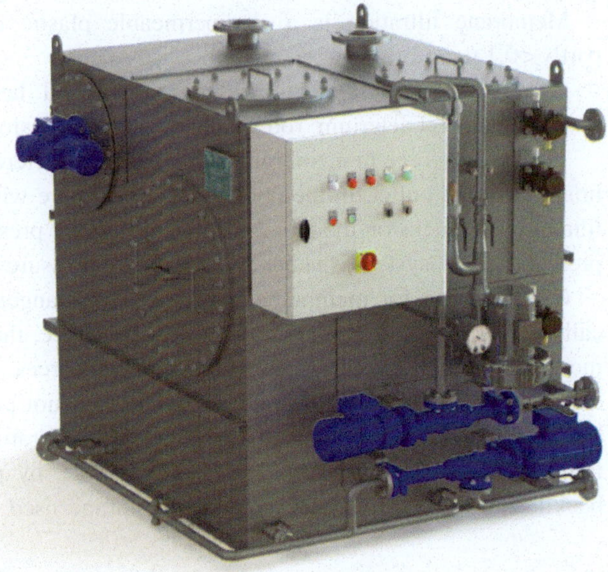

Fig. 8.21 Biological-physical wastewater treatment plant with membrane technology "UltraC-2." *Photo:* Company Ocean Clean, Rostock

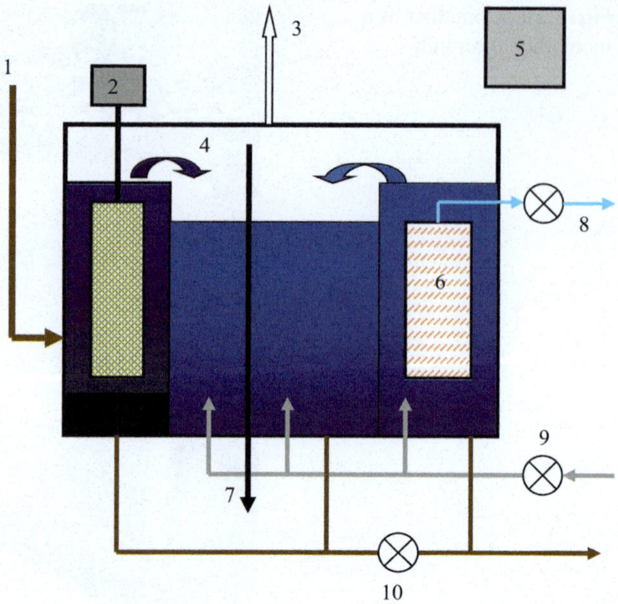

Fig. 8.22 Principle of the MBR process: *1* wastewater inlet, *2* drive fine screen, *3* venting, *4* overflow from one step to another, *5* control and operating panel, *6* membrane filter, *7* emergency overflow, *8* filtrate pump for purified water, *9* ventilation fan, *10* circulation and sludge pump

the filtrate pump. As a result, the mixture of wastewater and activated sludge flows constantly from the membrane chamber back to the activation (overflow).

In the third stage the membrane filtration[53] takes place in the membrane tank to physically separate the purified water from the activated sludge and bacteria.

Membrane filtration is a semipermeable plastic membrane with a pore size of partly < 0.1 µm [29].

Even bacteria and some viruses are retained here. In comparison: Size of an Escherichia coli bacterium: rod shape with the dimensions 1.1–1.5 × 2.0–6.0 µm [71].

The driving force in membrane filtration is the differential pressure between the membrane inlet and outlet. In the process described here with a so-called flat membrane, the filtrate pump sucks on the membrane. Atmospheric pressure prevails inside the AWA (no pressure vessel, system is vented). The suction pressure is max. 0.25 bar.

Characteristic for membrane filtration is the tangential overflow of the membrane, called cross-flow filtration. With this flow guidance, the build-up of a filter cake on the membrane is minimized by tangentially acting forces, since such a flow pattern—as in classical filtration—is not desired. However, it is not possible to avoid the formation of layers completely. Since most micro- and ultrafiltration membranes are self-supporting, the membrane can be cleaned, for example, by periodic backwashing. However, backflushing depends on the type of membrane used and is not recommended by all

[53]The processes differ in the pore size of their membranes: 0.5–0.1 µm is microfiltration, 0.1–0.01 µm is ultrafiltration [76]; 10–1 nm is nanofiltration.

manufacturers. In the plant described here, for example, backwashing is not carried out, but chemical in situ cleaning is carried out approximately once a year. This is possible because the membrane chamber is decoupled from the activation, so that chlorine bleach can be used in this tank to clean the biofilm on the membrane surface.

The micro- and ultrafiltration membranes are produced in the form of flat membranes, tubular membranes or hollow fibres and are installed in pressure pipes. A ready-to-connect unit consisting of pressure tube and membrane is called a module. Tubular modules with a tube diameter of 3–24 mm are generally used for the purification of wastewater [76].

The combination of **M**embrantechnology and **B**ioreaktor is generally referred to as the "MBR process."

In the following, the compact "UltraC-2" plant shown in Fig. 8.21 is described in more detail as an example of this wastewater treatment technology. It is designed and also approved by BG-Transport for a hydraulic load of 2.7 m^3/d and a biological load with a BOD_5 value of 1.9 kg/d. In addition to the process engineering, it contains the necessary electrical system (power supply 3×400 V/50 Hz or 3×460 V/60 Hz), in which, among other things, the supply voltage is transformed down to 24 V AC for the control voltage. The system is designed with protection class IP66. The supply and discharge pumps are controlled by means of level switches. An alarm is triggered when defined levels in the individual chambers are exceeded.

Tanks and pipelines are made of stainless steel. The following pipe connections are required:

• Entrance	DN 100,
• Ventilation	DN 100,
• Overflow	DN 25,
• Procedure	DN 25,
• Backwash	DN 25,
• Emptying	DN 25

A water pressure of 2–5 bar is required for the fresh water supply for backwashing.

Biological and biological-physical wastewater treatment plants are sensitive to wastewater quality. Intensive use of strong detergents or the addition of chemicals (e.g., Chlorine) destroys the microorganisms in the activated sludge. Insufficient "feeding" with organic matter can greatly reduce the concentration of microorganisms, which has a direct influence on the effectiveness of such plants [28].

Figure 8.22 shows the basic structure of an MBR plant.

Possible malfunctions, their causes and ways of eliminating them at an MBR plant are described below [29]:

- Fine screen system blocked: Wastewater entering the plant cannot pass through the fine screen and is retained in the wastewater inlet. The coarse material tank must be emptied and the fine sieve cleaned. Replace the brush if necessary.
- The membrane is clogged: purified water cannot be pumped out of the system. The vacuum is increased and the vacuum switch stops the filtrate pump. Otherwise the membrane filter would be damaged. (Never backwash the membrane filter!)
  This can be caused by poor biology: Either too little or too much activated sludge. Check the quality of the activated sludge and, if possible, take measures to improve the activated sludge quality.
  Carry out a chemical cleaning of the membrane filter.
- Aeration does not work: Aeration is crucial for bacterial life in the MBR plant. If no air or not enough air enters the activated sludge stage, the biology suffers. This leads to a deterioration of the water quality and, thus possibly, also to a blockage of the membrane filters. If the problem is caused by a defective blower, repair or replace it.
  Unusually high- or low-air flows can be caused by defective or dirty ventilation pipes. This can be determined by visually checking the appearance of the bubbles on the surface of the liquid in the aeration stage. If there are only a few bubbles, the ventilation pipes may be blocked. If very large bubbles appear, this may indicate damaged ventilation pipes. For a detailed inspection of the aeration pipes, the activation tank must be emptied and the aeration pipes must be removed by dismantling the flanges. Replace with new pipes if necessary.
- The circulating pump does not work: As long as the circulating pump is not working, the membrane tank is not supplied with "fresh" activated sludge, the liquid in the membrane tank thickens. Result: Either the membrane is clogged or it dries out—this damages both the membrane filter and the filtrate pump.
- The filtrate pump is not working: the level of the water in the sewage plant is controlled by the filtrate pump. When the liquid level reaches a high level, the filtrate pump starts to pump the cleaned water out of the system (via the membrane filter). If the filtrate pump cannot empty the apparatus, the liquid level rises up to the "high" level alarm, which causes the wastewater inlet to be closed. Therefore, the system must be run in bypass if the filtrate pump is not working.
- The activated sludge quantity is too high: danger of clogging the membrane! Ensure proper plant function. Find the right balance between activated sludge content and water content. In order to reduce the sludge quantity, a certain sludge volume is pumped off via the sludge pump. Attention: Do not empty the entire system! It is recommended to empty the filtrate container and refill it with clean water. In this way, sufficient sludge is removed and the remaining sludge inside the plant is sufficiently diluted.
  Another possibility: Refill the membrane tank by simply restarting the circulation. In this way the membrane tank is filled with activated sludge from the activation tank. The remaining liquid is sufficiently diluted by adding "fresh" wastewater.

- The filtrate is neither clear nor clean: damage to the membrane filter; replace. Before replacing the membrane, check each pipe connection/hose connection of the filtrate line inside the unit. A faulty connection can also lead to unclean filtrate.
- High-foam formation in the plant: Foam is a reaction of the plant biology to the supply of heavily polluted water (e.g., cleaning chemicals) or high-peak loads. The aeration supports the foam buildup, so switch off briefly. Caution: Without aeration over a longer period of time, the biology loses its ability to catabolize the wastewater.
- Numerous high-level alarms occur: High-level alarms can be caused by:
  - Peak load of the wastewater inlets,
  - Too high wastewater inflow,
  - Too little discharge via the filtrate pump,
  - Heavy ship movements,
  - Dirty sensors,
  - Foaming.

Alarms caused by peak loads or excessive wastewater inflow should not occur frequently; if they do, the cause is usually not in the plant itself. Rather, the reason is often too little filtrate discharge: a defect in the filtrate pump can be the cause. Therefore: Check and replace if necessary.

Similarly, strong ship movement due to sea conditions can cause the liquid in the system to move vigorously. It can happen that the alarm level sensor detects too high a level in response to the moving water in the system, even though the system is working perfectly.

Dirty sensors can also send false signals. In this case, remove, clean, and reinstall the sensor.

Excessive foaming can cause the sensors to detect liquid where there is none.

8.4.6 Assessment of Existing Technologies with Regard to Their Achievable Discharge Values

The discharge values required today according to the limit value table of IMO Resolution MEPC. 227(64) (Table 8.12) cannot be achieved by mechanical pretreatment with disinfection alone. The COD and BOD_5 values cannot be maintained due to insufficient biological degradation processes caused by a nonexistent biological treatment stage, nor can the filterable substances, as these are not retained sufficiently due to the size of the screen hole. Likewise, the value for free chlorine cannot be maintained if disinfection is carried out by chlorination. The only way to reliably eliminate the coliform germs is with sufficient disinfection. In this respect, such wastewater treatment is only suitable for discharging the wastewater outside the 3 nm zone outside special areas.[54]

With a combination of mechanical pretreatment and biological treatment, it should be possible to achieve compliance with COD and pH values according to the current limit values. This can be explained by the fact that a certain environment is required for the bacterial culture. If this is present (e.g., including a pH value between 6 and 8, which also corresponds to the currently valid pH limit value), the bacterial culture will also be "healthy." In combination with a sufficient oxygen input for chemical oxidation, a sufficient conversion of the organic loads will then take place in the cleaning stages. Through disinfection, the wastewater also satisfies theabovementioned values with regard to germ reduction. Nevertheless, the value for free chlorine cannot be maintained here either, as long as chlorination is carried out again; likewise, the filterable substances cannot be sufficiently limited. However, free chlorine could be obtained by adding sodium thiosulfate (1 g/m^3 Cl_2 consumed 0.8 g/m^3 sodium thiosulfate) to bind the free chlorine to the sodium atom to form NaCl; according to information from BG Transport, this technology is also used in part.[55]

Against the background of the current discharge values to be complied with within the 3-nm zone in German waters, the MBR process technology with upstream mechanical treatment, described in more detail above, has now become established on new cruise ship constructions. Due to the very small pore size of the micro and nanofiltration, there is no need for disinfection, as the limited number of coliform bacteria can be reliably maintained due to the size of the rods. The limit value for filterable substances is also achievable due to the pore size of the membrane technology.

In this respect, ships equipped with such systems and as far as they are approved by the BG Transport, can discharge their wastewater treated in this way into German waters within the 3 nm zone.

8.5 Bilge Water Treatment

A **bilge** is the lowest space on a ship. The bilge collects water from accidents and leaks that have penetrated into the hull of the ship, as well as condensation water with various impurities (especially hydrocarbons). This water is called *bilge water*. In motor-driven ships, the propulsion engine is usually equipped with its own bilge, the motor bilge.

The amount of bilge water produced depends on the type and size, but also on the condition and age of the ships. The IMO **calculates on average** with about 2–3 cbm per ship and day [20, p. 730].

[54]See MARPOL Annex IV Regulation 9 (para. 1.2).

[55]BG Transport, Marine Environmental Protection, Ship Safety Division, May 23, 2011.

Bilge water may only be pumped out at sea if it is pumped through an approved *bilge water separator*.

The legal basis for the discharge of bilge water at sea and for the establishment of suitable bilge oil separators is IMO Resolution MEPC.107(49). It follows that within the 12-nm zone, bilge water with an oil content of maximum 15 ppm (= 15 g of oil per 1000 L of bilge water) may be discharged. In special areas the limit value is 5 ppm (see also MARPOL Annex I).

A problem in the treatment of bilge water is its constantly varying and difficult to predict composition. In addition to pure mineral oils and synthetic oils, there are also impurities due to solid dirt particles that make cleaning more difficult. The following substances may be present in the bilge water in varying proportions as sweat, leakage, or wash water [14]:

- Fuels,
- Lube oils,
- Hydraulic oils,
- Water additives (mainly for cooling and boiler water),
- Solvent,
- Refrigerant,
- Cold cleaner,
- Various other dirt particles (e.g., rust, sand, paint residues, metal abrasion).

There are two possible methods for the disposal of bilgewater: In inland navigation, bilgewater is collected in a tank on board until it is transferred from a *bilge oil disposal boat* is bled (pumped off). The bilgewater is then separated into water and waste oil directly on the disposal boat. After filtering, the water is returned to the watercourse, while the waste oil is collected in tanks and later delivered to land for further disposal.

The second method of bilge water disposal is treatment on board in **bilge water treatment plants**. This process is also used on board bilge oil disposal vessels, but is unavoidable in maritime shipping. Water with a residual oil content of max. 15 ppm (in accordance with IMO Resolution MEPC.107(49)), or 5 ppm in some special areas, can be discharged into the sea, while the oil sludge is either incinerated with the heavy fuel in the propulsion engine or is incinerated in an oil sludge incineration plant on board or has to be disposed of on land.

8.5.1 Bilge Water Treatment Plants

The simplest way of separating oil–water mixtures is the phase separation based on the different densities—the *gravity separation*: The oil, fuel, cold cleaner, and similar droplets contained in the emulsion with a lower density than water are separated according to their buoyancy. The buoyancy corresponds to the weight of the displaced liquid. So the bigger it is, the stronger the buoyancy. Under the idealized assumption of spherical oil droplets and negligible friction, Stok's law gives a quadratic dependence of the buoyancy velocity on the droplet diameter ($v_A \sim d^2$). This means that a doubling of the droplet size results in a fourfold higher buoyancy speed. It also follows that if the droplets to be separated are reduced by half, the residence time in the settling tank must be quadrupled or the flow rate reduced to a quarter.

In the simplest case of gravity separation, the mixture is pumped into a tank, where the relatively lighter oil and other substances collect on the surface and can be skimmed off; the water, which is largely oil-free, can be drawn off in the lower part of the tank. The limits of this process are based on the density difference required for separation; this must be at least 15%. In case of smaller differences in density or even very small oil droplets, there is no clear separation into an oil and water phase. With small drops to be separated, the residence time is relatively long due to the low-buoyancy speed.

This is where the coalescence process gains importance. This is also based on the fact that the relatively lighter phase of liquid hydrocarbons rises to the surface. In the coalescence separator it is achieved that such small droplets, which are below the critical size for separation in the gravity field, flow together to form larger droplets and can thus be separated. For this purpose, the oil–water mixture is passed through variously designed internals in the coalescence separator, depending on the manufacturer. Due to the modified flow guidance, increased contact between the smallest droplets and thus a confluence is made possible. At the same time oil is deposited on the internals, for example, Shaft or angle profiles. Through openings in the upper part of the profiles, separated hydrocarbons can rise to the next row of profiles and "catch" further drops on their way. Due to the small distance between the individual profiles—and the correspondingly low rising height—oil droplets are separated more quickly than in a large vessel without installations. Alternatively, so-called fixed beds can be used for fine separation. These can consist of, for example, plastic foams, nonwoven, or granulate.

The lighter hydrocarbon phase collects in the dome of the separator and can be extracted [14].

The separation of oil and solid particles from the bilge water is all the more demanding the more these particles form a stable emulsion or suspension with the water. The structure and stability of these emulsions and suspensions, therefore, plays an important role in bilge water treatment.

Emulsion is defined as a mixture of two immiscible liquids. An oil-in-water emulsion contains small oil drops distributed in a coherent water phase. A suspension is a mixture containing solid particles in the continuous water phase.

Emulsions are created when small drops of one liquid (secondary phase) are trapped in the other liquid (main phase) under pressure, for example during pumping operations or when valves are throttled. Small drops of oil usually melt into larger drops in water, as oil is not water soluble.

Surfactants contained in the bilge water, such as detergents, soaps or other surface-active compounds, still contribute to the stabilization of the small drops. This makes separation difficult.

In order to safely maintain the currently required oil content of max. 15 ppm in the bilge water, bilge water deoiling is therefore usually carried out as a combination of the two principles "separation by gravity" and "unification of smallest droplets to larger ones" (coalescence). The minimum throughput of bilge-water separators depends on the size of the ship ([5, Annex 1]; see Table 8.16).

Experiment for phase separation by gravity separation
Some colored lamp oil is put into a beaker for better visualization, then water is added. The glass is shaken well to obtain an oil–water emulsion. After a short period of standing, the system separates: the oil phase floats on the water, separated by a relatively exact phase boundary layer (see Fig. 8.23).

In the following, bilge water de-oiling is described using the oil–water separator "OCEAN CLEAN EB" as an example (Fig. 8.24).[56]

This is a three-stage oil–water emulsion separation system with a throughput capacity of between 0.25 and 5.0 m^3/h, consisting of separation by.

1. Gravity separation,
2. Coalescence separation, and
3. A filter system.

The system can be operated either in manual or automatic mode.

The first stage is the *gravity separation stage*. Up to 100% oil can be separated from the water in the first stage without interruption for backwashing.

The oil–water separation is facilitated by a special inlet into the apparatus, which creates a hydrocyclone effect.

The second stage is the *coalescence stage*. Remaining fine oil droplets are separated from the water here. The coalescence unit consists of an oil and acid resistant polyether foam, which is cleaned by an automatic backwash. Depending on the operating conditions, the service life of these units can be between 2 and 5 years. The separated oil is detected by a measurement and automatically discharged through drain valves.

The first and second stages of this apparatus work without any moving components.

The third stage is the *emulsion breaking stage* (emulsion splitting).

The oil–water emulsion must pass through special filters with a hydrophobic surface. These filters are permeable for water, while oil particles are retained. After passing the "15 ppm alarm" (can be reduced to 5 ppm), the treated water is discharged outboard. The service life of the filter inserts is approx. 2 years.

[56]Bilge water oil separator of the company Ocean Clean GmbH, Rostock.

Table 8.16 Requirements for bilge water systems

	Minimum throughput or rather minimum content	<400 BRZ	400 BRZ to<1,600 BRZ	1,600 BRZ to<4,000 BRZ	4,000 BRZ to<15,000 BRZ	≥15,000 BRZ
15 ppm plant for special areas consisting of – 15 ppm plant[c] – 15 ppm alarm[b] – Stop device	m^3/h	0.25	0.5[a]	1.0	2.5	5
Sludge tanks[d]	m^3	2% of total content all fuel tanks in m^3				
Bilge water holding tanks [e,f]	m^3	–	–	2	6	15
Leakage oil tanks[f]	m^3	1	1	1	1.5	2.5
Dirty oil and waste oil tanks[f,g]	m^3	1	2	3	8	10
Delivery line to country	–	1 Land connection on port or starboard side or midship of the main deck according to MARPOL 73/78, Annex I, Regulation 13			2 Land connections on port and starboard side of the main deck according to MARPOL 73/78, Annex I, Regulation 13	

[a]When applying MARPOL regulation 14(5), a bilge water holding tank of approx. 1.5 m^3 should be provided

Legend to the "structural measures for the prevention of marine pollution by oil".

[b]15 ppm alarm = oil content measuring device with alarm device 15 ppm; stop device = automatic stop of the discharge of oil–water mixtures into the sea when the limit value is exceeded.

[c]The equipment table is designed in such a way that all the vessels are to be equipped with 15 ppm equipment for special areas. Thus, the requirement of MARPOL 73/78, Annex I, Regulation 14 is fulfilled.

On written application by the shipowner, the installation of the 15 ppm alarm and the stop device can be waived for ships up to 10,000 BRZ if the ship does not operate in special areas, such as Baltic Sea/North Sea. Depending on the type of construction or de-oiling process (membrane oil separator), a smaller 15 ppm system can also be approved by the BG Transport/Ship Safety Division.

[d]If an oil sludge incineration plant according to Annex 3 of document D.16 is present, the total capacity of sludge tank and dirty oil tank can be reduced by up to 30% of the specified tank volumes. The minimum throughput of this incinerator must be 1% of the daily fuel consumption.

[e]Bilge water holding tanks should be provided if the density of the fuel is higher than 0.94, that is, for heavy fuel oil operation.

[f]If the tanks are designed according to the contents of Table 8.16, the conditions of the IMO Guidelines for systems for handling oily wastes in machinery spaces of ships incorporating guidance note for an integrated bilge water treatments system (IBTS), MEPC.1/Circ. 511 of April 18, 2006 were fulfilled.

[g]If a complete exchange of the lubricating oil at sea is planned for main and auxiliary diesel engines, the size of the tanks—in deviation from the values in the table—should be at least 1.5 m^3 to be designed for a machine output of 1,000 kW each.

Fig. 8.23 Phase separation due to differences in density

Fig. 8.24 Bilge water oil separator. *Photo:* Company Ocean Clean GmbH, Rostock

Since the separation system is located on the discharge side of the system pump, there is no restriction on the suction height. If the suction lift is more than 6 m, the pump can be installed separately in a lower position.

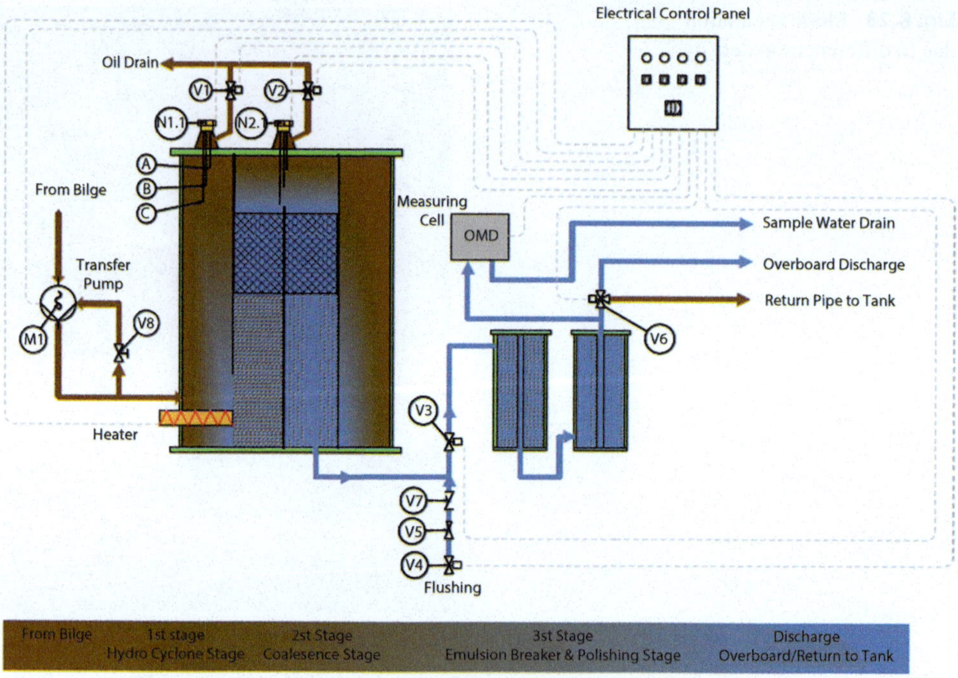

Fig. 8.25 Principle bilge water de-oiling with Ocean Clean EB. *Graphic:* Company Ocean Clean GmbH, Rostock

Description of the procedure

The oily water from bilge or bilge water storage tank (*from bilge*) is pumped by a pump (*transfer pump*) to the 1st stage (*1st Stage*), the coarse separation stage (*Hydro Cyclone Stage*; Fig. 8.25; [27]).

The inlet pipe is designed like a hydrocyclone, which makes the bilge water rotate. The lighter oil moves to the center of the tank where it is detected by a conductivity electrode (*N1.1*). This electrode measures the electrical resistance of oil and water between sensors (A, B, C) in the upper part of the tank. Water is a good electrical conductor, while the electrical resistance of oil goes against ∞. This means that if the two sensors are surrounded by oil, no electrical current can flow between them. This signal is accepted by the detection relay. The valve for discharging oil is opened (*V1*). While *V1* is open for the discharge of oil, *V3* remains closed.

As long as the said 1st stage sensor does not detect oil, the exhaust valve *V1* closed. The oil–water emulsion flows from below through the coalescence insert (*2nd stage*) to the top and forms larger oil drops by merging together the smallest oil drops. Oil separated here is removed by the sensor *N2.1* (operating on the same principle as described above) and is detected and controlled by the exhaust valve *V2* drained.

A backwashing of the coalescence separator (*flushing*) with sea or fresh water is done automatically. The pressure of the backwash water is controlled by means of a pressure control valve, which should be set on 2 bar.

After passing through the 1st and 2nd stages, the remaining predeoiled water flows through the emulsion breaking and filtering stage (*3rd Stage, emulsion breaker & polishing stage*), which are connected in series. The clean water—free of oil and emulsions—is then pumped through the valve *V6* to outboard (*overboard discharge*).

Between the last filter column and the outlet there is a bypass line back to the bilge or the bilge water tank (*return pipe to tank*). The 15 ppm alarm constantly detects the oil content (*messuring cell*) in the outlet; as soon as this rises to more than 15 ppm, the outlet valve *V6* is switched. It then closes, and the outboard discharge is stopped and the water flows via the bypass line back to the bilge or bilge water tank (*return pipe to tank*).

Table 8.17 shows the possible malfunctions and remedial measures of the presented plant.

Another process for bilge water treatment is coalescence separation (1st stage) followed by an adsorption stage (2nd stage). The bilge water separator OWS-COM from Veolia Water Technologies Germany GmbH works according to this principle (see Fig. 8.26).

Adsorption (from Latin "adsorptio," from "adsorbere" = "(suck in)") is the accumulation of substances from gases or liquids on the surface of a solid, more generally at the interface between two phases.

The function of this bilge water separator [37] is essentially based on the mode of action of the open-pored coalescer. In the downstream second stage the emulsion is broken and the oil is removed from the emulsion.

Table 8.17 Malfunction matrix bilge water separator *Ocean Clean EB*

Malfunction	Measures
No pressure or pumping of the pump	Check bilge-pump connection line. Damage or wear of the stator—replace. At the same time, check the condition of the rotor. Its surface must be smooth and free of grooves—replace if necessary. Check dry running, if free and clean—replace if necessary
The piston valves do not function properly	Check, whether air lines are clean and free—replace if necessary. Check the control valves—if dirty, replace if necessary. Check piston valve; if spring, sealing plates or O-rings are damaged, replace if necessary
15 ppm alarm is continuously in alarm position	Check emulsion gap and filter cartridges—replace if necessary. Check alarm system by flushing with fresh water—replace measuring cell if necessary
Oil–water separation is not satisfactory	Check sensor electrodes in the event of a ground short circuit—replace if necessary. Check the level relay

Fig. 8.26 Bilge water–oil separator. *Photo:* Veolia Water Technologies Germany GmbH

An eccentric screw pump (in Fig. 8.26 centrally between the two tanks) sucks from the bilge through the OWS-COM. This flow control prevents unnecessary additional mixing of oil and water due to pump turbulence upstream of the gravity separator.

In the 1st stage of the OWS-COM (left tank in Fig. 8.26) the coarse separation takes place by using the different density of water and oil (gravity separation). Due to its extremely oleophilic surface, an extremely open-pored coalescer causes the fine separation of the smallest oil droplets.

The adsorber bypass is controlled by the Bilge Alarm Monitor OMD. The adsorption elements of the second stage remove all kinds of hydrocarbons from the water. The adsorption capacity is mainly limited by the amount of dissolved or emulsified hydrocarbons in the water; however, large amounts of dirt particles can also block the adsorber (the right tank in Fig. 8.26). In order to extend the adsorber service life, heavy emulsions with high oil concentration should be avoided by using fast separating cleaners in the engine room. Emulsions from air conditioning systems or cleaner sludge should not be added to the bilge water.

To increase the service life of the adsorber cartridges, an automatic adsorber bypass is installed. The 15 ppm (or 5 ppm) oil alarm monitor periodically checks the water quality (every 5 min for 4 s) at the outlet of the first oil separator stage. If the value at this measuring point is then at 14 ppm (or 4 ppm), a bypass valve controls the purified water past the adsorber stage, directly outboard. However, if the value is via 14 ppm (or 4 ppm), the adsorber stage is activated. This process control guarantees a considerably extended service life of the adsorber cartridges.

During the short measuring time (4 s), a three-way valve directs the purified water back to the bilge/bilge tank for safety reasons. This type of control ensures that only purified water with less than 15 ppm (5 ppm) residual oil content can reach the outboard. Alarm limit value 2 is set at 15 ppm (or 5 ppm) and alarm limit value 1 is set at 14 ppm (4 ppm). Only if the residual oil content of 15 ppm (5 ppm) is exceeded for at least 2 s, an alarm message to the engine control room is activated. Alarm 1 must be set lower than Alarm 2.

Basically three control states are possible:

a) Measuring point falls below the limit value of 15 ppm (5 ppm) set for alarm 2 and measuring point falls below the limit value of 14 ppm (4 ppm) set for alarm 1:
Operation without adsorber, purified water goes outboard.
b) Measuring point exceeds the limit value of 14 ppm (4 ppm) set for alarm 1 and measuring point falls below the limit value of 15 ppm (5 ppm) set for alarm 2:
Operation with adsorber, purified water goes outboard.
There is no voltage at the solenoid valve of the three-way valve, the three-way valve directs the water through the adsorber.
c) Measuring point exceeds the limit value of 14 ppm (4 ppm) set for alarm 1 and measuring point exceeds the limit value of 15 ppm (5 ppm) set for alarm 2:
peration with adsorber, purified water goes back to the bilge or bilge water tank. There is no voltage at the solenoid valve of the three-way valve, the three-way valve directs the water through the adsorber and from there back to the bilge or bilge water tank. Additionally, an alarm message to the engine control room is activated. On the monitor, two red LEDs for both limit values indicate that they have been exceeded and therefore an alarm is triggered.

The separated oil collects in the upper calming zone of the oil separator. A sensor electrode measures the accumulated oil level. As soon as a predetermined amount of oil is collected, the automatic level control opens the oil drain valve and the flushing water inlet valve so that the existing oil is drained off by the flushing water pressure to the oil collection tank. The backwashing process starts after the shutdown. The high-performance coalescer is backwashed with clean water.

Deoiling interval and flushing cycle are coordinated in such a way that the OWS-COM operates largely maintenance-free.

The oil separator system is equipped with a 15 ppm (optionally also 5 ppm) oil alarm monitor OMD. The device has been tested in accordance with IMO Resolution MEPC.107(49).

Figure 8.27 shows the component overview of an OWS-COM system, and Fig. 8.28 shows an installation diagram for bilge water separators. Details, in particular pipe cross sections and required flange connections, can be found in the respective manufacturer's specifications.

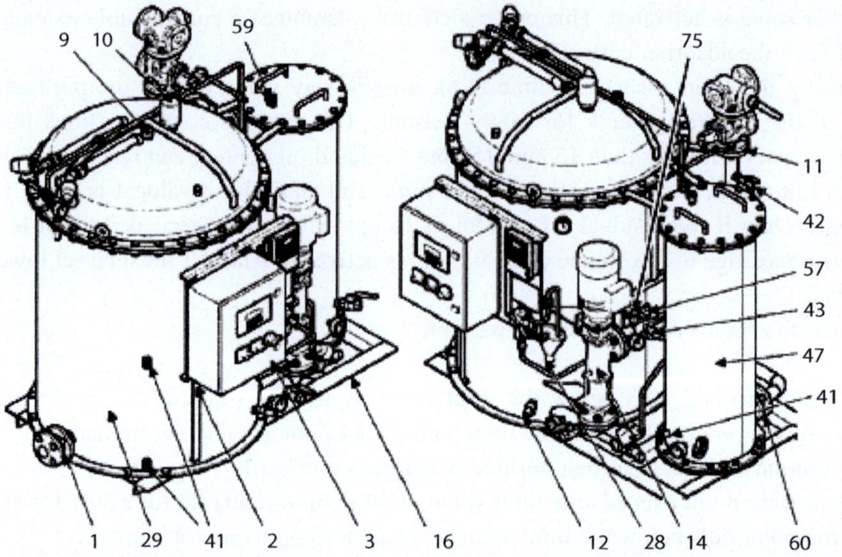

Fig. 8.27 Component overview of the OWS-COM system: *1* check valve, *2* feeding, *3* control box, *9* sensor electrode, *10* electric heating, *11* dust screen, *12* safety valve (1 bar), *14* monopump, *16* foundation, *28* funnel, *29* coalescer vessel (1st stage), *41* zinc anode, *42* sampling tap, *43* safety valve (3 bar), *47* adsorber (2nd stage), *57* test water valve, *59* venting, *60* swing check valve, *75* flow controller. *Picture:* Company Veolia Water Technologies Germany GmbH

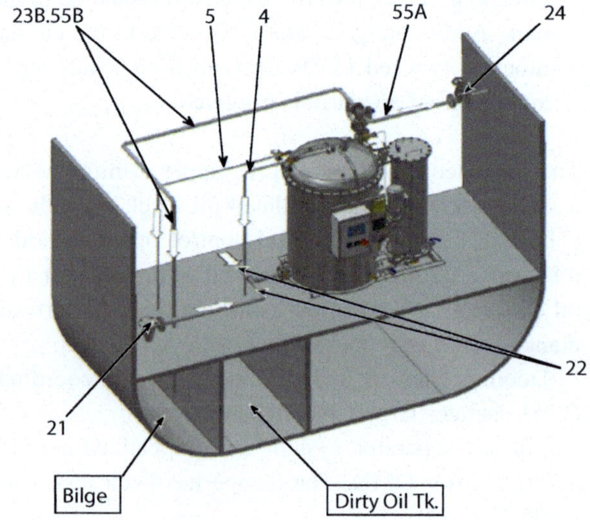

Fig. 8.28 Piping and instrument plan: *4* oil drain valve, *5* flushing valve (outlet), *21* sieve basket (optional), *22* suction line from bilge (bilge water tank), 23B/55B back to bilge/bilge water tank, *24* outboard valve (optional), 55A outboard line (outlet to outboard), *bilge*: Bilge water tank, *dirty oil tank*: Tank for separated oil. *Picture:* Company Veolia Water Technologies Germany GmbH

8.6 Ballast Water Treatment

8.6.1 Introduction

On August 21, 2014, the—"See-UmwVerhV" came into force. This Ordinance regulates requirements for environmentally sound behavior in shipping and the punishment of violations of environmentally sound behavior, in particular violations of the provisions of the Ballast Water Convention (see Sect. 1 SeeUmwVerhV).

Section 18 of the Ordinance regulates the discharge of ballast water. Under para 1, the discharge of ballast water into the sea is prohibited unless ballast water exchange in accordance with regulation D-1[57] or treatment in accordance with regulation D-2 or, in the cases referred to in Article 9(3) or Article 10(2) or (3) of the Ballast Water Convention, the BSH has issued a permit on request.

In this context, reference is made to the **Transition rule B-3** of the *Ballast Water Convention* referred [53]:

A ship built before 2009.

- with a ballast water capacity of 1,500–5,000 cm^3 including **may only be used until 2014** ballast water exchange[58]—D-1 procedure—perform; **after that date it should at least meet the standard described in regulation D-2 (ballast water treatment);**
- with a ballast water holding capacity of $<$ 1,500 or $>$ 5,000 cm^3 must carry out ballast water treatments from 2016 onward that at least meet the standard described in D-2.
  A ship built in 2009 or later with a ballast water capacity of $<$ 5,000 cm^3, should provide ballast water treatment complying at least with the standard described in regulation D-2.
- A ship constructed in 2009 or later, but before 2012 with a ballast water capacity of $\geq$ 5,000cm^3 should provide ballast water treatment at least equal to that specified in regulation D-1 or D-2, and thereafter at least equal to that specified in regulation D-2.
- A ship constructed in 2012 or later with a ballast water capacity of 5,000 cm^3 or more should provide ballast water treatment at least equal to that specified in regulation D-2.

According to the Ballast Water Convention, "treatment of ballast water" means mechanical, physical, chemical, and biological processes, which can be individual or in combination, remove or render harmless harmful aquatic organisms and pathogens contained in ballast water and sediments or prevent their uptake or release.

---

[57]Note transition periods according to the Regulation B-3—see below.

[58]See Ballast Water Convention, Regulation B-4: 200 nm from the nearest land, in exceptions 50 nm, and water depth of 200 m.

In the Ballast Water Convention, the procedure according to **Rule D-1** *standard for the exchange of ballast water* is described as follows:

1. Ships exchanging ballast water under this regulation must achieve an efficiency of 95% of the volume of ballast water exchanged.
2. For ships exchanging ballast water by the method of pumping through, pumping through the volume of each ballast water tank three times should be considered as complying with the standard referred to in paragraph 1. Pumping less than three times may be accepted, provided that the ship concerned can demonstrate that 95% of the quantity measured by volume has been replaced.

The **Rule D-2** *standard for the composition of ballast water* describes the following procedure:

1. Ships which treat ballast water in accordance with this regulation must discharge less than 10 viable organisms per cubic meter of 50 μm or more and less than 10 viable organisms per millilitre of 50 μm or less and 10 μm or more, and the rate of discharge of pilot microbes referred to in paragraph 2 should not exceed the concentrations specified.
2. The pilot microbes considered safe for human health at the concentrations indicated include those listed below:
   - Toxigenic *Vibrio cholerae* (O1 and O139) at a concentration of less than cfu[59] per 100 ml or of less than 1 cfu per 1 g zooplankton (wet weight),
   - *Escherichia coli* in a concentration of less than 250 cfu per 100 ml,
   - Intestinal enterococci in a concentration of less than 100 cfu per 100 ml.

From the above it is clear that two ballast water management procedures are currently common: ballast water exchange and ballast water treatment on board.

8.6.2 Ballast Water Exchange

In *sequential ballast water exchange* a ballast water tank is first completely emptied and then refilled with fresh seawater. In the *flow-through process* a filled ballast water tank is constantly flushed with fresh seawater. In the *dilution method* fresh seawater is filled into the ballast water tank and simultaneously the same amount of ballast water is pumped out of the tank so that the filling level remains constant.

[59]The abbreviation “cfu” stands for “colony-forming unit”.

8.6.3 Ballast Water Treatment

The challenge is that the plants—which must meet high requirements according to the above mentioned D-2 regulations with regard to the separation or killing of organisms—have high-water flow rates (in the range of 200–5,000 m^3/h) and at the same time must be very compact and robust [66]. In general, a high level of operational safety must be ensured. Membrane filter plants which could ensure the retention of all relevant organisms (as in drinking water or wastewater treatment), do not meet the requirements in terms of space requirements at the required throughput. Therefore, filters are used down to a separation limit in the range between 20 and 60 μm. The organisms that are not separated in this way must be killed by additional measures. For this reason, a ballast water treatment plant usually comprises two main stages: (1) a*mechanical separation stage* for the separation of microorganisms and sediments with a separation limit in the range of 20–50 μm and (2) a disinfection stage for killing microorganisms, which are not separated in the first stage. Determining the separation limit of the first stage is like a straight line, since a small separation limit improves separation, but reduces the throughput in relation to the filter area. This is also significantly influenced by a deposit layer on the filter medium. Small separation limits therefore usually require a large filter area, which means that a certain size cannot be undercut. One way to minimize the filter area is to carefully select the filter medium and to clean it periodically. This optimization is a demanding development task.

In the second stage *physical or chemical disinfection processes* are to be applied. One pure *physical processing* is present at this stage when UV oxidation is used. If necessary, this can be further intensified in combination with ultrasound or a biocidally active substance. Ozone can be used, for example, in a similar way to other water purification processes. This has the advantage that it does not burden the environment and can be generated on site [35].

The *physical procedure* used today is predominantly the *UV irradiation* of the ballast water. Here, ballast water flows through a UV disinfection unit. Such a UV disinfection system essentially consists of several quartz tubes containing UV lamps.

Another physical process is the *thermal processing* of the ballast water. This involves exposing the organisms contained in the water to a temperature that is sufficient to kill them. There are essentially three possible methods for thermal ballast water treatment: mixed preheating of ballast water with simultaneous flushing of a ballast water tank, heating of ballast water in the tanks, and heating of ballast water during intake or discharge. Heating of the ballast water is mainly done with engine cooling water or other heat sources, such as auxiliary boilers.

In the *ultrasonic disinfection* the ballast water is irradiated with a sound field (20–400 kHz), so that the alternating negative and positive pressure areas are formed in the fluid. In a negative pressure phase of the sound field, vapor bubbles form in the water, which collapse implosively in the subsequent positive pressure phase (cavitation). The sudden collapse of the vapor bubbles produces shock waves that cause the cell membranes of the organisms to rupture, thus killing them.

With a *chemical treatment* one or more chemical substances are added to the ballast water directly or indirectly by electrolysis, so that a toxic reaction takes place, which leads to the killing of water organisms, bacteria, and viruses. Ozone, chlorine, chlorine derivatives, and biocides can be used here.

However, chlorine, chlorine derivatives and biocides are not harmless for occupational safety and environmental protection reasons, so that the *ozone disinfection* is widely used. As ozone is unstable, it is generally produced locally from air or oxygen in ozone generators or by irradiating air with UV lamps.

8.6.3.1 Process of Electrochemical Disinfection

The process is a two-stage process consisting of filtration and electrochemical disinfection ([36]; see also Fig. 8.29).

In the first filtration stage, particles, sediments, and organisms are retained when ballast water is absorbed. Optionally, this is done by the so-called disc-filter depth filtration or by a proven *screen filter* with significantly lower space requirements compared to disc-filter technology.

In the second stage, the electrochemical disinfection unit, the remaining smaller organisms and bacteria are eliminated before the water reaches the ballast water tanks.

During the voyage to the next port of call, renewed growth of the organisms in the tank may occur. To ensure a constant quality of the water to be dispensed, the system is also used during the dispensing process. For this purpose the filtration stage is bypassed by means of a bypass line and only the electrochemical disinfection is used.

Description of the filtration process

During ballast water intake, the raw ballast water is pumped into the *disc filters,* which operate in parallel. Each disc filter is equipped with a series of grooved filter discs

Fig. 8.29 Ballast water treatment with electrochemical disinfection. *Photo:* K.-H. Hochhaus, CC BY 3.0

stacked on several spikes. As filtered water passes through the discs, particles, fibers, algae, etc. are retained in the grooves on the outside of the discs.

When a preset differential pressure is reached, the fully automatic backwash mode starts. The spring that compresses the filter discs is automatically released. Rinsing water then flows from the inside through the filter spikes to the outside, which guarantees effective backwashing of the filter in just a few seconds.

The RWO screen filter solution is based on a self-cleaning multiscreen filter, and is an alternative to the disc filter.

Description of electrochemical disinfection

The disinfection technology used here is an extremely efficient and robust system that works both in seawater and in low-salt water. In this respect it differs greatly from standard chlorine electrolysis. By applying electrical voltage to the special electrodes located in the cell, they generate very short-lived and reactive hydroxyl (OH) radicals directly in the pipeline, killing bacteria, and the other microorganisms. These radicals have the highest oxidation potential (2.8 mV vs. 1.36 V for chlorine) and thus a very high-disinfection efficiency. Due to their short service life, they do not cause corrosion of subsequent pipelines and coatings.

The advantages of this ballast water treatment process are summarized as follows:

- Low-energy consumption (0.006–0.06 kWh/m^3),
- Can also be used in low-saline water,
- Meets all requirements of the approval authorities,
- Minimum space requirement (< 1 m^2 per 500 m^3/h),
- No special contact time of the water, no additional chemicals in the system,
- Safe for crew and ship,
- No danger of corrosion.

8.6.3.2 Method of UV Disinfection

The second common method of disinfecting ballast water is disinfection by means of ultraviolet radiation.

This is also a two-stage process: first stage filtration, second stage UV disinfection. Whereas, when ballast water enters via the sea chests, the water passes through both the filter and disinfection stages before entering the ballast water tank, once the tanks are emptied, the water then only passes via the disinfection stage to kill the organisms which may have formed in the tanks during the voyage; it bypasses the filter stage through a bypass line (Fig. 8.30).

Description filter stage

The filter unit usually consists of a filter element with mesh sizes of 20–40 μm in stainless steel or duplex to retain suspended solids and zooplankton. An automatic

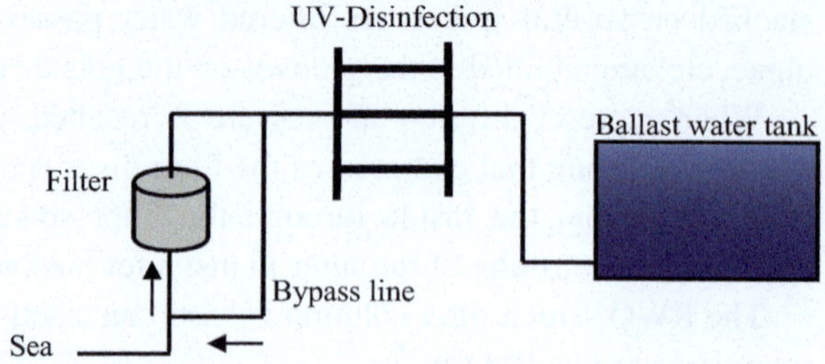

Fig. 8.30 Collection and discharge of ballast water

backwashing of the filter is carried out by a pressure switch, which is activated at a set differential pressure.

The filter can be installed horizontally or vertically, depending on the available space.

Description of UV disinfection

After pre-cleaning in the filter stage, the ballast water flows through the UV reactor unit. This consists of metal tubes containing monochromatic UV-C[60] -quartz glass lamps. The ultraviolet light with a wavelength of about 255 nm destroys the cell membranes of the microorganisms and inactivates them in this way.

The contamination of the quartz glass lamps is counteracted by various automatic cleaning measures. The UV light intensity (in Watt per m^2) is monitored by UV light sensors. If the intensity decreases to a certain value, the lamp must be replaced.

References

Print media

1. Achternbosch, M., Richers, U.: Stoffstromanalysen zur abwasserfreien und Abwasser erzeugenden Verfahrenskonzeption von nassen Rauchgasreinigungssystemen. Forschungszentrum Karlsruhe, Karlsruhe (1996)
2. Cruises, A.I.D.A.: AIDA cares—Nachhaltigkeitsbericht 2009. AIDA Cruises, Rostock (2009)
3. Arnold, C.: Beurteilung der chemischen Verhältnisse bei der Auswahl der besten Abluftreinigungsmethode. Z. Wlb **2002**(4), 51–54 (2002)
4. Bahadir, M., Parlar, H., Spitteler, M.: Springer Umweltlexikon. Springer, Heidelberg (2000)
5. BG Verkehr—Dienststelle Schiffssicherheit, Dokument D.16 „Anforderungen Bauliche Maßnahmen auf Seeschiffen zur Verhütung der Meeresverschmutzung durch Öl, Abwasser, Müll und Abgasemissionen", Stand 01.08.2014, Hamburg
6. Börger, G.: Umweltschutztechnik in der chemischen Industrie. Druckschrift der Bayer AG, Leverkusen (1985)

[60]Spectrum of UV light with a wavelength between 100 and 280 nm.

7. Büchner, H.-P., Goedecke, H.: Modernes MVA Anlagendesign aus der Sicht eines Betreibers. Z. Müll Abfall **5/2010**, 213 (2010)
8. Chafik, E.: Skriptum der Vorlesung ANLAGENBAU. Ruhr-Universität Bochum, Lehrstuhl für thermische Stofftrennverfahren, Bochum (1986)
9. Couple Systems, Brochure: „The very new exhaust gas cleaning system. There is no longer an excuse!“. Bardowick, Deutschland (2012)
10. Dreyhaupt, F.-J. (ed.): VDI-Lexikon Umwelttechnik. VDI-Verlag, Düsseldorf (1994)
11. Dubbel: DUBBEL Bd. I. Springer, Berlin, Heidelberg, New York (1974)
12. Fratzscher, W., Picht, H.-P.: Stoffdaten und Kennwerte der Verfahrenstechnik. VEB Deutscher Verlag für Grundstoffindustrie, Leipzig (1979)
13. Fonds der Chemischen Industrie: Umweltbereich Wasser—Textheft 13. Fonds der Chemischen Industrie, Frankfurt/M. (1990)
14. GAUSS mbH: Entölertechnik an Bord von Seeschiffen—Bestandsaufnahme. GAUSS mbH, Bremen (1998)
15. GAUSS-Workshop „Aufbereitung von Schwarz- und Grauwasser in der Seeschifffahrt“, Vortrag „Recht und rechtliche Entwicklung im Zusammenhang mit Schiffsabwasser“, Bremen 9.12.99, Autor unbekannt
16. GESTRA-Wegweiser: Bremen (2005)
17. Hansmann, K.: In: Landmann, R., Rohmer, G. (Hrsg.) Umweltrecht, Bd. I, 14. EL, C. H. Beck, München (1990)
18. Hellwig, K., Pfaff, M.: Zur Erlaubnispflicht nach dem WHG für Einleitungen von Schiffsabwasser. ZfW **3**(2005), 154–165 (2005)
19. Hellwig, U., Kölling, A., Hochhaus, K.-H.: Energetische Abfallnutzung auf Kreuzfahrtschiffen. In: Mensch & Technik, Bd. Nr. IV/2008, Verein Deutscher Ingenieure, Hamburg (2008)
20. Hochhaus, K.-H., Rachow, M.: Umwelt. In: Meier-Peter, H., Bernhardt, F. (Hrsg.) Handbuch Schiffsbetriebstechnik. Seehafen Verlag, Hamburg (2008)
21. International Maritime Organization: Resolution MEPC. 76(40), „Normspezifikation für bordseitige Verbrennungsanlagen“ vom 23. Februar 2007, (VkBl. Nr. 6 vom 31. Mar. 2007 S. 174)
22. Jarass, H.-D.: Kommentar BImSchG, 6th edn. C. H. Beck, München (2005)
23. Kaiser, A.: Vortragsskript „Umwelt: Ver- und Entsorgung auf Schiffen“, Prüf- und Entwicklungsgesellschaft für Abwassertechnik, RWTH Aachen e. V.
24. Landesumweltamt Nordrhein-Westfalen (Hrsg.), „Materialien Nr. 60—Untersuchungen zur Effizienz von unterschiedlichen Systemen zur Rauchgasreinigung bei Müllverbrennungsanlagen“, Essen (2001)
25. Marinekommando (Hrsg.): Jahresbericht (2016)
26. Nethe, L.-P.: Kalkhydrat vs. Natriumbicarbonat. Vortrag auf der 4. Fachtagung „Trockene Abgasreinigung“, Haus der Technik, Essen, 13.–14. Nov. 2008. (2008)
27. Ocean Clean GmbH Rostock: Bedienungsanleitung „Instruction Manual OWS 1.0 EB“
28. Ocean Clean GmbH Rostock: Prospekt „Technical Specification Biological Physical Sewage Treatment Plant—Ocean Clean UltraC-2“
29. Ocean Clean GmbH Rostock: Trainingsmanual „Training data and information for Ocean Clean sewage treatment plant OCS-compact“
30. Pfaff, M.: Ökologische Schiffsmüllverbrennung—neuartiges Abgasreinigungssystem mit sehr geringer Schadstoffemission. Z. Mar. Forum **11/2007**, 27 (2007)
31. Pfaff, M.: Ökologische Schiffsmüllverbrennung. Zeitschrift Hansa **9/2007**, 43 (2007)
32. Pfaff, M.: Schiffsmüllverbrennungsanlagen—rechtliche Anforderungen. Immissionsschutz **10**(3), 72–77 (2005)

33. Redaktion der Zeitschrift „Immissionsschutz“: Heft 4, 7. Jhrg., Erich Schmidt Verlag, Berlin (2002)
34. Reimann, D.O.: Rostfeuerungssysteme und Möglichkeiten der Emissionsminderung durch Feuerleistungsregelungen. In: Müllverbrennung und Umwelt Bd. 5. EF-Verlag für Energie- und Umwelttechnik, Berlin (1987)
35. Ripperger, S.: Anwendung der Filtration zur Behandlung von Ballastwasser. Fachjournal F&s Filtr. Sep. **23**(6), 314 (2009)
36. RWO GmbH—Marine Water technology, Firmenprospekt „CleanBallast—The RWO Ballast Water Treatment System“, Bremen
37. RWO GmbH—Marine Water technology, „Installations-, Betriebs- und Wartungsanleitung OWS-COM“, Bremen
38. Schwacke, P.: Recht der Ordnungswidrigkeiten. Deutscher Gemeindeverlag, W. Kohlhammer, Köln (1988)
39. Sieder, F., Zeitler, H., Dahme, H., Knopp, G.H.: WHG, Kommentar, 29. Ergänzungslieferung, Stand: 12/2004. C.H. Beck, München (2004)
40. Thier, B.: Apparate: Technik—Bau—Anwendung. Vulkan-Verlag, Essen, S. 362 (1997)
41. Thomé-Kozmiensky, K.J., Borchers, H.-W., Faulstich, M.: Filternde Abscheider. In: Müllverbrennung und Umwelt Bd. 2. EF-Verlag für Energie- und Umwelttechnik, Neuruppin (1991)
42. Thomé-Kozmiensky, K.J., Borchers, H.-W., Faulstich, M.: Maßnahmen zur Schadstoffreduzierung bei der Abfallverbrennung. In: Müllverbrennung und Umwelt, 2. Aufl. EF-Verlag für Energie- und Umwelttechnik, Neuruppin (1991)
43. Verband der chemischen Industrie (VCI): Chemie und Umwelt—Wasser. Verband der chemischen Industrie (VCI), Frankfurt/M. (1982)
44. Verein Deutscher Ingenieure: VDI-Richtlinie „Abgasreinigung durch Adsorption; Prozessgas- und Abgasreinigung“—VDI 3674. VDI, Düsseldorf (1998)
45. Verein Deutscher Ingenieure: VDI-Richtlinie „Massenkraftabscheider“—VDI 3676. VDI, Düsseldorf (1999)
46. Witthöft, H.-J.: Köhlers Flottenkalender. Koehlers Verlagsgesellschaft, Hamburg (2015)
47. Witthöft, H.-J.: Köhlers Flottenkalender. Koehlers Verlagsgesellschaft, Hamburg (2017)
48. Witte, U.: Steinmüller Taschenbuch Dampferzeugertechnik. Vulkan-Verlag, Essen (1984)

Internet

49. Köster, S., Westhof, L., Keller, L.: Stand der Technik der Abwasserreinigung an Bord von Kreuzfahrtschiffen“, Fachberichte Abwasserreinigung (2016). https://www.gwf-wasser.de/fileadmin/GWF/01_fb_Koester.pdf. Accessed: 2. Apr. 2017
50. https://www.bielefeld.de/de/un/stadtreinigung/wertabfall/wert/in/ Accessed: 22. Mar. 2017
51. http://www.bmu.de/redenarchiv/14/altmann/doc/1876.php Accessed: 3. Apr. 2017
52. http://www.bsh.de/de/Meeresdaten/Umweltschutz/Ballastwasser/index.jsp Accessed: 16. Mar. 2017
53. http://www.bsh.de/de/Meeresdaten/Umweltschutz/Ballastwasser/G-Ballastwasser.pdf Accessed: 16. Mar. 2017
54. http://www.bsh.de/de/Schifffahrt/Sportschifffahrt/Berichtigungsservice_NfS/Schifffahrtsvorschriften/2012/Beilage46-2012-172.pdf Accessed: 20. Apr. 2017
55. http://www.bund.net/lab/reddot2/eu_umweltpolitik.htm Accessed: 27. Mar. 2017
56. http://www.cargoforum.de/seefracht/dictionary/f_d.htm Accessed: 19. Mar. 2017

57. http://www.dammann.de/Produkte/Produktfamilie/Umweltschutz/Luftreinhaltung/Flour-Kaskaden-Absorber Accessed: 24. Mar. 2017
58. http://www.daserste.de/information/wissen-kultur/w-wie-wissen/sendung/2010/umweltfreundliche-schiffe-100.html Accessed: 17. Mar. 2017
59. http://www.deutsche-flagge.de/de/umweltschutz/marpol/abwasser Accessed: 29. Mar. 2017
60. https://www.dieselnet.com/standards/inter/imo.php Accessed: 18. Mar. 2017
61. http://www.energiewelten.de/elexikon/lexikon/seiten/htm/010607_Rauchgasentschwefelung_in_Verbrennungskraftwerken.htm Accessed: 24. Mar. 2017
62. http://www.fh-bochum.de/fb2/personen/bracke/flaechenentwicklung/leistungsbuch_bodenluftbehandlung.pdf Accessed: 23. Mar. 2017
63. http://www.gutmbh.de/AktivkohleVortrag3.htm Accessed: 26. Mar. 2017
64. http://www.gutmbh.de/AktivkohleVortrag4.htm Accessed: 23. Mar. 2017
65. http://www.gutmbh.de/AktivkohleVortrag5.htm Accessed: 23. Mar. 2017
66. https://hochhaus-schiffsbetrieb.jimdo.com/ballastwassermanagement-264/ Accessed: 21. Apr. 2017
67. http://www.imo.org/en/OurWork/environment/pollutionprevention/airpollution/pages/nitrogen-oxides-(nox)-regulation-13.aspx Accessed: 18. Mar. 2017
68. http://www.jacob-rohre.de/berechnu/abwei10/abweich1.htm Accessed: 27. Mar. 2017
69. http://www.patent-de.com/20030313/DE19613376C2.html ; 23. Mar. 2017
70. http://www.siclaro.ch/de/produkte/downloads/Maritim/ProspektRechen2004.pdf Accessed: 3. Apr. 2017
71. http://www.spektrum.de/lexikon/biologie/escherichia-coli/22571 Accessed: 5. Apr. 2017.
72. http://sundoc.bibliothek.uni-halle.de/diss-online/02/02H140/t5.pdf Accessed: 22. Mar. 2017
73. http://www.umweltbundesamt.de/themen/wasser/gewaesser/meere/nutzung-belastungen/schifffahrt#textpart-2 Accessed: 27. Febr. 2017
74. http://www.unhcr.de/recht/ii-staatenlosigkeit.html?L=0 Accessed: 19. Mar. 2017
75. http://www.wasser-wissen.de/abwasserlexikon/f/filtration.htm Accessed: 3. Apr. 2017
76. http://www.wasser-wissen.de/abwasserlexikon/m/mikrofiltration.htm Accessed: 5. Apr. 2017
77. http://de.wikipedia.org/wiki/AIDAdiva Accessed: 27. Juli 2017
78. http://www.zeit.de/2010/38/Kreuzfahrten Accessed: 17. Mar. 2017
79. www.cdni-iwit.org Accessed: 17. Mar. 2017
80. www.gutmbh.de Accessed: 17. Mar. 2017
81. www.imo.org Accessed: 27. Juli 2017
82. www.michaelis-umwelttechnik.de Accessed: 17. Mar. 2017
83. www.vonroll.ch Accessed: 17. Mar. 2017

Tables, Diagrams, and Overviews

A.1 Important regulations to be observed in the construction of german seagoing vessels[1]

Employer's Liability Insurance Association for Transport and Traffic Economy
Ship Safety Division
Reimerstwiete 2
20,457 Hamburg

1. Accident Prevention Regulations for Shipping Companies (UVV See) of January 1, 1981 (as amended by the 1st to 17th Supplement, as of January 10, 2003)
2. International Convention of 1974/88 for the Safety of Life at Sea—Ordinance of January 11, 1979 (Bundesgesetzblatt (BGBl., Federal Law Gazette) 1979 p. 141) as amended by the Protocol of 1988—Ordinance of September 20, 1994 (BGBl. 1994 II p. 2458), as last amended by 20th SOLAS—Amendment of November 27, 2009 (BGBl. 2009 II p. 1226 of December 8, 2009)
3. Announcement of the International Rescue Equipment (LSA) Code of June 4, 1998 in the currently valid version
4. Marine Equipment Directive 96/98/EC, as amended
5. Ship Safety Act (SchSG) of September 9, 1998 (BGBl. 1998 I p. 2860), last amended by the Tenth Ship Safety Adjustment Ordinance of March 11, 2009 (BGBl. I p. 507)

[1] http://www.deutsche-flagge.de/de/download/bau-und-ausruestung/neu-und-umbau/uebersicht/rechtsvorschriften. (January 20, 2016).

M. Pfaff, *Ship Operation Technology*, https://doi.org/10.1007/978-3-658-32729-3

6. Ship Safety Ordinance (Schiffssicherheitsverordnung, SchSV) (BGBl. 1998 I p. 3023)—last amended by the Tenth Ship Safety Adjustment Ordinance of March 11, 2009 (BGBl. I p. 507)
7. Ship Manning Ordinance of August 26, 1998 (BGBl. I p. 2577), last amended by Article 524 of the Ordinance of October 31, 2006 (BGBl. I p. 2407)
8. Act on the International Convention on Load Lines of 1966 (BGBl. II p. 249 of February 20, 1969) in the version of the 6th Regulation on Amendments to the International Convention on Load Lines of 1966 of June 24 24, 2008 (BGBl. II p. 668)
9. Ordinance on Sickness Benefits on Merchant Ships (BGBl. I No. 38 of April 29, 1972), last amended by the 3rd Amendment Ordinance of September 5, 2007 (BGBl. I No. 46, p. 2221)
10. Ordinance on the Accommodation of Crew Members on Board Merchant Ships and amendments to this Ordinance (BGBl. I No. 10 of February 10, 1973 and BGBl. I No. 109 of August 31, 1976)
11. Technical Rules for the Construction and Equipment of Accommodation Spaces on Seagoing Vessels (May 1975 edition) with 1st to 11th Amendment
12. Act on the International Convention for the Prevention of Pollution from Ships, 1973 and the Protocol of 1978 to this Convention (MARPOL 73/78; BGBl. II No. 1 of January 7, 1982), last amended by the 15th Ordinance on Environmental Protection at Sea of August 24, 2009 (BGBl. II p. 995)
13. Guidelines and leaflets of Maritime Employer's Liability Insurance Association (See-Berufsgenossenschaft, See-BG), status June 2003
14. Regulations of the respective classification society, which have a ship survey agreement with the See-BG, in the respective valid version
15. Requirements for the type approval of compasses, radio direction finding equipment, navigation lights and sound signaling devices, and conditions for the installation or location of the aforementioned devices
16. Ship Safety Regulations Handbook (as of June 2004). This book contains the regulations listed above under 2, 5, 6, 8, and 12 as well as the Ordinance on Dangerous Goods at Sea.
17. International Code of 2000 for Safety of High-Speed Craft of December 5, 2000 (Transport Gazette document No B 8128—Vers. 06/02), as amended
18. International Fire Safety Systems Code (FSS Code) of December 5, 2000—Resolution MSC.98(73), as amended

The above rules can be obtained from

Documents for No. 1	Print shop Paul Moehlke GmbH Hohenfeld Avenue 17–19, 22087Hamburg
Documents for No. 2, 5–10, and 12	Publisher Bundesgesetzblatt (Federal Law Gazette) PO Box 13 20, 53,003 Bonn
Documents for No. 11 and 13	BG Transport –Ship Safety Division Reimerstwiete 2, 20457Hamburg
Documents for No. 14	Respective classification society for example, DNV·GL Brooktorkai 18, 20457Hamburg
Documents for No. 15	Federal Maritime and Hydrographic Agency (BSH) PO Box 30 12 20, 20305Hamburg
Documents for No. 16	Print shop Paul Moehlke GmbH Hohenfeld Avenue 17–19, 22087Hamburg
Documents for No. 3, 4, and 17	Bundesanzeiger Verlagsges. mbH, Breite Str. 78–80, 50,667 Cologne or PO Box 10 05 34, 50445Cologne
Documents for No. 18 and 19	Verkehrsblatt-Verlag Borgmann GmbH & Co KG Hohe Street 39, 44139Dortmund

A.2 Values for the density of seawater in kg/m³ for approximate calculations

Fresh water	≈ 1000
Baltic Seawater	up to 1015
North Seawater	up to 1025
Red Seawater	up to 1044
Caspian Seawater	up to 1060
Great Salt Lake (USA)	up to 1230
Dead Seawater	up to 1278

A.3 Selection of vibration dampers[2]

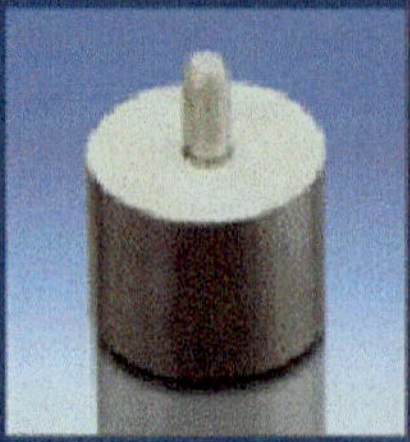

Megi® - Buffer

are suitable as simple and inexpensive standard elements in many cases for the elastic mounting of light and medium-heavy aggregates, engines and compressors.

Megi® - rails

Recommended for the elastic mounting of larger aggregates, motors, lathes, vibrating and vibrating machines, elevator machines, etc. (delivery length up to 2,000 mm).

Megi® - machine feet

are preferably used when high horizontal rigidity with good vertical deflection is required for the elastic mounting of eccentric presses, planing machines, printing and textile machines, etc.

Megi® - Cones

are particularly suitable for elastic mounting of engines, body superstructures on chassi frames, compressor units on railcars, etc., whereby shocks are progressively absorbed by stop plates.

[2]Source: sd-dresden.de.

Megi® - Bearings

Megi bearings are used as bearing elements with high vertical softness and sufficient horizontal stiffness for engine and compressor mounts etc.

Megi® - U-V-W parts

are special bearing elements for elastic mounting of sensitive apparatus, measuring devices and instruments.

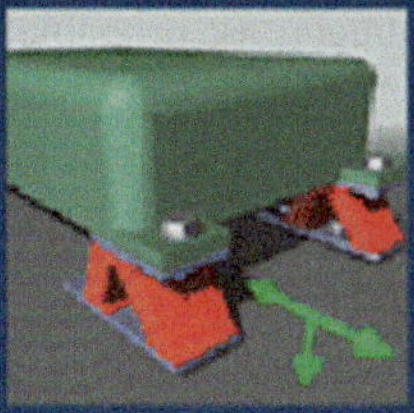

Megi® - bushings and rings

are used as elastic joints in vehicle construction and all branches of mechanical engineering. They work completely maintenance-free, noiseless and vibration-isolating with high fatigue strength.

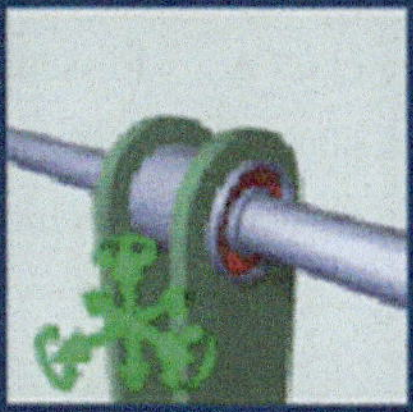

Megi® - ring elements

Megiflex washers are suitable for traction and shock absorbers, as torque supports, for the suspension of pendulum supports, leaf spring ends, etc.
Megi ring buffers provide an elastic mounting of machines which are relatively hard. The main task of the element is to separate the structure-borne sound bridge.

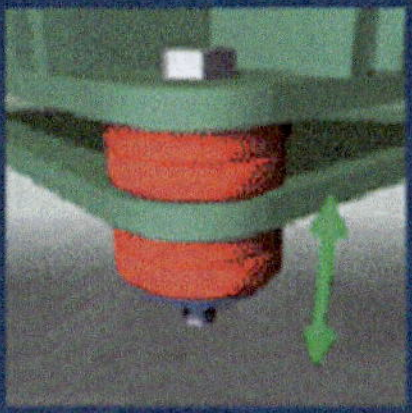

A.4 Total frequency Weighted RMS value for frequencies from 1 to 80 hz as guidelines for the habitability of different areas of a ship according to GL Technology Ship Vibration, 2001

	Area classification					
	A		B		C	
	mm/s^2	mm/s	mm/s^2	mm/s	mm/s^2	mm/s
Values above which adverse comments are probable	143	4	214	6	286	8
Values below which adverse comments are not probable	71.5	2	107	3	143	4

NOTE: The zone between upper and lower values reflects the shipboard vibration environment commonly experienced and accepted.

Area class A	Passenger cabins
Area class B	Crew accommodation
Area class C	Machines and work areas

Standard values for evaluated vibration severities in mm/s^2 and mm / s on floors and floor panels

The standard values in KB according to VDI Guideline 2057, as of October 24, 1987, are shown in brackets. Note: The standard values in the KB evaluation correspond to the values of the vibration severity in mm/s.

	DIN ISO 6954		(VDI Guidelines 2057, sheet 4.3)
	mm/s^2	mm/s	(KB)
Work area			
Bridge area incl. radio room	140	4.0	(4.0)
Engine rooms	221	6.3	(6.3)
Engine control rooms, washrooms, and workshops	175	5.0	(5.0)
Kitchens	175	5.0	(5.0)
Office space	140	4.0	(4.0)
Accommodation area			
Living and bedrooms	112	3.2	(3.2)
Mess and other common rooms	123	3.5	(3.5)
Hospital and other treatment rooms	112	3.2	(3.2)
Hobby - and scopes	140	4.0	(4.0)

A.5 Equipment with anchors, chains, and cables in accordance with DNV·GL Construction Regulations

The following table presents an **extract** from the DNV·GL Construction Regulation "Rules for Classification - Ships, Part 3 Hull, Chap. 11 Hull equipment, supporting structure and appendages."

Equipment number	Stockless bower anchors		Stud-link chain cables				Towline (guidance)		Mooring lines (guidance)		
	Number	Measure via anchor (kg)	Total length (m)	Diameter and steel just			Steel or fiber ropes				
				VL K1 (mm)	VL K2 (mm)	VL K3 (mm)	Minimum length (m)	Minimum breaking strength (kN)	Number	Length of each (m)	Minimum breaking strength (kN)
…	…	…	…	…	…	…	…	…	…	…	…
3040 to 3209	2	9300	660	97	84	76	280	1471	6	200	520
…	…	…	…	…	…	…	…	…	…	…	…
3800 to 3999	2	11,700	687.5	107	95	84	300	1471	6	200	647
4000 to 4199	2	12,300	687.5	111	97	87	300	1471	7	200	647
…	…	…	…	…	…	…	…	…	…	…	…
4600 to 4799	2	14,100	715	120	105	92	300	1471	7	200	677
4800 to 4999	2	14,700	742.5	122	107	95	300	1471	7	200	686
5000 to 5199	2	15,400	742.5	124	111	97	300	1471	8	200	686
…	…	…	…	…	…	…	…	…	…	…	…

A.6 Breaking loads for ropes[3]

Approximate minimum breaking loads of different ropes Ø 10 mm in daN (exact details can be found in the relevant manufacturer and supplier catalogs).

	Material	Minimum breaking load (daN)
Natural fiber ropes	Cotton	400
	Manila	690
	Hemp	700
Synthetic fiber ropes	PE	1020
	PP (turned)	1530
	Polyester (turned)	1700
	Polyamide (turned)	2040
	Aramid (braided)	3700
	Dyneema (braided)	9000
Wire ropes	Wire 1 × 19	8250
	Wire 7 × 19	5210
	Wire 7 × 7	5630

Wire rope construction

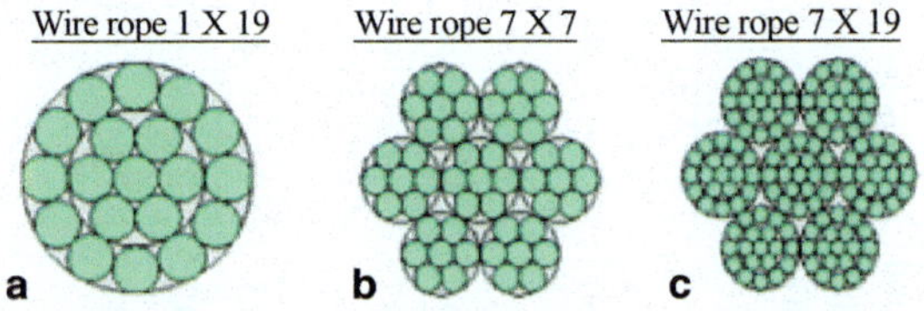

[3]After Rolf-Joachim Müller, PowerPoint presentation "Sports boat license inland/seemanlike works," DHH.

A.7 Strand ropes according to EN 12,385-4[4]

Rope type: stranded ropes according to EN 12385-4
Rope construction: 6x36WS-IWRC

Rope class/Number of outer strands
6X36 / 6
Surface available in:
U (ungalvanized) / B (galvanized)
Lay/Direction of lay available in
sZ/zS
Categories:
not rotation-resistant
Rope swivel/Twist catch insert: **not permitted**
Plastic insert:
no

Rope ø in mm	Weight in kg/ 100 m	Grade of rope classification 1770 N/mm² Minimum breaking force	Grade of rope classification 1960 N/mm² Minimum breaking force	Grade of rope classification 2160 N/mm² Minimum breaking force
8	26.2	40.30	44.70	49.20
9	33.1	51.00	56.50	62.30
10	40.9	63.00	69.80	76.90
11	49.5	76.20	84.40	93,00
12	58.9	90,70	100,00	111,00
13	69.1	106.00	118,00	130,00
14	80.2	124.00	137,00	151,00
16	105	161.00	179,00	197.00
18	133	204.00	226,00	249.00
20	164	252.00	279.00	308,00
22	198	305,00	338.00	372.00
24	236	363.00	402.00	443,00
26	276	426.00	472.00	520,00
28	321	494,00	547,00	603,00
32	419	645,00	715.00	787,00
36	530	817,00	904.00	997,00
40	654	1.010.00	1.120,00	1.230,00
44	792	1.220,00	1.360,00	1.490,00
48	942	1.450,00	1.610,00	1.770,00
52	1.110,00	1.700,00	1.890,00	2.080,00
56	1.280,00	1.980,00	2.190,00	2.410,00
60	1.470,00	2.270,00	2.510,00	2.770,00

Structure of stranded rope, type of construction 6×36 WS-IWRC. (Photo: Carl Stahl Hebetechnik GmbH).

[4]Extract from the catalog of company Carl Stahl Hebetechnik GmbH.

A.8 The most common knots[5]

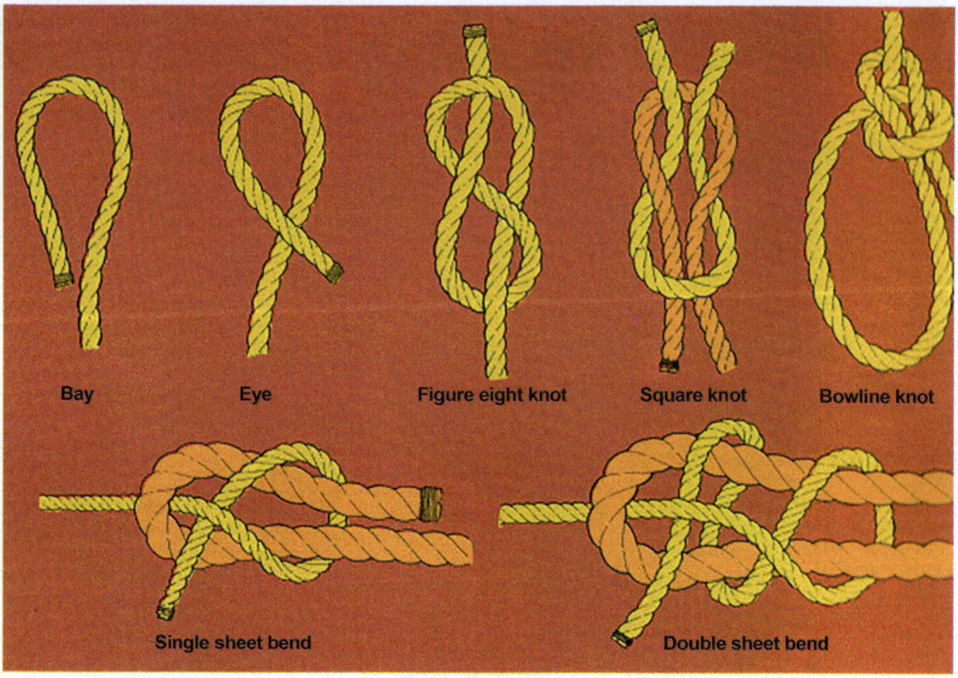

[5]From: Overschmidt/Gliewe, Sport boat license inland—Sail and Engine, Delius Klasing & Co, Bielefeld 1989.

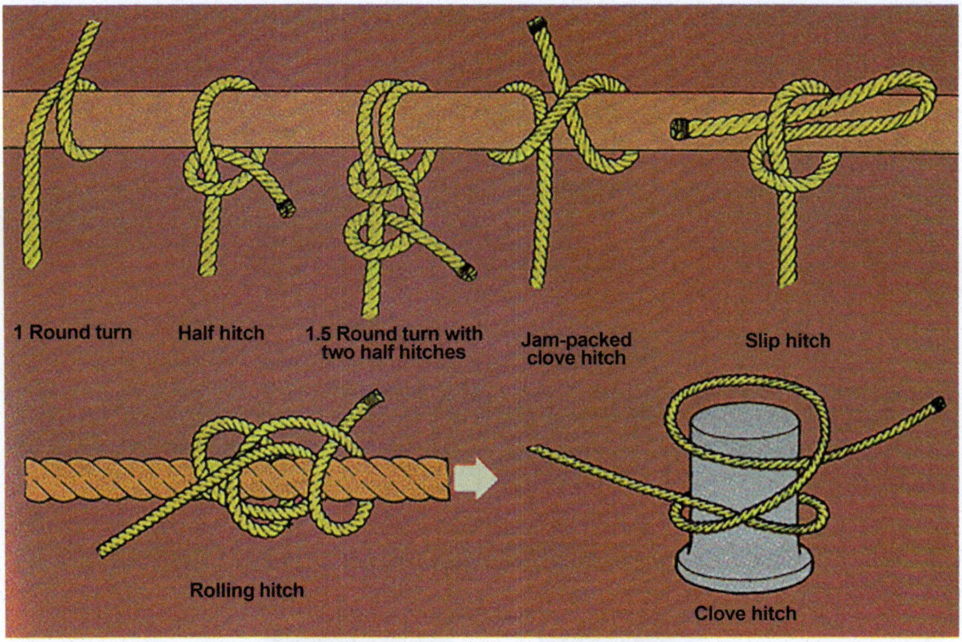

A.9 Marking of fiber ropes

The material is identified by weaving/inserting colored yarns into the rope.

Material	Identification color
Manila	Black
Sisal	Red
Hemp	Green
Polyamide	Green
Polyester	Blue
Polypropylene	Brown

A.10 Electrochemical series of various elements[6]

Gold	+ 1.42 V
Platinum	+ 1.20 V
Silver	+ 0.80 V
Carbon	+ 0.74 V
Copper	+ 0.34 V
Antimony	+ 0.14 V
Hydrogen	± 0.0 V
Lead	– 0.12 V
Tin	– 0.14 V
Nickel	– 0.25 V
Cadmium	– 0.40 V
Iron	– 0.44 V
Chromium	– 0.71 V
Zinc	– 0.76 V
Aluminum	– 1.67 V

– ← **0 V** → +

unnoble noble

A.11 Selection of static and dynamic friction coefficients

Material pairing	Coefficient of static friction μ_0		Coefficient of sliding friction μ	
	Dry	Lubricated	Dry	Lubricated
Steel/steel	0.15…0.30	0.10…0.12	0.10…0.12	0.04…0.07
Steel/cast iron	0.18…0.20	0.10…0.20	0.15…0.20	0.05…0.10
Steel/bronze	0.18…0.20	0.10…0.20	0.15…0.20	0.05…0.10
Leather/metal	0.30…0.50	0.12…0.16	0.30	0.15
Wood/metal	0.60…0.70	0.11	0.40…0.50	0.10
Wood/wood	0.40…0.60	0.16	0.20…0.40	0.08

[6] From: http://www.haustechnikdialog.de/shkwissen/images/elektrochemische_spannungsreihe, source: Bosy.

A.12 Selection of slip resistance characteristics for various rooms and areas according to Annex 2 of ASR A1.5/1,2

Number	Workrooms, areas, and operational traffic routes	Assessment group of the slip hazard (R-group)	Displacement area with code for the minimum volume
0	**General work rooms and areas**		
0.1	Entrance areas, inside	R 9	
0.2	Entrance areas, outside	R 11 or R 10	V 4
0.3	Stairs, inside	R 9	
0.4	Outdoor stairs	R 11 or R 10	V 4
0.5	Sloping ramps, inside (e.g., wheelchair ramps, compensating slopes, transport routes)	One R-group higher than required for the access flooring	V-value of access flooring, if applicable
0.6	Sanitary rooms		
0.6.1	Toilets	R 9	
0.6.2	Locker room and washrooms	R 10	
0.7	Break rooms (e.g., recreation room, company canteens)	R 9	
0.8	First aid rooms and similar facilities (see ASR A4.3)	R 9	
10	**Cold rooms, freezer rooms, cold storages, deep-freeze storages**		
10.1	For unpackaged goods	R 12	
10.2	For packed goods	R 11	
13	**Laundry**		
13.1	Rooms with continuous washing machines (washing tubes) or with washer extractors	R 9	
13.2	Rooms with washing machines, where the laundry is removed dripping wet	R 11	
13.3	Rooms for ironing and mangling	R 9	
30	**Operational traffic routes in outdoor areas**		
30.1	Walkways	R 11 or R 10	V 4

A.13 Beaufort scale

Strength in Bft	Designation Wind force	km/h	m/s	kn	Designation of sea state	Effect at the sea
0	Calm	0–<1	0–0.2	0–<1	Completely quiet, smooth sea	Mirror-smooth sea
1	Quieter draught	1–5	0.3–1.5	1–3	Quiet, crimped see	Light crimped waves
2	Light breeze	6–11	1.6–3.3	4–6	Lightly moving sea	Small, short waves, surface glassy
3	Weak breeze	12–19	3.4–5.4	7–10		Beginning of foam formation
4	Moderate breeze	20–28	5.5–7.9	11–15	Lightly moving sea	Small, longer waves, foam heads everywhere
5	Fresh wind	29–38	8.0–10.7	16–21	Moderately moving sea	Moderate waves of greater length, foam heads everywhere
6	Strong wind	39–49	10.8–13.8	22–27	Rough sea	Larger waves with breaking heads, white foam stains everywhere
7	Stiff wind	50–61	13.9–17.1	28–33	Very rough sea	White foam from the breaking wave heads settled into foam strips in wind direction
8	Stormy wind	62–74	17.2–20.7	34–40	High sea	Quite high crests of waves whose heads are blown away, foam strips everywhere
9	Storm	75–88	20.8–24.4	41–47		High waves with windblown spray, breakers begin to form
10	Severe storm	89–102	24.5–28.4	48–55	Very high sea	Very high waves, white spots on the water, long, breaking ridges, heavy breakers
11	Hurricane-like storm	103–117	28.5–32.6	56–63	Exceptionally heavy sea	Roaring sea, water is blown away horizontally, strong reduction of visibility
12	Hurricane	From 118	From 32.7	From 64		Sea completely white, air filled with foam and spray, no visibility

A.14 Selection of rigging of some tall ships

Images from: http://inmaris.de/unsere-seekiste/takelung-schiffstypen.html (retrieval date: July 1, 2017).

Four-masted barque

Sail of a four-masted barque

01 Outer jib (Jager)
02 Jib
03 Inner jib
04 Fore topmast staysail
05 Foresail
06 Lower fore topsails
07 Upper fore topsail
08 Lower fore topgallant sail
09 Upper fore topgallant sail
10 Fore royal
11 Main topmast staysail
12 Main topgallant staysail
13 Main royal staysail
14 Main sail
15 Lower main topsails
16 Upper main topsail
17 Lower main topgallant sail
18 Upper main topgallant sail
19 Main royal
20 Crossjack topmast staysail
21 Crossjack topgallant staysail
22 Crossjack royal staysail
23 Square sail
24 Sower crossjack topsails
25 Upper crossjack topsail
26 Lower crossjack topgallant sail
27 Upper crossjack topgallant sail
28 Crossjack royal
29 Mizzen staysail
30 Mizzen topmast staysail
31 Mizzen topgallant staysail
32 Lower mizzen
33 Upper mizzen
34 Mizzen topsegel

Three-masted full-rigged ship

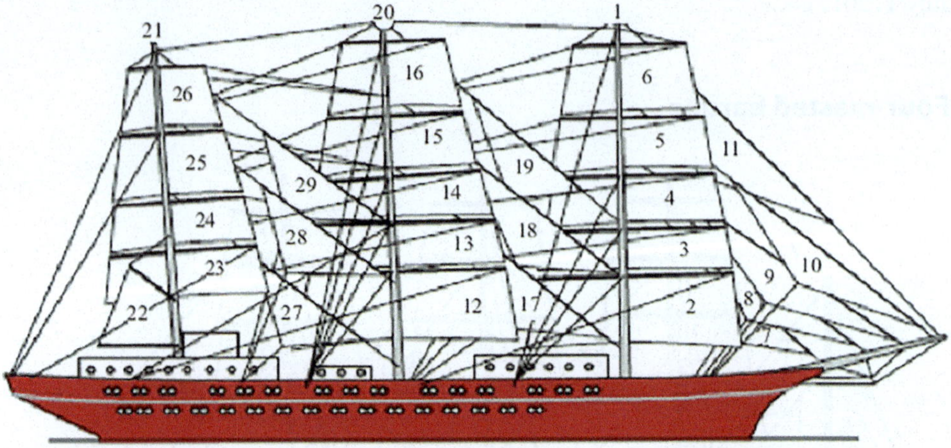

21 Crossjack mast
22 Mizzen
23 Lower crossjack topsails
24 Upper crossjack topsail
25 Lower crossjack topgallant sail
26 Crossjack royal
27 Crossjack topmast staysail
28 Crossjack topgallant staysail
29 Crossjack royal staysail

11 Flyer
12 Main sail
13 Lower main topsails
14 Upper main topsail
15 Lower main topgallant sail
16 Main royal
17 Main topmast staysail
18 Main topgallant staysail
19 Main royal staysail
20 Main mast

01 Foremast
02 Foresail
03 Lower fore topsails
04 Upper fore topsail
05 Lower fore topgallant sail
06 Fore royal
07 Fore topmast staysail
08 Inner jib
09 Jib
10 Outer jib

Three-masted barque

01 Mizzen
02 Mizzen topsegel
03 Mizzen topmast staysail
04 Main sail
05 Lower main top
06 Mizzen topmast staysail
07 Upper main top
08 Main topgallant
09 Main topmast staysail
10 Main topgallant staysail
11 Main royal staysail
12 Foresail
13 Lower fore top
14 Upper fore top
15 Fore topgallant
16 Fore topgallant staysail
17 Inner jib
18 Outer jib (Jager)

A.15. Coefficients of linear expansion of selected materials

Material	Linear expansion coefficient α (10^{-6}/K)
Aluminum, Al	23.8
Bronze	17.5
Iron, Fe	12.2
Glass (quartz glass)	0.5
Cast iron	10
Carbide	60
Copper, Cu	16.5
Brass	18.4
Polyamide, PA	110
Polystyrene	75
Polyvinyl chloride, PVC	80
Silver, Ag	19.5
Steel	11.7

Material	Linear expansion coefficient α (10^{-6}/K)
Steel, high-alloy, here V2A	16
Tungsten, W	4.5
Zinc, Zn	29
Tin, Sn	26.7

These values apply to the temperature interval $0^\circ\text{C} < T < 100^\circ\text{C}$.

A.16 Matrix for identifying the cause of faults in the simplex-compact stern tube sealing system

Source: TKMS/Blohm + Voss Industries GmbH.

Observation on 20-l high tank	Shaft stationary		Shaft rotating	
	Cause	Measure	Cause	Measure
Level falls continuously[a]	Leakage in system	Seal leakage point, check tank layout	Seal worn out, sealing ring running not round	Replace sealing ring renew, check bearing clearance, if necessary realign seal or bushing
Level increases continuously	Oil flows through the overflow[b] Possibly sealing ring 3 worn or damaged Screwed pipe connection to 20-l high tank at stern tube leaking	Close shut-off valve in feed line to 20-l high tank	Sealing ring 3 deformed or hardened Stern tube bearing leaked or seal housing or rather bushing poorly aligned	Close shut-off valve in feed line to 20-l high tank Check stern tube bearing play Check alignment of seal and bushing
Water in 20-l high tank[c]	Sealing ring 1 + 2 worn or damaged, due to foreign matter seal housing leaking	Close shut-off valve in feed line Check whether water is penetrating the stern tube Check water content more often Possibly set 20-l tank temporarily higher	Sealing ring 1 + 2 deformed, hardened, or otherwise damaged	Close shut-off valve in feed line Check whether water is penetrating the stern tube Check water content more often Possibly set 20-l tank temporarily higher Check alignment of seal housing and bushing and improve if required

[a]Oil consumption of up to 1% of the seal size in liters/day is not an indication of a fault on the seal. If the oil consumption is constantly increasing, an oil of higher viscosity can be used. Alternatively, the shut-off valve in the supply line can be closed and opened twice a day for 1 h.

[b]Slight oil drops over the overflow of the 20-l high tank are no sign of a damaged seal as long as the quantity per day remains below approximately 1% of the seal size.

[c]Small amounts of water entering the 20-l high tanks are not a sign of a damaged seal as long as the quantity per day remains below approximately 1% of the seal size.

A.17 Design rudder designs

The following figures show various constructive designs of rudders. The designation of the rudder design is based on the type of bearing of the rudder blade and rudder shaft.

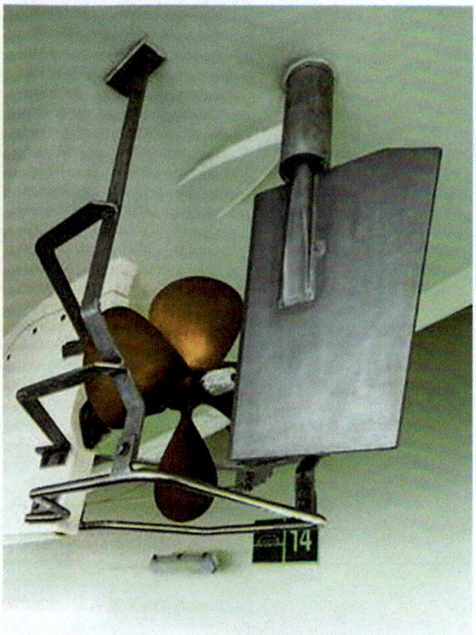

Fig. A.1 Hovering rudder

Fig. A.2 Semi-hovering rudder. (Photo: Brosen, CC BY-2.5)

Fig. A.3 Plate rudder mounted on bearings on both sides. (Photo: Cobija, CC0)

A.18 Marking of piping according to the flow substance[7]

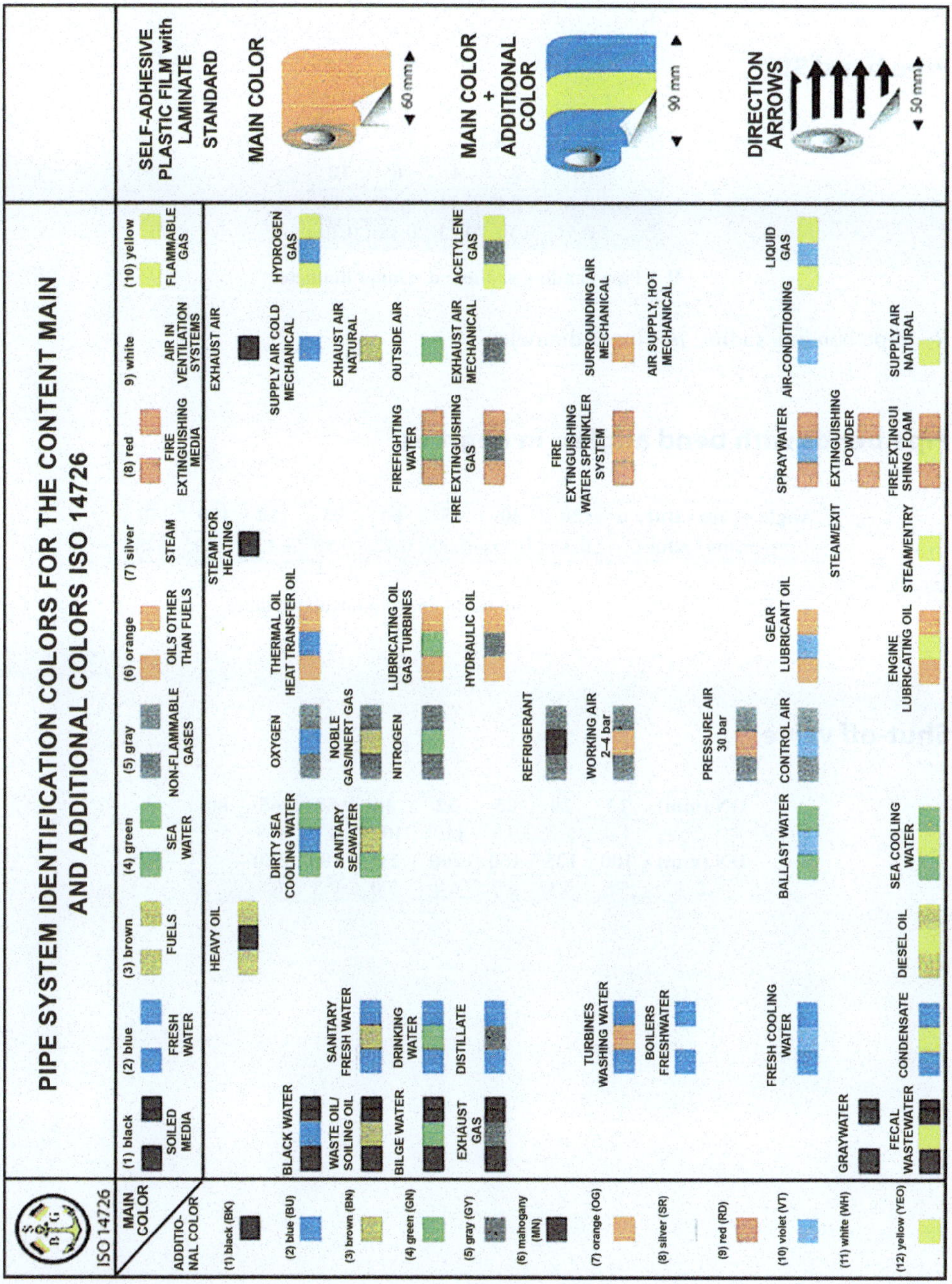

[7]BG Transport, Ship Safety Division, "Marking of pipelines on board - D.6," source: http://www.deutsche-flagge.de (July 7, 2017).

A.19 Table of selected resistance coefficients (mean values)[8]

Pipe bend 90°

	R / d				
	1	2	4	6	10
ξ_{smooth}	0.21	0.14	0.11	0.09	0.11
ξ_{rough}	0.51	0.30	0.23	0.18	0.20

R = Pipe bending radius, d = inner diameter

R = pipe bending radius, d = inner diameter.

Pipe bends with bend angle α in degrees

Angle of curvature α	20	30	45	60	75	110	130	150
Conversion factor f	0.31	0.45	0.60	0.78	0.90	1.13	1.20	1.28

$\zeta = \zeta_{90°} \cdot f$ $\zeta_{90°}$ can be taken from the above table

Shut-off valve

DN (mm)	15	20	25	32	40	50	65	80
ζ	1.4	5.2	9.8	11.7	10.2	8.7	6.8	5.9
DN (mm)	100	125	150	200	250	300	350	
ζ	5.5	5.5	5.7	6.4	7.0	7.3	6.8	

[8]Source: http://www.schweizer-fn.de/zeta/armaturen/armaturen.php (September 13, 2016).

Angle seat valve

Fig. A.4 Shut-off valve. (Photo: Heather Smith, CC BY 3.0)

DN (mm)	15	20	25	32	40	50	65	80	100	125	150
ζ	2.0	2.0	2.2	2.5	2.6	2.6	2.6	2.5	2.4	2.5	2.9

Butterfly valve

Fig. A.5 Butterfly valve. (Photo: Heather Smith, CC BY 3.0)

DN (mm)	20	25	32	40	50	65	80	100	125	150	200
ζ	0.79	0.86	0.70	0.54	0.42	0.33	0.28	0.25	0.23	0.22	0.21
DN (mm)	250	300	350	400	500	600	700	800	900	1000	
ζ	0.20	0.21	0.21	0.21	0.21	0.21	0.21	0.20	0.18	0.16	

Ball valve

DN (mm)	15	20	25	32	40	50	65
ζ	0.084	0.107	0.117	0.123	0.121	0.117	0.110
DN (mm)	80	100	125	150	200	250	
ζ	0.103	0.095	0.086	0.079	0.066	0.052	

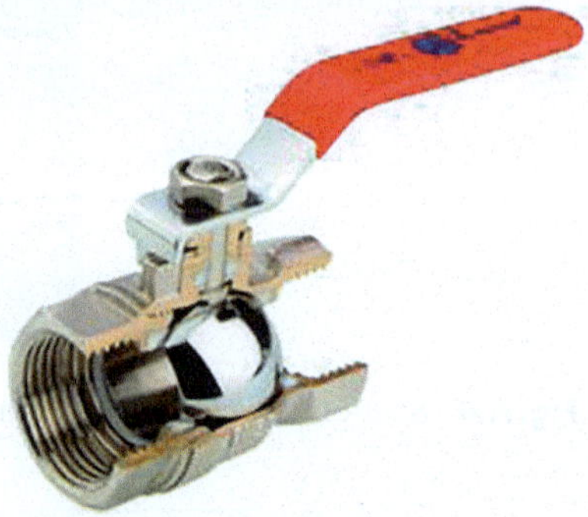

Fig. A.6 Ball valve. (Photo: Standard Hidráulica)

Angle valve

DN (mm)	15	20	25	32	40	50	65	80	100	125	150	200
ζ_{max}	3.1	3.1	3.1	3.1	3.4	3.8	4.1	4.4	4.7	5.0	5.3	5.7
ζ_{min}	2.0	2.0	2.0	2.0	2.0	2.0	2.0	2.0	2.0	2.0	2.0	2.0

Gate valve, open

DN (mm)	15	20	25	32	40	50	65	80	100	125	150	200
ζ_{max}	0.65	0.60	0.55	0.50	0.50	0.45	0.40	0.35	0.30	0.30	0.30	0.30
ζ_{min}	0.10	0.10	0.10	0.10	0.10	0.10	0.10	0.10	0.10	0.10	0.10	0.10

Fig. A.7 Gate valve. (Photo: Heather Smith, CC BY 3.0)

Throttle valve

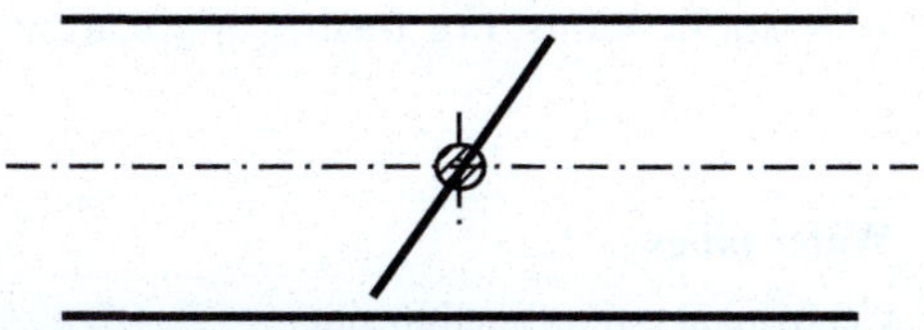

Fig. A.8 Schematic diagram of a throttle valve

The resistance coefficient depends on the opening angle α of the throttle valve (angle between the valve and the longitudinal axis of the component).

	α						
	10	20	30	40	50	60	70
ζ	0.5	1.5	4.0	11.0	33.0	120.0	250.0

Figure A.9 shows further resistance figures, especially for valves of larger nominal diameters

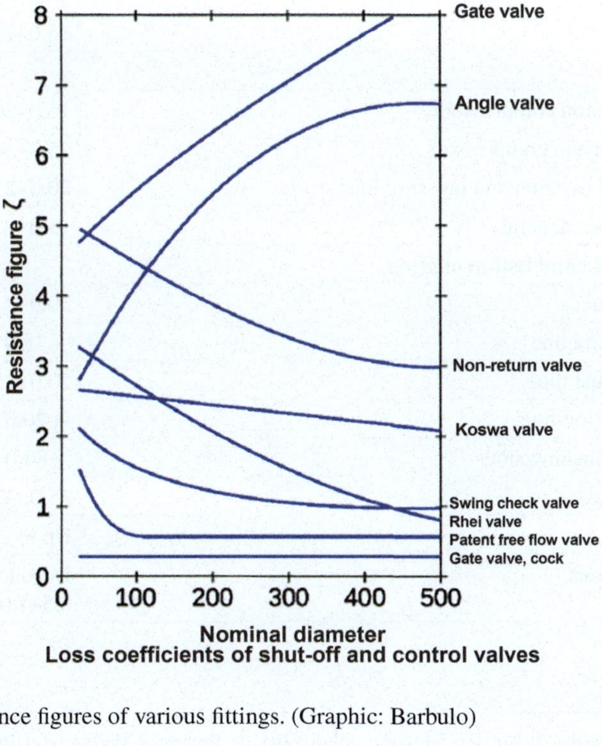

Fig. A.9 Resistance figures of various fittings. (Graphic: Barbulo)

A.20 Recommended flow velocities in pipes[9]

The following table gives a selection for recommended flow velocities in pipelines for some applications. For further applications, please refer to the relevant literature.

	Flow velocity (m/s)
Water pipes	
Centrifugal pump, suction side	0.7–1.5
Centrifugal pump, pressure side	1.0–5.5
Piston pump, suction side	0.8–1.0
Piston pump, pressure side	1.0–2.0
Pressurized water pressure lines	15.0–20.0
Hot water pressure line	2.0–3.0
Drinking and service water pipes	1.0–2.0
Steam lines	
Turbine, superheated steam, low performance	≈ 35.0
Turbine, superheated steam, average performance	40.0–50.0
Turbine, superheated steam, high performance	50.0–70.0
Saturated steam	≈ 25.0
Exhaust steam	15.0–25.0
Air ducts	
Suction line for piston compressors	16.0–20.0
Pressure line for piston compressors	25.0–30.0
Turbocompressors (suction and pressure line)	20.0–25.0
Compressed air line, general	15.0
Cables in internal combustion engines	
Gas engines, air duct	20.0
Gas engines, gas pipeline	≈ 35.0
Gas engines, exhaust pipe	20.0–25.0
Diesel engine, suction line	≈ 20.0
Diesel engine, in flushing slots	≥ 80.0
Blower supply lines	25.0–30.0
Fuel lines	Up to approx. 20.0
Lubricating oil lines	Dependent on toughness 0.5–1.0

[9]DUBBEL – Taschenbuch für den Maschinenbau; Bd. 1; Springer Verlag, Berlin, Heidelberg New York, 1974.

A.21 Specific heat capacity for selected substances

Liquids

The following values for the specific heat capacity apply to a temperature range of $0^{\circ}\mathrm{C} < t < 100^{\circ}\mathrm{C}$. For other temperatures, the values should be found in relevant tables or calculated according to the following equation:

$$c|_{T_1}^{T_2} = \frac{c|_{T_0}^{T_2} \cdot (T_2 - T_0) - c|_{T_0}^{T_1} \cdot (T_1 - T_0)}{T_2 - T_1}.$$

Substance	Specific heat capacity c (kJ/kgK)
Ethanol (ethyl alcohol)	2.38
Petrol	2.02
Benzene	1.70
Diesel fuel/marine diesel	2.05
Acetic acid	2.05
Fuel oil EL	2.07
Methanol (methyl alcohol)	2.51
Perchloroethylene	0.904
Petroleum ether	1.76
Petroleum	2.16
Lubricating oil	2.09
Transformer oil	1.88
Trichloroethylene	0.93
Toluene	1.67
Water	4.187

Gases

Specific heat capacity $c_{pm}|_0^t$ in kJ/kgK.

T °C	CO	CO_2	H_2	H_2O (vaporous)	Air	N_2
0	1.039	0.8205	14.38	1.858	1.004	1.039
100	1.041	0.8689	14.40	1.874	1.031	1.041
200	1.046	0.9122	14.42	1.894	1.013	1.044
300	1.054	0.9510	14.45	1.918	1.020	1.049
400	1.064	0.9852	14.48	1.946	1.029	1.057

A.22 Thermal conductivity, selection of some liquid, solid, and gaseous substances[10]

Fabric/material	Thermal conductivity λ W/mK
Aluminum	204
Diesel fuel	0.15
Ice	2.33
Iron	81
Gases	0.01–0.23
Plaster	0.45
Glass, window	0.81
Glass wool	0.04
Gray cast iron	58
Rigid foam	0.04
Wood	0.12–0.17
Scale	1.2–3.0
Carbon dioxide	0.015
Cork	0.05
Copper	3.84
Air	0.02
Brass	113
Polyamide	0.31

[10]**Further values, e.g., also GIECK, Technical formulae collection, Heilbronn 1981.**

Fabric/material	Thermal conductivity λ W/mK
Polyvinyl chloride	0.16
Red brass	38
Fireclay	1.2
Steel, NiRo	14
Steel, unalloyed	47–58
Transformer oil	0.13
Water	0.6
Water vapor (100 °C)	0.016
Soft rubber	0.14–0.24

A.23 Log *p*–*h* diagram for ammonia (NH_3)

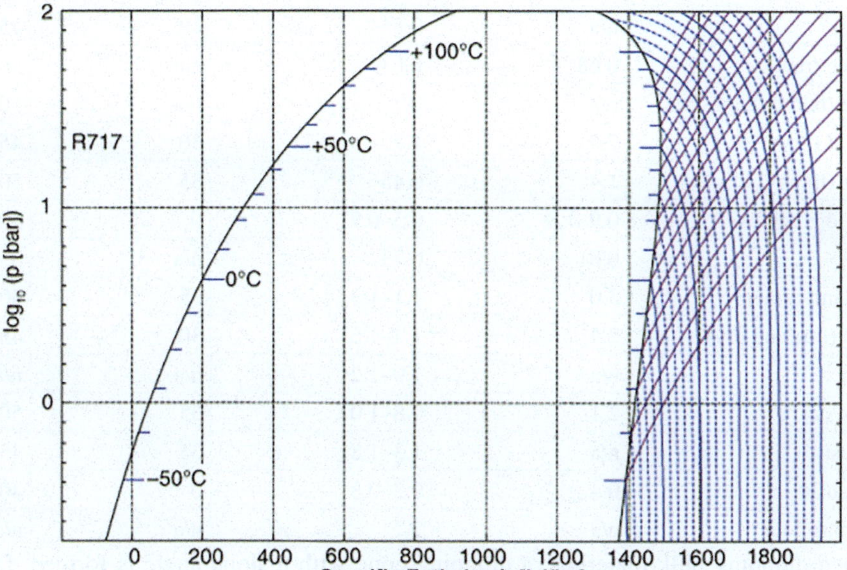

Source: Dohmann, J., Thermodynamik der Kälteanlagen und Wärmepumpen, Springer Verlag, Heidelberg 2016.

A.24 Bulk material table according to Professor Rüdiger

Selection of some bulk materials

Bulk material	Density	Bulk density	Slope angle [°]	
	(t/m^3)	(t/m^3)	Movement[a]	Rest[a]
Lignite	1.3	0.65–0.75	35	50
Iron ore	4–6	1.7–3.5	35	50
Barley	0.7	0.6	25	40
Plaster	1.81	1.4	n/a	40
Domestic waste	n/a	0.66	n/a	n/a
Wood (firewood)	n/a	0.32–0.42	n/a	n/a
Blast furnace slag, in pieces	2.5	0.6–1.5	35	50
Pulses	n/a	0.85	10	25
Coffee (raw)	0.68	0.63	n/a	n/a
Lime (burned), powder	n/a	0.5	n/a	50
Gravel (dry)	2.5	1.7	30	45
Coke (lumpy)	1.4	0.45	35	50
Plastic granules	0.9–1,8	0.5–0.9	0	30
Maize	0.81	0.73	n/a	35
Portland cement	3.0	1.1–1.3	35	50
Sand (fine and dry)	2.4	1.4–1.65	30	45
Saltpeter	n/a	1.0–1.2	n/a	n/a
Rock salt	2.1	0.8–1.0	35	50
Ballast	n/a	1.3–1.8	35	45
Sawdust	0.4	0.2–0.3	0	40
Thomas flour	n/a	2.2	n/a	n/a

[a] When dumping bulk materials, a dumping cone with a slope angle is formed. During tipping, the material slides down. In this respect, the slope angle is smaller than when the material has come to rest.

A.25 Security identifiers, security guidance system

Selection of the safety markings according to ISO 7010

Note: A comprehensive overview of the currently valid safety markings can be found at www.a-spe.com or also at www.brady.de.

Warning sign						
Hazardous area	Explosive substances	Radioactive substances/ ionizing radiation	Laser beams	Electromagnetic radiation	Dangerous electrical voltage	Slip hazard
Risk of tripping	Danger of falling	Biohazard	Suspended loads			
Rescue sign						
Knives for life raft	Lifeboat	Rescue boat	Life raft	Life jacket for infants	Life jacket	Closed slide for ship evacuation
Life belt with line and lamp	Life belt	Parachute signal rocket	Distress beacon	Embarkation ladder		
Prohibition sign						
Smoking prohibited	Fire, open light, and smoking prohibited	Climbing/ascending prohibited	No loading	Passenger transport in fire case prohibited	Do not switch	

 Use ear protection	 Use safety goggles	 Before opening power plug pull hands out	 Wash hands	 Use handrail	 Use safety harness	 Lower lifeboat during suspension procedure
 Use respiratory protection	 Put life raft in water during trigger sequence	 Lower rescue boat during at suspension procedure	 Release halyard during suspension procedure	 Release lifeboat lashings during suspension procedure	 Secure gas cylinders against falling over	 Wear self-contained breathing apparatus
Fire protection symbol						
 Fire extinguisher	 Fire hose	 Fire protection ladder	 Fire detector	 Fire alarm telephone	 Fixed built-in fire extinguisher battery	 Mobile fire extinguisher
 Firefighting monitor						

A.26 Radio equipment obligation depending on the areas of operation[11]

Equipment	A 1 USW	A 2 MW	A 3 Satellite	A 4 SW
USW radiotelephone system	X	X	X	X
USW DSC controller	X	X	X	X
USW watch receiver channel 70	X	X	X	X
Navtex receiver	X	X	X	X
EPIRB (406 MHz or 1.6 GHz)	X	X	X	X
Radar transponder [SART]	X	X	X	X
USW walkie-talkies	X	X	X	X
MW radiotelephone system		X	X	
MW DSC controller		X	X	
MW/SW radiotelephone system				X
MW/SW DSC controller				X
EGC receiver			X	X
INMARSAT-C or -B			X	

Explanations

USW **U** ltra **s** hort **w** ave (frequency range 30–300 MHz)

MW M arginal **w** ave (frequency band between 1605 and 3800 kHz, is at the "margin" between medium wave and shortwave)

SW **S** hort **w** ave (frequency range 3–30 MHz)

Sea area A 1 An area defined by the competent administration within the radio range of at least one USW coastal radio station, which is continuously available for DSC alerts (range 30 nm).

[11]See details: http://navigareberlin.de/onewebmedia/H%20Grundlagen%20Copy.pdf (retrieval date February 11, 2017).

Sea area A 2 An area defined by the competent administration (excluding sea area A 1) within the radio range of at least one MW coastal radio station, which is continuously available for DSC alerts (range 150 nm).

Sea area A 3 An area (excluding sea areas A 1 and A 2) within the coverage of a geostationary INMARSAT satellite, which is continuously available for alerts (range approximately 75° N to 75° S).

Sea area A 4 An area outside sea areas A 1 to A 3 (polar cap areas).
It is left to each administration to decide on the geographical boundaries it chooses for the establishment of sea areas A 1 and A 2. The boundaries of the individual sea areas can be taken from the "Master Plan for shore-based facilities" published by the IMO.

A.27 Lights guidance

a) According to CAR and SeeSchStrO[12]

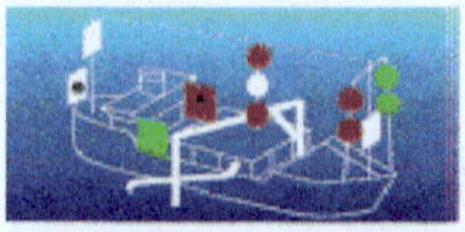

Maneuvering disabled vessel with motion through the water; passing on the side with two green lights possible.

Vessel of the public service in the performance of police duties.

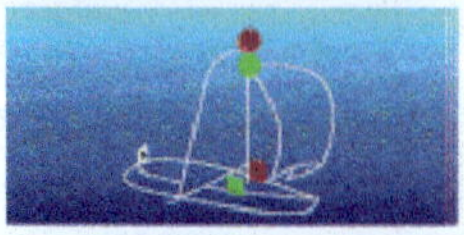

Sailing vessel in motion.

Hovercraft navigating in a non-water-displacing state.

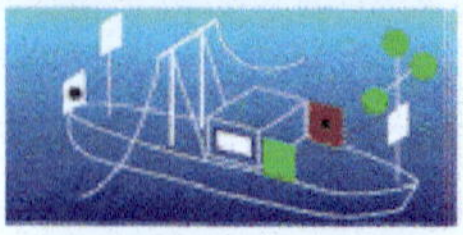

Vessel clearing mines; approach to less than 1,000 m is dangerous for other vehicles.

Pilot vessel in motion.

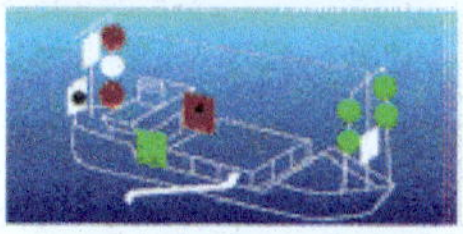

Vessel with limited manoeuvrability, dredging in the fairway or carrying out underwater work; no obstruction; passing on the right in the direction of travel (in the area covered by SeeSchStrO).

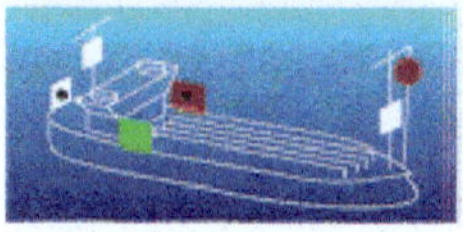

Vessels and pushing and towing units carrying certain dangerous goods and certain empty vehicles (within the area of validity of SeeSchStrO).

Vessels of the German Federal Armed Forces and the Federal Border Guard as well as machinery vessels towing targets (within the area of validity of SeeSchStrO).

Vessels under sail of less than 12 m length or vessels under rudder (within the area of validity of SeeSchStrO).

Non-free ferry in motion (in the area of validity of SeeSchStrO).

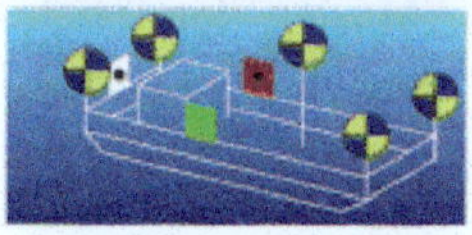

Free-moving ferry on the North-East Sec Canal, the Trave or the Warnow in motion (in the area of validity of SeeSchStrO).

Customs vessel within the area of application of SeeSchStrO).

Vessel moored on the shore of less than 50 m length (in the area of validity of SeeSchStrO).

Vessel moored on the shore of 50 m and more (in the area of validity of SeeSchStrO).

 Fixed all-round light.

 Fixed light over limited sector.

 Common mode light

 Light not visible to the observer.

[12] Taken from: Dreyer, R., Sportküstenschifferschein und Sportbootführerschein See, 14. Auflg., Verlag Delius Klasing, Bielefeld.

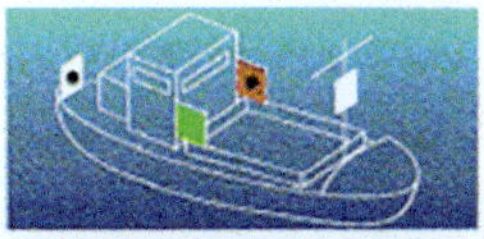
Machine vessel in motion of less than 50 m length.

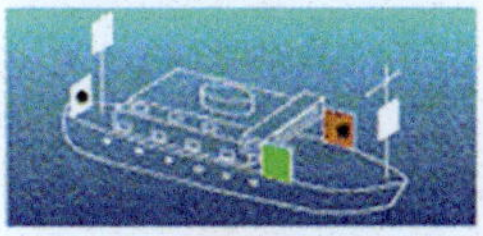
Machine vessel in motion (length cannot be specified; a vehicle less than 50 m length may carry a second toplight, but is not obliged to do so.)

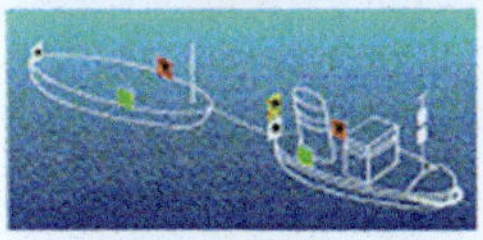
Towing unit in motion with attachment of 200 m length or less.

Towing unit in motion with attachment of 200 m or less. The towing unit cannot deviate from the course.

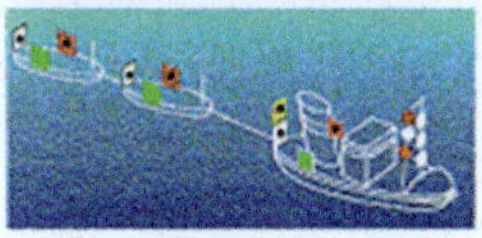
Towing unit in motion with an attachment of more than 200 m length. The towing unit cannot deviate from the course.

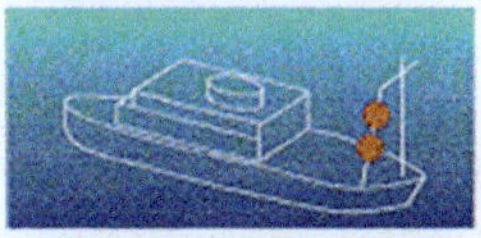
Non-maneuverable vessel without motion through water (not required for vessels less than 12 m length).

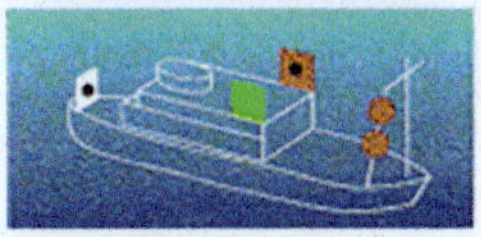
Non maneuverable vessel with motion through water (not required for vessels less than 12 m length).

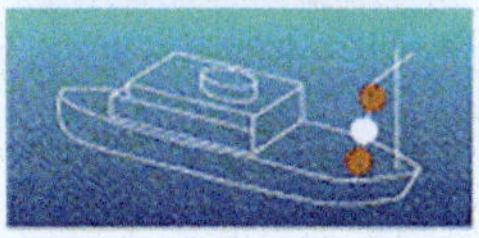
Maneuvering disabled vessel without motion through water (not required for vessels less than 12 m length).

Manoeuvring disabled vessel with motion through water (not required for vehicles less than 12 m length).

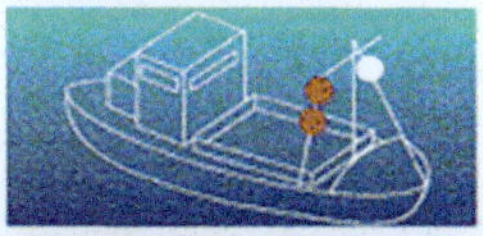
Vessel based on less than 50 m in length (not required for vehicles less than 12 m in length).

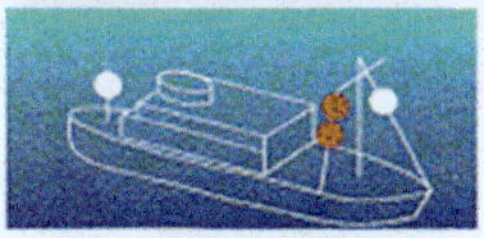
Vessel on ground (probably 50 m or longer; definitive indication of length not possible).

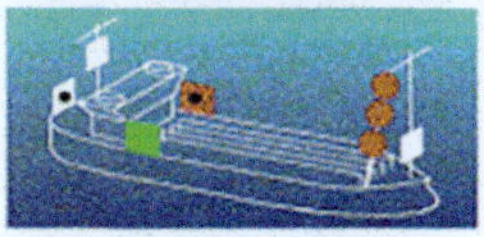
Draught restricted vessel in motion (length cannot be specified).

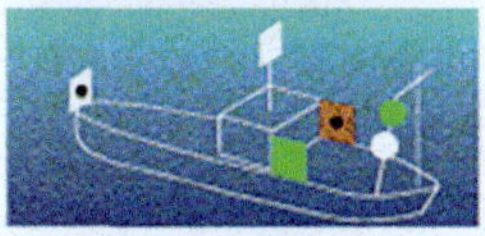
Fishing trawler with motion through the water (probably 50 m or longer; definitive length cannot be given).

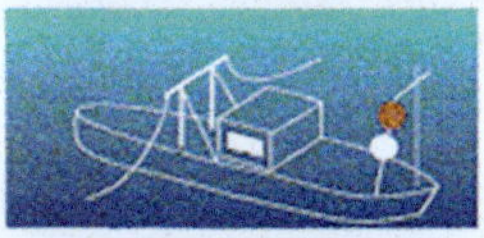
Fishing vessel in motion, not trawling, without motion through the water (see above for length).

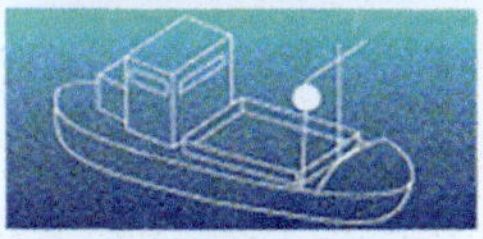
Vessel at anchor less than 50 m in length.

Vessel at anchor (see above for length).

b) According to BinSchStrO[13]

Commercial vessels in general.

Ship ≤ 110 m length

Ship > 110 m length

Towing formation; special feature → The last vessel has a white tail light, if it is a small vessel, then no light. Furthermore marking is shown during daytime

Push tow > 110 m length (pack of 2)

hinge tow > 110 m length

[13]Graphics taken from: Overschmidt, Gliewe, „Sportbootführerschein Binnen – Segel Motor “, Delius Klasing Verlag, Bielefeld.

Transport of hazardous goods (tags are guided by blue cones instead of blue lanterns).

Ship with fire hazardous goods

Ship with ammonia or equivalent goods

Ship with explosion-hazardous goods

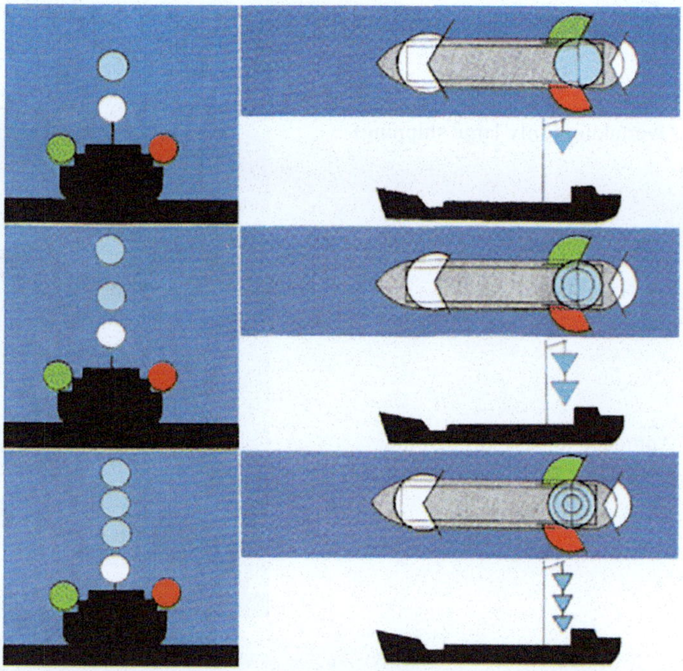

Authorities and organizations with security tasks (BOS) in operation, ferries.

Not free driving

Free driving

BOS vessel

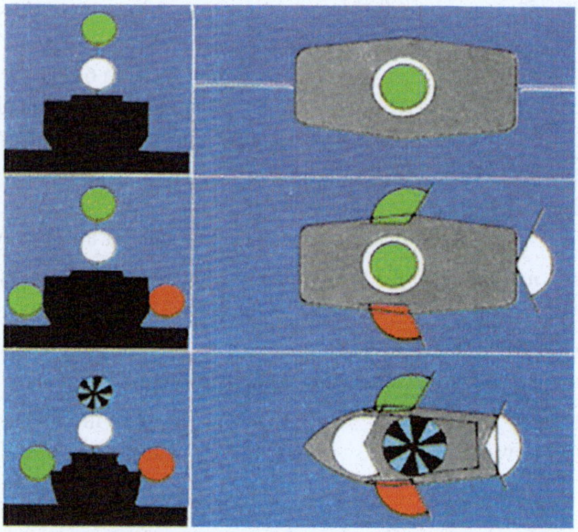

Overtaking and meeting.

Overtaking (only large shipping)

Approach to pass through the starboard side

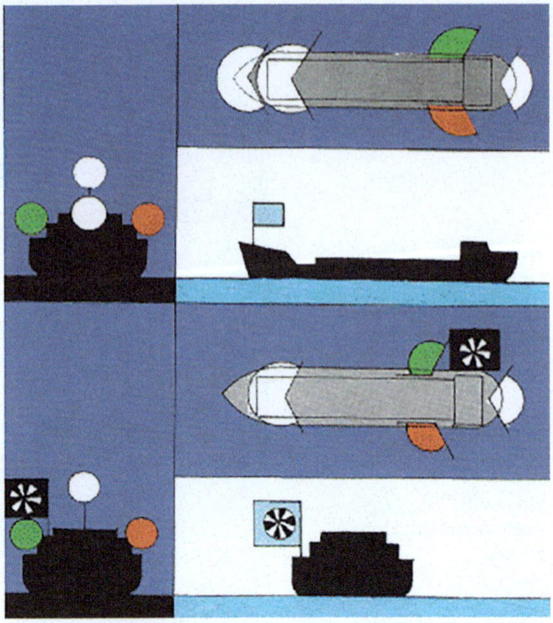

A.28 Overview of restrictions on the dumping or discharge of waste into the sea in accordance with MARPOL Annex V, Regulations 4–6

Type of waste[a]	All ships		Ships that are adjacent to or within 500 m of offshore platforms[d], more than 12 nm from nearest land (Rule 5)
	Outside from special areas (Rule 4) Distance to nearest land	**Within** special areas (Rule 6) Distance to nearest land or ice shelf	
Food waste, crushed or crushed[b]	$\geq$ 3 nm on course and as far away as possible	$\geq$ 12 nm on course and as far away as possible[c]	Dumping or discharge permitted
Food waste, not crushed or crushed	$\geq$ 12 nm on course and as far away as possible	Dumping or discharge prohibited	Dumping or discharge prohibited
Cargo residues[e,f], not contained in wash water	$\geq$ 12 nm on course and as far away as possible	Dumping or discharge prohibited	Dumping or discharge prohibited
Cargo residues[e,f], contained in wash water		$\geq$ 12 nm and as far away as possible (subject to the conditions in Rule 6 s. 1.2)	

Type of waste[a]	All ships		Ships that are adjacent to or within 500 m of offshore platforms[d], more than 12 nm from nearest land (Rule 5)
	Outside from special areas (Rule 4) Distance to nearest land	**Within** special areas (Rule 6) Distance to nearest land or ice shelf	
Detergents and additives[f], contained in wash water from cargo spaces	Dumping or discharge permitted	$\geq$ 12 nm and as far away as possible (subject to the conditions in Rule 6 s. 1.2)	Dumping or discharge prohibited
Detergents and additives, contained in wash water used on deck and external areas		Dumping or discharge permitted	
Animal carcasses (should be cut up or otherwise treated to ensure that the bodies sink immediately)	Ship must be on its course and as far away as possible from nearest land. If possible at $>$ 100 nm and maximum water depth	Dumping or discharge prohibited	Dumping or discharge prohibited
Other waste, including plastics, synthetic ropes, fishing gear, plastic waste bags, ashes from incinerators, slag, cooking oil, driving dunnage, cladding and packaging material, paper, rags, glass, metal, bottles, earthenware, and similar waste	Dumping or discharge prohibited	Dumping or discharge prohibited	Dumping or discharge prohibited

[a]If the waste is mixed or contaminated with other pollutants, which may not be introduced or discharged or to which other rules on introduction or discharge apply, the stricter rules shall apply.

[b]Crushed or ground food waste must be able to pass through a sieve with openings no wider than 25 mm.

[c]The introduction of imported bird products into the Antarctic area shall not be allowed unless such products have been incinerated, autoclaved, or otherwise treated to render them sterile.

[d]Offshore platforms located 12 nm from the nearest land and associated vessels shall include all fixed or floating platforms used for the exploration and exploitation and related processing of seabed resources and all vessels located adjacent to or within a radius of 500 m of such platforms.

[e]The term cargo residues refers only to those cargo residues, which cannot be resumed when using normally available discharge procedures.

[f]These substances must not be harmful to the marine environment.

A.29 Requirements for ship waste incinerators

	MEPC.76(40)	RL 2010/75/EU	17. BImSchV
Technical requirements to the firing	Combustion air at least 6% O_2; Exhaust gas temperature 850–1200 °C; Quenching of exhaust gas (to prevent the dioxin formation)		Requirement for the most extensive burnout of the input materials; Minimum temperature of the gases in after-burner with a residence time of at least 2 s and $O_{2\,min} = 6\%$ must 850 °C, and 1100 °C if halogen-containing waste is used; Auxiliary burner; Automatic locking of the waste supply if temperature falls below the minimum temperature
Emission limit values	Formation of dioxins must be prevented	**Daily average values:** Dust 10 mg/Nm³; HCl 10 mg/Nm³; HF 1 mg/Nm³; SO_2 50 mg/Nm³; NO_2 200 mg/Nm³ (plants with a nominal capacity > 6 t/h or new plants); NO_2 400 mg/Nm³ (plants with nominal capacity ≤ 6 t/h or old plants); CO 100 mg/Nm³ (half-hourly average); C_{tot} 10 mg/Nm³; Dioxin and furans 0.1 mg/Nm³	**Daily average values:** Total dust 10 mg/m³, C_{tot} 10 mg/m³, HCl 10 mg/m³, HF 1 mg/m³, SO_2 50 mg/m³, NO_2 200 mg/m³; Hg 0.03 mg/m³; CO 50 mg/m³ **Half-hourly averages:** Total dust 20 mg/m³, C_{tot} 20 mg/m³, HCl 60 mg/m³, HF 4 mg/m³, SO_2 200 mg/m³, NO_2 400 mg/m³, Hg 0.05 mg/m³; CO 100 mg/m³ **Reference O_2:** 11%
Heat utilization			Heat utilization commitment
Excluded waste	PCB, waste listed in MARPOL Annex V with more wastes of heavy metals, sophisticated petroleum products (the halogens included), PVC		

A.30 Diagrams for the determination of required adsorbent quantities

The following diagrams indicate the required mass of the adsorbent mixtures in each case as a function of the assumed volume of waste to be incinerated, ranging from 1.2 kg per person per day to 1.5 kg per person per day depending on the total number of persons on board.

Example for the application of the diagrams

An adsorbent mixture of hydrated lime and activated carbon was chosen. On board the cruise ship, there are 2,500 people in total. A waste volume of 1.3 kg per person and day is expected to be incinerated. How high is the adsorbent consumption per day? To do this, draw a vertical line upward on the abscissa at "2,500" in the diagram below, pointing upward to the dotted straight line. The required adsorbent mass - here 19.5 kg per day - can then be read off on the ordinate.

a) Diagram for an adsorbent mixture of 75% $Ca(OH)_2$ and 25% activated carbon

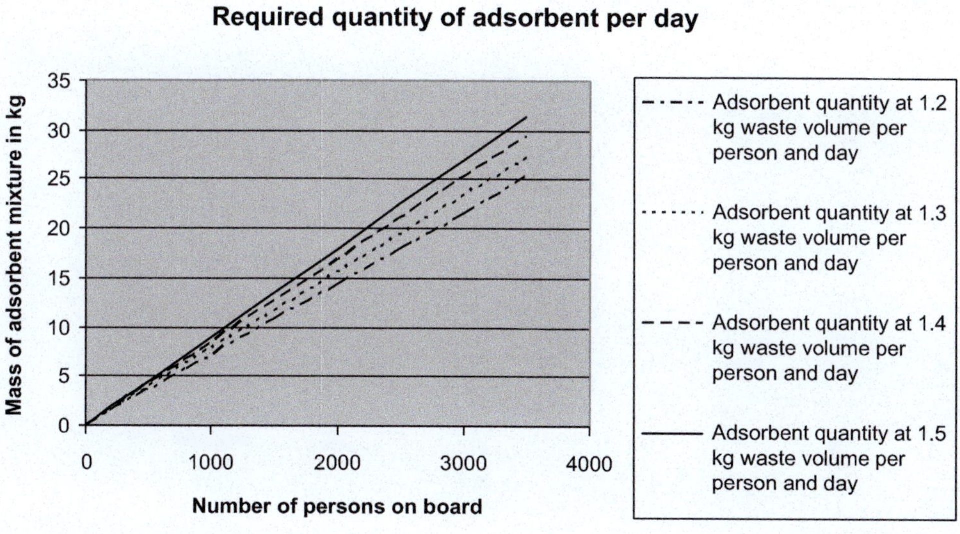

b) Diagram for an adsorbent mixture of 79% $CaCO_3$ and 21% activated carbon

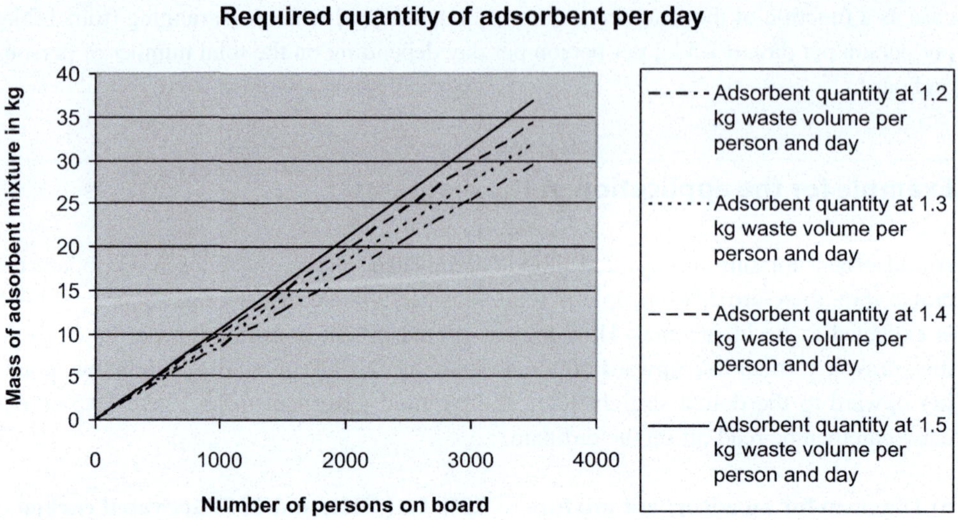

Index

M. Pfaff, *Ship Operation Technology,* https://doi.org/10.1007/978-3-658-32729-3

The manufacturer's authorised representative in the EU is Springer Nature Customer Service Centre GmbH, Europaplatz 3, 69115 Heidelberg, Germany. If you have any concerns regarding our products, please contact ProductSafety@springernature.com

Printed and bound by CPI Group (UK) Ltd, Croydon, CR0 4YY
30/09/2026
02225594-0006